Edited by
Tamara Bechtold, Gabriele Schrag, and
Lihong Feng

System-level Modeling of MEMS

Related Titles

Korvink, J. G., Smith, P. J., Shin, D.-Y. (eds.)

Inkjet-based Micromanufacturing

2012
ISBN: 978-3-527-31904-6

Saile, V., Wallrabe, U., Tabata, O., Korvink, J. G. (eds.)

LIGA and its Applications

2009
ISBN: 978-3-527-31698-4

Hierold, C. (ed.)

Carbon Nanotube Devices
Properties, Modeling, Integration and Applications

2008
ISBN: 978-3-527-31720-2

Tabata, O., Tsuchiya, T. (eds.)

Reliability of MEMS
Testing of Materials and Devices

2008
ISBN: 978-3-527-31494-2

Varadan, V. K., Vinoy, K. J., Gopalakrishnan, S.

Smart Material Systems and MEMS
Design and Development Methodologies

2006
ISBN: 978-0-470-09361-0

Edited by Tamara Bechtold, Gabriele Schrag, and Lihong Feng

System-level Modeling of MEMS

WILEY-VCH

WILEY-VCH Verlag GmbH & Co. KGaA

The Editors

Dr. Tamara Bechtold
IMTEK - Institute of Microsystem Technology
Laboratory for Simulation
Georges-Koehler-Allee 103
79110 Freiburg
Germany

Dr. Gabriele Schrag
Technische Universität München
Institute for Physics of Electrotechnology
Arcisstr. 21
80290 München
Germany

Dr. Lihong Feng
Max Planck Institute for Dynamics of Complex Technical Systems
Sandtorstr. 1
39106 Magdeburg
Germany

The front cover shows the finite element model and strain intensity distribution of a piezoelectric micro energy harvester. Mathematical order reduction was applied to this model enabling system level simulation as indicated in the block diagram. The modelling approach is described in the chapter "Towards system-level simulation of energy harvesting modules" of the present volume.

■ All books published by **Wiley-VCH** are carefully produced. Nevertheless, authors, editors, and publisher do not warrant the information contained in these books, including this book, to be free of errors. Readers are advised to keep in mind that statements, data, illustrations, procedural details or other items may inadvertently be inaccurate.

Library of Congress Card No.: applied for

British Library Cataloguing-in-Publication Data
A catalogue record for this book is available from the British Library.

Bibliographic information published by the Deutsche Nationalbibliothek
The Deutsche Nationalbibliothek lists this publication in the Deutsche Nationalbibliografie; detailed bibliographic data are available on the Internet at <http://dnb.d-nb.de>.

© 2013 Wiley-VCH Verlag GmbH & Co. KGaA, Boschstr. 12, 69469 Weinheim, Germany

All rights reserved (including those of translation into other languages). No part of this book may be reproduced in any form – by photoprinting, microfilm, or any other means – nor transmitted or translated into a machine language without written permission from the publishers. Registered names, trademarks, etc. used in this book, even when not specifically marked as such, are not to be considered unprotected by law.

Composition   Laserwords Private Limited, Chennai, India

Printing and Binding   Markono Print Media Pte Ltd, Singapore

Cover Design   Schulz Grafik-Design, Fußgönheim

Print ISBN: 978-3-527-31903-9
ePDF ISBN: 978-3-527-64712-5
oBook ISBN: 978-3-527-64713-2

Printed in Singapore
Printed on acid-free paper

Dedicated to our kids
Lalita, Max, Nils, Filip, Lukas, Sita, Yanning Zhang and Dorian Gopal

Contents

About the Editors *XIX*

Series Editor Preface *XXI*

Volume Editors Preface *XXIII*

List of Contributors *XXVII*

Part I Physical and Mathematical Fundamentals *1*

1 Introduction: Issues in Microsystems Modeling *3*
Gary K. Fedder and Tamal Mukherjee
1.1 The Need for System-Level Models for Microsystems *3*
1.2 Coupled Multiphysics Microsystems *4*
1.3 Multiscale Modeling and Simulation *6*
1.4 System-Level Model Terminology *7*
1.5 Automated Model Order Reduction Methods *9*
1.6 Handling Complexity: Following the VLSI Paradigm *10*
1.7 Analog Hardware Description Languages *11*
1.8 General Attributes of System-Level Models *12*
1.9 AHDL Simulation Capabilities *13*
1.10 Composable Model Libraries *14*
1.11 Parameter Extraction, Model Verification, and Model Validation *15*
1.12 Conclusions *16*
References *17*

2 System-Level Modeling of MEMS Using Generalized Kirchhoffian Networks – Basic Principles *19*
Gabriele Schrag and Gerhard Wachutka
2.1 Introduction and Motivation *19*
2.2 Generalized Kirchhoffian Networks for the Tailored System-Level Modeling of Microsystems *20*

2.2.1	Generic Modeling Approach for Microdevices and Systems 21
2.2.2	From Continuous-Field Level to Compact Models 23
2.2.3	Approaches to Compact Model Derivation 29
2.2.3.1	Finite Network (FN) Models 29
2.2.3.2	Compact Modeling of MEMS 30
2.2.3.3	Mixed-Level Modeling (MLM) 31
2.2.3.4	Mathematical Model Order Reduction (MOR) 32
2.3	Application 1: Physics-Based Electrofluidic Compact Model of an Electrostatically Actuated Micropump 32
2.3.1	Compact Model of the Membrane Drive 34
2.3.2	Compact Model of the Valves 36
2.3.3	System-Level Model of the Tubes 38
2.3.4	Physics-Based System-Level Model of the Micropump 39
2.3.5	Model Calibration and Parameter Extraction 40
2.4	Application 2: Electrostatically Actuated RF MEMS Switch 41
2.4.1	Conclusions 46
	References 48
3	**System-Level Modeling of MEMS by Means of Model Order Reduction (Mathematical Approximations) – Mathematical Background** 53
	Lihong Feng, Peter Benner, and Jan G. Korvink
3.1	Introduction 53
3.2	Brief Overview 55
3.3	Mathematical Preliminaries 56
3.3.1	Scalar, Vector, and Matrix 57
3.3.2	Vector Space, Subspace, Linear Independence, and Basis 57
3.3.3	Laplace Transform 60
3.3.4	Rational Function 60
3.3.5	Norms 61
3.3.5.1	Vector and Matrix Norms 61
3.3.5.2	Vector Function Norms and Matrix Function Norms 62
3.4	Numerical Algorithms 63
3.5	Linear System Theory 66
3.5.1	Transfer Function 66
3.5.2	Measure of the Difference between Any Two Different LTI Systems 67
3.5.3	Controllability and Observability 68
3.5.4	Realization Theory 69
3.5.5	Stability and Passivity of a System 70
3.6	Basic Idea of Model Order Reduction 71
3.7	Moment-Matching Model Order Reduction 73
3.7.1	Moments and Moment Vectors 73
3.7.2	Computation of the Projection Matrices W and V 74
3.7.3	Different Choices of the Expansion Points 75
3.7.4	Development of Moment-Matching MOR 76

3.8	Gramian-Based Model Order Reduction	77
3.8.1	Motivation of Balanced Truncation	78
3.8.2	Balancing Transformation	78
3.8.3	Truncation	80
3.8.4	Computation of the Balancing Transformation	81
3.8.5	Acceleration Methods of Computing the Gramians	82
3.8.6	Extension to More General Systems	83
3.9	Stability, Passivity, and Error Estimation of the Reduced Model	84
3.9.1	Stability, Passivity, and Error Bound of Moment-Matching MOR	84
3.9.2	Stability, Passivity, and Error Bound of the Gramian-Based MOR	84
3.10	Dealing with Nonzero Initial Condition	85
3.11	MOR for Second-Order, Nonlinear, Parametric systems	86
3.12	Conclusion and Outlook	86
	References	87

4 Algorithmic Approaches for System-Level Simulation of MEMS and Aspects of Cosimulation 95

Peter Schneider, Christoph Clauß, Ulrich Donath, Günter Elst, Olaf Enge-Rosenblatt, and Thomas Uhle

4.1	Introduction	95
4.2	Mathematical Structure of MEMS Models	96
4.2.1	Differential and Algebraic Equations	97
4.2.1.1	Multibody Systems	98
4.2.1.2	Fluidic Systems	99
4.2.1.3	Networks	101
4.2.2	Boolean Equations and Finite State Machines	102
4.3	General Approaches for System-Level Model Description	104
4.3.1	SPICE and Macromodeling	104
4.3.2	Model Description Languages	105
4.4	Numerical Methods for System-Level Simulation	107
4.4.1	Solution of Nonlinear DAEs	107
4.4.2	Mixed-Signal Simulation Cycle	109
4.4.3	Cosimulation	110
4.4.3.1	Coupling Algorithms	111
4.4.3.2	Cosimulation Interface	112
4.5	Emerging Problems and Advanced Simulation Techniques	113
4.5.1	Discontinuous Forcing Functions	114
4.5.2	Structural Changes in Model Equations	117
4.6	Conclusion	118
	References	118

Part II Lumped Element Modeling Method for MEMS Devices 123

5 System-Level Modeling of Surface Micromachined Beamlike Electrothermal Microactuators 125
Ren-Gang Li and Qing-An Huang
5.1 Introduction 125
5.2 Classification and Problem Description 127
5.2.1 In-Plane 127
5.2.1.1 U-Shaped Actuator 127
5.2.1.2 Bent-Beam Actuators 128
5.2.1.3 Long-Short Beam Actuator 128
5.2.2 Out of Plane 129
5.2.3 Material Properties 129
5.2.3.1 Thermal Conductivity 129
5.2.3.2 Specific Heat 130
5.2.3.3 Resistivity 130
5.2.3.4 Coefficient of Thermal Expansion 130
5.3 Modeling 131
5.3.1 Electrothermal Model of a Beam 131
5.3.2 Thermomechanical Model of the Beam 134
5.4 Solving 136
5.4.1 Equivalent Circuit of a Coupled Electrothermal Model 137
5.4.2 Equivalent Circuit of the Thermomechanical Model of a Beam 138
5.5 Case Study 139
5.6 Conclusion and Outlook 142
References 143

6 System-Level Modeling of Packaging Effects of MEMS Devices 147
Jing Song and Qing-An Huang
6.1 Introduction 147
6.2 Packaging Effects of MEMS and Their Impact on Typical MEMS Devices 148
6.2.1 Accelerometers 148
6.2.2 Gyros 149
6.2.3 Pressure Sensors 149
6.2.4 Thermal Actuators 149
6.2.5 Hall Sensors 150
6.3 System-Level Modeling 150
6.3.1 System Partitioning 151
6.3.2 Behavioral Modeling of Single Substructures 152
6.3.2.1 Microbeam Model 152
6.3.2.2 Chip/Adhesive/Package Model 153
6.3.2.3 Support Model 155
6.3.3 Element Integration 156
6.3.4 FEM and Experimental Validation 157

6.4	Conclusion and Outlook *160*	
	References *160*	

7 Mixed-Level Approach for the Modeling of Distributed Effects in Microsystems *163*
Martin Niessner and Gabriele Schrag

7.1	General Concept of Finite Networks and Mixed-Level Models *163*	
7.2	Approaches for the Modeling of Squeeze Film Damping in MEMS *165*	
7.2.1	Reynolds Equation-Based Modeling Strategies *166*	
7.2.2	Motivation for Using Mixed-Level Modeling *168*	
7.3	Mixed-Level Modeling of Squeeze Film Damping in MEMS *169*	
7.3.1	Finite Network-Based Evaluation of the Reynolds Equation *170*	
7.3.2	Physics-Based Lumped Element Models *173*	
7.3.2.1	Boundary Model *173*	
7.3.2.2	Hole Model *174*	
7.3.3	Gas Rarefaction Effects *175*	
7.3.4	Calculation of the Total Damping Force *177*	
7.3.5	Coupling with Mechanical Models *177*	
7.3.6	Automated Model Generation *178*	
7.4	Evaluation *179*	
7.4.1	Numerical Evaluation *179*	
7.4.2	Experimental Evaluation *180*	
7.4.3	Comparison with Alternative Damping Models *185*	
7.5	Conclusion *186*	
	References *187*	

8 Compact Modeling of RF-MEMS Devices *191*
Jacopo Iannacci

8.1	Introduction *191*
8.2	Brief Description of the MEMS Compact Modeling Approach *192*
8.3	RF-MEMS Multistate Attenuator Parallel Section *194*
8.4	RF-MEMS Multistate Attenuator Series Section *202*
8.5	Whole RF-MEMS Multistate Attenuator Network *205*
8.6	Conclusions *207*
	References *208*

Part III Mathematical Model Order Reduction for MEMS Devices *211*

9 Moment-Matching-Based Linear Model Order Reduction for Nonparametric and Parametric Electrothermal MEMS Models *213*
Tamara Bechtold, Dennis Hohlfeld, Evgenii B. Rudnyi, and Jan G. Korvink

9.1	Introduction *213*
9.2	Methodology for Applying Model Order Reduction to Electrothermal MEMS Models: Review of Achieved Results and Open Issues *213*
9.3	MEMS Case Study – Silicon-Based Microhotplate *220*

9.4	Application of the Reduced-Order Model for the Parameterization of the Controller *223*	
9.5	Application of Parametric Reduced-Order Model to the Extraction of Thin-Film Thermal Parameters *227*	
9.5.1	Parametric Model Order Reduction *228*	
9.5.2	Parameter Extraction Methodology *230*	
9.6	Conclusion and Outlook *232*	
	References *234*	

10 Projection-Based Nonlinear Model Order Reduction *237*
Amit Hochman, Dmitry M. Vasilyev, Michał J. Rewieński, and Jacob K. White

10.1	Introduction *237*
10.2	Problem Specification *238*
10.3	Projection Principle and Evaluation Cost for Nonlinear Systems *239*
10.4	Taylor Series Expansions *240*
10.4.1	Microfluidic Channel Example *241*
10.4.2	Model Reduction via Quadratic Taylor Expansion *242*
10.4.3	Stability Issues *244*
10.4.4	Taylor Series Expansions: Summary *245*
10.5	Trajectory Piecewise-Linear Method *245*
10.5.1	Nonlinear Transmission-line Model *246*
10.5.2	Stability Issues *247*
10.5.3	TPWL Summary *249*
10.6	Discrete Empirical Interpolation method *250*
10.6.1	Thermal Analysis *251*
10.6.2	Stability Issues *253*
10.6.3	DEIM Summary *254*
10.7	A Comparative Case Study of an MEMS Switch *255*
10.7.1	Pull-In Effect *256*
10.7.2	Generalizing from Training Inputs *258*
10.7.3	Harmonic Distortion *258*
10.7.4	A Note about CPU Times *259*
10.8	Summary and Outlook *260*
	Acknowledgment *260*
	References *261*

11 Linear and Nonlinear Model Order Reduction for MEMS Electrostatic Actuators *263*
Jan Lienemann, Emanuele Bertarelli, Andreas Greiner, and Jan G. Korvink

11.1	Introduction *263*
11.2	The Variable Gap Parallel Plate Capacitor *264*
11.2.1	Pull-In *266*
11.2.2	Regimes of the Trajectory *268*
11.3	Model Order Reduction Methods *269*

11.3.1	Representation of Nonlinearities	*270*
11.3.2	MOR Methods for Second-Order Linear Systems	*270*
11.3.3	MOR Methods for Nonlinear Systems	*272*
11.3.3.1	Polynomial Projection	*273*
11.3.3.2	Trajectory Piecewise-Linear (TPWL) Method	*274*
11.4	Example 1: IBM Scanning-Probe Data Storage Device	*275*
11.4.1	Cantilever Model	*277*
11.4.2	Model Order Reduction with Polynomial Projection	*277*
11.4.2.1	Extraction of Nonlinear System	*277*
11.4.2.2	Polynomial Approximation	*278*
11.4.2.3	Model Order Reduction of the Polynomial System	*278*
11.4.3	Results	*279*
11.4.4	Discussion	*281*
11.5	Example 2: Electrostatic Micropump Diaphragm	*281*
11.5.1	Plate Model Formulation	*282*
11.5.2	Model Order Reduction by TPWL and Arnoldi Methods	*284*
11.6	Results and Discussion	*285*
11.7	Conclusions	*286*
	Acknowledgments	*287*
	References	*287*

12 Modal-Superposition-Based Nonlinear Model Order Reduction for MEMS Gyroscopes *291*
Jan Mehner

12.1	Introduction	*291*
12.2	Model Order Reduction via Modal Superposition	*292*
12.3	MEMS Testcase: Vibratory Gyroscope	*293*
12.4	Flow Chart of the Nonlinear Model Order Reduction Procedure	*294*
12.5	Theoretical Background of Modal Superposition Technologies	*295*
12.5.1	Linear Mechanical Systems	*295*
12.5.2	Nonlinear Electromechanical Interactions and Body Loads of Capacitive Sensors	*297*
12.5.3	Parametric Reduced Order Models for Packaging Interactions	*298*
12.6	Specific Algorithms of the Reduced Order Model Generation Pass	*299*
12.6.1	Extraction of Body Load Contribution Vectors for Modal Superposition	*299*
12.6.2	Extraction of Capacitances for Comb Cell Conductors and Platelike Capacitors	*301*
12.6.3	Data Sampling and Function Fit Procedures for Multivariable Capacitances	*302*
12.7	System Simulations of MEMS Based on Modal Superposition	*304*
12.7.1	Behavioral Analysis of Vibratory Gyroscopes in Matlab/Simulink	*304*

12.7.2	Simulations with Reduced Order Models Based on Kirchhoffian Networks 306
12.7.3	Expansion of System Simulation Results to Full Order FEM Models 306
12.8	Conclusion and Outlook 307
	References 308

Part IV Modeling of Entire Microsystems 311

13	Towards System-Level Simulation of Energy Harvesting Modules 313
	Dennis Hohlfeld, Tamara Bechtold, Evgenii B. Rudnyi, Bert Op het Veld, and Rob van Schaijk
13.1	Introduction 313
13.1.1	Wireless Autonomous Sensor Nodes 314
13.1.2	Micropower Module 315
13.1.3	Vibrational Harvesters 316
13.2	Design and Fabrication of the Piezoelectric Generator 317
13.3	Experimental Results 318
13.4	Modeling and Simulation 318
13.4.1	Lumped Element Modeling 319
13.4.2	Finite Element Modeling 320
13.4.3	Model Order Reduction 322
13.4.4	Transient MEMS Circuit Cosimulation 323
13.4.5	Harmonic MEMS Circuit Cosimulation 325
13.5	Maximum Power Point for the Piezoelectric Harvester 327
13.5.1	Complex Matching 328
13.5.2	Real Matching 330
13.5.3	Consequence of Matching 331
13.6	Conclusions and Outlook 332
	References 333

14	**Application of Reduced Order Models in Circuit-Level Design for RF MEMS Devices** 335
	Laura Del Tin, Evgenii B. Rudnyi, and Jan G. Korvink
14.1	Model Equations for RF MEMS Devices 337
14.2	Extraction of the Reduced Order Model 340
14.2.1	Second Order ODE Systems 341
14.2.2	Handling Nonlinearities in the Input Function 341
14.2.3	Extraction Procedure 343
14.3	Application Examples 345
14.3.1	Vibrating Devices 345
14.3.2	Microswitch 351
14.4	Conclusion and Outlook 354
	References 355

15	**SystemC AMS and Cosimulation Aspects** *357*	
	François Pêcheux, Marie-Minerve Louërat, and Karsten Einwich	
15.1	Introduction *357*	
15.2	Heterogeneous Modeling with SystemC AMS *358*	
15.2.1	SystemC AMS Timed Dataflow (TDF) *358*	
15.2.2	Timed Dataflow Model of the Accelerometer *359*	
15.2.3	Electrical Linear Network Model of the MEMS Accelerometer *361*	
15.3	Case Study: Detection of Seismic Perturbations Using the Accelerometer *363*	
15.3.1	Seismic Stimuli and Seismic Sensors in SystemC AMS TDF *365*	
15.3.2	Digital Controller in SystemC *365*	
15.3.3	2.4 GHz RF Transceiver *366*	
15.3.4	Battery Modeling *368*	
15.3.5	Embedded Software, Cross-Compiled GNU GCC Application for MIPS *369*	
15.3.6	Simulation Results *369*	
15.4	Conclusion *370*	
	Appendix *371*	
	References *374*	
16	**System Level Modeling of Electromechanical Sigma–Delta Modulators for Inertial MEMS Sensors** *377*	
	Michael Kraft	
16.1	Introduction and Motivation *377*	
16.2	Second Order Electromechanical $\Sigma\Delta M$ for a MEMS Accelerometer *380*	
16.2.1	Basic Model *380*	
16.2.2	Advanced Model *386*	
16.3	Higher Order Electromechanical $\Sigma\Delta M$ for MEMS Accelerometer *391*	
16.3.1	Design Methodology for Higher Order EM$\Sigma\Delta M$ *391*	
16.3.2	Example Design *392*	
16.3.3	Feedback Linearization *395*	
16.3.4	Advanced Model *396*	
16.4	Higher Order Electromechanical $\Sigma\Delta M$ for MEMS Gyroscopes *397*	
16.5	Concluding Remarks *400*	
	References *401*	
	Part V Software Implementations *405*	
17	**3D Parametric-Library-Based MEMS/IC Design** *407*	
	Gunar Lorenz and Gerold Schröpfer	
17.1	About Schematic-Driven MEMS Modeling *407*	
17.2	A 3D Parametric Library for MEMS Design–MEMS+® *409*	
17.2.1	3D Design Entry for MEMS *409*	

17.2.2	MEMS Model Library	410
17.2.3	Integration with System Simulators	412
17.2.4	Integration with MATLAB and Simulink	413
17.2.5	Integration with EDA Tools	414
17.3	Toward Manufacturable MEMS Designs	415
17.3.1	Parameterization of Process and Material Properties	415
17.3.2	Process Design Kits	418
17.4	Micromirror Array Design Example	419
17.5	Conclusions	422
	References	423
18	**MOR for ANSYS**	**425**
	Evgenii B. Rudnyi	
18.1	Introduction	425
18.2	Practice-Oriented Research during the Development of MOR for ANSYS	426
18.3	Programming Issues	429
18.3.1	Obtaining System Matrices from ANSYS	430
18.3.2	Solvers	431
18.4	Open Problems	432
18.5	Conclusion	436
	References	437
19	**SUGAR: A SPICE for MEMS**	**439**
	Jason V. Clark	
19.1	Introduction	439
19.2	SUGAR	439
19.2.1	Equation of Motion	440
19.3	SUGAR-Based Applications	444
19.3.1	Library	444
19.3.2	Design/Simulation	445
19.3.3	Layout Generation	445
19.3.4	Common Ground Tracers	445
19.3.5	Etch Holes	447
19.3.6	Multilayer Pads	447
19.3.7	Parameterized Arrays	447
19.3.8	Optimization	448
19.3.9	NEMS	448
19.3.10	PSugar	450
19.3.11	GUI Configuration	450
19.3.12	DAEs	451
19.3.13	Simulation	453
19.4	Integration of SUGAR + COMSOL + SPICE + SIMULINK	454
19.4.1	Integration	456
19.4.2	Verification	457

19.5	Conclusion	457
	References	458

20 Model Order Reduction Implementations in Commercial MEMS Design Environment 461
Sandeep Akkaraju

20.1	Introduction	461
20.1.1	*Ab initio* (First Principles) Simulations	461
20.1.2	Technology CAD (TCAD)	462
20.1.3	Schematic or Component-Based Design (Top-Down Design)	462
20.1.4	Layout-Based Design (Bottom-Up Design)	463
20.1.5	System Model Extraction (SME)	464
20.1.6	Verification	465
20.1.7	System Simulation	467
20.2	IntelliSense's Design Methodology	467
20.2.1	From the Top Down	467
20.2.2	One Step at a Time	468
20.2.3	Closing the Loop	469
20.3	Implementation of System Model Extraction in IntelliSuite	470
20.3.1	High-Level Overview	470
20.3.2	Capturing Residual Stress and Film Damping Effects	471
20.3.3	Implementation	473
20.4	Benchmarks	474
20.4.1	Accelerometer	474
20.4.2	Inertial Gyroscope	474
20.4.3	Fluid Damping	477
20.4.4	Coupled Package-Device Modeling	478
20.5	Summary	480
	References	481

21 Reduced Order Modeling of MEMS and IC Systems – A Practical Approach 483
Sebastien Cases and Mary-Ann Maher

21.1	Introduction	483
21.2	The MEMS Development Environment	484
21.3	Modeling Requirements and Implementation within SoftMEMS Simulation Environment	485
21.3.1	Models and Inputs	487
21.3.2	Modeling Process	490
21.3.3	Model Output	492
21.3.4	Parameterization	492
21.4	Applications	494
21.5	Conclusions and Outlook	498
	References	498

22	**A Web-Based Community for Modeling and Design of MEMS** *501*
	Peter J. Gilgunn, Jason V. Clark, Narayan Aluru, Tamal Mukherjee, and Gary K. Fedder
22.1	Introduction *501*
22.2	The MEMS Modeling and Design Landscape *501*
22.3	Leveraging Web-Based Communities *502*
22.3.1	Concepts of Web-Based Design *503*
22.3.2	Design Community Constitution *504*
22.4	MEMS Modeling and Design Online *505*
22.4.1	System-Level Modeling of MEMS *506*
22.4.2	Web-Based Community Conventions *507*
22.5	Encoding MEMS Behavioral Models *508*
22.5.1	Inside the Black Box – Nonlinear Beam *509*
22.5.2	Behavioral Model Performance – Nonlinear Beam *511*
22.5.3	Behavioral Model Extensions *512*
22.5.4	Behavioral Model Composability *513*
22.6	Conclusions and Outlook *515*
	References *515*

Index *519*

About the Editors

Tamara Bechtold is currently an Interim Professor for microsystems simulation at the University of Freiburg. She obtained her MSc in microelectronics and microsystems engineering from the University of Bremen, Germany, in 2000 and her PhD in microsystem simulation from the University of Freiburg, Germany, in 2005. Between 2006 and 2010, Dr. Bechtold worked as an experienced researcher for Philips Research Laboratories and NXP Seminconductors in Eindhoven, The Netherlands. The objective of her research work was to enhance the standard IC design flow through model order reduction and optimization modules. From 2010 to 2011, Dr. Bechtold was with CADFEM GmbH in Stuttgart, Germany, supporting industry and academia in the application of advanced modeling and simulation tools for system-level simulation and electromagnetic device simulation. Dr. Bechtold is the author or coauthor of over 40 technical publications in the area of microsystem simulation and the lead author of the textbook Fast Simulation of Electro-Thermal MEMS: Efficient Dynamic Compact Models, published by Springer.

Her research interests cover applications of advanced mathematical methods of model order reduction and topology optimization to engineering problems and a multiphysic modeling on the system and device levels.

Gabriele Schrag is currently heading a research group at the Munich University of Technology, Germany, working in the field of MEMS modeling with a focus on virtual prototyping and predictive simulation methodologies, parameter extraction, and model verification for microdevices and microsystems. She studied physics at the University of Stuttgart and received her doctorate (with honors) from the Munich University of Technology in 2002, her thesis covering the "Modeling of Coupled Effects in Microsystems" with special emphasis on fluid-structure interaction and viscous-damping effects. Gabriele Schrag authored and co-authored more than 70 publications in technical journals and conference proceedings.

Lihong Feng is a team leader in the research group of Computational Methods in Systems and Control theory headed by Professor Peter Benner, Max Planck Institute for Dynamics of Complex Technical Systems in Magdeburg, Germany. After her PhD from Fudan University in Shanghai, China, she joined the faculty of the State Key Laboratory of Application-Specific Integrated Circuits (ASIC) & System, Fudan University, Shanghai, China. From 2007 to 2008 she was a Humboldt research fellow in the working group of Mathematics in Industry and Technology at the Technical University of Chemnitz, Germany. During 2009–2010, she worked in the Laboratory for Microsystem Simulation, Department of Microsystems Engineering, University of Freiburg, Germany. Her research interests are in the field of reduced order modelling and fast numerical algorithms for control and optimization in Chemical Engineering, MEMS simulation, and circuit simulation.

Series Editor Preface

The book you hold in your hands was planned and edited, upon my instigation, by three very capable young scientists: Dr. Tamara Bechtold, Dr. Lihong Feng, and Dr. Gabriele Schrag. Each of them has made a strong impact on the field, pushing physical-based and model order reduction (MOR)-based compact modeling into engineering applications, especially into MEMS applications. Each would certainly have been more than capable of editing the book alone. That they did so as a team is all the better for us readers and I wish you many pleasurable hours in perusing its content.

The book is very timely, because the broad area of compact modeling has considerably matured over the past decade, conquering a vast number of application-specific issues. On the one hand, MOR has the potential to completely revolutionise scientific computation, but this awareness will only come when evidence and good experience spreads around. On the other hand, as physical-based compact modeling gains a robust foundation, its application has the potential to excite theoreticians. For the newcomer to the area, the book will provide a valuable introduction to the state of the art and access to the relevant literature. For the experienced user or applicator, the book will show clever techniques from neighbouring areas. Because the techniques described in the book represent both, a means to considerably speed up equation solution and ways to create high-quality compact models, the book should also lower the acceptance barrier to computational design software developers. In this spirit, the book does not stop with the theory, but also addresses practical usage and implementation issues.

Drs. Bechtold, Feng and Schrag have collected together an impressive and substantial set of articles written by a fair proportion of the key players in the area of compact modeling as applied to MEMS, from academia, the major software vending industry, and industrial players. The authors come from eight countries, and cover a very broad and representative range of research application areas.

I want to take the opportunity to thank the numerous authors for their hard work.

There is much pressure today for authors to only publish in journal literature. From time to time it is good, I think, to write an overview that is accessible to a broader community, and that consolidates key results found only in highly specialized literature. It helps to advertise the key idea, to make the community aware of the results and demonstrate the benefit. This book would not have been possible if the top R&D authors had not taken some of their valuable time to do this work. And for this, I want to thank them most heartily and assure them that their work will do much good during its lifetime. I also want to thank the Wiley staff for their strong support for our writing projects, the excellent team work, all of which makes the writing process palatable. The final printed result speaks for itself!

Karlsruhe *Jan G. Korvink*
October 2012

Volume Editors Preface

The rapid progress in MEMS technology and the evolution from a limited set of well-established applications, example, in automotive industry, print heads, and digital light projection, to other fields mainly driven by consumer applications (such as image stabilization, smart phones, game consoles, etc.) opens a new market for these devices. The vast potential application areas for MEMS in the developing "Internet of Things" are still to be explored. On the other hand, it simultaneously creates the need for fast, efficient, and adequate design and optimization tools not only for stand-alone devices, but also for entire microsystems comprising the MEMS component, the attached control and read-out circuitry and the package, while additionally considering environmental impacts potentially affecting the system functionality.

Owing to their nature as transducing elements, an inherent feature of many MEMS devices is that multiple energy domains and their couplings determine their operation. Furthermore, they often exhibit complex geometrical structures. Modeling these features leads to coupled-domain and large-scale ordinary differential equation systems for each single device. In addition, taking into account that the optimal performance of a microsystem is generally achieved by optimizing the entire system including the interplay of all its components, there is a strong need for device models with reduced order of complexity. Various compact modeling methods are able to reduce the number of degrees of freedom drastically in order to make these device models tractable.

In the field of semiconductor devices and modules, comprehensive simulation environments and methodologies for the top-down and bottom-up design are well established since many years and environments for the virtual prototyping of devices and systems including their compact modeling are in a mature state. In contrast, this evolution for MEMS has just started within the last two decades and is still in progress. In the meanwhile, several approaches for system-level modeling have been established within the world-wide community and found their way into practical application and – to a certain extent – into commercially available software implementations.

This book intends to bring together worldwide experts in this field in order to give a broad overview on the state of the art in system-level modeling of MEMS, with special emphasis on the theoretical fundamentals of compact modeling, the application of different approaches to specific problem classes and, finally, already existing implementations of the presented methodologies into commercially available software environments.

In order to cover the numerous aspects of this topic adequately, this book is partitioned into five major parts.

The first part focuses on the theoretical fundamentals of compact modeling. A comprehensive introduction to modeling issues in microsystems in Chapter 1 is followed by a chapter on the theoretical fundamentals of generalized Kirchhoffian network theory. It is demonstrated that this method can be derived thoroughly from a generic, thermodynamical description, and, thus, provides a generic framework for system-level modeling, which is demonstrated for two typical MEMS case studies. The theoretical background of mathematical model order reduction (MOR) based on various projection techniques is given in Chapter 3 providing the basis for linear and also more advanced, that is, non-linear and parameterized MOR methods. The last contribution of Part I, Chapter 4, focuses on algorithmic approaches for system-level simulation and aspects of co-simulation, which is of specific interest with regard to the system-aspect in MEMS.

The second and third parts of this book are dedicated to the application of physics or lumped element-based compact models and MOR methods to specific examples and problem classes of MEMS. Part II contains contributions on compact modeling methodologies applied to electrothermal actuators taking into account also packaging effects, on RF MEMS devices and networks, and on a mixed-level approach for the simulation of distributed effects like viscous damping. Part III demonstrates the application of linear MOR methods as well as their extension to non-linear systems and parameterized problems. The applications range from parameter optimization of a micro hotplate system over microfluidic channels and non-linear transmission lines to electro-mechanical actuators like MEMS switches, where highly non-linear processes are involved in the operation. This part is concluded by a contribution on reduced order models of MEMS gyroscopes based on the modal superposition technique, but extended to non-linear phenomena.

The fourth part focuses completely on the system aspect, that means the transducer with attached control, read-out and supply circuitry. It answers the question "What can the compact model of the single transducer be used for?". The first system presented here is a MEMS-based energy harvesting module for application in wireless sensor networks. This module is composed of the piezoelectric transducer, the power management and the energy storage components. Second, a system model of an RF MEMS switch and a resonator with attached control circuitry is demonstrated and discussed. The third contribution focuses on system-level simulation applying SystemC and SystemC-AMS for model implementation, which is demonstrated on a wireless sensor network for the detection of seismic perturbations. Finally, a contribution on inertial sensors like accelerometers or gyroscopes demonstrates the benefit of compact modeling when designing closed-loop, force-feedback control systems for read-out and stabilization of mechanically movable microstructures.

Finally, an overview on the most-known software environments evolved specifically in the field of microsystem modeling during the last years is given in Part V. The individual software suppliers point out their respective views on the needs and requirements arising from a customer-oriented perspective and the modeling philosophies and methodologies lying behind each tool as well as its applicability with view to certain demonstrator systems are given.

The book is concluded by a contribution, which gives a vision on how experienced researchers, users from industry and university, as well as novices in the field may benefit from a web-based worldwide interactive platform including knowledge management, discussion fora, online simulation tools, model libraries, and so on. This closes the circle to the first contribution of the book and encourages all experts to bring their knowledge together in order to push this interesting and challenging field forward.

The wish and hope of the editors is that the broad collection of expertise accomplished by this book will make a considerable contribution in this respect. Finally, we would like to express our gratitude to all authors and collaborators, who supported this project by providing excellent surveys of their research topics, and to our families, who had to exercise a lot of patience during the composition of this volume.

August 2012 *Tamara Bechtold, Gabriele Schrag, and Lihong Feng*

List of Contributors

Sandeep Akkaraju
IntelliSense
175 Grove St
Wellesley, MA 02482
USA

Narayan Aluru
Mechanical Science and
Engineering
Department University of Illinois
at Urbana-Champaign
2140 Mechanical Engineering
Laboratory
1206 West Green Street
MC-244
Urbana, IL 61801
USA

Tamara Bechtold
CADFEM GmbH
Leinfelder Str. 60
70771 Leinfelden-Echterdingen
Germany

and

University of Freiburg
IMTEK
Georges-Köhler-Allee 103
79110 Freiburg
Germany

Peter Benner
Max Planck Institute for
Dynamics of Complex
Technical Systems
Sandtorstr 1
39106 Magdeburg
Germany

Emanuele Bertarelli
Politecnico di Milano
Department of Structural
Engineering
Piazza Leonardo da Vinci 32
20133 Milano
Italy

Sebastien Cases
SoftMEMS EURL
2 rue Eugène Sue
38100 Grenoble
France

Jason V. Clark
Purdue University
School of Electrical and
Computer Engineering
School of Mechanical
Engineering
Discovery Park
Nanotechnology Center
1205 West State Street
BRK-1289
West Lafayette, IN 94906-2057
USA

Christoph Clauß
Fraunhofer-Institute for
Integrated Circuits IIS
Design Automation Division EAS
Zeunerstraße 38
01069 Dresden
Germany

Laura Del Tin
Delft University of Technology
Department of Precision and
Microsystems Engineering
Mekelweg 2
2628 CD Delft
The Netherlands

Ulrich Donath
Fraunhofer-Institute for
Integrated Circuits IIS
Design Automation Division EAS
Zeunerstraße 38
01069 Dresden
Germany

Karsten Einwich
Fraunhofer Institute for
Integrated Circuits IIS
Design Automation Division EAS
Zeunerstraße 38
01069 Dresden
Germany

Günter Elst
Fraunhofer Institute for
Integrated Circuits IIS
Design Automation Division EAS
Zeunerstraße 38
01069 Dresden
Germany

Olaf Enge-Rosenblatt
Fraunhofer Institute for
Integrated Circuits IIS
Design Automation Division EAS
Zeunerstraße 38
01069 Dresden
Germany

Gary K. Fedder
Carnegie Mellon University
The Institute for Complex
Engineered Systems
The Robotics Institute and
The Department of Electrical and
Computer Engineering
5000 Forbes Avenue
Pittsburgh, PA 15213-3890
USA

Lihong Feng
Max Planck Institute for
Dynamics of Complex
Technical Systems
Sandtorstr 1
39106 Magdeburg
Germany

Peter J. Gilgunn
Carnegie Mellon University
The Institute for Complex
Engineered Systems
5000 Forbes Avenue
Pittsburgh, PA 15213-3890
USA

List of Contributors

Andreas Greiner
University of Freiburg
Department of Microsystem
Engineering (IMTEK)
Georges-Köhler-Allee 103
79110 Freiburg
Germany

Amit Hochman
Research Laboratory of
Electronics
Massachusetts Institute of
Technology
77 Massachusetts Avenue
Cambridge, MA 02139
USA

Dennis Hohlfeld
Reutlingen University
School of Engineering
Alteburgstr. 150
72762 Reutlingen
Germany

Qing-An Huang
Southeast University
Key Laboratory of MEMS of the
Ministry of Education
Nanjing 210096
China

Jacopo Iannacci
MEMS Research Group
Center for Materials and
Microsystems - CMM
Fondazione Bruno Kessler - FBK
via Sommarive 18
38123 Trento
Italy

Jan G. Korvink
University of Freiburg
IMTEK
Georges-Köhler-Allee 103
79110 Freiburg
Germany

and

Freiburg Institute for
Advanced Studies (FRIAS)
Albertstraße 19
79104 Freiburg
Germany

Michael Kraft
University of Southampton
Electronics and Computer
Science
Highfield Campus
Southampton SO17 1BJ
UK

Ren-Gang Li
Southeast University
Key Laboratory of MEMS of the
Ministry of Education
Nanjing 210096
China

Jan Lienemann
University of Freiburg
Department of Microsystem
Engineering (IMTEK)
Georges-Köhler-Allee 103
79110 Freiburg
Germany

and

Schmid & Partner
Engineering AG
Zeughausstrasse 43
8004 Zürich
Switzerland

Gunar Lorenz
European Headquarter and
Development Center
Coventor Sarl
3 Avenue du Quebec
91140 Villebon sur Yvette
France

Marie-Minerve Louërat
Université Pierre et Marie Curie
LIP6 Laboratory
4 place Jussieu
75005 Paris
France

Mary-Ann Maher
SoftMEMS LLC
2391 Nobili Avenue
Santa Clara, CA 95051
USA

Jan Mehner
Chemnitz University of
Technology
Department of Microsystems and
Precision Engineering
Faculty of Electrical Engineering
and Information Technology
Reichenhainer Str. 70
09107 Chemnitz
Germany

Tamal Mukherjee
Carnegie Mellon University
The Department of Electrical and
Computer Engineering
5000 Forbes Avenue
Pittsburgh, PA 15213-3890
USA

Martin Niessner
Institute for Physics of
Electrotechnology
Technische Universität München
Arcisstraße 21
80290 München
Germany

Bert Op het Veld
Philips Research Laboratories
High Tech Campus 31
5656 AE Eindhoven
The Netherlands

François Pêcheux
Université Pierre et Marie Curie
LIP6 Laboratory
4 place Jussieu
75005 Paris
France

Michał J. Rewieński
Research Laboratory of
Electronics
Massachusetts Institute of
Technology
77 Massachusetts Avenue
Cambridge, MA 02139
USA

Evgenii B. Rudnyi
CADFEM GmbH
Leinfelder Str. 60
70771 Leinfelden-Echterdingen
Germany

Peter Schneider
Fraunhofer-Institute for
Integrated Circuits IIS
Design Automation Division EAS
Zeunerstraße 38
01069 Dresden
Germany

Gabriele Schrag
Institute for Physics of
Electrotechnology
Technische Universität München
Arcisstraße 21
80290 München
Germany

Gerold Schröpfer
European Headquarter and
Development Center
Coventor Sarl
3 Avenue du Quebec
91140 Villebon sur Yvette
France

Jing Song
Southeast University
Key Laboratory of MEMS of the
Ministry of Education
Nanjing 210096
China

Thomas Uhle
Fraunhofer-Institute for
Integrated Circuits IIS
Design Automation Division EAS
Zeunerstraße 38
01069 Dresden
Germany

Rob van Schaijk
IMEC/Holst Centre
High Tech Campus 31
5656 AE Eindhoven
The Netherlands

Dmitry M. Vasilyev
Research Laboratory of
Electronics
Massachusetts Institute of
Technology
77 Massachusetts Avenue
Cambridge, MA 02139
USA

Gerhard Wachutka
Institute for Physics of
Electrotechnology
Technische Universität München
Arcisstraße 21
80290 München
Germany

Jacob K. White
Research Laboratory of
Electronics
Massachusetts Institute of
Technology
77 Massachusetts Avenue
Cambridge, MA 02139
USA

Part I
Physical and Mathematical Fundamentals

1
Introduction: Issues in Microsystems Modeling
Gary K. Fedder and Tamal Mukherjee

1.1
The Need for System-Level Models for Microsystems

Multiphysics microsystems are having an increasing practical impact on our lives as the industry creates a wealth of new products based on microelectromechanical systems (MEMS) such as accelerometers, gyroscopes, resonant timers, microphones, radio-frequency (RF) switches, tunable RF passives, micro-optical displays, microvalves, and microfluidic total analysis systems. System-level modeling and simulation are essential tools in designing such complex multiphysics microsystems. This book provides an overview of system-level modeling methodologies and tools, drawing from experts in microsystems and tapping into their multiple perspectives. It acts as a resource for future researchers who wish to build on this foundation.

As the field continues to mature, more diverse and integrated microsystems will evolve from exploratory prototypes into tomorrow's product offerings. This high-level commercial activity is motivating the research on methodologies of accurate multiphysics system-level simulations that provide rapid analysis of iterative design and allow the transference and archiving of design knowledge. Simultaneous with these advances in multiphysics aspects, an increasing number of microsystems comprise a multiplicity of devices that are integrated with electronics. Such integrated microsystems require cosimulation of electronics and MEMS, further stimulating the need for system-level models in support of rapid and efficient development.

There is ample motivation to refine and to automate system-level modeling methodologies of multiphysics microsystems [1–3]. One can turn to the semiconductor industry to put into perspective the complexities of device modeling. In advanced complementary metal–oxide semiconductor (CMOS) electronics, sophisticated transistor models for each emerging technology node must be created before commencement of analog and digital circuit design. Foundries can justify employing hundreds of engineers and technicians to create these models in a relatively short time frame. The microsystems field does not have the ability to apply such a large amount of resources to complete multiphysics device models.

System-level Modeling of MEMS, First Edition. Edited by T. Bechtold, G. Schrag, and L. Feng.
© 2013 Wiley-VCH Verlag GmbH & Co. KGaA. Published 2013 by Wiley-VCH Verlag GmbH & Co. KGaA.

To exacerbate the issue, even in a fixed process, the multiplicity of MEMS devices renders it impractical to replicate the enormous effort given to transistor modeling. Instead, microsystems modeling efforts must leverage the CMOS design infrastructure while continuing to advance automated methodologies in order to meet the challenges.

A particularly important recent trend is the growth of the number of foundries offering custom MEMS process services and the emergence of CMOS MEMS process offerings within CMOS foundries. The "one process, one product" mantra that was common during the past two decades of MEMS commercial development must become a relic of the past for these latter foundry processes to be successful. A significant roadblock to process reuse is the lack of adoption of common modeling methodologies that enable designers to exploit existing MEMS processes for rapid creation of new products. Part of the solution, and the core mission of this book, is to make available a near comprehensive overview of the state of the art in corresponding system design and modeling tools and methodologies for microsystem developers.

This chapter first provides an introduction to coupled multiphysics phenomena and to multiscale modeling and simulation of microsystems. A concise glossary of system-level model terminology is next presented, followed by a short description of model order reduction (MOR) methods. The concepts of very-large-scale integration (VLSI) hierarchy and views are next presented as a means to handle complexity in the microsystem design process. Modern analog hardware description languages (AHDLs) for system-level model implementation are then introduced, followed by general attributes of AHDL models and capabilities of AHDL simulators. The chapter ends with consideration of multiphysics model libraries for microsystems design and the need for parameter extraction, model verification, and model validation to produce trusted models.

1.2
Coupled Multiphysics Microsystems

Microsystems are generally less than a cubic centimeter in size and have one or more critical aspects of operation that are dependent on micron-scale, or even smaller, dimensions. The small scale leads to an extremely tight coupling of multiphysics aspects arising from the processes and devices composing a microsystem. This intimate coupling sets microsystem design apart from most macroscale system design and presents subsequent challenges and opportunities for the modeling community. Microsystems pose complex problems that are at best difficult and time consuming to solve with continuum field analysis. Especially, in the case of time-stepping analyses, many problems are intractable using the currently available continuum field analysis software running on the fastest computers. Layered over these issues is the desire to perform iterative design that requires multiple sequential parametric analyses. Practical realization of rapid time-domain

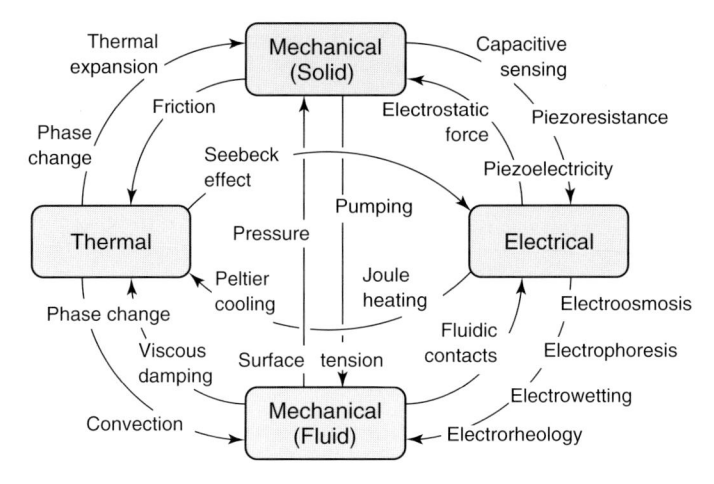

Figure 1.1 A sampling of coupled multiphysics in microsystems.

analyses in support of the design process for complex coupled-physics systems necessitates the formation of system-level models.

The list of coupled physics in microsystems is truly inexhaustible; a short summary here of some highlight areas helps underscore this observation. Electronic and active materials used in sensors and actuators have inherent coupling between energy domains, exemplified by electromechanical, electrothermal, magnetostrictive, piezoelectric, piezoresistive and shape memory (phase change) effects. A seminal paper by Middlehoek and Hoogerwerf [4] in 1986 categorized the coupling in solid-state sensors into six signal domains – radiant, mechanical, thermal, electrical, magnetic, and chemical – with physical sensing effects classified according to the domains of the input signal, the output signal, and the auxiliary energy source that modulates the output signal. A sampling of coupling between a subset of physical energy domains of major importance in microsystems is illustrated in Figure 1.1. Examples of actuator models are given in Chapters 5, 6, and 11 and modeling of energy harvester systems are overviewed in Chapter 13. Effects of induced stress in packaging represent a very important aspect of overall system simulation, with one modeling approach outlined in Chapter 6.

MEMS inertial sensors and resonant devices have complex interactions between inertial excitations, mechanical stresses in flexures, electrostatic fields with moving walls, thermal interactions with material properties, viscous losses from the surrounding ambient, and intrinsic losses within structural materials [5]. Examples of modeling and simulation of inertial microsystems are given in Chapters 12, 15, and 16.

RF microswitches and tunable capacitors add interactions of impedance matching, wiring and substrate loss at RF frequencies, long-term creep in metallic materials, and tribology of surfaces including dielectric charging phenomena and the physics of electrical contacts [6]. Modeling and cosimulation of RF MEMS with circuits is presented in Chapters 8, 10, and 14.

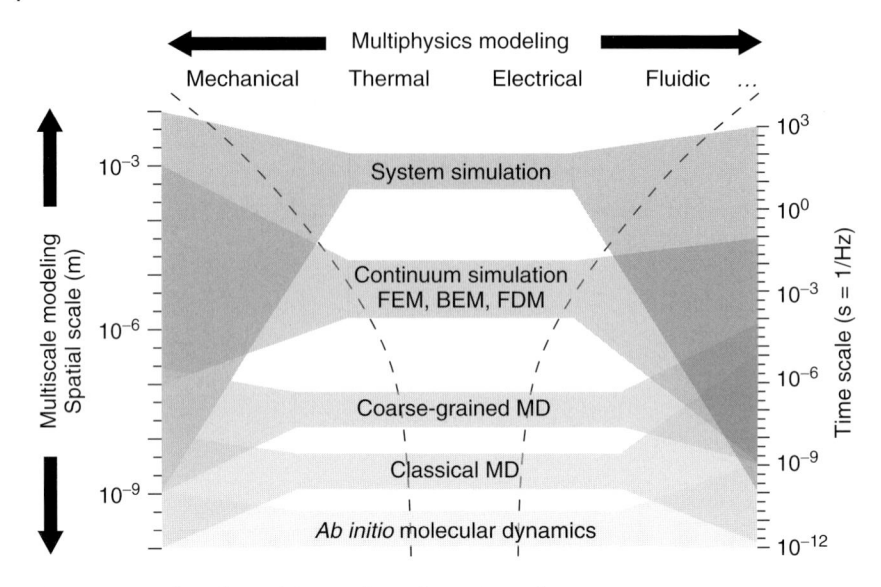

Figure 1.2 Multiscale modeling and simulation hierarchy.

Optical-based and probe-based microsystems are two emerging areas poised for future commercialization. Optical microsystems may include arrayed micromirror, lens, and waveguide components, [7] as well as, in more complex cases, exploit interactions of optical, thermal, and mechanical forces to create optically coupled microcavities [8] and chip-scale atomic clocks and sensors [9]. In nanoprobe and nanorelay systems, scaling of mechanics down to 50 nm and below brings the need to model atomic-scale forces, such as the van der Waals and the Casimir forces [10].

Microfluidic systems are governed by a complementary set of physics: incompressible flow, diffusion, convection, two-phase flow, electroosmotic and electrophoretic forces, surface tension, electrowetting, and fluid–particle interaction [11, 12]. Chapters 10 and 11 introduce some examples of microfluidic device modeling. An approach to model fluidic damping is covered in Chapter 7. A final extensive category comprises the innumerable chemical, biological, and material interactions exploited in chemical and bioMEMS devices [13]. The modeling procedures demonstrated in this book are applicable to these systems.

1.3
Multiscale Modeling and Simulation

A hierarchy of multiscale modeling and simulation is illustrated in Figure 1.2. System-level tools lie at the top of the simulation hierarchy and rely on behavioral models to describe the underlying physics. A system's behavior can be simulated across a very wide range of spatial scale and timescales; however, this occurs at the expense of the granularity of the system representation. Continuum simulation, which includes the finite-element method (FEM), boundary-element method (BEM), and finite-difference method (FDM), handles physics expressed by continuum partial differential equations. Molecular dynamics (MD) simulations represent interactions between atoms in the most fine-grained *ab initio* approaches. Classical MD approaches represent interactions between molecules and coarse-grained MD deals with interactions of larger molecular units, for example, grains in polycrystalline materials.

As briefly discussed in Section 1.2, simulation of a complete system with software that solves all physics at a continuum level (or at one of the MD levels) are generally impractical when run over timescales of interest for most microsystems (e.g., over billions of time steps). It is theoretically conceivable to embed continuum numerical analyses for devices into system simulation (this is sometimes called *mixed-mode simulation*); however, the resulting simulation speed is not acceptable for design. Furthermore, the continuum approach becomes even less practical for complex microsystems with numerous interacting components that require simulation over multiple spatial scales. The flexibility in abstracting behavior over multiple time and spatial scales in system-level simulation allows for the analysis of extremely complex coupled phenomena.

A key step in implementing system-level simulation is the translation of the physical behavior of the constitutive components in a system from the more fine-grained continuum level to more abstract coarse-grained models. Building this relationship between different tools in the simulation hierarchy, where each tool has viable utility at different spatial and time scales, is known as *multiscale modeling*. Information is passed up the hierarchy from fine-grained simulation to fill in physical parameter values, material property functions, or other behavioral relationships in a more coarse-grained model.

An important challenge in system-level modeling is the preservation of accuracy from fine-grained simulation to a degree that is deemed adequate. For the simulator to run in a reasonable time, the system-level model should only include the degrees of freedom (DOFs) necessary to capture the relevant physics. Very handy in this sense are the mathematical methods of MOR, which under certain conditions enable almost automatic transfer from the continuum level simulation up to the behavioral models with minimal loss of accuracy. These methods are described in Chapter 3, with more detail on specific approaches in Chapters 9 and 10, examples of nonlinear MOR in Chapters 11 and 12, and three commercial implementations of MOR in Chapters 18, 20, and 21.

1.4
System-Level Model Terminology

Identifying relevant multiscale and multiphysics phenomena in a device or system and encoding the interactions appropriately are usually a significant challenge that is very application specific. Combinations of techniques are required to build most multiphysics models for microsystem simulation. Subsequently, there is a long list of terminology used to describe system-level models and modeling approaches.

A *component*, in the context of computer-aided design tools, generally describes a functional part of a system that is represented by the combination of its system-level model and its symbol for use in system-level schematics.

A *behavioral model* is described, in whole or in part, mathematically through relationships between the model's terminals and external parameters. At least some part of a behavioral model is defined by differential and algebraic equations, but it can also incorporate interconnected subcomponents. A *primitive behavioral model* (or just *primitive model*) has no interconnected subcomponents; it is a model defined solely by differential and algebraic equations.

A *structured model* is described solely in terms of interconnected subcomponents (i.e., a schematic). At the lowest possible level in a system-level modeling hierarchy, these components must be described by primitive models.

A *circuit model* or *network model* is a structured model with potential and flow variables assigned to each terminal connection and that obeys the conservative Kirchhoffian network laws.

A *signal-flow model* (or *block-diagram model*) is a structured model where only potential variables, and no flow variables, are defined for each terminal connection.

A *physics-based model* incorporates equations derived from the physics of the problem. This is in contrast to most MOR techniques that fit to nonphysical basis functions. One potential benefit of physics-based models is their use for scaling studies and extrapolative studies.

A *compact model* is a well-established name for accurate transistor device models used for the representation and simulation in schematics. Compact models may be formed from physics-based equations, fitted to basis functions, or built from lumped components. A mix of these approaches is used for compact models of advanced-technology-node transistors. The term is often applied to complex microsystem behavioral models and, in general, encompasses both primitive and structured models.

A *reduced-order model* is a behavioral model formed by reducing the order of a high-DOF model, typically from numerical continuum simulation (e.g., using MOR).

A *lumped-element model* specifies a reduced-order modeling approach where spatially distributed physical behavior is "lumped" into a finite set of "elements" that approximate behavior at discrete points in space. Typically, a lumped-element model is implemented as a structural model with "elements" comprising physics-based primitive models. A common microsystem example is a 1-DOF mass-spring-damper model. A common electrical example is an inductor–capacitor network approximation of a transmission line.

The term *macromodel* has its origins from SPICE circuit modeling, where it is used to describe a "subcircuit" (i.e., a structured model embedded in SPICE code) comprising some combination of available primitive models such as transistors, idealized dependent sources, inductors, capacitors, and resistors [14]. In SPICE, all primitive models are built into the simulator and thus constrain designers to creating structured models (i.e., macromodels) based on the limited inventory of primitive models. The term is sometimes used to convey a compromise of accuracy somewhere between a circuit made solely of transistors and a more abstract model having no transistors. Common macromodel examples exist for operational amplifiers, comparators, timers, and other high-level analog electronic components. In the MEMS field, the term macromodel is often interchanged with reduced-order modeling; however, this association is not always synonymous with the older SPICE usage and is discouraged.

1.5
Automated Model Order Reduction Methods

Continuum field solvers are important tools that are used to verify system-level models that are created manually. A large swath of coupled multiphysics can now be solved for modest-sized device-level problems with commercially available software on desktop computers or on parallel computing clusters. Increasingly, these tools are also being used to generate system-level models automatically (Chapters 18, 20, and 21). Advancements in algorithms to speed up finite-element and boundary-element computations, and their greatly increased ease of use within commercial multiphysics tools has enabled their practical use in automated parameterized model generation.

Often, system-level models are created from fine-grained simulation results by optimally adjusting coefficients to a basis function, which may be polynomials or rational functions of polynomials. Dynamic models use the fundamental differential equations governing the system, but with a reduced order. For example, the number of DOFs in a continuum model with 10 000 elements may be reduced to 10 DOFs by extracting the lower 10 eigenmodes of the simulated system. An issue with this approach can be the decision of where to truncate the matrix so that all important modes are included while excluding as many insignificant modes as possible to speed up system simulation. One example of high-order modes that could be significant are self-resonances of comb finger beams in electrostatic sensors and actuators. Modulation frequencies applied to comb transducers could conceivably stimulate resonance, yet these frequencies may be far above the first 10 modes of the overall system. A generalized automated modeling algorithm would have no way of knowing that these modes were essential to consider for proper system design. Also, systems with high nonlinearity or dynamic parametric effects may result in modal modifications that are difficult for automated modeling algorithms to predict without manual intervention. For these reasons, the Krylov-subspace methods, the accelerated grammian methods, and the parametric and nonlinear projection methods are being developed by mathematicians and increasingly used by engineers (Chapters 9–12).

1.6
Handling Complexity: Following the VLSI Paradigm

The VLSI design paradigm uses both hierarchy and views to handle complexity [15]. Hierarchy is implemented by building a system or component comprising smaller subcomponents. The components have their behavior wholly encapsulated by their time-varying terminal relations and the fixed values of external parameters. This behavioral encapsulation allows components to be instantiated anywhere in the system to implement their function instead of being replicated from scratch. Partitioning within a system hierarchy may continue until the subsystems comprise only primitive behavioral models. For example, an inertial microsystem can be partitioned into a MEMS subcomponent and an electronics subcomponent. The MEMS subcomponent can be partitioned further into individual accelerometer and gyroscope components. Each accelerometer can be further partitioned into mass-spring-damper and electrostatic transduction subcomponents. As a possible final layer in the hierarchy, the mass-spring-damper component can be modeled as interconnecting beams and plates. Alternatively, mechanical and electrical subcomponents of the system can be represented by reduced-order models, gained by mathematical MOR methods and inserted into the hierarchical representation. For example, the springs in an accelerometer may be represented by an accurate reduced-order model.

However, if hierarchy was the only advantage of VLSI design then there would be little difference from the many other areas in engineering that exploit system partitioning. VLSI design is truly unique because of the combination of hierarchy with parallel model views. A *view* is a representation of a component, which may take the form of a component symbol, a layout, or any of the various model types; for example, a primitive behavioral model, a signal-flow model, a structured model (i.e., a schematic), or a schematic that includes parasitic elements extracted from layout information. The powerful implication for design is the ability to explore the hierarchy from the perspective of any of these views, as illustrated in Figure 1.3. Different model abstractions (i.e., model views) of particular components may be swapped in for evaluation at any level in the hierarchy, which allows for top-down conceptual thinking and "what-if" experimentation. If a subcomponent is specified by a primitive behavioral model then its detailed implementation at a lower level in the hierarchy can be decoupled from the rest of the system. It is not necessary for all views to be filled in order to evaluate performance at each level of the hierarchy. This feature allows refinement of modeling and design to occur in parallel and iteratively at any level in the hierarchy.

Microsystem design benefits directly from the VLSI paradigm; however, the supporting infrastructure for multiphysics systems is in its infancy relative to analog/digital electronic systems. The increased complexity of interacting energy domains provides a huge incentive to design microsystems with simultaneous access to both hierarchy and multiple model views. Key infrastructure needs in support of this design paradigm are fast, designer-friendly model-generation tools and comprehensive libraries of trusted system-level models.

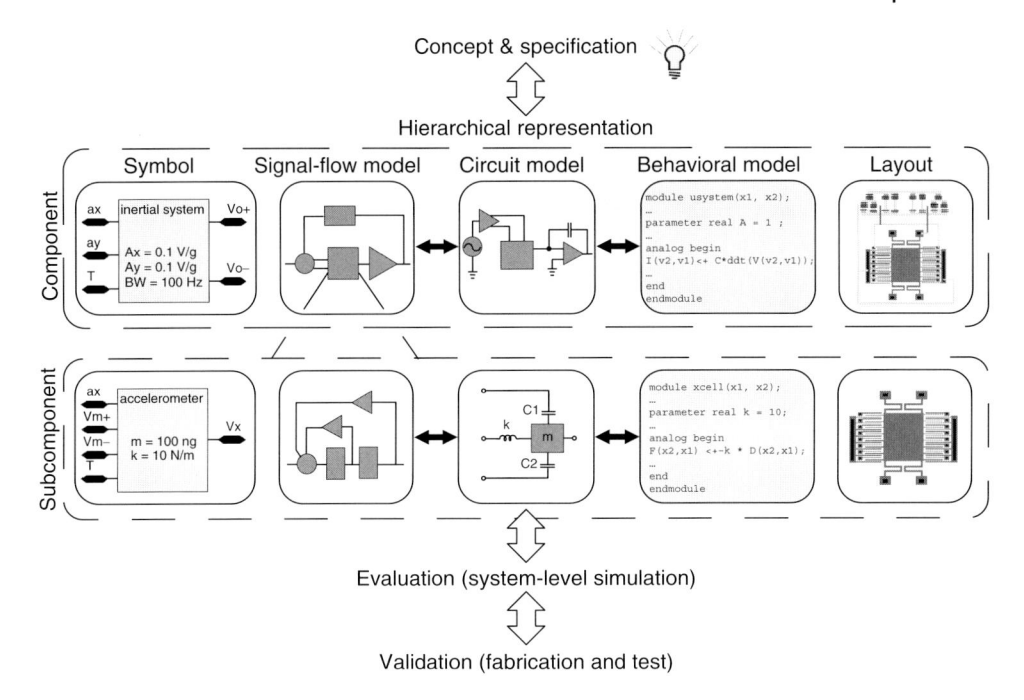

Figure 1.3 VLSI design methodology illustrating two levels in the hierarchical representation, where each component has multiple views. Several possible model views are shown, although only one model per component is used during simulation.

1.7
Analog Hardware Description Languages

Several AHDLs are available for model specification of physics governed by differential and algebraic equations. AHDLs are supported by commercial circuit simulators and include the open-source standards OpenMAST® [16], Verilog®-AMS [17], VHDL-AMS [18], and SystemC-®AMS [19], where AMS stands for analog mixed signal. AHDL models are decoupled from the simulation software, thereby providing users access to the algorithmic capabilities of modern simulators directly through their custom model code. System-level models written in AHDLs coupled with modern commercial electronic design frameworks provide a path to encode and document all design aspects: system architecture, device topology and sizing, process settings, multiphysics behavior with complex interactions, signal timing, and external stimuli emulating the application. Cosimulation of interconnected electronic and multiphysics devices is enabled, as the circuit simulators support system-level modeling across all energy domains.

These AHDLs allow designers to create modules that encode component models as behavioral (mathematical) descriptions and also support structural descriptions of systems by interconnecting components. Various terms used in describing multiphysics circuits are illustrated in Figure 1.4. Each terminal or *port* on a

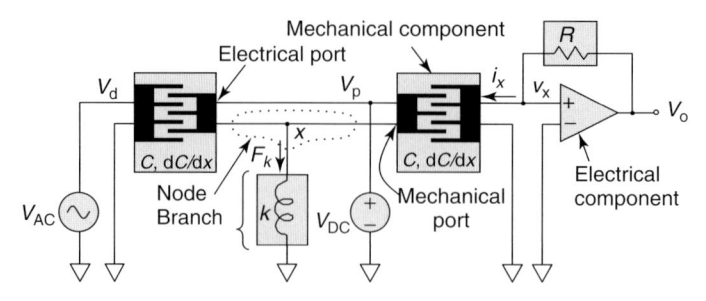

Figure 1.4 An example of a multiphysics circuit schematic: a mechanical spring connected to two electrostatic interdigitated comb capacitors with a motional current readout circuit. Components are designated by the gray symbols. Potential variables include v_d, v_p, v_x, v_o, and x. Flow variables include F_k and i_x. The ground nodes denote zero potential in all energy domains. Parameters include voltages V_{AC} and V_{DC}, resistance, R, spring constant, k, air-gap capacitance, C, and motional sensitivity, dC/dx.

component is associated with a potential variable and a flow variable (also known as *across* and *through* variables, respectively) that are used within the model. AHDLs support both *conservative* and *signal-flow* definitions for the component ports. In a signal-flow port, only the terminal potentials are defined and used within the model. Conservative circuit interconnections follow the generalized Kirchhoff's potential law (KPL) and Kirchhoff's flow law (KFL). These laws are more generally known as *Kirchhoff's mesh rule* and *Kirchhoff's node rule*, respectively. A detailed treatment of multiphysics modeling with Kirchhoffian networks is given in Chapters 2 and 4. The definitions of KPL and KFL relate to nodes and branches in a generalized circuit. A *node* is most generally defined as any contiguous equipotential connection (i.e., wiring) between two or more component ports. A *branch* is a path of flow through a component from one of its ports to another. In KPL, the potential variable values associated with the branches around a closed loop of a circuit sum to zero. In KFL, the flow variable values flowing through all branches out of a common node of a circuit sum to zero. Most circuit simulators allow mixing of conservative and signal-flow components in a single system. In these cases, signal-flow outputs are treated as dependent potential sources and signal-flow inputs are treated as potential probes (i.e., as infinite-impedance inputs). As described in Section 1.6, this flexibility to mix component model views at any hierarchical level is especially handy for top-down design where details of a component's potential to flow relationship may be deferred to a later stage in the design process.

1.8
General Attributes of System-Level Models

Interoperability is a critically important attribute of models to enable their use with other models to form systems. The potential and flow variables must be consistent in definition and in their associated reference direction. Interoperability is taken for granted in electronic circuit design. Electrical terminal standards – current as flow, voltage as potential, and the associated reference direction of current flow into the positive potential terminal to mean that power is delivered to the component – are well known and followed in device modeling efforts. Standards for macrosystem terminal relations in many energy domains have been established for AHDLs. However, multiphysics terminal relations particularly suitable for microsystems do not yet exist as accepted standards. Exacerbating adoption of standards are often significant differences in notation and approach between various microsystem designers.

Several challenges exist in appropriately and accurately translating the results of lower-level simulations to populate system-level models. Microsystems almost always involve conversion of energy across physical domains (e.g., mechanical to electrical, thermal to mechanical). Therefore, of foremost importance is the attribute that system-level models abide by principles of conservation of energy to preserve the physical integrity of the simulations in which they are used. The energy flowing into the model plus any sources internal to the model must equal the energy leaving the model plus the thermodynamic losses internal to the model.

Parameterization is a third important model attribute that enables iterative design and is supported within AHDLs. External parameters may be created to set materials properties and geometric sizing during system simulation. For example, Young's modulus and structural thickness parameter values may be set for a particular micromechanical process and then modified to allow exploration of process sensitivities. Full specification within parameterized models includes design constraints (e.g., geometric design rules) and nominal and variation values for the layer thicknesses and materials properties within a process. CMOS foundries supply this information in documents and files that comprise *physical design kits* that customize the computer-aided design environment. In an analogous manner, system-level model parameters can be formulated for compatibility with MEMS physical design kits. Examples of constraints set by MEMS design rules include the maximum proof mass size in an accelerometer, the minimum beam width in flexures, and the minimum gap in electrostatic actuators.

1.9
AHDL Simulation Capabilities

MEMS models in AHDLs exploit the availability of fast circuit simulators that incorporate DC, AC, and transient analysis. In general, simulations will converge if the corresponding behavioral models are formed from physical principles and if the structural models follow physically realizable interconnect rules. Simulation time is often dependent on the specific form of the model code and so can be optimized for speed. Chapter 4 addresses aspects of system-level cosimulation that have to be understood and solved if different physical energy domains are to be coupled and simulated successfully.

Most commercial circuit simulators support several additional analyses that are useful for the evaluation of microsystems designed with AHDL-based models.

Periodic ac analysis and periodic steady-state analysis provide mechanisms to simulate highly nonlinear systems that result in asymmetric periodic waveforms arising either from AC excitation or from intrinsic oscillatory behavior. In contrast, basic AC analysis linearizes around an operating point and only includes the fundamental harmonic term. Use of transient analysis for nonlinear problems may take very long to settle to a steady-state solution that displays all of the frequency content of interest. In particular, periodic analyses are helpful in simulating microsystems that employ modulation and nonlinear resonant conditions. For example, capacitive accelerometer systems that use chopper stabilization or correlated double sampling at megahertz frequencies are challenging to simulate in regular transient analysis over periods on the order of seconds needed to capture response to low-frequency input acceleration signals. Noise analyses are available to analyze the effects of stochastic disturbances originating from sources internal and external to the system. The amplitude and nature of the noise must be included in the constitutive behavioral models. In many design frameworks, these various simulation analyses are overlaid with parametric sweeps to automate exploration of the design space and with the Monte Carlo capabilities to estimate statistical distributions arising from parameter and process variations. Model support for incorporating and propagating process variations is needed to fully exploit the Monte Carlo capabilities and is essential for estimating manufacturing yield.

1.10
Composable Model Libraries

In MEMS, along with most other fields, the overwhelming majority of system-level designers rely on experts in device physics and in materials processing to create behavioral models. As there are few device modelers relative to system designers, a bottleneck in time and efficiency ensues. One way to help alleviate this bottleneck is to create models automatically from fine-grained simulation through MOR as described earlier. An alternate approach is to create system-level models at a low enough granularity that allows their reuse for a significantly broad design space. As long as terminal relations remain interoperable, this approach remains compatible and can be synergistic with automated MOR techniques.

Such a model reuse paradigm has existed successfully for over 40 years for circuit simulation (i.e., SPICE [20]). Widespread and free access to the evolving models in SPICE led to its adoption as a *de facto* standard. The long-term success of SPICE occurred because (i) high-fidelity models of CMOS transistors were developed, (ii) the infrastructure was created to extract model parameters from experimental process and device data, (iii) the education of designers centered on use of these models, (iv) a financial incentive for modeling activities within foundries arose to enable their propagation to external designers, and (v) there was dedication and incentive to revise models and insert new device physics as the CMOS technology advanced. It is now taken for granted by the electronics designer that models of transistors, resistors, capacitors, and inductors provide exquisite predictive accuracy regardless of how they are used in a circuit. These primitive behavioral models are *composable*, which in this context means that they are interoperable and of a basic elemental nature to enable the creation of a large number of useful structured models at higher levels in the design hierarchy.

In the mid-1990s, a series of composable and parameterized primitive models were introduced for MEMS design using handcrafted physics-based techniques. Common electromechanical composable models include straight and curved beams, plates with various geometric features, electrostatic gaps, and contact points. Examples of composable model libraries are given in Chapters 17, 19, 21 and 22, with some versions supported commercially.

Physics related to distributed effects are particularly difficult to capture in composable models. For example, electrostatic fields in principle depend on the location of all conductive and dielectric components in a microsystem (unless the components are completely shielded). This is a key reason why creation of accurate general-purpose electromechanical "gap" models remains an open challenge. Approximations or abstractions to far-field effects must be incorporated in the system-level models to eliminate the need for computations involving all components. Other examples of distributed effects that are challenging to model include stress, covered in Chapter 6, and damping, covered in Chapter 7. The latter damping example uses a "mixed-level" modeling approach that takes advantage of the hierarchical nature of structured modeling.

Current composable model libraries for MEMS are extremely useful, but they are not comprehensive; there are plenty of general-purpose models that could, should, and probably will be added in the future. Much effort is required to expand and improve composable models, as they are currently handcrafted. Model libraries will naturally evolve and improve through incorporation of additional physical phenomena, such as piezoresistance, piezoelectricity, loss, and thermal properties. Other microsystem design domains, most notably microfluidics, will benefit from analogous composable libraries.

1.11
Parameter Extraction, Model Verification, and Model Validation

Microsystem models can be accurate only if the parameter values incorporated within the simulation reflect the actual outcomes of the manufacturing process. Process parameters include material properties, layer thicknesses, and feature offsets from layout. Determining accurate values for these parameters is a particular challenge for MEMS. Procedures to provide parameter values from experimental test data are collectively known as *parameter extraction*. Parameter extraction for multiphysics systems is currently performed with painstaking custom test structure design and measurement. An example is described in Chapter 9, where determination of materials thermal properties is based on fitting a reduced-order model to experimental measurements. Standardizing and streamlining these kinds of activities would greatly reduce the time to market for new microsystem technologies.

Model verification comprises procedures to check on the mathematical and structural form of a system-level model. Models created automatically through MOR techniques are verified by construct, as they are directly derived from continuum simulations. However, handcrafted microsystem models are subject to errors in form and to errors in physical assumptions. Behavior of these kinds of models must be verified by using a system-level simulation "test bench" and comparing results against continuum simulation using identical boundary conditions, materials property values, and geometric sizing values in both simulations. Identifying the appropriate suite of verification problems to cover all aspects of model form is challenging. To be comprehensive, the problem set must cover static and dynamic behavior up to maximum frequencies of interest; it must cover the dynamic range of interest for the input stimuli; and it must cover the design space over all external parameters. Also, suitable criteria must be defined when comparing the simulations. An uncertainty criterion that is set too small may capture effects of numerical error that have nothing to do with the model form. Too large of an uncertainty criterion may mask detection of modeling errors.

Model validation describes procedures that check accuracy of a system-level model when compared with experimental results. Model simulations for use in validation exercises should be performed with extracted parameter values. Subsequent experimental validation within process uncertainty is an essential requirement for trusted models. Most of the same issues that occur in verification arise when identifying an appropriate suite of validation tests. Coverage of static responses, frequency responses, dynamic range, and external parameter space are all required for complete model validation.

The microsystem designer should be wary in trusting a system-level model to provide accurate performance predictions when its operation regime is extrapolated beyond its verified and validated range. However, physics-based composable models that operate in their verified and validated range potentially provide a basis for accurate predictive behavior when used to create new structured models. Verification of the design space for structural models that comprise composable components is an open research area. Validating individual composable models is an additional challenge, as practical test structures may need to be formed from a combination of components. For example, testing the electrostatic force generated from an air gap probably requires a flexure to be connected to the gap. Discrepancies between the simulation and experiment may then arise from the flexure and not from the gap.

1.12
Conclusions

System-level modeling is a core activity in support of microsystems design. Various case studies on multiphysics modeling and their use in systems design are given in various chapters throughout this book. Owing to the availability of analog hardware description languages such as Verilog-AMS, coupled multiphysics and multiscale behavior can be readily incorporated within system-level models that run on fast commercial circuit simulators that support DC, AC, periodic, noise, and the Monte Carlo analyses. AHDLs allow model formation in any combination of mathematical and structured descriptions, including primitive behavioral models, signal-flow models, and circuit macromodels (Chapters 2 and 4). Accurate translation of results from continuum field simulations into system-level models is best accomplished through model order reduction techniques (Chapters 3, 9, 10, 18, 20, and 21). Progress being made in nonlinear MOR is broadening the impact of these automated approaches to system-level modeling (Chapters 10–12). To be trusted, models for systems design constructed in whole or in part with manual techniques must be verified through comparison to continuum simulation. Regardless of their origin, all models to be considered trusted must be validated through comparison with experiment using process parameter values extracted from experimental test structures.

Leveraging the principles of the VLSI design hierarchy, multiphysics system-level modeling and simulation provides inherent capability to handle ever increasing amounts of microsystem complexity. The fast simulation speed, relative to fine-grained continuum simulation, powers iterative design by enabling simultaneous exploration of trade-offs in system architecture along with topology and sizing of multiphysics components. These capabilities become increasingly important as microsystems trend toward greater integration of multiphysics components with digital and analog electronic subsystems.

Useful models intended for rapid iterative design of multiphysics systems must be interoperable and parameterized. While interoperable model libraries for microsystem design exist (Chapters 17, 19, and 22), a significant amount of resources and effort must be mustered to build comprehensive libraries of trusted models that include the plethora of physics of interest to microsystem designers. Growing a worldwide microsystem modeling community is perhaps the most practical approach to addressing the major issues of education, adoption of standards, availability of accurate generalized models, process parameter extraction, and model verification and validation (Chapter 22).

References

1. Wachutka, G. (1995) *Sensors and Actuators A: Physical*, **47** (1–3), 603–612.
2. Senturia, S. (1998) *Proceedings of the IEEE*, **86**, 1611–1626.
3. Mukherjee, T., Fedder, G.K., Ramaswamy, D., and White, J. (2000) *IEEE Transactions on Computer-Aided Design of Integrated Circuits and Systems*, **19** (12), 1572–1589.
4. Middelhoek, S. and Hoogerwerf, A.C. (1986) *Sensors and Actuators*, **10**, 1–8.
5. Senturia, S. (2001) *Microsystem Design*, Springer, New York, NY.
6. Rebeiz, G. (2003) *RF MEMS. Theory, Design, and Technology*, John Wiley & Sons, Inc., Hoboken, NJ.
7. Solgaard, O. (2009) *Photonic Microsystems: Micro and Nanotechnology Applied to Optical Devices and Systems*, Springer, New York, NY.
8. Vahala, K. (2004) *Optical Microcavities*, World Scientific Publishing, Singapore.

9. Kitching, J., Knappe, S., and Donley, E.A. (2011) *IEEE Sensors Journal*, **11** (9), 1749–1758.
10. Lin, W.H. and Zhao, Y.P. (2005) *Microsystem Technologies*, **11** (2–3), 80–85.
11. Nguyen, N.T. and Wereley, S. (2002) *Fundamentals and Applications of Microfluidics*, Artech House, Norwood, MA.
12. Karniadakis, G., Beskok, A., and Aluru, N. (2002) *Micro Flows: Fundamentals and Simulation*, Springer, New York, NY.
13. Ferrari, M. (ed.) (2007) *BioMEMS and Biomedical Nanotechnology* (4 volume set), Springer, New York, NY.
14. Boyle, G.R., Cohn, B.M., Pederson, D.O., and Solomon, J.E. (1974) *IEEE Journal of Solid-State Circuits*, **9** (6), 353–364.
15. Weste, N.H.E. and Harris, D.M. (2011) *CMOS VLSI Design: A Circuits and Systems Perspective*, 4th edn, Chapter 14, Addison-Wesley, Boston.
16. Synopsys OpenMAST Language Reference Manual, version 1.0, (2004) http://www.openmast.org/home.html (accessed 28 July 2012).
17. Accelera (2009) Verilog-AMS Language Reference Manual: Analog and Mixed-Signal Extensions to Verilog HDL, Version 2.3.1, http://www.eda.org/verilog-ams/htmlpages/public-docs/lrm/2.3.1/VAMS-LRM-2-3-1.pdf (accessed 28 July 2012).
18. IEEE (2011) Standard 1076.1.1-2011. *IEEE Standard for VHDL Analog and Mixed-Signal Extensions*, http://ieeexplore.ieee.org/servlet/opac?punumber=5752647 (accessed 28 July 2012).
19. Accelera and Open SystemC Initiative, (2005) SystemC Analog/Mixed-Signal Extensions, Release 1.0. http://www.accellera.org/downloads/standards/systemc/ams (accessed 28 July 2012).
20. Nagel, L.W. and Pederson, D.O. (1973) *SPICE (Simulation Program with Integrated Circuit Emphasis)*, Memorandum No. ERL-M382, University of California, Berkeley, http://www.eecs.berkeley.edu/Pubs/TechRpts/1973/22871.html (accessed 28 July 2012).

2
System-Level Modeling of MEMS Using Generalized Kirchhoffian Networks – Basic Principles

Gabriele Schrag and Gerhard Wachutka

2.1
Introduction and Motivation

As already explicated in Chapter 1, microelectromechanical systems (MEMS) are, because of their nature as transducers, inherently dominated by the coupling between different physical energy domains [1, 2] and, in general, exhibit a high degree of physical and geometrical complexity, which originates also from their integration into standard IC frameworks and packages. In order to obtain the best possible performance, the entire system consisting of transducer, electronics, and package has to be optimized as a whole. Although it might still be possible to model the single transducer and its operation on the level of partial differential equations ("device level," continuous-field models), for example, by finite element simulations, this becomes prohibitive when its interplay with the package, the surrounding electronics or other transducers have to be taken into account [3]. Thus, models with an extremely reduced order of complexity on a higher level of abstraction than the continuous-field description, the so-called system-level models, are inevitable in order to describe the system as a whole (see Figure 1.2 in Chapter 1).

The term "system-level model" or "system model" is used in general and also within this book as a synonym for a reduced order model (ROM) of the entire system under consideration, which comprises all electrical and nonelectrical parts of it and, thus, enables their fast and efficient cosimulation, preferably within the same simulation environment. In general, the derivation of such a system model starts with partitioning the entire system into tractable subsystems, for which the so-called compact models have to be derived. These are models with a drastically reduced number of degrees of freedom compared to continuous-field models, and their complexity ranges from simple "behavioral models" extracted by pure curve fitting of measured or calculated characteristics to "physics-based compact models" containing a multitude of design parameters and their respective dependencies on internal and external quantities (Chapters 5–8) or "ROMs," which are derived by applying mathematical projection techniques to discretized, large-scale dynamical systems (Chapters 3 and 9–12). An overview of different approaches for the derivation of compact models is given in Section 2.2.3.

System-level Modeling of MEMS, First Edition. Edited by T. Bechtold, G. Schrag, and L. Feng.
© 2013 Wiley-VCH Verlag GmbH & Co. KGaA. Published 2013 by Wiley-VCH Verlag GmbH & Co. KGaA.

A proper framework for the final assembly of the subsystem models to a system model and, thus, for the cosimulation of all electrical and nonelectrical parts is provided by Kirchhoffian network theory, which is a well-established methodology for the mathematical representation of electrical networks, but can be generalized to other energy domains [4]. The system is described by discrete components, which are interlinked by interconnects with negligible resistance appearing as terminal nodes between the single submodels (Chapter 4). Between these single blocks, energy is exchanged via these terminals, controlled by balance equations as expressed by the Kirchhoffian network rules for currents and driving potentials.

Chronologically speaking, a first approach to extend this originally electrical concept to other energy domains was to describe nonelectrical elements (mechanical, thermal, and hydraulic problems) by electrical network analogies (e.g., resistances, inductances, and capacitances) [5–9]. The next step was to formulate the governing physical equations and dependencies from design and external parameters directly in a hardware description language (HDL) and to implement them into analog network simulators as functional model blocks [10, 11]. Examples for the application of such physics-based system-level models are given in Refs [12, 13], Sections 2.3 and 2.4 of this chapter, and Chapters 5–8 of this book.

In the following list, it will be explained:

- how Kirchhoffian network theory can be extended to the simulation of nonelectrical problems;
- how this approach enables the easy coupling of different energy domains leading to the so-called generalized Kirchhoffian networks, in which not only electrical energy is exchanged; and
- how this approach can be successfully applied to system-level modeling of MEMS.

The underlying theoretical fundamentals of generalized Kirchhoffian network theory are introduced on the basis of irreversible thermodynamics (Section 2.2.1), and the way from continuous-field to compact model-based simulation is sketched (Section 2.2.2). Finally, a short overview of different compact modeling approaches and their advantages and disadvantages as well as two exemplary applications of physics-based compact modeling are given (Sections 2.3 and 2.4).

2.2
Generalized Kirchhoffian Networks for the Tailored System-Level Modeling of Microsystems

In this section, a thermodynamics-based generic modeling approach for microsystems and their respective subcomponents is presented, which serves as a basic framework for system macromodeling on the basis of generalized Kirchhoffian networks. The transition from a continuous to a lumped system description applying this generic methodology is sketched, and finally, a short overview on alternative compact modeling approaches is given.

2.2.1
Generic Modeling Approach for Microdevices and Systems

The first step in setting-up a MEMS system-level model consists mostly in decomposing the system into feasible and tractable subsystems. This can be achieved by considering functional, structural, or behavioral aspects according to the given problem. The models that are derived for the subsystems are then linked via properly chosen, consistent flux-conserving interface conditions (e.g., balance of forces, energy, and electrical flux,) according to the physical quantities exchanged between the subsystems under consideration. This is schematically shown in Figure. 2.1.

For instance, the electrostatic RF MEMS switch described in Section 2.4 can be decomposed according to the different energy domains involved in the operation, which leads to four submodels: a mechanical model representing the membrane and the four suspension beams, an electrostatic model that is needed for the actuation of the switch, a fluidic submodel that takes into account the damping forces caused by the ambient air, and a contact model that describes the closing of the switch (Figure 2.16).

For the micropump introduced in Section 2.3, a decomposition into functional units is more appropriate, which are in that case the driving unit, the valves, and the tubes (Figure 2.6).

Applying the basic principles of thermodynamics [14–16], each of the so-defined subsystems can be described in general by the following quantities and relations:

1) A subset $\mathcal{J} = \{Y_1, Y_2, ..., Y_M\}$ of the intensive state variables Y, which might consist, for example, of the mechanical stress tensor σ, the hydrostatic pressure p, the electric field $\mathbf{E}$, the temperature T, and the electrochemical potential ϕ_k governing drift and diffusion of carrier species k; here, the concrete selection of variables has to be "tailored" for the specified system.
2) A subset of thermodynamic equations of state that connect the intensive state variables $Y \in \mathcal{J}$ through material-specific relations with respective conjugate extensive state variables $X \in \mathcal{E}$, where $\mathcal{E} = \{X_1, X_2, ..., X_M\}$ comprises, for

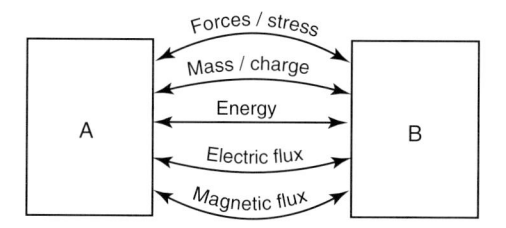

Figure 2.1 Partition of a system into two subsystems A and B. Along the interfaces, each of the quantities (fluxes) is exchanged in a physically consistent manner (subject to balance equations).

example, the mechanical strain tensor ϵ, the volume V, the dielectric displacement field $\mathbf{D}$, the entropy density s, or the particle concentrations c_k of carrier species k.

3) A set of governing equations that in most cases exhibit the generic structure of a balance equation

$$\frac{\partial n_X}{\partial t} + \mathrm{div} \mathbf{j}_X = \Pi_X \quad \text{for } X \in \mathcal{E} \tag{2.1}$$

where n_X, $\mathbf{j}_X$, and Π_X denote the density, the current density, and the production rate of the respective extensive quantity X.

4) And a set of constitutive current relations, by which, according to Onsager's principles of irreversible thermodynamics [15, 17], the current densities ("fluxes") $\mathbf{j}_X$ are related to the gradients of generalized potentials as "driving forces"

$$\mathbf{j}_X = -\sum_{Y \in \mathcal{J}} \lambda_{XY} \nabla \varphi_Y \quad \text{for} \quad X \in \mathcal{E} \tag{2.2}$$

These generalized potentials $\varphi_Y(\vec{r}, t)$ are one-to-one functions of the associated intensive state variables Y, for example, the temperature T, the hydrostatic pressure p, the electric potential ϕ_{el}, or the electrochemical potentials Φ_k.

Equation (2.2) describes the transport phenomena within a device, where particles, charge, or energy flows[1]. The matrix of generalized conductivities λ_{XY} (or transport coefficients) describes various transducer effects – direct effects such as the heat flux driven by a temperature gradient as well as cross-linked transducer effects, notably thermoresistivity, thermoelectricity, piezoresistivity, piezoelectricity, and galvanothermomagnetism. These transport coefficients, in turn, are functions of the respective state variables; in particular, they are temperature-, stress- and magnetic-field dependent, which is the reason why Eq. (2.2) is termed a *pseudo-linear* relation.

As an illustrative example, how the above-described generic description of a MEMS component works, let us consider a simple thermoelectric (sub)structure, which exhibits electric and thermal conductivity and, in addition, shows the thermoelectric effect. Table 2.1 gives an overview of the governing quantities and the respective equations. Couplings occurring between different energy domains can be easily incorporated through extending the single domain model equations by the coupling terms in the constitutive current relation (2.2) (i.e., the nondiagonal entries of $\lambda_{X_k Y_l}, k \neq l$). For an electrothermal structure made of an n-type and/or p-type semiconductor, the constitutive current relations take the form Refs [15, 16]

$$\vec{j}_\alpha = -\sigma_\alpha (\nabla \Phi_\alpha + P_\alpha \nabla T_\alpha) \text{ with } (\alpha = n, p) \tag{2.3}$$

$$\vec{j}_{th} = -\kappa \nabla T + P_n T \vec{j}_n + P_p T \vec{j}_p \tag{2.4}$$

---

1) Currents of the extensive quantities, later also referred to as *through quantities* or *generalized fluxes* are driven by the gradients of the intensive state variables (generalized potentials), later also referred to as *across quantities*

Table 2.1 Generic relations for modeling electrical and thermal conductors, respectively (with $\mathbf{E}$ = electric field, $\mathbf{j}_{el}, \mathbf{j}_{th}$ = electrical or thermal current densities, respectively, σ, κ = electrical or thermal conductivities, respectively, Π_{el}, Π_{th} = electrical/thermal generation term).

	Electrical domain	Thermal domain
Intensive quantity	Electrical potential ϕ_{el}	Temperature T
Extensive quantity	Electrical charge Q_{el}	Heat Q_{th}
Driving force	$\mathbf{E} = -\nabla \phi_{el}$	$-\nabla T$
Constitutive current relation	$\mathbf{j}_{el} = \sigma \mathbf{E}$	$\mathbf{j}_{th} = -\kappa \nabla T$
Balance equation	$\frac{\partial \rho_{el}}{\partial t} + div \mathbf{j}_{el} = \Pi_{el}$	$\frac{\partial Q_{th}}{\partial t} + div \mathbf{j}_{th} = \Pi_{th}$

Here, $\vec{j}_\alpha$ denotes the electric current density of the electrons and holes, respectively, and P_α the Seebeck coefficients (or thermopowers), which describe the effect that an electrical voltage is generated by a temperature gradient.

The system of all relations subsumed under items 1) to 4) in this section represents a generic model of a microelectromechanical device, system, or subsystem, which obeys general generic construction principles for any given configuration of energy domain couplings. From a mathematical point of view, we face a more or less large system of coupled partial differential-algebraic equations, which can be solved – in not only the most rigorous and most exact but also the most expensive way – by discretizing these equations on continuous-field level, using numerical methods such as finite elements, finite boxes, and finite networks. It is worthwhile to make sure that it is the particular structure of the generic model, which provides a natural bridge to a system-level description, as it will be explained in the following section.

2.2.2
From Continuous-Field Level to Compact Models

Given that the entire system is properly decomposed into suitable subsystems – as, for example, the system depicted in Figure 2.2, which is divided into two subsystems A and B – the generic model presented in the previous section can be specified and, under certain conditions and assumptions, the distributed field variables can be lumped and mapped onto the so-called terminals (or nodes) along the interfaces of the subsystems. The device operation is then described by exchanging the values of these lumped variables via these terminals.

In general, the transformation (or better: projection) of a flux-conserving equation system to a description in terms of compact models is realized by discretizing the partial differential equation system by means of finite boxes, finite volumes, or finite networks, and lumping the state variables, until, after iterating the process, the subsystem is described by a few lumped variables only. These are then exchanged between the subsystems via their node values as depicted in Figure 2.2.

Technically, this can be done by meshing the spatial domain of the system and constructing a box around each mesh point K, which is defined as the volume

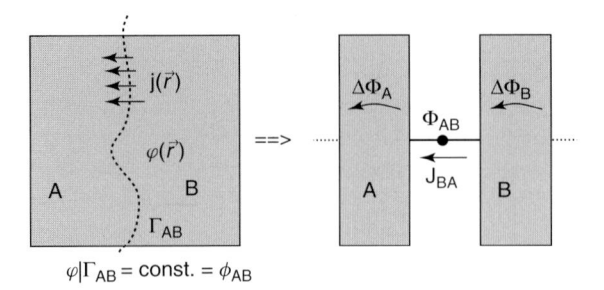

Figure 2.2 Schematic view of a system separated into two subsystems A and B along the common boundary Γ_{AB}. Under the assumption that Φ_{AB} can be regarded as constant along Γ_{AB}, the system can be separated and the interaction can be described by the lumped generalized potentials Φ_A and Φ_B as well as by the lumped flux (= branch current or edge current) J_{BA}.

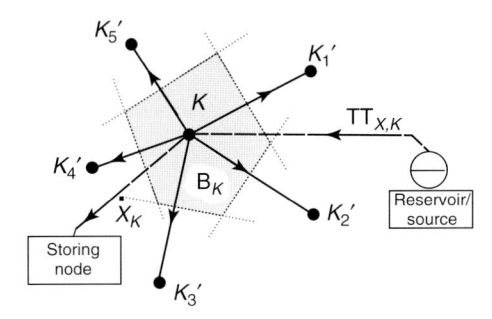

Figure 2.3 Finite box discretization around mesh node K: transition from continuous to lumped element system description.

enclosed by the perpendicular bisectors (or bisectorial symmetry planes) of all lines connecting K with its nearest neighbors K'. In this way, the polyhedron cell B_k depicted in Figure 2.3 around node K is formed.

The total current density of the extensive variable $X \in \mathcal{E}$ inside B_k is then obtained by integrating Eq. (2.1) over this cell

$$\int_{B_K} \frac{\partial n_X}{\partial t} d^3r + \int_{B_K} \vec{\nabla} \vec{j}_X d^3r = \int_{B_K} \Pi_X d^3r \tag{2.5}$$

Applying the Gauss divergence theorem (or Gauss' theorem), the second term is converted into a flux integral along the boundary ∂B_K. Introducing the discrete sum of the contributions originating from each of the polyhedra side faces $S_{KK'}$ between K and nearest neighbor K', we obtain

$$\int_{B_K} \frac{\partial n_X}{\partial t} d^3r + \int_{\partial B_K} \vec{j}_X d\vec{a} = \int_{B_K} \Pi_X d^3r \tag{2.6}$$

$$\int_{B_K} \frac{\partial n_X}{\partial t} d^3r + \sum_{K'} \int_{S_{KK'}} \vec{j}_X d\vec{a} = \int_{B_K} \Pi_X d^3r$$

2.2 Generalized Kirchhoffian Networks for the Tailored System-Level Modeling of Microsystems

Finally, we define the total amount of current flowing through the polyhedron side face $S_{KK'}$ in the direction of the connection line ("branch" or "edge") $\overrightarrow{KK'}$ from K to K' as

$$J_X(\overrightarrow{KK'}) := \int_{S_{KK'}} \vec{j}_X d\vec{a} \tag{2.7}$$

and we define the amount of the extensive quantity X stored on the node K as

$$X_K := \int_{B_K} n_X d^3r \tag{2.8}$$

Finally, we define a current source attached to node K as

$$\Pi_{X,K} := \int_{B_K} \Pi_X d^3r \tag{2.9}$$

In this notation, the box-integrated version of Eq. (2.1) reads

$$\frac{dX_K}{dt} + \sum_i J_X(\overrightarrow{KK'_i}) = \Pi_{X,K} \tag{2.10}$$

It has to be pointed out that Eqs. (2.7–2.9) constitute the projection of the continuous-field model onto a lumped element description, which has the distinguished property that it is flux conserving. The box-integrated formulation can be conceived as a Kirchhoffian network description: The mesh points K are the nodes, the connection lines between K and K' are the directed branches (or edges), and the continuous current $\mathbf{j}_X(\vec{r})$ distributed over the side face $S(KK')$ is concentrated in the branch current $J_X(\overrightarrow{KK'})$ flowing through these edges. Thus, Eq. (2.10) represents the lumped formulation of the balance equation (2.1) on network level, which expresses a node rule at node K (generalized "Kirchhoff's current law," "GKCL") for any extensive quantity that obeys a balance equation. In the special case of an electrical network, Eq. 2.10 states the well-known node rule for electric currents I ("Kirchhoff's current law," "KCL")

$$\sum_i I(\overrightarrow{KK'_i}) = I_{\text{source}} \tag{2.11}$$

where it is assumed that no charge is stored on a node (i.e., only one-port capacitive elements are considered).

The currents or fluxes $\mathbf{j}_X$ defined in the generic model are driven by the gradients of generalized potentials φ_Y as expressed by the phenomenological current relations (Eq. 2.2). The integration of a gradient field $\nabla \varphi_Y$ along a closed loop (or "mesh") $\mathcal{M} = \{\overrightarrow{K_0 K_1}, \overrightarrow{K_1 K_2}, \ldots, \overrightarrow{K_{N-1} K_N}, \overrightarrow{K_N K_0}\}$ composed of the $N+1$ branches $\overrightarrow{K_j K_{j+1}}$ ($j = 0, \ldots, N$) yields zero (Figure 2.4)

$$\oint_{\mathcal{M}} \nabla \varphi_Y d\vec{r} = 0 \tag{2.12}$$

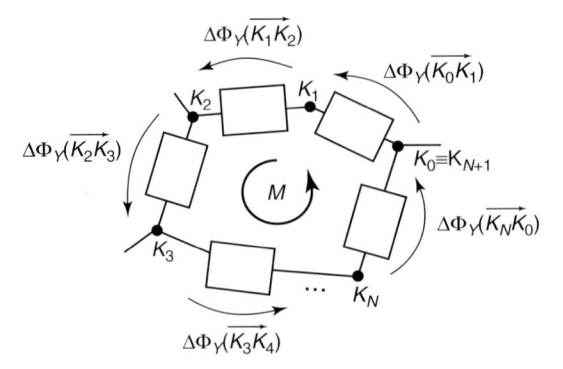

Figure 2.4 Closed-loop integration of the gradient of a generalized potential $\nabla\varphi_Y$ leads to the generalized Kirchhoff's voltage law.

Evaluating the closed-loop integral as the sum over the contributions from the individual branches connecting two neighboring nodes yields

$$0 = \oint_{\mathcal{M}} \nabla\varphi_Y d\vec{r} = \sum_{i=0}^{N} \int_{K_i}^{K_{i+1}} \nabla\varphi_Y \cdot d\vec{r} = \sum_{i=0}^{N}(\varphi_Y(K_{i+1}) - \varphi_Y(K_i)) \quad (2.13)$$

where we set $K_{N+1} \equiv K_0$. Hence, we obtain for the generalized potential drops

$$\Delta\Phi_Y(\overrightarrow{K_i K_{i+1}}) := \varphi_Y(K_i) - \varphi_Y(K_{i+1}) \quad (2.14)$$

the sum rule

$$\sum_{i=0}^{N} \Delta\Phi_Y(K_i, K_{i+1}) = 0 \quad (2.15)$$

Equation (2.15) generalizes the Kirchhoff's mesh rule (or "voltage law," "KVL") for the generalized potential drops $\Delta\Phi_Y$ (later also referred to as *across quantities*); its well-known electrical analog is the Kirchhoff's voltage law for electric voltages U

$$\sum_{i=0}^{N} U(\overrightarrow{K_i K_{i+1}}) = 0 \quad (2.16)$$

Therefore, we term the relation (Eq. 2.15) "generalized Kirchhoff's law" ("GKVL"). The interaction between neighboring nodes K and K' is now described by a pair of conjugate lumped variables $J_X(\overrightarrow{K_i K_{i+1}})$ and $\Delta\Phi_Y(\overrightarrow{K_i K_{i+1}})$ that can be mapped onto the branches of the generalized Kirchhoffian network, which is defined by the directed graph of nodes and branches together with these lumped variables as dynamic state variables. The difference to a conventional Kirchhoffian network consists in that it may include more than one extensive quantity $X \in \mathcal{E}$ and the respective intensive quantities $Y \in \mathcal{J}$, which allows us to describe the coupling of different energy domains through interactions on the nodes (expressed by the source term $\Pi_{X,K} = \Pi(X_1(K), ..., X_M(K), \varphi_Y(K), ..., \varphi_M(K), ...)$ in the GKCL, Eq. (2.10)) and through interactions along the branches (expressed by the discretized lumped version of the constituent current relations (Eq. 2.2), see Eq. (2.18) below). We may imagine such a network as a colored circuit, where each of the involved extensive quantities $X \in \mathcal{E}$ (i.e., each involved energy domain) is marked by a different color. This circuit may contain single-colored subcircuits with color-conserving lumped elements and single-colored nodes (e.g., a purely electrical resistor network), but it will also contain color-converting lumped elements (e.g., a thermoelectric element, cf. Eqs. (2.3) and (2.4)) as well as multicolored nodes representing the transducer properties of the MEMS component described by the network. The goal of compact modeling is now, to continue the lumping process by eliminating a majority of the network nodes, until the entire system is decomposed into a few subsystems only so that it can be very efficiently described by a low (or at least tractable) number of lumped variables. This can be achieved not only by using the algebraic methods of network theory but also (and preferably) by the intuitive view of the model maker. A prerequisite for describing a system successfully by a small set of lumped parameters without significant loss of accuracy is that the system can be partitioned into subsystems in a way that its operation is not essentially affected by the spatial distribution of the generalized potential $\varphi_Y(\vec{r})$ along the interface between adjacent subsystems. Referring to Figure 2.2, the potential $\varphi_Y(\vec{r})$ on the continuous-field level must satisfy the condition

$$\varphi_Y|\Gamma_{AB} \approx \text{constant} =: \Phi_{Y,AB} = \text{lumped value on terminal node "AB"} \quad (2.17)$$

along the interface Γ_{AB} between the two subsystems A and B.

The functional behavior of the subsystems A and B is then controlled by the generalized potential drops $\Delta\Phi_{Y,A}$ and $\Delta\Phi_{Y,B}$ and/or the fluxes $J_{X,AB}$, which are obtained by integrating the distributed current densities $\mathbf{j}_x(\vec{r})$ along the interface Γ_{AB}. As derived before, Φ_Y and J_X constitute pairs of conjugate lumped variables, which satisfy the respective balance equations and conservation laws (GKCL and GKVL) on system level. Hence, the network analysis of the generalized "colored" Kirchhoffian network describing the system follows the same methodologies as in electric circuit analysis. This means that all commonly available electric circuit simulators, which rely on the above-described physical principles, can also be employed to nonelectrical devices and systems, such as MEMS. To this end, the system-level models have to be formulated in terms of generalized fluxes (or "through quantities") and generalized potentials (or "across quantities"), which are, in general, chosen so that their products have the dimension of a measurable physical quantity, namely, power, power density, or similar (instead of using the entropy production as in the original thermodynamical formulation [15, 16]). Table 2.2 lists commonly used pairs of conjugate variables for some exemplary energy domains.

Accordingly, the constitutive current relations (Eq. 2.2) have to be reformulated in terms of the lumped variables $J_X(\overrightarrow{KK'})$ and $\Delta\Phi_Y(\overrightarrow{KK'})$ (branch currents and potential drops)

$$J_X(\overrightarrow{KK'}) = \sum_{Y \in \mathcal{J}} L_{XY}(K, K') \Delta\Phi_Y(\overrightarrow{KK'}) \quad (2.18)$$

Table 2.2 Pairs of conjugate variables for exemplary energy domains (with U is the electrical voltage, I_{el} is the electrical current, Q_{el} is the electrical charge, $\vec{v}$ is the velocity, $\vec{F}$ is the mechanical force, $\vec{p}$ is the momentum, p is the hydrostatic pressure, w is the volume flow, m is the mass, T is the temperature, I_{th} is the heat flow, and Q_{th} is the heat).

Energy domain	Across quantity Φ	Through quantity J	Conserved quantity
Electrical	U	I_{el}	Q_{el}
Mechanical	$\Delta \vec{v}$	$\vec{F}$	$\vec{p}$
Fluidic	Δp	w	m
Thermal	ΔT	I_{th}	Q_{th}

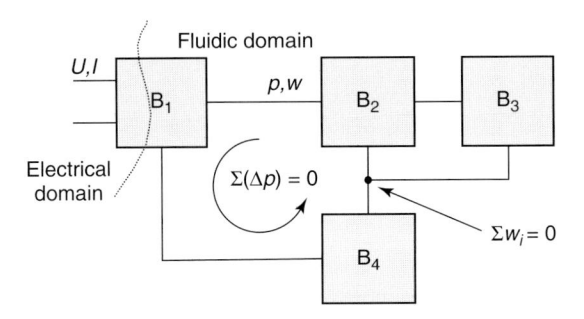

Figure 2.5 Schematic view of an electrofluidic Kirchhoffian network. The conservation laws for volume flow w and hydrostatic pressure p are reflected in sum rules at the nodes and along closed loops (meshes).

where the generalized conductances L_{XY} correspond to the generalized conductivities λ_{XY} in Eq. (2.2) and describe the transport properties in compact formulation (i.e., the compact model that determines the functional dependence between through and across quantities).

Figure 2.5 illustrates the application of the generalized Kirchhoffian network theory schematically for an electrofluidic transducer (e.g., electrostatically actuated micropump). The fluidic domain is described by the hydrostatic pressure p and the volume flow w, and the electrical domain is described by the electrical voltage U and current I. The coupling between electric and fluidic domain is realized via the transducer element B_1, which is a four-terminal element with the respective terminals in each domain. The transduction mechanism between electric and fluidic domain can be realized by a mechanical transducer, for example, an electrostatically actuated membrane. The membrane is deflected by an electrical voltage, causing a volume change and, thereby, driving the fluid flow w. A more detailed description of an electromechanical fluidic system-level model based on generalized Kirchhoffian networks is discussed in Section 2.3.

Advantages of a generalized Kirchhoffian network formulation for microsystems are as follows:

- It constitutes a physics-based description of the system, as it can be directly derived from the continuous-field formulation.
- The couplings between different energy domains are realized in a natural, generic (thermodynamics-based) way (e.g., by the matrix of transport coefficients L_{XY}, see Eq. (2.18)). By implementing the generalized Kirchhoffian network model of the microsystem in a system-level simulation tool, these couplings can be calculated simultaneously and consistently within the same simulation environment.
- Kirchhoffian network models can be incorporated in any standard circuit simulator that allows for model implementation in terms of a hardware description language (HDL, Verilog A, VHDL-AMS, etc.). This capability enables the easy cosimulation with the electric circuitry. In addition, all analysis options available in system simulators (e.g., static, transient, and small signal analysis) can be used with the fully coupled system model. Also, the global optimization of the system performance including transducers and electric circuitry becomes feasible with an acceptable computational expense. The aspects of cosimulation and the arising numerical issues will be addressed in more detail in Chapter 4.

The real challenge in system-level modeling consists in setting-up adequate compact models for the chosen subsystems, that is, the problem, how to fill the boxes in Figure 2.5. Here, a universal approach that could be applicable for any kind of MEMS structure and subsystem cannot be identified, as the best choice of the available methodologies depends strongly on specified function and application, the context, in which the model is supposed to be used, the system architecture, the required accuracy, and the desired scalability, which is equivalent to the reusability of the model for different design variants. In general, modeling methodologies should be preferred, which guarantee that physical transparency will be maintained in the transformation process from continuous-field to compact model, so that the models remain physics based and include all physical dependencies in a consistent manner.

In the following section, a short overview on different approaches to the compact modeling of microsystems is given. Most of them will be addressed in more detail within the following chapters of this book.

2.2.3
Approaches to Compact Model Derivation

2.2.3.1 Finite Network (FN) Models
FNs represent a spatially discretized description of systems by (generalized) Kirchhoffian networks. A system of physical equations can be transferred into an FN by box discretization as sketched in Section 2.2.2. FNs are, by their nature, distributed models and, to a certain extent, equivalent to finite element models concerning their complexity and their number of degrees of freedom. The advantage of using FNs is, however, that they constitute a flux-conserving discretization, which can be easily realized within the framework of analog circuit simulators, either by directly coding them in an HDL (e.g., VHDL-AMS, Verilog A, and Spectre HDL) or, alternatively, if the problem can be regarded as quasi-linear, translating them into an equivalent network description using electric network components. Thus,

an FN model can be combined with lumped element models within the same software environment (homogeneous simulation approach, mixed-level models, see Chapter 7).

An FN description should be favored, if it is not possible to derive lumped compact models for a certain subsystem, either because the geometry or the underlying physics is too complex to derive a physics-based compact model or the accuracy losses due to the simplifications made during the lumping procedure are not tolerable for the given problem or application. Especially, if the physical effects are distributed and cannot be described adequately by a single lumped parameter relation, an FN approach can be useful. As described below, FNs can be successfully combined with compact models to form a physics-based mixed-level model, for example, for the modeling of viscous damping effects in perforated microstructures (Chapter 7). In addition, an FN discretization of a subsystem can serve as a starting point for further order reduction, applying one of the projection techniques as described in Chapters 3 and 9–12 of this book.

2.2.3.2 Compact Modeling of MEMS

Compact models ("lumped element models," sometimes also called "macro models" [18, 19]) describe the operation of devices or systems through a small set of few concentrated variables ("lumped variables"). Among the existing approaches for setting-up compact models, basically three can be distinguished in the field of MEMS simulation: equivalent linear electrical network models, behavioral models, and physics-based compact models. In the following, these three approaches and their advantages and disadvantages are briefly discussed.

Equivalent Network Models In the equivalent network approach, electrical standard elements such as resistors, inductors, and capacitors are employed in order to emulate the nonelectrical behavior by equivalent RLC networks. This method has a long tradition in engineering and has been successfully applied to mechanical or fluidic systems [7, 20, 21]. The disadvantage of this method is that it is restricted to linear effects. In order to model also nonlinear effects, new network elements have to be implemented. This is why we may follow one of the two following approaches.

Behavioral Compact Models "Behavioral compact models" reproduce the operation of the original system typically by relatively simple mathematical relations or transfer characteristics that are extracted from static and dynamic FEM simulations or from measured data by means of curve fitting procedures. Often polynomials are used. The advantage of this method is that it is relatively easy to automate, for example, as a part of an FEM simulation platform [22], but, in general, the resulting models are inherently not able to reproduce the correct dependencies on the relevant design, material, and ambient parameters, so that a reliable extrapolation to design variants is not possible. Hence, the modeling procedure has to be repeated for varying device geometry, which implies a relatively high computational expense for design and optimization processes. An alternative approach to circumvent this problem is the application of physics-based compact models.

Physics-Based Compact Models Physics-based compact models describe the device operation by applying physics-based, analytical expressions as far as possible in order to enable extrapolation on different design variants and, thus, the predictive simulation of the device operation. The number of internal model parameters used for model adaption and fine-tuning should be kept as low as possible; instead, design, material, and ambient parameters should serve as input parameters. Ideally, no fit parameters should be used at all. However, the device geometry and operational principle of realistic devices may be so complex that it cannot be reproduced exactly by employing idealized, analytical expressions solely, so that, in general, fit parameters have to be introduced in order to account for these deviations from the "real-world" behavior. In order to maintain the predictive power of these models, those fit parameters should be introduced in such a way that their impact on the device operation is well understood and that they control the device operation in dedicated and well-defined parts of the characteristics. Then it is possible to derive systematic parameter extraction strategies for these fit parameters from FEM simulation or measurements, maybe by applying already existing parameter extraction tools from the semiconductor industry [12, 23]. An example is given in Section 2.3.

The strengths of physics-based compact models are their efficiency with respect to computational expense and speed, and their scalability, hence, their potential for extrapolation to other device geometries. Therefore, they are very well suited for design and parameter studies as well as for optimization processes. However, an automated model derivation is not possible for this kind of models. An experienced engineer or expert user with profound knowledge on the given problem has always to be involved in the model derivation process in order to properly incorporate all important physical effects adequately.

2.2.3.3 Mixed-Level Modeling (MLM)

The mixed-level modeling (MLM) approach combines compact models with a distributed discretization of a subsystem by FNs. It was suggested for the simulation of damping effects in perforated microstructures in Refs [24, 25]. In a mixed-level model, parts of the system are discretized and transferred into an FN formulation, a process that can be automated to a large extent [26, 27], while other parts of the system are described by a small number of concentrated variables using compact models. This method combines the advantages of both approaches, as the system can be described with distributed models in regions, where parameter lumping would come along with an unacceptably high loss of accuracy, while other regions can be modeled using computationally high-efficient compact models. This allows for a dedicated tailoring of the model complexity according to given requirements. Formulated in an HDL, MLMs can be implemented directly in a standard system-level simulator and combined with other subsystem models and the electronic circuitry for readout and control. Owing to its modularity, this method proved to be very powerful and flexible. Hence, a detailed overview of it is given in Chapter 7 by Niessner *et al.* with a special emphasis on its application to the modeling of viscous damping effects.

2.2.3.4 Mathematical Model Order Reduction (MOR)

Another approach to derive models with a highly reduced number of degrees of freedom is based on numerical order reduction methods. These concepts can be applied to any spatially discretized model and reduce the order of the underlying equation system by applying specific mathematical order reduction algorithms. All methods reduce the system complexity by applying projection methods such as modal superposition, Krylov-subspace techniques, moment matching, and proper orthogonal decomposition. The goal is to project the original system onto a properly chosen subspace, where the operation of the system can be described with a highly reduced number of degrees of freedom without loosing too much of accuracy. The resulting ROMs are compatible with the structure of a generalized Kirchhoffian network and, therefore, can easily be integrated and combined with other system-level models.

As ROMs are based on straight-forwardly and universally applicable mathematical algorithms, the derivation procedure can be standardized, automated, and integrated in a consisting simulation framework. In contrast to lumped element models, which are frequently "hand-built" by an experienced engineer or physicist, ROMs may then be derived using a standard tool of a simulation platform and, for that reason, potentially faster and easier to apply than analytical compact models. However, they are not as transparent and as intuitive as physics-based compact models, although approaches exist that provide the possibility to vary parameters to a certain extent (Chapter 9). The basics of mathematical model order reduction are addressed in detail in the following chapter and specific applications, in Chapters 9–12.

Which of the above-proposed modeling methods should be applied for the practical implementation depends on the given application, the intended use of the final system-level model, the desired accuracy, the available simulation tools and platforms and, of course, the user and his working environment itself. Regardless of which approach has been chosen, one should always bear in mind that a reduced model is the projection of a continuous-field (i.e., a much more accurate) model and, thus, is inherently restricted to a certain range of validity in the operating area of the device. Thus, it should be checked carefully, if the model complies with the given prerequisites before it is employed in the analysis of a specific problem. Considering this basic aspect, one should be aware of the two requirements a model maker should account for, namely, transparency and physical consistency in the design of models [15, 16].

2.3
Application 1: Physics-Based Electrofluidic Compact Model of an Electrostatically Actuated Micropump

An illustrative and instructive example for physics-based, lumped element system-level modeling is the electrostatically driven micropump, which was developed at Fraunhofer EMFT (formerly Fraunhofer IZM) in Munich in the late

1990s [28–30] and is nowadays established in various applications, the electrostatic driving unit being replaced by a piezoelectric one [31, 32]. The electromechanical fluidic system model has been derived in Refs [12, 33–35] and constitutes one of the first system-level models based on generalized Kirchhoffian network theory for this class of devices, which is not restricted to an on-purpose developed simulation tool but is implemented through an HDL and simulated in a standard system simulator [36].

A schematic view of the micropump is depicted in Figure 2.6. It consists of a pump chamber with an electrostatically displaceable membrane as driving element, passive inlet and outlet valves, and externally attached tubes. The membrane is separated from a solid counterelectrode by an air gap and an isolating oxide, which prevents electrical short circuiting when the membrane touches the counterelectrode. The pump is fabricated by assembling several silicon wafers, which were structured by anisotropic etching [28, 29]. When a voltage is applied to the driving unit, the membrane deflects and causes an underpressure inside the pump chamber, which opens the passive inlet valve and induces a fluid flow into the pump chamber. When the voltage is turned off, the electrostatic driving unit is discharged and the membrane releases causing an overpressure inside the pump chamber, which opens the outlet valve and pushes fluid out of the chamber. Then, the membrane is charged again and the process is repeated. The described pump cycle illustrates that the micropump represents a complex system, whose operation is governed by an intricate coupling of various physical energy domains, namely, the mechanical, the electrical, and the fluidic domain, while, at the same time, exhibiting a large number of degrees of freedom. So, computations on continuous-field level of the entire system, for example, by means of coupled three-dimensional finite element analysis, become prohibitive in this case, even if we keep in mind that the operation of the pump, by its nature, has to be modeled transiently. A system-level model with a highly reduced number of degrees of freedom is therefore inevitable for the design and optimization of such devices. Accordingly, we apply the above-developed methodology and divide the system

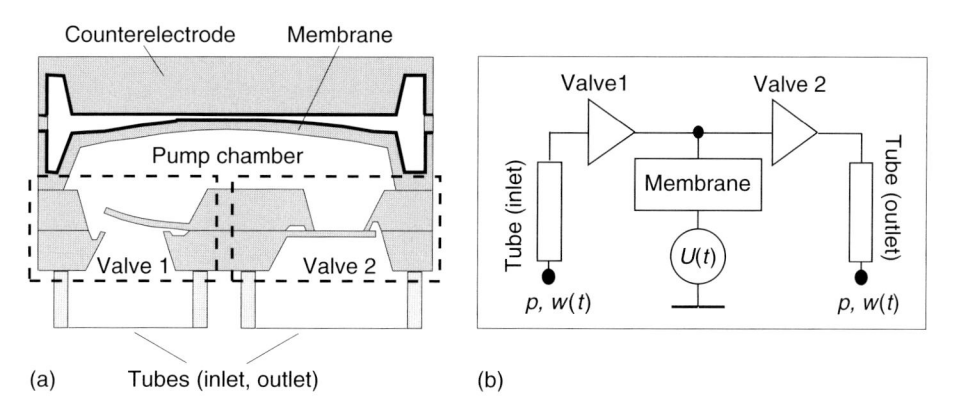

Figure 2.6 Schematic view of the electrostatically driven micropump (a) and its division into functional blocks for the derivation of the generalized Kirchhoffian network model (b).

first into manageable parts. Fluid mechanically coupled quasi-static finite element simulations reveal that the main pressure drop (and the maximum fluid velocity) occurs across the valve openings, while the pressure distribution inside the pump chamber can be regarded as spatially uniform. This justifies to divide the system along the dashed lines in Figure 2.6 into the functional units "tubes," "valves," and "pump chamber with membrane drive" and to describe the operation of the system parts by the flow rates through the interfaces (i.e., the surface integrals of the flow density along the interfaces). The constitutive set of conjugate variables for the fluidic components consists of the hydrostatic pressure p and the volume flow w. The resulting generalized Kirchhoffian network with its subsystems represented by simple "boxes" is sketched in Figure 2.6 (right). Next, compact models for each of the basic functional units have to be derived. A strictly physics-based modeling strategy ensures the models to be scalable for predictive simulation, viz, for their application in design and optimization studies. In order to illustrate this, the derivation of the single subsystem models is described shortly in the following sections. For more details, the interested reader is referred to [12, 33–35], from where most figures and graphs are also taken.

2.3.1
Compact Model of the Membrane Drive

The driving unit consists of a flexible silicon membrane and a rigid counterelectrode, both coated by a thin silicon dioxide layer in order to avoid electric shortage (Figure 2.7).

The compact model of the drive unit exhibits four terminals, two within the fluidic domain for the hydrostatic pressure p acting on the membrane and the driven fluid volume w and two in the electrical domain for the voltage V and the current I. The volume flow w per pump cycle results from the fluid displaced by the actuated membrane. Thus, the deflection of the membrane middle h_{mid} and an effective lateral length of the membrane l_{eff} are introduced as additional internal parameters of the model in order to calculate the membrane bending as a function of the applied pressure. Before the membrane touches the counterelectrode (pull-in), the effective length l_{eff} is constant and equal to the lateral length of the membrane l_m. Then the deflection of the membrane center h_{mid} and the position-dependent bending $h(x, y)$ can be determined according to the following analytic formula for

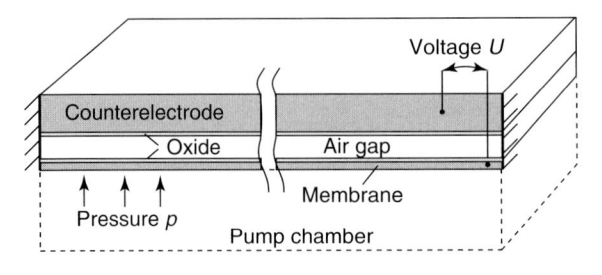

Figure 2.7 Schematic view of the membrane drive.

a clamped membrane [37]

$$h_{mid} = 0.00126 \frac{p \cdot l_m^4}{D \cdot d_m^3} \text{ with } D = \frac{E}{12(1-v)} \quad (2.19)$$

$$h(x,y) = h_{mid} \cdot \cos\left(\frac{\pi x}{l_m}\right) \cos\left(\frac{\pi y}{l_m}\right) \quad (2.20)$$

Here, p stands for the hydrostatic pressure, d_m stands for the thickness and l_m for the lateral length of the membrane, E for the Young's modulus, and v for the Poisson number.

After pull-in of the membrane occurs, h_{mid} remains constant and the free standing length of the membrane l_{eff} decreases according to the enlarging fraction of the membrane area touching the counterelectrode (Figure 2.8). As a consequence, an approximately square segment in the middle of the membrane is excluded from the force balance. The bending of the remaining rectangular membrane areas at each of the four sides can then be calculated applying a 2D approximation for long rectangular membranes [37]

$$h_{max} = 0.00126 \frac{p \cdot l_{eff}^4}{D \cdot d_m^3} \quad (2.21)$$

$$h(x) = h_{mid} \cdot \cos\left(\frac{\pi(x - x_t)}{l_{eff}/2}\right) \quad (2.22)$$

In contrast to the situation before pull-in, now the maximum deflection h_{mid} is fixed by the contact to the counterelectrode while l_{eff} decreases. Effects due to the corner areas of the membrane, which cannot be covered by the geometrical approximation made in Eq. (2.21), have been taken into account by introducing a fit factor extracted from 3D FEM simulations.

The electrostatic force, viz, the electrostatic pressure acting on the membrane due to the applied voltage, is calculated according to

$$p_{el}(x,y) = \epsilon \cdot V^2 / (d_{gap} - h(x,y))^2 \quad (2.23)$$

inserting the bending line $h(x,y)$ of the membrane. p_{el} is averaged over the membrane area and added to the hydrostatic pressure. Then, the bending shape

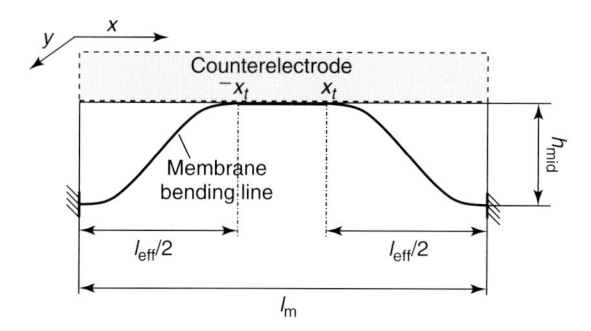

Figure 2.8 Geometry of membrane model (not to be scaled).

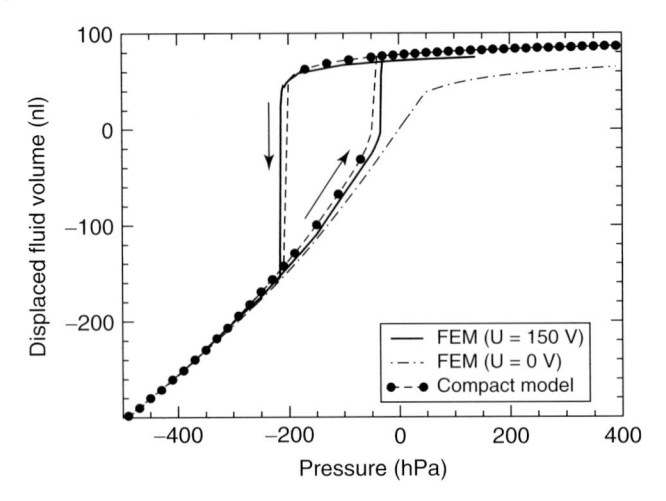

Figure 2.9 Quasi-static characteristics of the membrane drive: comparison between compact model and FEM simulation. A voltage of 150 V is applied and the pressure inside the pump chamber is varied.

of the membrane and, thus, the displaced fluid volume w is determined and given as through variable to the respective node of the fluidic part of the Kirchhoffian network. Figure 2.9 shows the quasi-static characteristics of the membrane drive. A voltage of 150 V is applied and the pressure inside the pump chamber is varied. The membrane shows the typical pull-in behavior as observed for electrostatically actuated devices, with the sudden pull-in, when the electrostatic force exceeds the repulsive mechanical force, and the release from the counterelectrode for a much smaller pressure value (electromechanical hysteresis). The compact model is in very good agreement with the FEM simulations concerning the pull-in as well as the release point of the membrane.

2.3.2
Compact Model of the Valves

The passive flap valves are fabricated by etching and subsequent wafer bonding of three differently structured silicon wafers (Figure 2.6a) [28, 29]. A rectangular cantilever-like structure forms the flap and closes the valve opening when touching the valve seat. The third, upper wafer constitutes the upper part of the pump chamber and, at the same time, the connection to the driving unit.

The compact model of the valve exhibits two terminals in the fluidic domain and is formulated using the hydrostatic pressure p and the volume flow w as across and through quantity, respectively. The flow through the valve depends on the opening of the valve, namely, the size of the slid between flap and valve seat. Coupled FEM simulations of the hydrostatic pressure distribution justify to lump the acting force in the middle of the flap and to calculate the resulting displacement of the flap center y_{mid} from the analytical formula for a beam [37]

$$y_{mid} = \frac{\Delta p \cdot A \cdot l_{mid}^3}{3D}$$ (2.24)

$$\text{with } D = \frac{Ed_{fl}^3}{12(1-\nu)}$$

where p is the hydrostatic pressure, A the cross section of the valve, l_{mid} the distance between the center of flap and flap suspension, E the Young's modulus, d_{fl} the thickness of the flap, and ν the Poisson number.

The mechanical quantity y_{mid} has been introduced as an internal parameter of the compact model in order to calculate the opening of the valve and, thus, the resulting volume flow through it by applying the analytical formula for a rectangular slit given in Ref. [38]. Figures 2.10a,b show that the quasi-static flow rate through the valve and the mechanical displacement of the flap obtained by this compact model are in very good agreement to FEM simulations and measurements.

The transient behavior of the flap is further dominated by damping forces and the inertia of the surrounding fluid. In addition, a displacement flow has to be taken into account, which results from the fluid displaced by the moving flap and constitutes the main effect causing the frequency dependence of the pump characteristics.

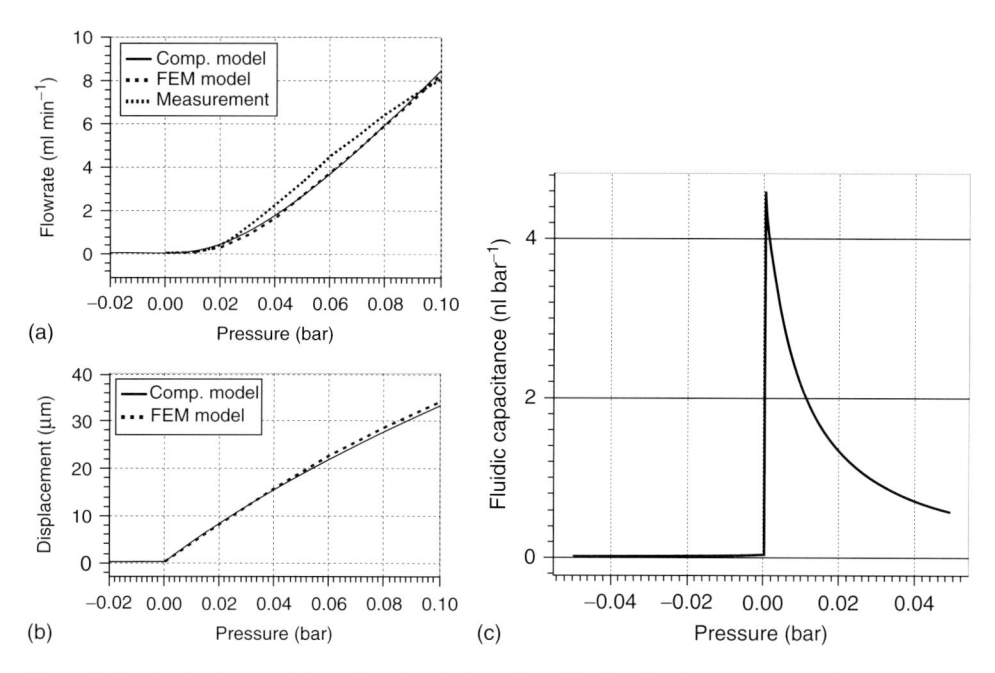

Figure 2.10 Static characteristics of a valve (volume flow (a), flap displacement (b), and fluidic capacitance versus pressure (c)). The simulation results of the compact model are represented by solid lines.

These effects can be taken into account by applying inertial forces of the flap and the surrounding fluid layer dragged by its motion to the center of the flap area. The thickness of this additional fluid layer has been determined as a fit parameter from FEM simulation data (fit of the resonance frequency) to about 10 times the thickness of the flap. This showed to be a rather reasonable and more likely result than the values of 1020 – 2360 given in the model of [39] obtained by analytical estimations.

The displacement flow w_d due to the flap motion is calculated according to

$$w_d = A \cdot dy_{\mathrm{mid}}/dt \tag{2.25}$$

with the corresponding fluidic capacitance defined as

$$C_f = dV/d(\Delta p) \tag{2.26}$$

which is shown in Figure 2.10c.

Finally, the damping force is extracted from FE simulations and added to the force balance. The entire flow rate through the valve is then obtained by summing up the quasi-static flow rate and the displacement flow w_d. The compact model so achieved describes the transient volume flow through the valve as a function of the hydrostatic pressure and the internal mechanical parameter y_{mid} of the flap and is connected to the other submodels of the pump via the two terminals in the fluidic domain.

A difficulty with this model is its discontinuity, when the pressure changes from positive to negative values and the flap hits the valve seat. This has its correspondence in a discontinuity of the fluidic capacitance at $\Delta p = 0$ (Figure 2.10), which could cause numerical instabilities. In order to avoid them, a (realistic) bending of the valve flap for negative values of $\Delta p = 0$ is admitted, thus storing the kinetic energy in the reverse (much stiffer) bending of the closed flap. This is another example that demonstrates how a physics-based modeling strategy can lead to adequate (here even more stable) numerical models.

2.3.3
System-Level Model of the Tubes

Connecting tubes have a decisive impact on the performance of a microfluidic system, especially, when they consist of an elastic material and are compliant as it is the case for the considered micropump. Reliable tube models are therefore inevitable in order to reproduce the system operation properly.

In other disciplines (mainly medicine), various approaches have been applied to derive analytic compact models for elastic tubes [20, 40, 41]. In Reference [12], a Kirchhoffian network-based tube model has been derived, which finally exhibits the same structure as an electrical transmission line. In the following, only the basic steps of the model derivation will be recalled, and for more details the reader is referred to [12].

Starting from the general Navier-Stokes equation for cylindrical symmetry and taking into account the elasticity of the tubes, the propagation of pressure

Figure 2.11 Finite network model for the tubes. The model is implemented as an electrical transmission line model.

waves inside the tubes can be derived. After introducing some geometry- and problem-specific assumptions and simplifications, the following expression for the pressure propagation is obtained straightforwardly

$$\frac{\partial^2 p}{\partial t^2} = \frac{\partial^2}{\partial x^2}\left(\frac{R_0 k}{2}p + \eta_k \frac{\partial p}{\partial t}\right) \tag{2.27}$$

where R_0 denotes the radius of the tube, k a constant factor that combines the tube elasticity and the fluid density, and η_k the kinematic viscosity. It arises that the fluidic problem described through Eq. (2.27) is equivalent to the electric circuit depicted in Figure 2.11. Each element represents a piece of the tube with a certain length. By relating the fluidic parameters to the parameters of the electric analogon, the tube can be modeled by a certain number of these subnetworks connected in series thus forming a one-dimensional FN model of the tube. The circuit model so derived is very similar to an electrical transmission line model and can be implemented directly in an analog network simulator.

2.3.4
Physics-Based System-Level Model of the Micropump

For the system-level model of the entire micropump, the above-derived submodels are implemented in a standard circuit simulator (here: Spectre [36]) using an HDL and connected according to the generalized Kirchhoffian network depicted in Figure 2.6. The comparison of the simulated frequency-dependent pump rate to measurements is shown in Figure 2.12. A very good agreement is obtained not only for the design of the pump, which has been used for model calibration and validation (Figure 2.12a), but also for a second design with different geometrical dimensions (Figure 2.12b). This agreement has been achieved without any readjustment of the implemented fit parameters and demonstrates the power and strength of physics-based compact models, namely, their ability to extrapolate to design variants, which is a prerequisite for optimization and design studies.

Physics-based models further offer a deep insight into the device operation. Figure 2.13, for example, shows the opening and closing of the valve flaps, the resulting flow rate, and the averaged pressure drop during different pump cycles. This is highly complex, as transient and fluid mechanically coupled process with mechanical contact at the valve seat, and can be hardly modeled by three-dimensional simulations on continuous-field level. The results obtained with the system-level model show that the phase shift, which occurs between the pressure inside the

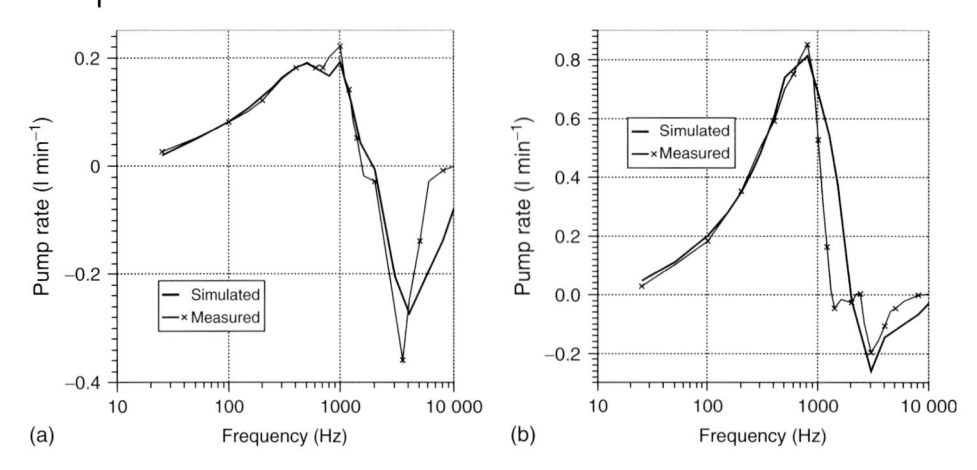

Figure 2.12 Frequency-dependent pump rate of the electrostatically operated micropump: comparison between measured and simulated data for the design under consideration (a) and a second design variant (b). The calibrated models were not recalibrated for the second device.

pump chamber and the flap displacement, reveals as the decisive impact on the resulting flow rate of the pump and on the frequency-dependent pump characteristics. For a frequency lower than the resonance frequency of the valve flaps the pump operates in forward direction, and for frequencies above this value the phase shift between pressure and flap displacement increases (Figure 2.13, d-f) and the resulting net flow becomes negative. This insight makes the frequency-dependent pump characteristics of Figure 2.12 interpretable from a physical point of view and, thus, accessible and controllable by dedicated design measures.

2.3.5
Model Calibration and Parameter Extraction

The basic feature of physics-based models is that they (ideally) contain all relevant parameters explicitly and reproduce their dependencies correctly. As a consequence, all those parameters are directly accessible for dedicated parameter extraction and calibration strategies by measurements or with respect to continuous-field simulations (e.g., FEM), which offers the possibility to derive high-fidelity models for predictive simulation.

Of course, analytical formulas are, in general, abstracting and, thus, idealizing and simplifying the physical reality to a certain extent. Hence, in most cases, fit parameters have to be introduced that take these nonidealities into account in order to reproduce the operation of the devices not only qualitatively but also quantitatively. Preferably, these fit parameters are introduced in a way that their action is well understood with respect to the device operation, so that dedicated parameter extraction strategies can be derived and applied. For the model of the membrane drive, for example, we basically introduced three fit parameters. The first affects only the linear region (small deflections of the membrane) and takes

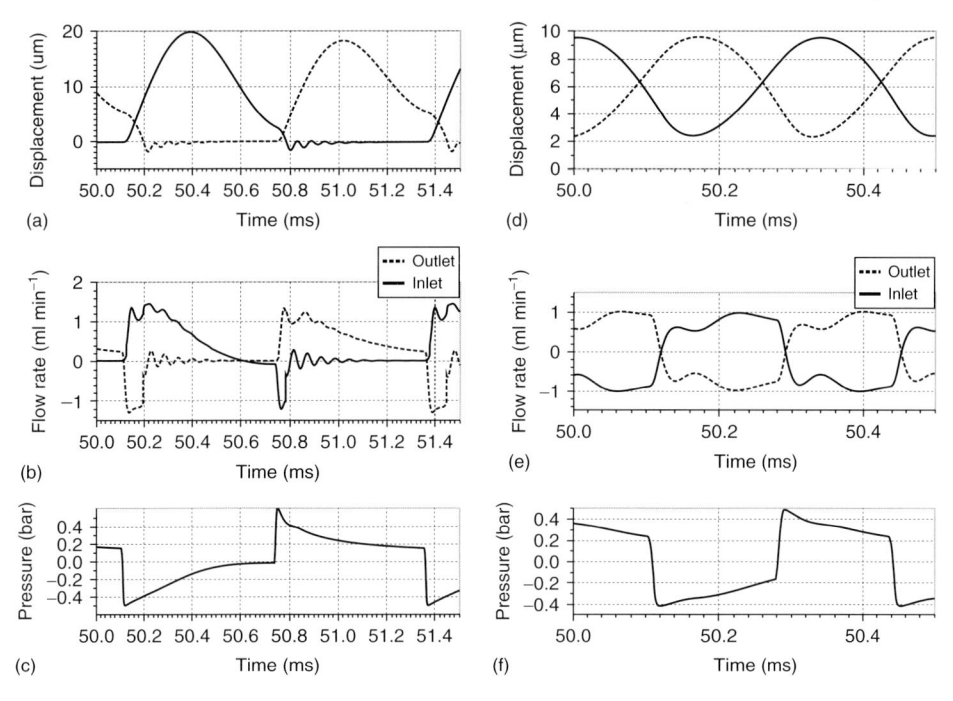

Figure 2.13 Opening and closing of the inlet and the outlet valve near the fundamental resonance frequency of the flap (a–c) and at a much higher frequency (d–f): displacement of both flaps, flow rate, and mean pressure inside the pump chamber are displayed.

into account manufacturing tolerances such as variation in geometry and material parameters. The second parameter is introduced to model nonlinearities of the mechanical deformation in the region of large deflections. Finally, the third fit parameter accounts for corner and edge filling effects and comes into play for large pressure values when the membrane is in touch with the counterelectrode. Each of the three parameters has its main impact on a specific region of the operation range. Thus, they can be adjusted by a sequential fitting procedure, in this case with respect to coupled electromechanical FEM simulations of the membrane drive [23]. As the effect of these fit parameters is well known and understood, it is also possible to decide up to what range of extrapolation their calibration is valid and under which circumstances a recalibration of them is necessary (cf. Figure 2.12 and related text).

2.4
Application 2: Electrostatically Actuated RF MEMS Switch

The second demonstrator is a MEMS switch, which has been fabricated at Fondazione Bruno Kessler and is intended for radio frequency (RF) applications [42, 43]. It will be addressed only shortly in order to demonstrate that there are, in

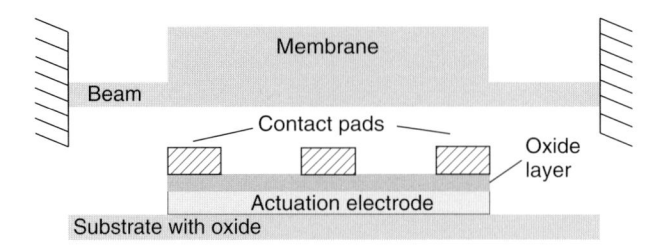

Figure 2.14 Schematic cross section of the RF MEMS switch.

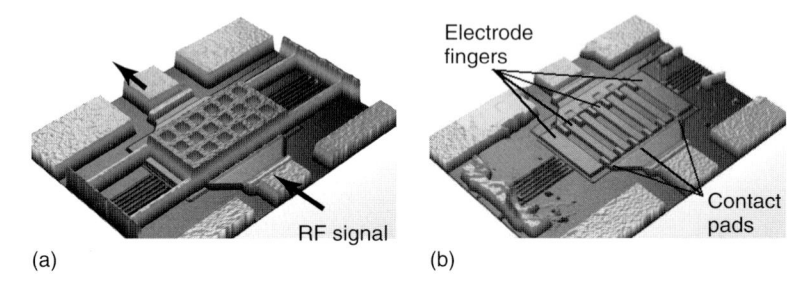

Figure 2.15 (a) Micrograph of the RF MEMS switch taken by a white-light interferometer. (b) Switch without membrane in order to observe the electrodes and the contact pads.

general, several alternatives to realize a system-level model of a device. Further and more detailed aspects of this demonstrator or similar devices will be picked up in Chapters 7 and 8.

The device structure and the operation principle of this actuator is shown schematically in Figure 2.14. The switch consists of a movable perforated gold membrane suspended above a fixed ground electrode by four straight beams. The fixed ground electrode acts as an actuation electrode of the switch and consists of several lateral fingers, which are connected in parallel. By applying a voltage, the suspended membrane is pulled toward the ground electrode, collapses onto 12 elevated contact pads, and closes an ohmic contact so that an RF signal path is closed. For more details on the technology and the fabrication process of the switch, refer to [42, 43]. The topography of the switch has been analyzed by applying a white-light interferometer. In Figure 2.15, the 3D topographies of the switch and the underlying structures are displayed.

As mentioned already in Section 2.2.1, the system model is partitioned into four subsystems according to the dominant forces and energy domains (Figure 2.16): the mechanical subsystem represents the inertia and elastic forces of the perforated membrane and the four flat suspension springs, the electrostatic subsystem accounts for the electric field between the membrane and the actuation electrode, the fluidic subsystem comprises the ambient air that exerts damping forces on the moving parts of the structure, and finally, the contact model describes the closing phase of the switch and the contact between the membrane and the counterelectrode.

2.4 Application 2: Electrostatically Actuated RF MEMS Switch

The mechanical forces F_i and the mechanical deformations are introduced as through and across quantities, respectively, and the complete system model of the RF MEMS switch is obtained by interlinking the contributions of all single domain submodels to form a generalized Kirchoffian network. The node law in the mechanical domain then reads as follows:

$$F_{\text{mech}} + F_{\text{damp}} + F_{\text{electro}} + F_{\text{cont}} = 0 \tag{2.28}$$

Here, F_{mech} denotes the mechanical force of the switch, F_{damp} the viscous damping force, F_{electro} the electrostatic actuation, and F_{cont} the force that occurs when the membrane is in contact with the underlying structure.

In principal, there are several choices how the "boxes" in Figure 2.16 could be filled, that is, which approach may be chosen to model the single subsystems. A simple approach would be to implement $F_{\text{mech}} = -M_{\text{pl}}\ddot{z} - Kz$ and $F_{\text{damp}} = -D\dot{z}$ with M_{pl} being the equivalent mass of the perforated plate, D the damping coefficient due to the surrounding air, and K the stiffness of the mechanical springs. Then, Eq. (2.28) becomes

$$M_{\text{pl}}\ddot{z} + D\dot{z} + Kz = F_{\text{elmech}} + F_{\text{contact}} = F_{\text{ext}} \tag{2.29}$$

This relation represents a one-dimensional spring-mass-damper model excited by external forces F_{ext}. The coefficients can be calculated by analytical equations, for example, K by the one-dimensional beam bending formula or the electrostatic force by applying a simple plate capacitor approach with varying gap

$$F_{\text{elmech}} = \frac{1}{2}\varepsilon_0\varepsilon_r A_{\text{eff}} \frac{U^2}{(d(U))^2} \tag{2.30}$$

where $\varepsilon_0, \varepsilon_r$ being the electric permittivity, U the electric voltage, $d(U)$ the variable gap height between movable plate and counterelectrode, and A_{eff} the effective plate area taking into account the perforations in the structure. This model could be refined by introducing additionally the fringing fields occurring around the perforation holes, maybe by a fit factor extracted for a perforated substructure from three-dimensional FEM simulations. The damping constant D can be calculated from distributed continuous-field models or extracted from measurements. Equation (2.29) then constitutes the energy-coupled compact model of the switch with the mechanical forces as through and the displacement z of the mechanical

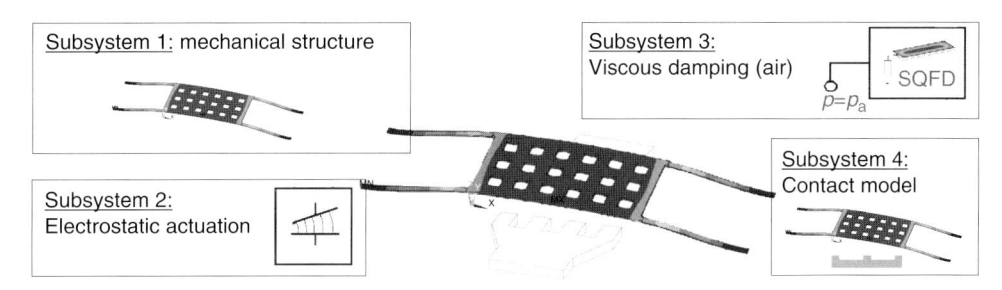

Figure 2.16 Subsystems of the RF MEMS switch, for which compact models have to be derived.

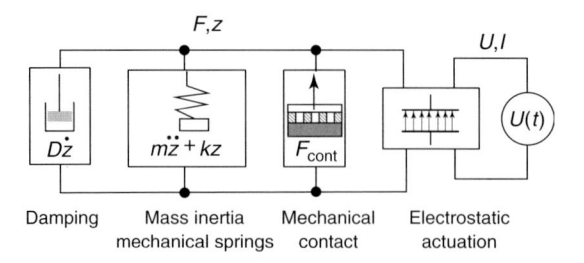

Figure 2.17 Generalized Kirchhoffian network model of the RF MEMS switch according to the system decomposition of Figure 2.16. The switch is represented as a simple, damped, one-dimensional spring-mass system; through and across variables are chosen to be the force F and the displacement z.

mass as across variable in its generalized Kirchhoffian network formulation. A schematic representation of this simple spring-mass-damper model is given in Figure 2.17.

If, however, a more elaborated model is desired that reflects the underlying physics in a better way, the single subsystems of Figure 2.16 have to be described by more sophisticated methods (as for example demonstrated in Refs [24, 44–46]). This is illustrated in Figure 2.18.

For modeling the viscous damping forces, the mixed-level approach proposed in Ref. [24] can be followed, which leads to a much more physics-based model than introducing a simple damping constant extracted from experiments or FEM simulations. The damping forces are calculated by a distributed network model of the Reynolds equation – a simplified version of the Navier-Stokes equation, which can be applied for a wide variety of MEMS geometries – combined with analytical compact models (basically fluidic resistances) containing all relevant geometrical quantities and physical dependencies from ambient parameters (pressure, viscosity, etc.). The fluidic mixed-level model is formulated in terms of the volume flow and the pressure of the surrounding air when transferring it to a Kirchhoffian network representation. The coupling to the mechanical and electrostatic submodels is

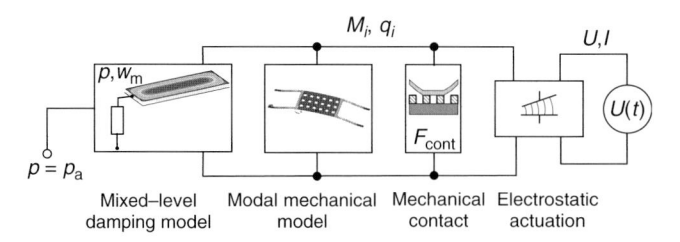

Figure 2.18 Generalized Kirchhoffian network model of the RF MEMS switch applying more elaborated compact models than in Figure 2.17. Through and across variables are the modal moments M and the modal amplitudes q_i, a mixed-level model is applied to describe the damping, the mechanical part is modeled by using the modal superposition technique as described in Chapter 12.

realized via the damping force acting on the moving membrane (Eq. (2.28)) and the varying gap height underneath the moving structure (see also Chapter 7, where this approach is described in detail).

In order to take into account the flexibility of the suspended membrane and, thus, its bending, a model based on the modal superposition technique described in Chapter 12 or in Refs [47, 48], for example, might be applied. To this end, the eigenmode shapes and frequencies of the suspended membrane are calculated and serve as a set of basis functions. The vector of displacements **u** can then be approximated by a superposition of a number m of weighted and discretized eigenmode shape functions $\mathbf{\Phi}_i$

$$\mathbf{u}(t) \approx \mathbf{u}_0 + \sum_{i=1}^{m} q_i(t) \mathbf{\Phi}_i \tag{2.31}$$

$\mathbf{u}_0$ accounts for the displacement in the equilibrium state and $q_i(t)$ denotes the modal amplitude that scales the corresponding shape function $\mathbf{\Phi}_i$. The most significant modes are identified – in the case of the considered switch the fundamental and the next higher completely symmetric eigenmode – and used to formulate a compact model in terms of the modal amplitudes consisting of only one second-order differential equation per included eigenmode

$$\ddot{q}_i + \omega^2 q_i = M_{\text{external},i} \tag{2.32}$$

Here, ω_i denotes the angular eigenfrequency of the ith eigenmode and $M_{\text{external},i}$ an external modal moment, for example, due to the electrostatic actuation. This set of equations constitutes the mechanical reduced order model with the modal amplitudes q_i and the modal moments M_i acting as across and through variables in a generalized Kirchhoffian network, respectively.

A more elaborated submodel for the electrostatic actuation is derived by first determining the electrostatic energy stored between a single electrode finger and the membrane in terms of the modal amplitudes and then deriving the respective capacitance functions for the respective eigenmode i. In a second step, Lagrangian energy functionals, and from these, the electrostatic moments $M_{\text{el},i}$, are calculated in terms of the capacitance functions and included in the right-hand side of Eq (2.32), realizing straightforwardly the coupling between mechanical and electrostatic domain. In the electrical energy domain, the electrostatic submodel is represented by a variable capacitor (depending on the modal amplitude q_i) that is described by the relation between the total electrical current I through the capacitor formed by the electrode fingers (through variable) and the voltage V (across quantity) at the electrode fingers (for more details, see Ref. [46]). The contact model represented by $M_{\text{contact},i}$ is more complex and is not discussed here. Details are explained in Ref. [46].

The entire macromodel of the switch is then obtained by formulating all submodels in terms of the modal coordinates using Eq. (2.32)

$$\ddot{q}_i + \omega^2 q_i = M_{\text{el},i}\left(\underline{q}, V\right) + M_{\text{reynolds},i}\left(\underline{q}, \underline{\dot{q}}, P_0\right) + M_{\text{contact},i}\left(\underline{q}\right) \tag{2.33}$$

In Eq. (2.33), the mechanical equations in their modal form are no longer independent from each other. Their interaction, which means the coupling between

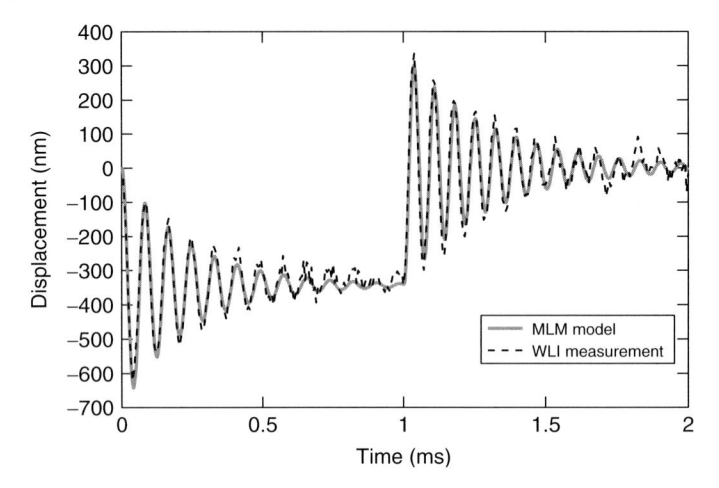

Figure 2.19 Transient response of the RF MEMS switch to the actuation by a voltage step (voltage below pull-in voltage). Comparison between simulated measured displacement (Source: Taken from [49]).

the single energy domains, enters this equation via the right-hand side, where the electrostatic, fluidic, and contact submodels are included.

Replacing the simple mass–spring model as well as the damping constant in the Kirchhoffian network in Figure 2.17 by a more sophisticated mechanical model based on modal superposition and a distributed mixed-level damping model implies, of course, a higher investment in the model derivation process. However, this pays off with view to the accuracy and the scalability and, hence, the predictive power and the reusability of the model. This becomes obvious in the experimental validation for different device geometries and for varying ambient conditions (see results for damping model demonstrated in Chapter 7).

Figure 2.19 shows the simulated response of the RF MEMS switch to a rectangular voltage step compared to measured data. The perforated membrane is actuated, displaced to a new equilibrium position, and released again to its original position, which is both accompanied by damped oscillations. It can be seen that the frequency shift due to electrostatic spring softening as well as the increased damping due to the smaller gap in the displaced state of the switch are reproduced accurately due to the physics-based character of the underlying models.

2.4.1
Conclusions

Generalized Kirchhoffian networks offer a generic and comprehensive framework for the modeling of microdevices and systems. Following the basic principles of irreversible thermodynamics, this theory, originally established for electrical circuits, can be extended to general microsystems including electrical and non-electrical components and their respective energy domains and, thus, offers the possibility to describe the considered system on a physical basis and to formulate and calculate the couplings between the involved subsystems and energy domains in a natural way within a uniform, comprehensive simulation framework.

Kirchhoffian network models can be incorporated in any standard circuit simulator that offers model implementation via a hardware description language (HDL, Verilog A, VHDL-AMS, etc.). Thus, the cosimulation of one or more transducers together with the attached electrical circuitry is easy to realize, which enables the investigation and optimization of full microsystems within a homogeneous simulation environment (Chapter 4). As the generalized Kirchhoffian network theory constitutes a general theoretical framework for the modeling on system level, it supports the assembly of any system model from submodels that are derived in a flux-conserving formulation. The overview of different compact modeling approaches given in Section 2.2.3 is subsumed in Figure 2.20, where the

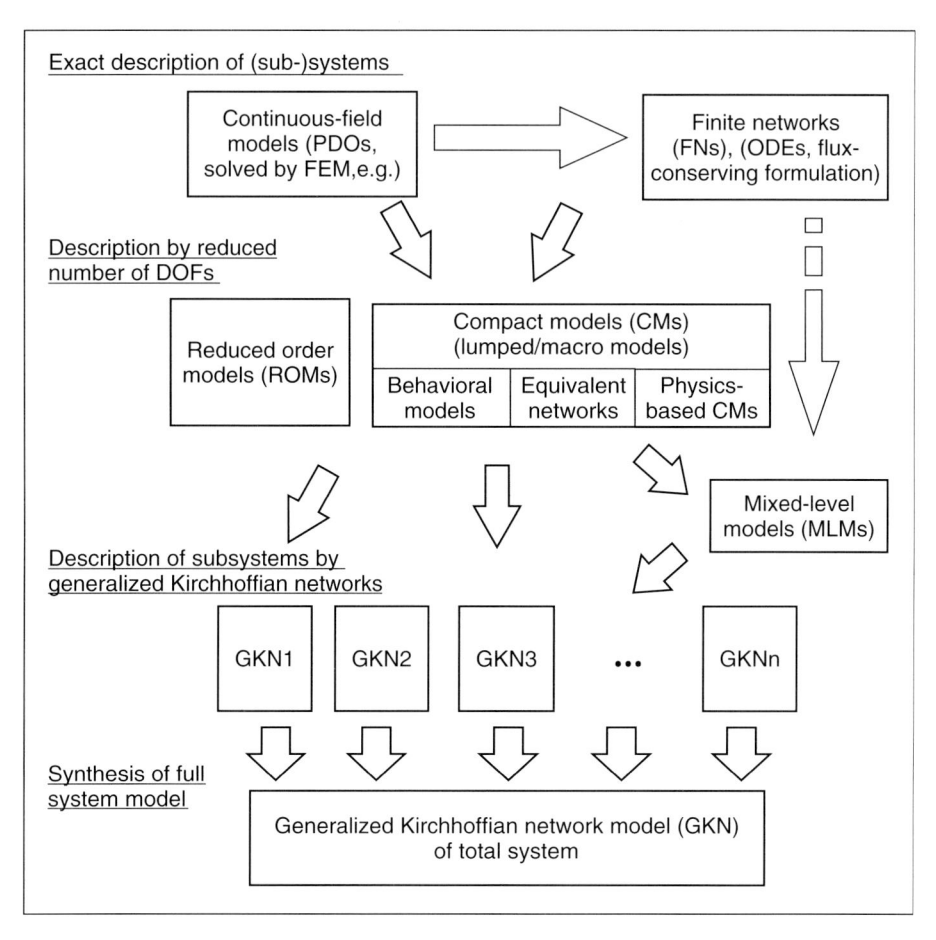

Figure 2.20 Overview over model hierarchies and modeling approaches within the framework of generalized Kirchhoffian networks.

different hierarchies in model complexity are also indicated. As already mentioned, it depends on the given application and the requirements on accuracy, scalability, and computational efficiency, which approach is feasible and to what extent model simplifications (or model complexity) can be still accepted.

The application examples given in this chapter demonstrate that physics-based compact models containing the relevant material and design parameters and their physical dependencies correctly offer a deeper insight in the system operation and performance than simple models based on parameter fitting or one-dimensional spring-mass-dampers. In general, however, their derivation requires a higher effort and deeper insight in device and system functionality – in general gained by profound investigations – as it cannot be automated like the mathematics-based ROM generation or the parameter fitting for purely behavioral compact models. This investment however pays off at the time, when the models are reused for design and optimization studies where extrapolation within a given design space is required. Nevertheless, it depends on the given problem, the operation of the system, the intended application, and the model requirements arising from the design process, which approach might be suitable and is, hence, chosen individually by the engineer or system designer. However, the above-mentioned approaches offer the basis to tailor system models according to the nature of the given problem and the needs arising from the specific application field.

Regardless of which of the approaches depicted in Figure 2.20 is finally applied, generalized Kirchhoffian networks provide the framework for the physics-based model tailoring of system-level models for microsystems with view to the given needs and practicalities. A bundle of examples in the sequel of this book will demonstrate the power and generic character of this method.

References

1. Middelhoek, S. and Hoogerwerf, A.C. (1986) Classifying Solid-State Sensors: the "Sensor Effect Cube". *Sensors and Actuators*, **10**, 1–8.
2. Middelhoek, S. (1998) The Sensor Cube Revisited. *Sensors and Materials*, **10** (7), 397–404.
3. Schwarz, P. (1998) Microsystem CAD: From FEM to System Simulation, in *Simulation of Semiconductor Processes and Devices (Sispad'98)* (eds K. DeMeyer and S. Biesemans), Springer Verlag, Leuven, pp. 141–148.
4. Schwarz, P. (2000) Physically oriented modeling of heterogeneous systems. Proceedings of 3rd IMACS Symposium of Mathematical Modeling (MATHMOD), 2000, Wien, Austria, pp. 309–318.
5. Koenig, H.E. and Blackwell, W.A. (1961) *Electromechanical System Theory*, McGraw-Hill, New York.
6. Klein, A. and Gerlach, G. (1996) System modeling of microsystems containing mechanical bending plates using an advanced network description method. Proceedings of MICROSYSTEMS Technologies, 1996, Potsdam, Germany, pp. 299–304.
7. Lenk, A., Pfeiffer, G. and Werthschuetzky, R. (2000) *Elektromechanische System. Mechanische und akustische Netzwerke, deren Wechselwirkung und Anwendungen*, Springer Verlag, Berlin.
8. MacNeal, R.H. (1951) The solution of elastic plate problems by electrical analogies. *Journal of Applied Mechanics*, **18**, 59–67.

9. Tilmans, H. (1996) Equivalent circuit representation of electromechanical transducers: I. Lumped-parameter systems. *Journal of Micromechanics and Microengineering*, **4**, 157–176.
10. Fedder, G. and Jiang, Q. (1999) A hierarchical circuit-level design methodology microelectromechanical systems. *IEEE Transaction on Circuits and Systems II (TCAS)*, **46** (10), 1309–1315.
11. Jing, Q., Mukherjee, T., and Fedder, G. (2002) Schematic-based lumped parameterized behavioral modeling for suspended MEMS. Techn. Digest of the ACM/IEEE International Conference on Computer Aided Design (ICCAD '02), 2002, San Jose, CA, pp. 367–373.
12. Voigt, P. (2003) Compact modeling of microsystems, in *Selected Topics of Electronics and Micromechatronics* (eds G. Wachutka and D. Schmitt-Landsiedel), Shaker Verlag, Aachen, Germany.
13. Schrag, G. (2003) Modellierung gekoppelter effekte in mikrosystemen auf kontinuierlicher feldebene und systemebene, in *Selected Topics of Electronics and Micromechatronics* (eds G., Wachutka and D. Schmitt-Landsiedel), Shaker Verlag, Aachen, Germany.
14. Callen, H.B. (1985) *Thermodynamics and an Introduction to Thermostatistics*, John Wiley & Sons, New York.
15. Wachutka, G. (1994) Problem-oriented modeling of microtransducers: state of the art and future challenges. *Sensors and Actuators A*, **41**, 279–283.
16. Wachutka, G. (1995) Tailored modeling: a way to the 'virtual microtransducer fab'? *Sensors and Actuators A*, **46-47**, 603–612.
17. Onsager, L. (1931) Reciprocal Relations in Irreversible Processes. *Physical Review*, **37**, 405–426.
18. Senturia, S.D. (1998) CAD Challenges for Microsensors, Microactuators and Microsystems. *Proceedings of the IEEE*, **86**, 1611–1626.
19. Senturia, S.D. (2001) *Microsystem Design*, Kluwer Academic Press, Norwell, MA.
20. Womersley, J. (1957) Oscillatory flow in arteries: the constrained elastic tube as a model of arterial flow and pulse transmission. *Physics in Medicine and Biology*, **2**, 178–187.
21. Veijola, T., Kuisma, H., and Lahdenperä, J. (1998) Dynamic modelling and simulation of microelectromechanical devices with a circuit simulation program. Proceedings of 1st International Conference on Modeling and Simulation of Microsystems (MSM'98), 1998, Santa Clara, CA. pp. 245–250.
22. Swart, N.R., Bart, S.F., Zaman, M.H., Mariappan, M., Gilbert, J.R., and Murphy, D. (1998) AutoMM: automatic generation of dynamic macromodels for MEMS devices. Proceedings of MEMS'98, Heidelberg. pp. 178–183.
23. Voigt, P. and Wachutka, G. (2000) Compact MEMS modeling for design studies. Proceedings of 3rd International Conference on Modeling and Simulation of Microsystems (MSM'00), 2000 Mar 27-28, San Diego, CA. pp. 134–137.
24. Schrag, G. and Wachutka, G. (2002) Physically-based modeling of squeeze film damping by mixed level simulation. *Sensors and Actuators A*, **97–98**, 193–200.
25. Schrag, G. and Wachutka, G. (2004) Accurate system-level damping model for highly perforated micromechanical devices. *Sensors and Actuators A*, **111**, 222–228.
26. Bedyk, W., Niessner, M., Schrag, G., Wachutka, G., Margesin, B., and Faes, A. (2008) Automated extraction of multi-energy domain reduced-order models demonstrated on capacitive MEMS microphones. *Sensors and Actuators A*, **145–146**, 263–270.
27. Niessner, M., Schrag, G., Iannacci, J., and Wachutka1, G. (2011) COMSOL API based toolbox for the mixed-level modeling of squeeze-film damping in MEMS: simulation and experimental validation. Proceedings of COMSOL Conference 2011, Stuttgart.
28. Zengerle, R., Ulrich, J., Kluge, S., Richter, M., and Richter, A. (1995) A bidirectional silicon micropump. *Sensors and Actuators A*, **50**, 81–86.

29. Zengerle, R. (1994) Mikromembranpumpen als komponenten für Mikro-fluidsysteme. PhD thesis, Univ. der Bundeswehr, Verlag Shaker, Aachen.
30. Richter, M., Linnemann, R., and Woias, P. (1998) Robust design of gas and liquid micropumps. *Sensors and Actuators A*, **68**, 480–486.
31. Herz, M., Askamp, N., and Richter, M. (2009) An industrialised silicon micropump for precise liquid dosing. Proceedings of European Meeting on Microflow Metrology, Braunschweig, PTB Braunschweig.
32. Richter, M., Kruckow, J., and Drost, A. (2003) A high performance silicon micropump for fuel handling in DMFC systems. Proceedings of Fuel Cell Seminar, Miami Beach, FL. pp. 272–275.
33. Voigt, P., Schrag, G., and Wachutka, G. (1996) Micropump macromodel for standard circuit simulators using HDL-A, in *Proceedings of the 10th European Conference on Solid-State Transducers (EuroSensors X)* (eds R. Puers), Timshel BVBA, Leuven, pp. 1361–1364.
34. Voigt, P., Schrag, G., and Wachutka, G. (1998) Electrofluidic full-system modeling of a flap valve micropump based on Kirchhoffian network theory. *Sensors and Actuators A*, **66**, 9–14.
35. Voigt, P., Schrag, G., and Wachutka, G. (1998) Microfluidic system modeling using VHDL-AMS and circuit simulation. *Microelectronics Journal*, **29**, 791–797.
36. Cadence, San Jose, CA, http://www.cadence.com. *SpectreHDL Reference manual*.
37. Timoshenko, S. (1959) *Theory of Plates and Shells*, McGraw-Hill, New York.
38. Bohl, W. (1984) *Technische Strömungslehre*, VEB Fachbuchverlag Leipzig, Leipzig.
39. Ulrich, J. and Zengerle, R. (1996) Static and dynamic flow simulation of a KOH-etched microvalve using the finite-element method. *Sensors and Actuators A*, **53**, 379–385.
40. Prud'homme, R., Chapman, T., and Bowen, J. (1986) Laminar compressible flow in a tube. *Applied Science Research*, **43**, 67–74.
41. Atabek, H. and Lew, H. (1966) Wave propagation through a viscous incompressible fluid contained in an initially stressed elastic tube. *Biophysical Journal*, **6**, 481–503.
42. Rangra, K., Giacomozzi, F., Margesin, B., Lorenzellia, L., Mulloni, V., Collini, C., Marcelli, R., and Soncini, G. (2004) Micromachined low actuation voltage RF MEMS capacitive switches, technology and characterization. Proceedings of the IEEE 2004 International Semiconductor Conference, Sinaia, Romania. pages 165–168.
43. Mulloni, V., Giacomozzi, F., and Margesin, B. (2010) Controlling stress and stress gradient during the release process in gold suspended micro-structures. *Sensors and Actuators A*, **162**, 93–99.
44. Schrag, G., Niessner, M., and Wachutka, G. (2011) Reliable system-level models for electrostatically actuated devices under varying ambient conditions: modeling and validation, in *Proceedings of the Symposium on Design, Test, Integration & Packaging of MEMS/MOEMS (DTIP 2011)*, Aix-en-Provence, France, pp. 8–13.
45. Niessner, M., Schrag, G., Wachutka, G., and Iannacci, J. (2010) Modeling and fast simulation of RF-MEMS switches within standard IC design framework. Proceedings of the 15th International Conference on Simulation of Semiconductor Processes and Devices (Sispad 2010), Bologna, Italy. pp. 317–320.
46. Niessner, M., Schrag, G., Iannacci, J., and Wachutka, G. (2011) Macromodel-based simulation and measurement of the dynamic pull-in of viscously damped RF-MEMS switches. *Sensors and Actuators A*, **172**, 269–279.
47. Gabbay, L.D. and Senturia, S. (1998) Automatic generation of dynamic macro-models using quasistatic simulations in combination with modal analysis. Proceedings Solid-State Sensor & Actuator Workshop, Hilton Head, SC. pp. 197–220.

48. Gabbay, L.D., Mehner, J.E., and Senturia, S. (2000) Computer-aided generation of nonlinear reduced-order dynamic macromodels– I: non-stress-stiffened case. *Journal of Microelectromechanical Systems*, **9** (2), 262–269.
49. Niessner, M., Schrag, G., Wachutka, G., Iannacci, J., and Margesin, B. (2009) Automatically generated and experimentally validated system-level model of a microelectromechanical RF switch. Proceedings of NSTI Nanotechnology Conference and Trade Show, May 3-7 (NSTI Nanotech 2009), Houston, USA. pp. 655–658.

3
System-Level Modeling of MEMS by Means of Model Order Reduction (Mathematical Approximations) – Mathematical Background

Lihong Feng, Peter Benner, and Jan G. Korvink

3.1
Introduction

Modeling and numerical simulation are unavoidable for MEMS design because of the very small size and the high complexity of the devices. MEMS devices can be modeled by partial differential equations (PDEs). To simulate such models, spatial discretization via, for example, finite element discretization is necessary, which results in a system of ordinary differential equations (ODEs) or differential algebraic equations (DAEs).

After spatial discretization, the number of degrees of freedom usually is very high. Therefore, it is time consuming to simulate the large-scale systems of ODEs or DAEs. Developed from well-established mathematical theories and robust numerical algorithms, model order reduction (MOR) has been recognized as being very efficient in reducing the simulation time of large-scale systems. Through MOR, a small system of ODEs with reduced number of equations (reduced model) are derived. The reduced model is simulated instead, and the solution of the original PDEs or ODEs can then be recovered from the solution of the reduced model. As a result, the simulation time of the original large-scale system can be shortened by several orders of magnitude. The reduced model as a whole can also replace the original system and be reused for many times during the design process, which can save much time further. To date, MOR has been widely applied to simulation of MEMS and has achieved much success in enhancing traditional simulation tools [1–3].

As an example, we illustrate in Figure 3.1 a process of electrothermal simulation. We show in Figure 3.2 how MOR can be applied to enable a fast simulation. Electrothermal simulation at system level is a joint simulation of electrical and thermal parts of the system (as schematically shown in Figure 3.1). The circuit produces power dissipation, which is used by the thermal subsystem to evaluate the temperatures. Temperatures in return influence the circuit parameters, which results in a two-way coupling. The thermal model in Figure 3.1 is a simple, lumped element model. For general complex geometries, a more accurate, physical model

System-level Modeling of MEMS, First Edition. Edited by T. Bechtold, G. Schrag, and L. Feng.
© 2013 Wiley-VCH Verlag GmbH & Co. KGaA. Published 2013 by Wiley-VCH Verlag GmbH & Co. KGaA.

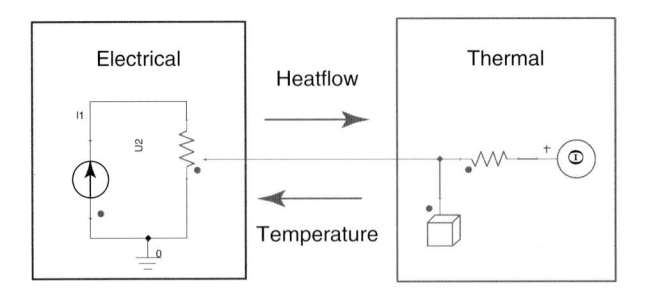

Figure 3.1 Simple electrothermal simulation [4].

in the form of heat transfer PDE is required,

$$\nabla \cdot (\kappa \nabla T) + Q - \rho c_p \frac{\partial T}{\partial t} = 0, \tag{3.1}$$

where $\kappa(r)$ is the thermal conductivity in Watts per meter per Kelvin at the position r, $c_p(r)$ is the specific heat capacity in Joules per kilogram per Kelvin, $\rho(r)$ is the mass density in kilogram per cubic meter, and $T(r,t)$ is the temperature distribution. Spatial discretization (via, e.g., the finite element method) of Eq. (3.1) results in a large-scale system of ODEs in the form of

$$E \frac{dT(t)}{dt} + KT(t) = F, \tag{3.2}$$

where E is the heat capacity matrix, K is the heat conductivity matrix, and T is a vector of nodal temperatures varying with time. The model in Eq. (3.1) can also be considered as an electrical network where the vector T is equivalent to unknown voltages, the matrix E is a capacity matrix, and the matrix K is the resistance matrix. In both cases, Eq. (3.1) is not compatible with system-level simulation, as the vector T usually contains several hundred thousands of degrees of freedom. The remedy is to apply the technique of MOR, which enables a formal transformation of Eq. (3.2), to a system in the same form but with much less equations, as illustrated in Figure 3.2. Here, the model of the thermal domain is replaced by a reduced model, which enables the efficient system-level simulation with the standard simulation tools.

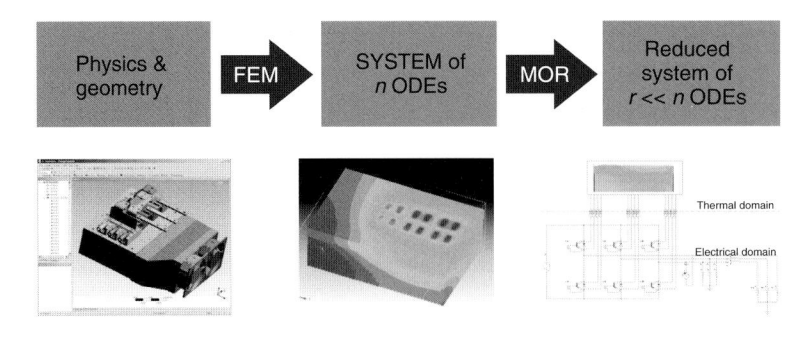

Figure 3.2 MOR applied to the electrothermal simulation [4].

3.2
Brief Overview

The technique of MOR can be traced back to the 1980s [5, 6] or even earlier [7–13]. Besides its application in MEMS simulation, MOR has already been applied to various research areas. There are different kinds of MOR methods in the literature. The modal truncation methods [7–9] are among the earliest developed methods and are mainly applied in structural dynamics. The Gramian-based MOR methods are mostly used in the area of electrical and control engineering [5, 14–16], etc. Reduced basis methods [12, 17, 18] are known in mechanical engineering, chemical engineering, etc. Proper orthogonal decomposition (POD) is widely used in fluid dynamics [19–23], etc. MOR based on moment matching [24–27] is popular in integrated circuit (IC) design and MEMS simulation. With the increasing communication among scientists in different areas, more interdisciplinary application opportunities have been found for the above-mentioned model reduction methods. At the same time, more and more hybrid methods that try to combine the advantages of different methods are being developed to meet the new requirements of the more complex problems.

In this chapter, we consider system-level modeling of MEMS, where MOR methods based on moment matching play an important role. Furthermore, we also introduce the basic technique of the Gramian-based MOR methods because of their increasing potential in solving large-scale MEMS models.

To make the book self-contained, we first introduce the basic mathematical definitions and concepts in Section 3.3, which are necessary to understand all the model reduction methods to be introduced in Sections 3.6–3.9. Some numerical algorithms are the core of the moment-matching MOR methods, which are explained in Section 3.4 for a better and easier application of moment-matching MOR. To understand the Gramian-based MOR methods, we need some knowledge on system theory, which is preliminarily introduced in Section 3.5. Afterward, we present the basic idea of MOR in Section 3.6. The motivation and implementation for each kind of MOR method are explored in Sections 3.7 and 3.8, respectively. Advanced computing issues such as choosing proper expansion points for the moment-matching MOR methods, and more efficient numerical algorithms for the Gramian-based MOR methods are also addressed in each corresponding section. The error estimation, stability, and passivity properties of the reduced models are analyzed in Section 3.9. An issue about nonzero initial condition, which is rarely considered in the literature but is of interest and importance, is discussed in Section 3.10. MOR for more complex systems is briefly mentioned in Section 3.11 to introduce the related MOR methods in Part III. Conclusions and outlook are given in Section 3.12, where some important opening problems and possible solutions are addressed.

It should be pointed out here that Sections 3.3–3.5 can be seen as mathematical preparations for the MOR methods to be introduced in the following sections. Those who are familiar with the content in these sections may directly start from Section 3.6. However, Sections 3.3–3.5 are not simply the copy of the known mathematical theories, we add many related explanations and remarks that illustrate how each of them is connected with the MOR methods to be introduced later.

MOR methods for nonlinear systems or parametric systems are introduced in Part III. In this chapter, we only introduce moment-matching MOR methods and the Gramian-based MOR methods for nonparametric linear time-invariant (LTI) systems as below

$$E\frac{d\mathbf{x}}{dt} = A\mathbf{x} + B\mathbf{u}(t),$$
$$\mathbf{y}(t) = C\mathbf{x} + D\mathbf{u}(t). \quad (3.3)$$

Here, $\mathbf{x}(t)$ is usually called the state vector, and its entries are called *state variables*. The system in Eq. (3.3) is called the *state-space representation of the system*. For example in Figure 3.1, $\mathbf{x}(t)$ represents the vector of nodal temperatures T in Eq. (3.2).

For most model order reduction methods, the term $D\mathbf{u}(t)$ remains unchanged during the process of MOR. Therefore, it is not affected by MOR. For simplicity, we assume that $D = 0$, a zero matrix. $E \in \mathbb{R}^{n \times n}$, $A \in \mathbb{R}^{n \times n}$, $B \in \mathbb{R}^{n \times m_1}$, and $C \in \mathbb{R}^{m_2 \times n}$ are the system matrices. There are m_1 input terminals and m_2 output terminals. When $m_1 = m_2 = 1$, the system is called *single-input single-output (SISO) system*. Otherwise, if both are larger than 1, then it is called multiple-input multiple-output system (MIMO). Accordingly, if only $m_1 > 1$, we call it multiple-input single-output (MISO), or if only $m_2 > 1$, we have a single-input multiple-output (SIMO) system. Here, "nonparametric linear time-invariant" means that all the system matrices are constant matrices, which are linearly related to the state vector, and are independent of time, with no physical, material, or geometrical parameters included.

3.3
Mathematical Preliminaries

In this section, mathematical preliminaries are introduced in order to understand the MOR methods that are introduced later. We first review the notations of scalar, vector, and matrix, which are frequently used not only in this chapter but also in the following chapters, especially in Part III of the book. After introducing the vectors, it is necessary to introduce the concepts of subspace, linear independence, and basis of a subspace, which are employed in the moment-matching MOR methods to construct the projection matrices W and V for the reduced model (Section 3.6). The norm of a vector or a matrix, the norm of a vector function, and the norm of a matrix function are used to estimate the accuracy of the MOR methods. In other words, they are used to estimate the accuracy of the reduced model in terms of the error between the output of the original system and that of the reduced model (reduced system) or in terms of the error between the transfer functions of the two systems. Especially, the vector function norm and the matrix function norm are essential for the inference of the computational error bound of the Gramian-based MOR methods, which make the methods automatic.

3.3.1
Scalar, Vector, and Matrix

- A variable x is a complex scalar, if $x \in \mathbb{C}$. When the imaginary part of x is zero, that is, $Im(x) = 0$, then x is a real number, and can be denoted as $x \in \mathbb{R}$.
- A vector is denoted in bold face $\mathbf{x} \in \mathbb{C}^n$, which means $\mathbf{x}$ is a complex vector with length n. A real vector $\mathbf{x}$ with length n is denoted as $\mathbf{x} \in \mathbb{R}^n$. Usually, we write $\mathbf{x} = (x_1, x_2, \ldots, x_n)^T$, where each x_i is a scalar in $\mathbb{C}$ or $\mathbb{R}$.
- A matrix A with m rows and n columns is denoted as $A \in \mathbb{C}^{m \times n}$. If A is a real matrix, then $A \in \mathbb{R}^{m \times n}$. A is written as

$$A = \begin{pmatrix} a_{11} & a_{12} & \cdots & a_{1n} \\ \vdots & \vdots & \vdots & \vdots \\ a_{m1} & \cdots & \cdots & a_{mn} \end{pmatrix} \tag{3.4}$$

where $a_{ij} \in \mathbb{R}$ or $\mathbb{C}$, the entry of A. It can be seen that the model in Eq. (3.3) includes scalars, vectors, and matrices.

Remark

Here, $\mathbb{R}$ is the set of real numbers and $\mathbb{C}$ is the set of complex numbers.

$$\mathbf{x} = (x_1, x_2, \ldots, x_n)^T = \begin{pmatrix} x_1 \\ x_2 \\ \vdots \\ x_n \end{pmatrix} \tag{3.5}$$

means the transpose of the row vector $(x_1, x_2, \ldots, x_n)$.

3.3.2
Vector Space, Subspace, Linear Independence, and Basis

We first define the vector space, then the concept of subspace is defined accordingly. An important property of the basis of the subspace is that they are linearly independent, such that any elements in the subspace can be represented by the basis. All these are the basic mathematical theories for MOR.

Definition: Vector space A vector space U over $\mathbb{R}$ is a nonempty assembly on which a scalar multiplication and an addition

- $\mathbb{R} \times U \to U$, denoted by $a \cdot \mathbf{x}$ or $a\mathbf{x}$ for all real numbers a and elements $\mathbf{x}$ in U,
- $U \times U \to U$, denoted by $\mathbf{x} + \mathbf{y}$ for all elements $\mathbf{x}$ and $\mathbf{y}$ in the Vector space U,

are defined with the following rules:

- There is an element $\mathbf{0}$ in U, such that $\mathbf{0} + \mathbf{x} = \mathbf{x}$ for all $\mathbf{x} \in U$.
- For each $\mathbf{x} \in U$, there is an element $(-\mathbf{x}) \in V$ with $\mathbf{x} + (-\mathbf{x}) = \mathbf{0}$.
- For all $\mathbf{x}, \mathbf{y} \in U$, $\mathbf{x} + \mathbf{y} = \mathbf{y} + \mathbf{x}$.

- For all $\mathbf{x}, \mathbf{y}, \mathbf{z} \in U$, $(\mathbf{x} + \mathbf{y}) + \mathbf{z} = \mathbf{x} + (\mathbf{y} + \mathbf{z})$.
- For all $a \in \mathbb{R}$ and all $\mathbf{x}, \mathbf{y} \in U$, $a(\mathbf{x} + \mathbf{y}) = a\mathbf{x} + a\mathbf{y}$.
- For all $a, b \in \mathbb{R}$ and all $\mathbf{x} \in U$, $a(b\mathbf{x}) = (a \cdot b)\mathbf{x}$.
- For all $a, b \in \mathbb{R}$ and all $\mathbf{x} \in U$, $(a + b)\mathbf{x} = a\mathbf{x} + b\mathbf{x}$.
- For all $\mathbf{x} \in U$, $1 \cdot \mathbf{x} = \mathbf{x}$.

Remark

For example, $\mathbb{R}^n$ is the vector space of all the vectors $\mathbf{x} = (x_1, \ldots, x_n)^T$ with entries $x_i \in \mathbb{R}, i = 1, \ldots, n$. If $x_i \in \mathbb{C}, i = 1, \ldots, n$, and the two scalars $a, b \in \mathbb{C}$, then it is the vector space $\mathbb{C}^n$ defined over $\mathbb{C}$. These two vector spaces are mostly used in engineering simulation. Sometimes, vector space is also called *linear space* in the community of numerical linear algebra.

Definition: Subspace A nonempty subassembly S from the vector space U is called a *subspace*, if it is closed for addition and scalar multiplication, that is, $\forall \mathbf{x}, \mathbf{y} \in S, \forall a \in \mathbb{R}$, the following is true: $\mathbf{x} + \mathbf{y} \in S, a\mathbf{x} \in S$.

Definition: Linear combination All possible linear combinations of vectors $\mathbf{x}_1, \ldots, \mathbf{x}_m$ from U constitute a subspace S, which we call the subspace spanned by $\mathbf{x}_1, \ldots, \mathbf{x}_m$. The linear combination is defined as

$$\sum_{k=1}^{m} a_k \mathbf{x}_k, \; a_k \in \mathbb{R} \text{ or } \mathbb{C}, k = 1, \ldots, m \tag{3.6}$$

S is denoted by $S = \text{span}\{\mathbf{x}_1, \ldots, \mathbf{x}_m\}$.

Definition: Linear independence The vectors $\mathbf{x}_1, \ldots, \mathbf{x}_m \in U$ are linearly independent, if none of them can be represented as a linear combination of others. That is, if $a_1 \mathbf{x}_1 + \cdots + a_m \mathbf{x}_m = 0$, then $a_1 = \ldots = a_m = 0$.

Orthogonality of the vectors A group of vectors $\mathbf{x}_1, \ldots, \mathbf{x}_m \in \mathbb{R}^n$ are said to be orthogonal with each other, if they satisfy

$$\mathbf{x}_i^T \mathbf{x}_j = \begin{cases} 0, & \text{if } i \neq j \\ \sigma_{ij}, & \text{if } i = j \end{cases}, \; i, j = 1, 2, \ldots, m \tag{3.7}$$

Remark

For any two vectors $\mathbf{x}, \mathbf{y} \in \mathbb{R}^n$, $\mathbf{x}^T \mathbf{y}$ is the inner product or scalar product of two vectors, which is defined as $\mathbf{x}^T \mathbf{y} = x_1 y_1 + \ldots + x_n y_n$. It is known that the angel θ between two vectors $\mathbf{x}$ and $\mathbf{y}$ can be computed as $\cos \theta = \frac{\mathbf{x}^T \mathbf{y}}{\|\mathbf{x}\|_2 \|\mathbf{y}\|_2}$. If $\mathbf{x}$ and $\mathbf{y}$ are orthogonal with each other, it means $\theta = \frac{\pi}{2}$, therefore, we have $\mathbf{x}^T \mathbf{y} = 0$. If $\mathbf{x}_1 \ldots, \mathbf{x}_m \in \mathbb{R}^n$ are orthogonal with each other, then it is not difficult to prove that they are linearly independent. If $\sigma_{ij} = 1$, $\mathbf{x}_i, i = 1, 2, \ldots, m$ are said to be orthonormal with each other.

Definition: Basis A set of vectors $\mathbf{x}_1, \ldots, \mathbf{x}_m \in U$ form a basis for the subspace S if given any vector $\mathbf{x}$ in S, there are unique scalars $a_1, \ldots, a_m$, such that $\mathbf{x}$ can be linearly represented by $\mathbf{x}_1, \ldots, \mathbf{x}_m$, that is, $\mathbf{x} = a_1 \mathbf{x}_1 + \cdots + a_m \mathbf{x}_m$.

Remark
The MOR methods introduced in this chapter try to find an approximation $\tilde{\mathbf{x}}$ of the state vector $\mathbf{x}$ from a subspace S of the vector space $\mathbb{R}^n$ or $\mathbb{C}^n$. If one knows the basis of the subspace S, for example, $\mathbf{v}_1, \ldots, \mathbf{v}_m$, then $\tilde{\mathbf{x}} \in S$ can be represented by the basis $\tilde{\mathbf{x}} = z_1 \mathbf{v}_1 + \ldots + z_m \mathbf{v}_m$. Therefore, the key for MOR is to find a proper subspace S and the basis of it. It seems not easy to construct the basis simply from its definition. However, we have the following equivalent definition of basis.

Equivalent Definition of Basis If a set of vectors $\mathbf{x}_1, \ldots, \mathbf{x}_m$ span the subspace S, and if they are linearly independent, then $\mathbf{x}_1, \ldots, \mathbf{x}_m$ form a basis of S.

Remark
By using the equivalent definition of the basis, it is easier to find the basis of the subspace. For example, if one knows that $\mathbf{x}_1, \ldots, \mathbf{x}_m$ span the subspace S, what is left is to check if these vectors are linearly independent or not. Usually, it is also difficult to check the linear independence. However, the modified Gram-Schmidt process in Section 3.4 can help us to derive a group of linearly independent (orthogonal) vectors $\tilde{\mathbf{x}}_1, \ldots, \tilde{\mathbf{x}}_q$ from an initial group of vectors $\mathbf{x}_1, \ldots, \mathbf{x}_m$. This means, whatever the initial group of vectors $\mathbf{x}_1, \ldots, \mathbf{x}_m$ are (linearly independent or not), one can always obtain a group of linearly independent vectors $\tilde{\mathbf{x}}_1, \ldots, \tilde{\mathbf{x}}_q$ (using the modified Gram-Schmidt process), which is the basis of the subspace spanned by $\mathbf{x}_1, \ldots, \mathbf{x}_m$.

Krylov Subspace and Block Krylov Subspace The moment-matching MOR method tries to find a basis of a (Block) Krylov subspace, which is closely related to the moments of the transfer function (The concept of transfer function and moments are introduced later in Sections 3.5 and 3.7, respectively). As a result, the reduced model derived matches the moments of the transfer function. The number of matched moments shows the degree of accuracy of the reduced model. Therefore, we need to introduce the Krylov subspace and the Block Krylov subspace as below.

Krylov Subspace A subspace $K_q(A, \mathbf{r})$ is called *Krylov subspace* if it is generated by a vector $\mathbf{r} \neq \mathbf{0} \in \mathbb{C}^n$ and a square matrix $A \in \mathbb{C}^{n \times n}$

$$K_q(A, \mathbf{r}) = \mathrm{span}\{\mathbf{r}, A\mathbf{r}, A^2\mathbf{r}, \ldots, A^{q-1}\mathbf{r}\} \tag{3.8}$$

Here, q is called the *order of the Krylov subspace*. The largest q that may keep $\mathbf{r}, A\mathbf{r}, A^2\mathbf{r}, \ldots, A^{q-1}\mathbf{r}$ linearly independent is n. It could happen that $A^{q_0}\mathbf{r}$ is zero for $q_0 \leq n$. Then, only the first q_0 vectors are linearly independent, and we have $K_q(A, \mathbf{r}) = K_{q_0}(A, \mathbf{r})$ for all $q > q_0$.

Block Krylov Subspace A block Krylov subspace is generated by a group of nonzero vectors $\mathbf{r}_1, \mathbf{r}_2, \ldots, \mathbf{r}_m$ and a square matrix $A \in \mathbb{C}^{n \times n}$. If we denote the group of vectors by a matrix, $R = (\mathbf{r}_1, \mathbf{r}_2, \ldots, \mathbf{r}_m) \in \mathbb{C}^{n \times m}$, then the block Krylov subspace $K_q(A, R)$ is defined as

$$K_q(A, R) = \mathrm{span}\{R, AR, A^2R, \ldots, A^{q-1}R\}. \tag{3.9}$$

Generally, the vectors $R, AR, A^2 R, \ldots, A^{q-1} R$ become linearly dependent quite fast. Therefore, the dimension of the subspace is usually smaller than the total number of vectors $m \times q$.

3.3.3
Laplace Transform

Laplace transform has a wide range of applications in mathematics, physics, and engineering. It is an integral transform, which transfers a function $f(t), t \geq 0$ (assuming $f(t)$ is locally integrable[28]), defined on the real domain into another function defined on the complex domain

$$F(s) = \int_0^\infty e^{-st} f(t) dt \qquad (3.10)$$

Here, $t \in \mathbb{R}$ is a real number, and $s = \sigma + j\omega \in \mathbb{C}$ is a complex number, with real part σ and imaginary part ω, and j is the imaginary unit $j = \sqrt{-1}$. ω is the angular frequency, in radians per unit time. It is also common to write $\omega = 2\pi f$, where f has the unit of cycles per second, or Hertz (Hz). In engineering, the real variable t mostly represents time. Therefore, the Laplace transform is often seen as a transform from the time domain to the frequency domain.

In the following Section 3.3.5.2, Laplace transform is used to explain the matrix function norm $||\cdot||_{\mathcal{H}_\infty}$. It is also used to get the transfer function of the system in Eq. (3.3) (Section 3.5), and the frequency domain expression of the state vector $\mathbf{x}$ in Eq. (3.3) (Section 3.7.1) as well. The transfer function can be used to measure the accuracy of the reduced model (see Theorem 4 in Section 3.7.2) and Eq. (3.44) in Section 3.9). The expression of the state vector in frequency domain is employed to find the projection matrix V for the moment-matching model reduction methods.

3.3.4
Rational Function

In the case of one variable s, a rational function $R_{p,q}(s)$ is the quotient of two polynomials $P_p(s)$ and $Q_q(s)$ of degree p and q, respectively

$$R_{p,q}(s) = \frac{P_p(s)}{Q_q(s)}$$

Remark
Moment-matching MOR methods are often seen as Padé or Padé-type approximations. This may be due to the fact that the moment-matching MOR methods are originally motivated by the observation that the transfer function of the system in Eq. (3.3) is a rational function. A good approximation of the rational function is Padé approximation. One of the very early MOR methods, the method of asymptotic waveform evaluation (AWE) in [24, 29] uses the Padé approximation of the transfer function (to be introduced in Section 3.3) of the system in Eq. (3.3) as the transfer

function of the reduced model, whereby the reduced model in state space can be constructed. More information can be seen in Section 3.7.4.

3.3.5
Norms

3.3.5.1 Vector and Matrix Norms

Vector and matrix norms are used to measure the distance between two vectors or the error of the numerical solution. Models of MEMS are described by PDEs, which are usually spatially discretized (e.g., by finite element discretization) for the derivation of the numerical solution. The discretized system can be solved numerically with manipulations involving matrices and vectors. Therefore, it is necessary to know how to compute and manipulate the vector norms as well as the matrix norms, such that one can estimate the accuracy of the numerical solution. The definitions of the vector and matrix norms below can be found in [30].

Definition: Vector norm Let U be a vector space. Here, we are mostly concerned with $\mathbb{R}^n$ (or $\mathbb{C}^n$). It is normed if there is a function $||\cdot||: U \to \mathbb{R}$, which we call a norm, satisfying all of the following:

1) $||\mathbf{x}|| \geq 0$, and $||\mathbf{x}|| = 0$ if and only if $\mathbf{x} = \mathbf{0}$ (positive definiteness).
2) $||\alpha \mathbf{x}|| = |\alpha| \cdot ||\mathbf{x}||$ for any real (or complex) scalar α (homogeneity).
3) $||\mathbf{x} + \mathbf{y}|| \leq ||\mathbf{x}|| + ||\mathbf{y}||$ (the triangular inequality).

Remark
The function $||\cdot||$ assigns a real value to each vector in the vector space U, for example, in $\mathbb{R}^n$ or in $\mathbb{C}^n$. The most commonly used norms are $||\mathbf{x}||_p = (\Sigma_i |x_i|^p)^{1/p}$ for $1 \leq p < \infty$, which we call p-norms, as well as $||\mathbf{x}||_\infty = \max_i |x_i|$, which we call the ∞-norm or infinity-norm.

Definition: Matrix norm $||\cdot||$ is a matrix norm on the space of $m \times n$ matrices if it is a vector norm on the corresponding $m \cdot n$ dimensional space:

1) $||A|| \geq 0$ and $||A|| = 0$ if and only if $A = 0$.
2) $||\alpha A|| = |\alpha| \cdot ||A||$.
3) $||A + B|| \leq ||A|| + ||B||$.

Remark
For example, $\max_{ij} |a_{ij}|$ is called the *max norm* and $||A||_F = (\Sigma |a_{ij}|^2)^{1/2}$ is called the *Frobenius norm*. Here, a_{ij} is the entry on the ith row and jth column of A. If an $m \times n$ matrix A is viewed as a vector in $\mathbb{R}^{mn}$, then the Frobenius norm is just the p-norm with $p = 2$ in the vector space $\mathbb{R}^{mn}$, which is a natural extension of the vector norm to the matrix norm.

Definition: Induced norm Let A be an $m \times n$ matrix and $||\cdot||$ be a vector norm on $\mathbb{R}^n$. Then

$$||A|| = \max_{x \neq 0, x \in \mathbb{R}^n} \frac{||Ax||}{||x||}$$

is called an *operator norm*, *induced norm*, or *subordinate matrix norm*.

Remark
Induced matrix norms are defined in terms of the behavior of a matrix as an operator, which "stretches" a vector $\mathbf{x}$ by a maximum factor $||A||$.

$||A||_p, p = 1, 2, \infty$ are the induced norms that are often used for error estimation of the numerical solution of PDEs, ODEs, as well as an algebra equation like $g(x) = f$. They are defined as

$$||A||_1 = \max_{x \neq 0,\, x \in \mathbb{R}^n} \frac{||Ax||_1}{||x||_1} = \max_j \Sigma_i |a_{ij}| = \text{maximum absolute column sum}.$$

$$||A||_2 = \max_{x \neq 0,\, x \in \mathbb{R}^n} \frac{||Ax||_2}{||x||_2} = \sqrt{\lambda_{\max}(A^*A)}, \text{ where } \lambda_{\max}(M) \text{ denotes the largest eigen-}$$

value of any matrix $M \in \mathbb{R}^{n \times n}$.

$$||A||_\infty = \max_{x \neq 0,\, x \in \mathbb{R}^n} \frac{||Ax||_\infty}{||x||_\infty} = \max_i \Sigma_j |a_{ij}| = \text{maximum absolute row sum}.$$

3.3.5.2 Vector Function Norms and Matrix Function Norms

To obtain the error bound for the reduced system, we need function norms. These norms are used to calculate the difference between the outputs of two different systems (introduced in Section 3.5). For example, a computable error bound for the difference between the output of the original system of ODEs and the output of the reduced model is derived in Section 3.9 by using the function norms. With this computable error bound, the reduced model can be automatically obtained.

In the following equation, we define the $\mathscr{L}_p$ norm of a real valued vector function $\mathbf{f}(t) = (f_1(t), f_2(t), \ldots, f_m(t))^T : \mathscr{I} \to \mathbb{R}^n$ as [31]:

Definition: $\mathscr{L}_p$ norm of a real valued vector function

$$||\mathbf{f}||_p = \left(\int_{t \in \mathscr{I}} ||\mathbf{f}(t)||_p^p\right)^{\frac{1}{p}} = \sqrt[p]{\sum_i \int_{t \in \mathscr{I}} |f_i(t)|^p dt}, \quad 1 \le p < \infty \tag{3.11}$$

where $\mathscr{I} = \mathbb{R}, \mathbb{R}_+, \mathbb{R}_-$, or some finite interval $[a, b]$.

Let $\mathbb{C}_+ \subset \mathbb{C}$ denote the right half of the complex plane: $s = \delta + j\omega$. Consider the complex matrix-valued function $F : \mathbb{C}_+ \to \mathbb{C}^{m_1 \times m_2}$, which is analytic in $\mathbb{C}_+$. The $\mathscr{H}_p$ norm of F is defined as [15]

Definition: $\mathscr{H}_p(p < \infty)$ norm of a complex matrix-valued function

$$||F||_{\mathscr{H}_p} = \left(\sup_{\delta > 0} \int_{-\infty}^{\infty} ||F(\delta + j\omega)||_p^p d\omega\right)^{\frac{1}{p}}, \quad 1 \le p < \infty. \tag{3.12}$$

For a particular value of s, say s_0, $F(s_0) \in \mathbb{C}^{m_1 \times m_2}$ is a matrix. If we denote $F(s_0)$ as $\tilde{F} \doteq F(s_0)$, then $||F(s_0)||_p = ||\tilde{F}||_p = \left[\sum_{i=1}^{m_2} \sigma_i^p(\tilde{F})\right]^{\frac{1}{p}}$ is also called the *Schatten p-norm* of the matrix $F(s_0)$ [15], where $\sigma_i(\tilde{F})$ is the ith singular value of $F(s_0)$. Assuming $m_1 > m_2$, and when $m_2 = 1$, the above $\mathcal{H}_p$ norm also defines the norm of a complex vector-valued function $\mathbf{f}: \mathbb{C}_+ \to \mathbb{C}^{m_1}$. When $p = 2$, the corresponding $\mathcal{H}_2$ norm of the vector function $\mathbf{f}$ can be used to evaluate the $\mathcal{H}_\infty$ norm of the complex matrix-valued function F (defined below), which is used for the error estimation of the reduced model (see the analysis in Section 3.9.2).

The $\mathcal{H}_\infty$ norm of a complex matrix-valued function F is defined as

Definition: $\mathcal{H}_\infty$ **norm of a complex matrix-valued function**

$$||F||_{\mathcal{H}_\infty} = \sup_{\omega \in \mathbb{R}} \sigma_{\max}(F(j\omega)), \tag{3.13}$$

where $\sigma_{\max}$ is the largest singular value of $F(j\omega)$. It can be shown that the $\mathcal{H}_\infty$ norm is the $\mathcal{L}_2$ induced norm, as well as the $\mathcal{H}_2$ induced norm [15], that is

$$||F||_{\mathcal{H}_\infty} = \sup_{X \neq 0} \frac{||FX||_{\mathcal{H}_2}}{||X||_{\mathcal{H}_2}} = \sup_{x \neq 0} \frac{||\mathbf{f} * \mathbf{x}||_2}{||\mathbf{x}||_2}. \tag{3.14}$$

Here, X is the Laplace transform of $\mathbf{x}: \mathbb{R} \to \mathbb{R}^{m_2}$ and F is the Laplace transform of $\mathbf{f}: \mathbb{R} \to \mathbb{R}^{m_1 \times m_2}$. The term $\mathbf{f} * \mathbf{x}$ is the convolution of $\mathbf{f}$ and $\mathbf{x}$, which is defined as

$$\mathbf{f} * \mathbf{x}(t) = \int_{-\infty}^{\infty} \mathbf{f}(\tau)\mathbf{x}(t-\tau)d\tau. \tag{3.15}$$

The relation between $\mathcal{H}_\infty$ norm and $\mathcal{L}_2$ norm, and the relation between $\mathcal{H}_\infty$ norm and $\mathcal{H}_2$ norm can be used to derive the difference between the output responses of two different systems (Section 3.5), from which the error estimation of the output response of the reduced system can be obtained.

3.4 Numerical Algorithms

In this section, we discuss some well-known numerical algorithms in numerical linear algebra, which are the computational cores of the moment-matching MOR methods. By using the modified Gram-Schmidt process, an orthogonal basis of a subspace spanned by a group of vectors, for example, $\tilde{\mathbf{r}}_1, \tilde{\mathbf{r}}_2, \ldots, \tilde{\mathbf{r}}_q$ can be derived. The vectors of basis $\mathbf{r}_1, \mathbf{r}_2, \ldots, \mathbf{r}_q$ are not only linearly independent and orthogonal but also orthonormal with each other (see the definition in Section 3.3.2).

The modified Gram-Schmidt process (Algorithms 1 and 2) helps us to understand the Arnoldi algorithm (Algorithm 3) and the Band Arnoldi process (Algorithm 4). These algorithms can be directly used to compute the projection matrix V for the reduced model in Section 3.7.

Algorithm 1 *Modified Gram-Schmidt process*

INPUT: $\tilde{\mathbf{r}}_1, \tilde{\mathbf{r}}_2, \ldots, \tilde{\mathbf{r}}_q$.
OUTPUT: $\mathbf{r}_1, \mathbf{r}_2, \ldots, \mathbf{r}_q$.
1) For $i = 1$ to q
2)   For $j = 1$ to $i - 1$

3) $\quad \tilde{\mathbf{r}}_i = \tilde{\mathbf{r}}_i - \dfrac{\mathbf{r}_j^T \tilde{\mathbf{r}}_i}{\mathbf{r}_j^T \mathbf{r}_j} \mathbf{r}_j$

4) $\quad$ end

5) $\quad \mathbf{r}_i = \tilde{\mathbf{r}}_i / \|\tilde{\mathbf{r}}_i\|_2$

6) end

Notice that the vectors $\mathbf{r}_i, i = 1, 2, \ldots, q$ are actually normalized, that is the 2-norm of each vector $\mathbf{r}_i$ is 1. Therefore, $\mathbf{r}_i, i = 1, 2, \ldots, q$ are orthonormal with each other. If after orthogonalization with all the previous vectors $\mathbf{r}_j$, the 2-norm of the current vector $\tilde{\mathbf{r}}_i$ is zero or very close to zero, then it shows that $\tilde{\mathbf{r}}_i$ can be linearly represented by the previous vectors, and it does not provide any new information to the subspace. Therefore, $\mathbf{r}_i$ should be deleted once $\|\mathbf{r}_i\|_2 < tol$, where tol is a suitably chosen small number. This process is called *deflation*, which is a must in any practical implementation of the modified Gram-Schmidt process. Finally, the above-modified Gram-Schmidt process becomes Algorithm 2 as below, which is the one implemented in practice.

Algorithm 2 *Modified Gram-Schmidt process with deflation*

INPUT: $\tilde{\mathbf{r}}_1, \tilde{\mathbf{r}}_2, \ldots, \tilde{\mathbf{r}}_{\tilde{q}}$.

OUTPUT: $\mathbf{r}_1, \mathbf{r}_2, \ldots, \mathbf{r}_q$ (orthonormal).

1) For $i = 1$ to q

2) $\quad$ For $j = 1$ to $i - 1$

3) $\quad\quad \tilde{\mathbf{r}}_i = \tilde{\mathbf{r}}_i - \dfrac{\mathbf{r}_j^T \tilde{\mathbf{r}}_i}{\mathbf{r}_j^T \mathbf{r}_j} \mathbf{r}_j$

4) $\quad$ end

5) $\quad$ If $\|\tilde{\mathbf{r}}_i\|_2 > tol$

6) $\quad\quad \mathbf{r}_i = \tilde{\mathbf{r}}_i / \|\tilde{\mathbf{r}}_i\|_2$

7) $\quad$ else

8) $\quad\quad$ delete $\mathbf{r}_i$

9) $\quad$ end

10) end

Next, we introduce the Arnoldi algorithm (Algorithm 3), which has close relationship with the modified Gram-Schmidt process. One of the properties of the Arnoldi algorithm is that it generates an orthonormal basis for the Krylov subspace $K_q(A, \mathbf{r})$, that is, it transforms the vectors $\mathbf{r}, A\mathbf{r}, \ldots, A^q\mathbf{r}$ into a group of orthonormal vectors, which forms a basis of the Krylov subspace. We can say that the Arnoldi algorithm is the modified Gram-Schmidt process used to orthogonalize the vectors $\mathbf{r}, A\mathbf{r}, \ldots, A^q\mathbf{r}$, which span the Krylov subspace $K_q(A, \mathbf{r})$. In Algorithm 3, the output is a group of orthonormal vectors $\mathbf{v}_1, \mathbf{v}_2, \ldots, \mathbf{v}_q$, which form a basis of the Krylov subspace $K_q(A, \mathbf{r})$.

Algorithm 3 *Arnoldi algorithm*

INPUT: $A, \tilde{\mathbf{r}}$.

OUTPUT: $\mathbf{v}_1, \mathbf{v}_2, \ldots, \mathbf{v}_q$ and a Hessenberg matrix H.

1) $\mathbf{v}_1 = \frac{\tilde{\mathbf{r}}}{\|\tilde{\mathbf{r}}\|_2}$
2) For $i = 1$ to $q - 1$
3) $\mathbf{w} = A\mathbf{v}_i$
4)   For $j = 1$ to i
5) $h_{ji} = \mathbf{v}_j^T \mathbf{w}$
6) $\mathbf{w} = \mathbf{w} - h_{ji}\mathbf{v}_j$
7)   end
8)   If $\|\mathbf{w}\|_2 > tol$
9) $h_{i+1,i} = \|\mathbf{w}\|_2$
9) $\mathbf{v}_i = \mathbf{w}/h_{i+1,i}$
10)   else
11)   stop
12)   end
13) end

When the system in Eq. (3.3) is a MISO or MIMO system, the Arnoldi algorithm cannot be used. Instead, the block Arnoldi algorithm or the so-called Band Arnoldi process is needed to derive the reduced model. This is because the input matrix B of a MISO or MIMO system is not a vector but a matrix; hence, a basis of the block Krylov subspace $K_q(A, B)$ (Subsection 3.3.2) rather than a Krylov subspace needs to be computed. More details about applying these algorithms can be found in Section 3.7. The block Arnoldi algorithm proposed in [32] and the Band Arnoldi process proposed in [33] can be used to generate an orthonormal basis of a block Krylov subspace. They are used by moment-matching MOR methods [26, 34]. Actually, the block Arnoldi algorithm 6.22: Block Arnoldi with Block MGS in [32] is used by the MOR method PRIMA in [26], which is a popular MOR method in the field of circuit simulation. The Band Arnoldi process [33] is widely used in MEMS simulation; therefore, we describe it in Algorithm 4. Besides the Band Arnoldi process, there are also global Arnoldi algorithms, which can also be used for MOR of MIMO systems and are more efficient for certain systems [35, 36]. Owing to space limitation, we do not introduce them in this book.

The output of Algorithm 4 is a group of orthonormal vectors, which form a basis of the block Krylov subspace $K_q(A, R)$. Usually, the number q_B (the number of the orthonormal vectors) is smaller than $q \times m$, the number of the total input vectors due to the deflation.

Algorithm 4 *Band Arnoldi process in [33]*

INPUT: A matrix $A \in \mathbb{C}^{n \times n}$;
    A block of m right starting vectors $R = (\mathbf{r}_1, \mathbf{r}_2, \ldots, \mathbf{r}_m) \in \mathbb{C}^{n \times m}$.
OUTPUT: The $n \times n$ Arnoldi matrix $G_{q_B}^{(pr)}$.
    The matrix $V_{q_B} = [\mathbf{v}_1, \mathbf{v}_2, \ldots, \mathbf{v}_{q_B}]$ containing the first q_B Arnoldi vectors,
    and the matrix $\rho_{q_B}^{(pr)}$.
0) For $k = 1, 2, \ldots, m$, set $\hat{\mathbf{v}}_k = \mathbf{r}_k$.
  Set $m_c = m$ and $\mathcal{I} = \emptyset$ (an empty set).
  For $i = 1, 2, \ldots$, until convergence or $m_c = 0$ do:

1) (If necessary, deflate $\hat{\mathbf{v}}_i$.)
Compute $\|\hat{\mathbf{v}}_i\|_2$ and check if the deflation criterion ($\|\hat{\mathbf{v}}_i\|_2 < tol$) is fulfilled.
If yes, do the following:
Set $\hat{\mathbf{v}}_{i-m_c}^{defl} = \hat{\mathbf{v}}_i$ and store this vector. Set $\mathscr{I} = \mathscr{I} \cup \{i - m_c\}$.
Set $m_c = m_c - 1$. If $m_c = 0$, set $i = i - 1$ and stop.
For $k = i, i+1, \ldots, i + m_c - 1$, set $\hat{\mathbf{v}}_k = \hat{\mathbf{v}}_{k+1}$.
Repeat all of step 1).
2) (Normalize $\hat{\mathbf{v}}_i$ to obtain $\mathbf{v}_i$.)
Set $g_{i,i-m_c} = \|\hat{\mathbf{v}}_i\|_2$ and $\mathbf{v}_i = \hat{\mathbf{v}}_i / g_{i,i-m_c}$.
3) (Orthogonalize the candidate vectors against $\mathbf{v}_i$)
For $k = i+1, i+2, \ldots, i + m_c - 1$, set:
Set $g_{i,k-m_c} = \mathbf{v}_i^H \hat{\mathbf{v}}_k$ and $\hat{\mathbf{v}}_k = \hat{\mathbf{v}}_k - \mathbf{v}_i g_{i,k-m_c}$.
4) (Advance the block Krylov subspace to get $\hat{\mathbf{v}}_{i+m_c}$.)
(a) Set $\hat{\mathbf{v}}_{i+m_c} = A \mathbf{v}_i$.
For $k = 1, 2, \ldots, i$ set:
$g_{k,i} = \mathbf{v}_k^H \hat{\mathbf{v}}_{i+m_c}$ and $\hat{\mathbf{v}}_{i+m_c} = \hat{\mathbf{v}}_{i+m_c} - \mathbf{v}_k g_{k,i}$.
5) (a) For $k \in \mathscr{I}$, set $g_{i,k} = \mathbf{v}_i^H \hat{\mathbf{v}}_k^{defl}$.
(b) Set
$G_i^{(pr)} = [g_{j,k}]_{j,k=1,2,\ldots,i}$,
$k_\rho = m + \min\{0, i - m_c\}$,
$\rho_i^{(pr)} = [g_{j,k-m}]_{j=1,2,\ldots,i; k=1,2,\ldots,k_\rho}$.
6) Check if i is large enough. If yes, set $q_B = i$, stop

3.5
Linear System Theory

In this section, we introduce some basic knowledge on linear system theory, which is necessary for the analysis of MOR methods. Most part of this section is prepared for the Gramian-based MOR methods introduced in Section 3.8. We also give the definitions of stability and passivity of an LTI system. Usually, it is preferred that the reduced model should preserve the stability and passivity of the original system, such that it is a physically proper substitute for the original system.

3.5.1
Transfer Function

The transfer function of the system in Eq. (3.3) is the input/output relation of the system in frequency domain. By applying Laplace transform to both sides of the equations in Eq. (3.3), we get

$$sEX(s) - E\mathbf{x}(0) = AX(s) + BU(s),$$
$$Y(s) = CX(s) \tag{3.16}$$

Here, $X(s)$ is the Laplace transform of $\mathbf{x}(t)$ and $\mathbf{x}(0)$ is the value of $\mathbf{x}(t)$ at $t = 0$. We point out here that in system theory [37, 38], the state vector $\mathbf{x}(t)$ is defined as the state of the system and $x(0)$ is the *initial state of the system*. The term *state* is used frequently in Section 3.8 where the balanced truncation method is introduced.

Assuming $x(0) = 0$, we have the expression of the transfer function $H(s)$

$$H(s) = Y(s)/U(s) = C(sE - A)^{-1}B \qquad (3.17)$$

Many MOR methods, for example, the moment-matching MOR methods, the Gramian-based MOR methods, assume the zero initial condition $x(0) = 0$. Although little work has been done for $x(0) \neq 0$, there are cases with nonzero initial condition that need to be efficiently dealt with by MOR as well. We discuss in Section 3.10 about MOR for systems with $x(0) \neq 0$.

3.5.2
Measure of the Difference between Any Two Different LTI Systems

In this section, we describe the measure of the difference between the output responses of any two different LTI systems, based on which a computable error bound of the Gramian-based MOR methods is established in Section 3.9. Since this measure is derived with no individual MOR method taken into consideration, that is it is independent of the MOR method, it could also be the bases for the error estimation of all other MOR methods, for example, the moment-matching MOR methods.

It is known [39] that the time domain output response $y(t)$ of the system in Eq. (3.3) can be expressed as the convolution integral of input $u(t)$ and the system's impulse response $h(t)$, that is

$$y(t) = h * u \doteq \int_{-\infty}^{\infty} h(\tau) u(t - \tau) d\tau$$

Now we can measure the difference between the output responses of two different systems driven by the same input signal $u(t)$. Assuming $y_1(t)$ and $y_2(t)$ are the corresponding output responses of two different LTI systems, then we have

$$||y_1 - y_2||_2 = ||h_1 * u - h_2 * u||_2 = ||(h_1 - h_2) * u||_2.$$

Here, $h_1(t)$ and $h_2(t)$ are the impulse responses of the two LTI systems, respectively. We also know that the transfer function is the Laplace transform of the impulse response of $h(t)$ [39]. Therefore, from Eq. (3.14), we immediately get

$$||y_1 - y_2||_2 \leq ||H_1 - H_2||_{\mathcal{H}_\infty} ||u||_2 \qquad (3.18)$$

and the difference between the output responses in frequency domain

$$||Y_1 - Y_2||_{\mathcal{H}_2} = ||H_1 U - H_2 U||_2 \leq ||H_1 - H_2||_{\mathcal{H}_\infty} ||U||_{\mathcal{H}_2}. \qquad (3.19)$$

From Eqs. (3.18) and (3.19), we see that being driven by the same input signal $u(t)$, the difference between the output responses of two different systems can be bounded by $||H_1 - H_2||_{\mathcal{H}_\infty}$. The model reduction methods try to derive a reduced system, such that $||H - H_r||_{\mathcal{H}_\infty}$ is very small, where H_r is the transfer function of the reduced system. If $||H - H_r||_{\mathcal{H}_\infty}$ can be kept small, it means both $||Y - Y_r||_{\mathcal{H}_2}$

and $||\mathbf{y} - \mathbf{y_r}||_2$ are very small. Here, Y_r and $\mathbf{y_r}$ are the output response of the reduced system in frequency domain and time domain, respectively. A computable bound of $||H_1 - H_2||_{\mathcal{H}_\infty}$ for all $s \in \mathbb{C}_+$ can be derived for the reduced model obtained by the Gramian-based MOR methods. With this global error bound, the reduced system can be computed automatically according to the required accuracy.

3.5.3
Controllability and Observability

Controllability Gramian and Observability Gramian are the basic concepts in system theory, and they are used to study some important properties of the system, for example, controllability and observability of the states. They play a crucial role in the Gramian-based MOR methods.

Controllability (of a state) Definition of controllability [38]: Given an LTI system as in Eq. (3.3), a nonzero state $\mathbf{x}$ is controllable if there exists an input $\mathbf{u}(t)$ with finite energy such that the state of the system goes to zero from state $\mathbf{x}$ within a finite time $\bar{t} < \infty$.

Remark
Controllability is the measure for how well a given nonzero state is steered by the input to the zero state. It depicts the possibility of steering the state from the input.

Observability (of a state) Definition of observability [38]: Given any input $\mathbf{u}(t)$, a state $\mathbf{x}$ of the system is observable, if starting with the state $\mathbf{x}$ (i.e., $\mathbf{x}(0)=\mathbf{x}$), and after a finite period of time $\bar{t} < \infty$, $\mathbf{x}$ can be uniquely determined by the output $y(\bar{t})$.

Remark
Observability is the measure for how well internal states of a system can be estimated by knowledge of its external outputs. It depicts the possibility of estimating the state from the output.

Controllability Gramian For a controllable, observable, and stable LTI system in Eq. (3.3), the controllability Gramian (matrix) P is defined as

$$P = \int_0^\infty e^{At} BB^T e^{A^T t} dt \tag{3.20}$$

Observability Gramian For a controllable, observable, and stable LTI system in Eq. (3.3), the observability Gramian (matrix) Q is defined as

$$Q = \int_0^\infty e^{A^T t} C^T C e^{At} dt \tag{3.21}$$

Remark
The Controllability Gramian and Observability Gramian make the controllability and observability of the states measurable, such that one can analyze the states

by means of the two Gramians. This makes the analysis of the states theoretically accurate and mathematically computable, which also make the truncation of those unimportant states during MOR practically implementable. In system theory, the states that are difficult to be observed and controlled are unimportant for the system and could be ignored. The Gramian-based MOR methods are proposed based on this principle.

3.5.4
Realization Theory

Realization In system theory, a realization of an LTI system is a set of the matrices

$$(E, A, B, C) \in \mathbb{R}^{n\times n} \times \mathbb{R}^{n\times n} \times \mathbb{R}^{n\times m_1} \times \mathbb{R}^{m_2 \times n}$$

corresponding to Eq. (3.3). In general, an LTI system has infinitely many realizations. This is because its transfer function is invariant under state-space transformations or, in other words, invariant under coordinate transformations.

The state-space transformation is defined as below

$$\mathcal{T}: \begin{cases} \mathbf{x} & \mathbf{Tx}, \\ (E, A, B, C) & (TET^{-1}, TAT^{-1}, TB, CT^{-1}) \end{cases} \quad (3.22)$$

It means that if we replace $\mathbf{x}$ with $\tilde{\mathbf{x}} = \mathbf{Tx}$, then the system in Eq. (3.3) is transformed into the following system

$$\begin{aligned} TET^{-1}\tfrac{d\tilde{\mathbf{x}}}{dt} &= TAT^{-1}\tilde{\mathbf{x}} + TB\mathbf{u}(t), \\ \mathbf{y}(t) &= CT^{-1}\tilde{\mathbf{x}} \end{aligned} \quad (3.23)$$

The invariance of the transfer function under state-space transformation can be proved by the following simple calculation

$$\tilde{H}(s) = (CT^{-1})(sTET^{-1} - TAT^{-1})^{-1}(TB) = C(sE - A)^{-1}B = H(s) \quad (3.24)$$

Here, $\tilde{H}(s)$ is the transfer function of the transformed system in Eq. (3.23) and $H(s)$ is the transfer function of the original system in Eq. (3.3).

But this is not the only nonuniqueness associated to LTI system representations. Any addition of state variables that does not influence the input–output relation, meaning that for the same input $\mathbf{u}$ and the same output $\mathbf{y}$ is achieved, leads to a realization of the same LTI system [16]. Therefore, the order n of a system can be arbitrarily enlarged without changing the input–output mapping. On the other hand, for each system, there exists a unique minimal number of state variables $\hat{n}$, which is necessary to describe the input– output behavior completely. This number is called the *McMillan degree* of the system, and a realization $(\hat{E}, \hat{A}, \hat{B}, \hat{C})$ of order $\hat{n}$ is a *minimal realization* of the system.

Note that only the McMillan degree is unique; any state-space transformation (Eq. 3.22) leads to another minimal realization of the same system.

Remark

The state-space transformation in Eq. (3.22) implicates that the transformed system in Eq. (3.23) and the original system in Eq. (3.23) represent one system. They are only two different realizations of the same system under a state-space (coordinate) transformation. This idea is used in the Gramian-based MOR methods to get the balanced transformation of the system in Eq. (3.3) (Subsection 3.8.2).

Balanced realization Informally speaking, a balanced realization is a minimal realization in which the controllability and observability Gramians are equal and diagonal.

More formally: Let $(\hat{E}, \hat{A}, \hat{B}, \hat{C})$ be a minimal realization of an LTI system, then $(\hat{E}, \hat{A}, \hat{B}, \hat{C})$ is called *balanced* if the controllability Gramian P and the observability Gramian Q satisfy $P = Q = \text{diag}(\sigma_1, \sigma_2, \ldots, \sigma_n) \triangleq \Sigma$. Lemma 5.6 in [15] tells us that the entries σ_i in Σ are equal to the positive square roots of the eigenvalues of the products of the Gramians PQ, that is

$$\sigma_i = \sqrt{(\lambda_i(PQ))}, \quad i = 1, \ldots, n$$

Here, $\sigma_i, i = 1, \ldots, n$ are the so-called Hankel singular values (HSVs), which are used to give the global error bound of the reduced system obtained by the Gramian-based MOR methods in Section 3.9.

In Section 3.8, it is shown that in order to obtain the final reduced model, the original system must be balanced first. This means a balanced realization of the system must be firstly found, and the reduced model is then derived from the balance realization of the system, rather than directly from the original system. The motivation to do so is to keep the states that are both easily controllable and easily observable into the reduced model. Only these states are important, and weight a lot in the presentation of the system, hence should be kept.

3.5.5
Stability and Passivity of a System

Stability The discussion of stability below originates from Section 3.4 in [33].

An LTI system in Eq. (3.3) is stable if its free-response, that is, the solution $x(t)$, $t \geq 0$, of

$$E\frac{dx}{dt} = Ax, \qquad (3.25)$$
$$\mathbf{x}(0) = \mathbf{x}_0$$

remain bounded as $t \to \infty$ for any possible initial vector x_0. Note that if the matrix E is singular, then there are certain restrictions on the possible initial vectors $\mathbf{x}_0$. See Section 2.1 in [33] for more explanations.

Stability of the system can be proved by the system matrices as below.

Theorem 1 *An LTI system is stable if and only if, the following two conditions are satisfied:*

(i) All finite eigenvalues $s \in \mathbb{C}$ of the matrix pencil $A - sE$ satisfy $Re(s) \leq 0$;
(ii) All finite eigenvalues of $A - sE$ with $Re(s) = 0$ are simple.

Stability can also be verified in terms of the transfer function in the following theorem.

Theorem 2 *If $H(s)$ is the transfer function of the minimal realization of system in Eq. (3.3), then the system is stable if and only if all finite poles s_i of $H(s)$ satisfy $Re(s_i) \leq 0$ and any pole with $Re(s_i) = 0$ is simple.*

Remark
In the Theorems above, $Re(s)$ means the real part of the complex variable s. Stability of the system can be understood as the property that the state $\mathbf{x}(t)$ of the system will never go to infinity with the increase of time t. It should be pointed out here that almost all the Gramian-based MOR methods require that the original LTI system in Eq. (3.3) is stable.

Passivity Here, we cite some sentences in [34, 40] to explain the importance of passivity in circuit simulation. Roughly speaking, a (linear or nonlinear) dynamical system is passive if it does not generate energy. The concept was first used in circuit theory [41]. For example, networks consisting of only resistors, inductors, and capacitors are passive. In circuit simulation, reduced order modeling mostly is applied to large passive linear subcircuits such as RLC networks. When reduced models of such subcircuits are used within a simulation of the whole circuit, stability of the overall simulation can be guaranteed only if the reduced models preserve the passivity of the original subcircuits [42, 43].

From the system's point of view, the passivity of an LTI system can be expressed in terms of its transfer function as below.

Theorem 3 *The LTI system in Eq. (3.3) is passive if, and only if, the associated transfer function $H(s) = C(sE - A)^{-1}B$ is positive real.*

The positive realness of a matrix-valued function is defined as below [41].

Definition A function $H : \mathbb{C} \mapsto (\mathbb{C} \bigcup \infty)^{m \times m}$ is called *positive real* if
  (i) H has no poles in $\mathbb{C}_+$;
  (ii) $H(\bar{s}) = \overline{H(s)}$ for all $s \in \mathbb{C}$;
  (iii) $Re(\mathbf{x}^H H(s)\mathbf{x}) \geq 0$ for all $s \in \mathbb{C}_+$ and $\mathbf{x} \in \mathbb{C}^m$.

3.6
Basic Idea of Model Order Reduction

After showing the above-mentioned mathematical basics, basic numerical algorithms, and backgrounds of system theory, we are ready to introduce the MOR techniques.

The general idea of almost all the MOR methods is to find a subspace S_1 that approximates the manifold where the state vector $\mathbf{x}(t)$ resides. Afterward, $\mathbf{x}(t)$ is approximated by a vector $\tilde{\mathbf{x}}(t)$ in S_1. The reduced model is produced by

Petrov-Galerkin projection onto another subspace S_2 or by Galerkin projection onto the same subspace S_1.

We use the system in Eq. (3.3) as an example to explain the basic idea. Assuming that an orthonormal basis $V = (v_1, v_2, \ldots, v_q)$ of the subspace S_1 has been found, then the approximation $\tilde{x}(t)$ in S_1 can be represented by the basis as $\tilde{x}(t) = Vz(t)$. Therefore, $x(t)$ can be approximated by $x(t) \approx Vz(t)$. Here, z is a vector of length $q \ll n$.

Once $z(t)$ is computed, we can get an approximate solution $\tilde{x}(t) = Vz(t)$ for $x(t)$. The vector $z(t)$ can be computed from the reduced model, which is derived by the following two steps.

1) By replacing x in Eq. (3.3) with Vz, we get

$$E\frac{dVz}{dt} \approx AVz + Bu(t),$$
$$y(t) \approx CVz \qquad (3.26)$$

2) Notice that the equations in Eq. (3.3) are not true anymore. Therefore, we can only use "$\approx$" in Eq. (3.26). In order to solve for z in Eq. (3.26), we should transform (3.26) into equations. (Petrov-) Galerkin projection provides us a way for transformation, the basic idea is as below. Since there is no equation in Eq. (3.26), there must be a difference between the left-hand side and right-hand side. We denote the difference as $e = AVz + Bu(t) - E\frac{dVz}{dt}$, which is called *residual*. From Eq. (3.26), we see $e \neq 0$ in the whole vector space $\mathbb{R}^n$. However, it is possible to force $e = 0$ in a properly chosen subspace S_2 of $\mathbb{R}^n$. If we have computed a matrix W, whose columns are the basis of S_2, then $e = 0$ in S_2 means e is orthogonal to each column in W, the basis vector of S_2, that is, $W^T e = 0$. As a result, we get the reduced model

$$W^T E \frac{dVz}{dt} = W^T AVz + W^T Bu(t),$$
$$\hat{y}(t) = CVz \qquad (3.27)$$

By defining $\hat{E} = W^T EV, \hat{A} = W^T AV, \hat{B} = W^T B, \hat{C} = CV$, we get the final reduced model

$$\hat{E}\frac{dz}{dt} = \hat{A}z + \hat{B}u(t),$$
$$\hat{y}(t) = \hat{C}z \qquad (3.28)$$

Notice that the approximation $\tilde{x}(t) = Vz(t)$ of $x(t)$ can be obtained from $z(t)$ by solving the system in Eq. (3.28) which has many less equations than the system in Eq. (3.3). Therefore, Eq. (3.28) is much easier to be solved, which is the so-called reduced model. In order to save the simulation time of solving Eq. (3.3), Eq. (3.28) can be used to replace the original large system in Eq. (3.3) to attain a fast simulation. Furthermore, the error between the two systems should be within acceptable tolerance. The error can be measured through the error between the output responses or the transfer functions of the two systems.

It can be seen that once the two matrices W and V have been obtained, the reduced model is derived. While the Gramian-based methods usually compute W different from V, some methods use $W = V$, for example, some of the moment matching methods, the reduced basis methods, and some of the POD methods. When $W = V$, Petrov-Galerkin projection becomes Galerkin projection. MOR methods differ in the computation of the two matrices W and V. The Gramian-based MOR methods compute W and V by the controllability and observability Gramians defined in Section 3.5. Reduced basis methods and POD methods compute V from snapshots of the state vector **x** at different time steps (also at the samples of the parameters if the system is parametric). The methods based on moment matching compute V from the moment vectors of the transfer function. One common goal of all methods is that the behavior of the reduced model should be sufficiently "close" to that of the original model. For example, given the input **u**(·) to both systems, the error between the output response of the reduced model **ŷ**(·) and that of the original model **y**(·) should be very small. The error between the transfer function of the reduced model and that of the original model is also often used to measure the accuracy of the reduced model because of Eqs. (3.18) and (3.19).

In the following sections, we introduce the moment-matching MOR methods and the Gramian-based MOR methods separately in Sections 3.7 and 3.8.

3.7
Moment-Matching Model Order Reduction

Moment-matching MOR methods try to derive a reduced model whose transfer function matches the moments of the transfer function of the original system. Generally speaking, the more moments matched the more accurate the reduced model will be. In the following section, we first introduce the definition of the moments and moment vectors, then we show how to compute the matrices W and V based on moment matching.

3.7.1
Moments and Moment Vectors

If we expand the transfer function $H(s)$ defined in Eq. (3.17) around an expansion point s_0 into its Taylor series

$$\begin{aligned} H(s) &= C^T[(s - s_0 + s_0)E - A]^{-1} B \\ &= C^T[(s - s_0)E + (s_0 E - A)]^{-1} B \\ &= C^T[I + (s_0 E - A)^{-1} E(s - s_0)]^{-1} (s_0 E - A)^{-1} B \\ &= \sum_{i=0}^{\infty} \underbrace{C^T[-(s_0 E - A)^{-1} E]^i (s_0 E - A)^{-1} B}_{:= m_i(s_0)} (s - s_0)^i \end{aligned} \qquad (3.29)$$

and if the system is a SISO system, then $m_i(s_0)$, $i = 0, 1, 2, \ldots$ are called the *moments of the transfer function* $H(s)$. If the system is a MIMO, SIMO or MISO system, then

$m_i(s_0), i = 0, 1, 2, \ldots$ are matrices and they are called *block moments* [26]. In the field of circuit design, the entry in the jth row, kth column of $m_i(s_0)$ is called the ith moment of the current that flows into port j when the voltage source at port k is the only nonzero source [26]. In this chapter, we only consider $m_i(s_0)$ as a whole and do not consider their entries individually. This means, when we talk about moments of the transfer function, we mean $m_i(s_0), i = 0, 1, \ldots$.

From Eqs. (3.16) and (3.29), it is straightforward to get the corresponding Taylor series expansion of $X(s)$

$$X(s) = \sum_{i=0}^{\infty} [-(s_0 E - A)^{-1} E]^i (s_0 E - A)^{-1} B U(s)(s - s_0)^i \qquad (3.30)$$

Here, we call the vectors $[-(s_0 E - A)^{-1} E]^i (s_0 E - A)^{-1} B, i = 0, 1, \ldots$, moment vectors, which are to be used for computation of the projection matrix V. Notice that when the system in Eq. (3.3) is a MISO or MIMO system, then $[-(s_0 E - A)^{-1} E]^i (s_0 E - A)^{-1} B, i = 0, 1, \ldots$ are matrices rather than vectors. For simplicity, we call them moment vector, even in the case of multiple-inputs.

3.7.2
Computation of the Projection Matrices W and V

Computation of V From Eq. (3.30), we see that the state vector $X(s)$ locates in the manifold spanned by all the moment vectors. Instead of using all the moment vectors, we only use a few of them to span a small subspace S_1. Then, $X(s)$ can be approximated by a vector in S_1, that is $X(s) \approx VZ(s)$. Here, V is an orthonormal basis of S_1. After inverse Laplace transform, we get the corresponding approximation of $\mathbf{x}(t)$ in S_1, that is, $\mathbf{x}(t) \approx V\mathbf{z}(t)$, where $\mathbf{z}(t)$ is the inverse Laplace transform of $Z(s)$. This means, $\mathbf{x}(t)$ in time domain can be approximated by $V\mathbf{z}(t)$. Usually, the more moment vectors included in S_1, the more accurate the approximation $V\mathbf{z}(t)$ will be. However, in order to keep the reduced model small, we usually choose a small number of moment vectors starting from $i = 0$, that is, the columns of the orthonormal matrix V span the subspace

$$\text{range}\{V\} = \text{span}\{\tilde{B}(s_0), \tilde{A}(s_0)\tilde{B}(s_0), \ldots, \tilde{A}^{q-1}(s_0)\tilde{B}(s_0)\} \qquad (3.31)$$

where $\tilde{A}(s_0) = (s_0 E - A)^{-1} E$, $\tilde{B}(s_0) = (s_0 E - A)^{-1} B$ and $q \ll n$.

Computation of W To obtain the reduced model, we also need to compute the (Petrov-)Galerkin matrix W. The method in [25] computes W and V by the unsymmetric Lanczos process, where $W \neq V$. The columns of the matrix W span the subspace below,

$$\text{range}\{W\} = \text{span}\{C, \tilde{A}^T(s_0) C, \ldots, (\tilde{A}^T(s_0))^{q-1} C\} \qquad (3.32)$$

and V satisfies Eq. (3.31). Owing to space limitation, we do not describe the unsymmetric Lanczos process and the corresponding algorithms in [25], see also [44]. It can be proved [25] that if the above-mentioned two matrices W and V are used to obtain the reduced model in Eq. (3.28), the transfer function of the reduced model matches the first $2q$ moments of the transfer function of the original system. We summarize this in the following theorem

Theorem 4 *If W and V span the subspaces in Eqs. (3.32) and (3.31), then the transfer function $\hat{H}(s) = \hat{C}(s\hat{E} + \hat{A})^{-1}\hat{B}$ of the reduced model in Eq. (3.28) matches the first $2q$ moments of the transfer function of the original system in Eq. (3.3), that is*

$$m_i(s_0) = \hat{m}_i(s_0), i = 0, 1, \ldots, 2q-1,$$

where $\hat{m}_i(s_0) = \hat{C}[-(s_0\hat{E} - \hat{A})\hat{E}]^{-i}(s_0\hat{E} - \hat{A})^{-1}\hat{B}, i = 0, 1, \ldots, 2q-1$ are the ith order moments of $\hat{H}$.

The moment-matching MOR method PRIMA [26] uses $W = V$, which is computed by the block Arnoldi algorithm 6.22 in [32]. In this case, only $q + 1$ moments of the transfer function are matched. The goal of the method is to compute an orthonormal matrix V that is the basis of the Krylov subspace in Eq. (3.31). Therefore, the block Arnoldi algorithm in [32] is not the only choice. Alternatively, V can be computed by the Arnoldi algorithm described in Algorithm 3 when the system in Eq. (3.3) is a single-input system (SISO or SIMO) whose input matrix B is actually a vector, $B = \mathbf{b}$. The right starting vector $\mathbf{r}$ in the Arnoldi algorithm is nothing but $\mathbf{r} = \tilde{B}(s_0)$, and the matrix A is actually $\tilde{A}(s_0)$. When the original system has multiinputs (MISO or MIMO), the Band Arnoldi process, Algorithm 4 can be used instead to generate V. The two inputs A and R of the algorithm are $R = \tilde{B}(s_0)$ and $A = \tilde{A}(s_0)$. It can be seen that Algorithm 4 includes the case of single input ($B = \mathbf{b}$) as a special case, and it is equivalent to the Arnoldi algorithm for single input systems. Only matrix–vector multiplications are used in the unsymmetric Laczos process, Arnoldi, or Band Arnoldi process, such that the algorithms are easily implemented and the complexity of the resulting methods is only $O(nq^2)$.

3.7.3
Different Choices of the Expansion Points

The accuracy of the moment-matching methods depends not only on the number of the moments matched or, in other words, the number of vectors included in the Krylov subspace in Eqs. (3.31) and (3.32), but also on the expansion points. Taylor series expansion in Eq. (3.29) tells us that the series is only accurate within a certain radius around the expansion point s_0, otherwise the reduced model will become inaccurate beyond the radius.

If the expansion point s_0 is chosen as zero, then the moments can be simplified to $m_i(0) = C^T/(A^{-1}E)^i(-A^{-1})B, i = 1, 2, \ldots$. If $s_0 = \infty$, the moments are also called *Markov parameters* [45], which are $C^T A^{i-1} B, i = 1, 2, \ldots$. To increase the accuracy of the single-point expansion, one may use more than one expansion points and match more than one moments for each expansion point. Moment-matching by multipoint expansion is also known as *rational interpolation* [27]. For example, if using a set of q distinct expansion points $\{s_1, \cdots, s_q\}$, the reduced order system obtained by, for example,

$$\text{range}\{V\} = \text{span}\{\tilde{B}(s_1), \cdots, \tilde{B}(s_q)\},$$
$$\text{range}\{W\} = \text{span}\{(\tilde{C}(s_1), \cdots, \tilde{C}(s_q)\}$$

matches the first two moments $m_0(s_i), m_1(s_i)$ at each s_i, $i = 1, \cdots, q$ [27]. Here, $\tilde{B}(s_i) = (s_i E - A)^{-1} B$, $\tilde{C}(s_i) = (s_i E - A)^{-T} C$, $i = 1, \ldots, q$. For this case, the standard Lanczos process, Arnoldi algorithm, or Band Arnoldi process cannot be used because the subspace spanned by either W or V is not a Krylov subspace anymore. Instead, the rational Krylov methods proposed (rational Lanczos algorithm and rational Arnoldi algorithm) in [27] would be the proper choice.

3.7.4
Development of Moment-Matching MOR

In order to have a deeper understanding of the moment-matching MOR methods, we further briefly review the history of this kind of methods, such that one can know the motivations at different stages of the development.

Among the early work of moment-matching MOR [24, 29, 46, 47], the method of AWE in [24, 29] is able to solve large-scale interconnect electrical circuit models that stimulated broad interest in this kind of methods. The AWE method tries to find a Padé approximation of the transfer function $H(s)$, which can be derived much more quickly than obtaining $H(s)$ itself. For a SISO system, the transfer function $H(s)$ is a scalar function. The Padé approximation of a scalar function can be defined as follows:

Padé Approximation The Padé approximation of a function $H(s)$ is a rational function $H_{p,q}(s)$ whose Taylor series at $s = 0$ agrees with that of $H(s)$ in at least the first $p + q + 1$ terms [25, 48].

For a MIMO system, the transfer function $H(s)$ is a matrix function and each entry of it can be approximated by the above-mentioned Padé approximation. For clarity of explanation, we use a SISO system as an example to briefly describe the method.

From the definition of the Padé approximation, we know that if $H(s) = H_{p,q}(s_0 + \sigma) = \frac{P_p(\sigma)}{Q_q(\sigma)}$ is a Padé approximation of the transfer function $H(s_0 + \sigma)$, then we have

$$H_{p,q}(s_0 + \sigma) = H(s_0 + \sigma) + O(\sigma^{p+q+1}). \tag{3.33}$$

Here, we introduce a new variable $\sigma = s - s_0$, because on the one hand, we get a Padé approximation of $H(s)$, on the other hand, we use a Padé approximation at s_0, rather than at 0. However, Padé approximation is defined on the derivatives of $H(s)$ at 0; therefore, we first need to do a variable transformation, $H(s) = H(s_0 + \sigma)$, such that we seek the Padé approximation of $H(s_0 + \sigma)$ with variable σ.

The derivatives of $H(s_0 + \sigma)$ at $\sigma = 0$ are actually the derivatives of $H(s)$ at s_0, they are also the moments $m_i(s_0), i = 0, 1, \ldots$ of the transfer function. By definition, the Padé approximation $H_{p,q}(s_0 + \sigma)$ matches the first $p + q + 1$ moments of the transfer function.

If the coefficients of the two polynomials $P_p(\sigma)$ and $Q_q(\sigma)$ in $H_{p,q}(s_0 + \sigma)$ are computed, then $H_{p,q}(s_0 + \sigma)$ is obtained. The coefficients can be obtained by solving two groups of equations that are derived from equating the coefficients of the Taylor series expansion (at $\sigma = 0$) on both sides of Eq. (3.33).

Since the moments $m_i(s_0), i = 0, 1, \ldots$, are the coefficients of the Taylor series expansion of $H(s_0 + \sigma)$ at $\sigma = 0$, they are involved in solving the equations to get the coefficients of $P_p(\sigma)$ and $Q_q(\sigma)$. However, in the AWE method, the moments are computed explicitly, which can cause serious numerical instability. The problem is that the vectors $B, \tilde{A}(s_0)B, \tilde{A}(s_0)^2 B, \ldots$ and $C, \tilde{A}^T(s_0)C, (\tilde{A}(s_0)^T)^2 C, \ldots$, which are used to compute moments, quickly converge to a right and, respectively, left, eigenvector corresponding to a dominant eigenvalue of A. As a result, the equations are heavily ill-conditioned, and therefore, the coefficients of the polynomials have very poor accuracy. For more detailed explanations and some illustrative results of AWE, refer to [25, 33].

In order to overcome the numerical instability of AWE, a more robust method Padé via Lanczos (PVL) [25] ([44]) is proposed. PVL also computes the Padé approximation of $H(s) = H(s_0 + \sigma)$; however, the moments of $H(s)$ do not have to be computed explicitly. Instead, an orthonormal basis of the subspace spanned by the moment vectors is computed, which constitutes the projection matrix V in Eq. (3.31), and the projection matrix W in Eq. (3.32) is also computed simultaneously. Both of them are computed by the unsymmetric Lanczos process. It is proved in [25] that the transfer function of the reduced model produced by W and V is the Padé approximation of the original transfer function $H(s)$. The PVL method avoids explicit computation of the moments and therefore avoids the possible numerical instability.

Unfortunately, PVL does not preserve passivity of the original system, which is not preferable in some engineering applications, especially in the IC design. For this target, the method PRIMA was proposed such that the reduced model is computed by only one projection matrix V. The resulting reduced model can preserve the passivity of the original system, particularly for those systems satisfying the conditions in Theorem 5 in Section 3.9.1. The trade-off is that only half the number of the moments can be matched by PRIMA as compared with PVL, if the matrix V in both methods expands the same Krylov subspace in Eq. (3.31). Such approximation of the transfer function is called *Pade-type approximation*. Many other moment-matching methods are also developed in parallel or based on the above-mentioned methods, among which the Band Arnoldi process in Algorithm 4 is proposed in [33], which has already been widely used in MEMS simulation [49]. For survey papers on moment-matching model reduction, see [15, 27, 33, 50].

3.8 Gramian-Based Model Order Reduction

Most of the Gramian-based MOR methods are proposed for LTI systems with $E = I$, the identity matrix. While most methods can be extended to the case of E being a nonsingular matrix, Gramian-based methods cannot be applied directly to the case when E is singular, that is, when the system is a descriptor system. Recently, some new theories and algorithms for computing the reduced model for a descriptor system (the E singular case) are proposed [51, 52]. More efficient algorithms are also developed, which can solve large-scale descriptor systems. Gramian-based methods for descriptor systems are outside the scope of this chapter as well as this book, refer to [51, 52] for detailed explanations. Since the theories of the Gramian-based methods are originally established for systems with $E = I$, when introducing the Gramian-based methods, we also assume $E = I$. We shortly address the case of $E \neq I$, but E being nonsingular at the end of this section.

The first proposed Gramian-based MOR method is balanced truncation [5], and on the basis of it, many more methods are developed, which are more efficient than balanced truncation in some aspects. All these methods compute the projection matrices W and V from the Gramians of the system, therefore, we call them Gramian-based MOR. We focus on the basic idea of balanced truncation and the derivation of the reduced model. For other more computationally efficient Gramian-based methods, we give a short introduction with corresponding references.

3.8.1
Motivation of Balanced Truncation

The name of this method implicates that the reduced model might be obtained by truncation after balancing. This is also what the method does. Balanced truncation tries to get the reduced model by truncating a balanced system, and the balanced system is derived by balancing the original LTI system with a balancing matrix T (note that in computational procedures for balanced truncation, the balancing transformation T is not computed explicitly, but only used implicitly in deriving the reduced model, see the discussion in Section 3.8.4). For survey papers on balanced truncation, refer to [53–55].

The goal of balanced truncation is to get a reduced model that keeps only those states $\mathbf{x}(t)$ that are both easily controlled and easily observed. Those states that are either difficult to be controlled or difficult to be observed are not important to the system and therefore can be ignored and truncated. Usually, the number of the state variables in the reduced model is much less than the number of the state variables in the original system. As a result, the reduced model is of much smaller size than the original system.

3.8.2
Balancing Transformation

The reason that we need the balancing transformation and cannot directly truncate the states from the original (realization of the) system relies on the degree of controllability and observability of the state $\mathbf{x}(t)$ of the system. In reality, there are many systems whose states are easily controlled and actually are hardly observed, and vice versa. This implicates that if we truncate the states that are difficult to be observed directly from the original system, then we may also truncate some states that are easily controlled. However, we need to remain those states with easy controllability. We have the same contradiction for those states that are difficult to be controlled but easily observed.

If we look at the degree of the controllability and observability mathematically, it is possible to find the solution of the above contradiction. It can be deduced from the Lemma 4.27 in [15] that for stable LTI systems, the states that are easily controlled can be approximately represented by the eigenvectors of the controllability Gramian (matrix) corresponding to the eigenvalues with the largest magnitudes. The states that are easily observed can be approximately represented by the eigenvectors of the observability Gramian (matrix) corresponding to the eigenvalues with the largest magnitudes [15].

In most cases, the controllability Gramian does not have the same eigenvectors as the observability Gramian. There are even extreme cases in which the eigenvectors of one Gramian corresponding to the largest eigenvalues span almost the same subspace spanned by the eigenvectors of the other Gramian corresponding to the smallest eigenvalues. For such typical cases, there exist states that are difficult to be controlled but are easy to be observed, or vice versa. As a result, we do not know how to keep those both easily controlled and easily observed states.

However, if we use the balanced realization, then the problem can be solved. This is because the balanced realization has two identical Gramians that must have the same eigenvectors.

It is clear now that for the balanced realization of the system, the states that are difficult to be controlled must also be hard to observe. If we truncate those hard to control states from the balanced realization of the system, we have truncated the hard to observe states at the same time. A realization of the system such that the Gramians are identical is called *balanced*. The transformation that transforms the original realization into a balanced realization is called *balancing transformation*. From the above-mentioned analysis, we see that in order to keep the important states (easily observed and easily controlled states) into the reduced model, the balancing transformation is an important step.

Definition: balancing transformation A balancing transformation is a nonsingular matrix T, such that $\tilde{P} = TPT^T$, $\tilde{Q} = T^{-T}QT$, and $\tilde{P} = \tilde{Q}$. Here, P and Q are given in Eqs. (3.20) and (3.21).

Remark
It can be readily verified that $T = \Sigma^{\frac{1}{2}} K^T U^{-1}$ is a balancing transformation and its inverse is $T^{-1} = U K \Sigma^{-\frac{1}{2}}$. Here, $P = UU^T$ is the Cholesky factorization of P and $U^T Q U = K \Sigma^2 K^T$ is the eigendecomposition of $U^T Q U$, with the eigenvalues σ_i^2 nonincreasingly ordered on the diagonal of Σ^2. The Gramians are symmetric matrices, so is the matrix $U^T Q U$. Therefore, in the eigendecomposition $U^T Q U = K \Sigma^2 K^T$, the matrix K is an orthonormal matrix, that is, $K^T K = I$, $K^{-1} = K^T$. By using $T = \Sigma^{\frac{1}{2}} K^T U^{-1}$, we actually have obtained two diagonal Gramians, that is,

$\tilde{P} = TPT^T = \Sigma$ and $\tilde{Q} = T^{-T}QT^{-1} = \Sigma$. The elements on the diagonals of Σ are exactly the square roots of the eigenvalues of the matrix $U^T QU$.

Definition: balanced system A controllable, observable, and stable LTI system is balanced, if its two Gramian matrices are equal $P = Q \in \mathbb{R}^{n \times n}$, it is principal-axis balanced if $P = Q = diag(\sigma_1, \ldots, \sigma_n)$, a diagonal matrix.

If we use $T = \Sigma^{\frac{1}{2}} K^T U^{-1}$ to transform P and Q into two identical Gramians, that is, $\tilde{P} = TPT^T$, $\tilde{Q} = T^{-T}QT^{-1}$ and $\tilde{P} = \tilde{Q}$, then it is not difficult to prove that the corresponding balanced system of Eq. (3.3) is

$$\frac{d\tilde{x}(t)}{dt} = TAT^{-1}\tilde{x}(t) + TBu(t) \tag{3.34}$$

$$y(t) = CT^{-1}\tilde{x}(t).$$

Remark

By using $T = \Sigma^{\frac{1}{2}} K^T U^{-1}$, the two Gramians are transformed to the same diagonal matrix Σ; therefore, the unit vectors $e_i = [0, \ldots, 0, 1, 0, \ldots, 0]$ (1 is the ith entry of e_i), $i = 1, 2, \ldots n$, are the eigenvectors of Σ whose eigenvalues are the entries on the diagonal. For the convenience of discussion, we define $\tilde{A} = TAT^{-1}$, $\tilde{B} = TB$, and $\tilde{C} = CT^{-1}$.

3.8.3
Truncation

After we get the balanced system, we can remove the states that are both difficult to be observed and difficult to be controlled from the balanced system and get the reduced model. For the balanced system, we see that if a state $\tilde{x}$ is easily observable and easily controllable, then it can be approximately represented by the eigenvectors of Σ corresponding to the largest eigenvalues.

Assuming that $\sigma_{q+1}, \ldots, \sigma_n$ are the smallest eigenvalues of Σ and $\sigma_1, \ldots, \sigma_q$ are the largest eigenvalues of Σ, if a state $\tilde{x}$ is a easily controllable and easily observable state of the balanced system, then it must almost locate in the subspace spanned by the eigenvectors $e_1, \ldots, e_q$ of Σ corresponding to $\sigma_1, \ldots, \sigma_q$, that is

$$\tilde{x}(t) \approx \alpha_1(t)e_1 + \ldots + \alpha_q(t)e_q \tag{3.35}$$

This means that $\tilde{x}$ can be approximately and linearly represented by $e_1, \ldots, e_q$. From Eq. (3.35), we see that $\tilde{x}(t)$ can be approximated by

$$\tilde{x}(t) \approx [\alpha_1(t), \ldots, \alpha_q(t), 0, \ldots, 0]^T = [z(t), 0]^T \tag{3.36}$$

where $z(t) = [\alpha_1(t), \ldots, \alpha_q(t)]^T$, $\mathbf{0} = [0, \ldots, 0]^T$. Therefore, if we only keep the easily controllable and easily observable states, then we can use the approximation in Eq. (3.36) to replace $\tilde{x}$ in the balanced system (Eq. 3.34) and get

$$\begin{bmatrix} \frac{d}{dt}z(t) \\ 0 \end{bmatrix} \approx \begin{bmatrix} \tilde{A}_{11} & \tilde{A}_{12} \\ \tilde{A}_{21} & \tilde{A}_{22} \end{bmatrix} \begin{bmatrix} z(t) \\ 0 \end{bmatrix} + \begin{bmatrix} \tilde{B}_1 \\ \tilde{B}_2 \end{bmatrix} u(t),$$

$$\tilde{y}(t) \approx \begin{bmatrix} \tilde{C}_1 & \tilde{C}_2 \end{bmatrix} \begin{bmatrix} z(t) \\ 0 \end{bmatrix}$$

(3.37)

The matrices $\tilde{A}$, $\tilde{B}$, and $\tilde{C}$ are partitioned according to the partition of $\tilde{x}(t)$ in Eq. (3.36). The part corresponding to 0 can be neglected, and the truncated system is obtained as

$$\frac{d}{dt}z(t) = \tilde{A}_{11}z(t) + \tilde{B}_1 u(t),$$

$$y(t) = \tilde{C}_1 z(t) \tag{3.38}$$

which is the reduced model we seek.

Notice that the approximation of $\tilde{x}$ in Eq. (3.36) is actually derived from the partition of the matrix Σ

$$\Sigma = \begin{bmatrix} \Sigma_1 & \\ & \Sigma_2 \end{bmatrix} \tag{3.39}$$

Here, Σ_1 contains the eigenvalues $\sigma_1, \ldots, \sigma_q$ of large magnitudes and Σ_2 contains the eigenvalues $\sigma_{q+1}, \ldots, \sigma_n$ of small magnitudes. Therefore, once we get the partition of Σ in Eq. (3.39), we can pick out the matrices $\tilde{A}_{11}$, $\tilde{B}_1$, and $\tilde{C}_1$, then the reduced model (Eq. 3.38) can be derived directly.

Remark

Partitioning of the matrix Σ, that is, the suitable number q can be obtained from the required accuracy of the reduced model thanks to the computable global error bound of the reduced model introduced in Section 3.9.

Remark

It seems that we have obtained the reduced model of the balanced system rather than that of the original system. However, we state that Eq. (3.38) is also the reduced model of the original system (Eq. 3.3). This is because that although we have transformed the original system into a balanced system, due to realization theory (Eqs. 3.22 and 3.24), they have the same transfer functions, and therefore are only different realizations of the same physical system.

3.8.4
Computation of the Balancing Transformation

To get the balanced system, we need to compute the two Gramians. It is difficult to compute them from their definitions. Fortunately, they can be equivalently obtained by solving the two Lyapunov equations as below [5, 15]

$$AP + PA^T = -BB^T,$$

$$A^T Q + QA = -C^T C \tag{3.40}$$

The balancing transformation matrix $T = \Sigma^{\frac{1}{2}} K^T U^{-1}$ and its inverse $T^{-1} = UK\Sigma^{-\frac{1}{2}}$ also need to be computed. However, the Gramians are usually ill-conditioned, it means that both the matrices U and Σ are nearly singular. Therefore, it will cause numerical inaccuracy if we explicitly compute T. In order to avoid this problem, a square-root algorithm (SR-method) [14, 56] is proposed, which computes the balanced system without explicitly computing the transformation matrix T.

Algorithm 5 *SR-method*

1) Do Cholesky factorization of the two Gramians: $P = Z_p Z_p^T$, $Q = Z_Q Z_Q^T$, where Z_p, Z_Q are lower triangular matrices.
2) Do singular value decomposition (SVD) of matrix $Z_p^T Z_Q$, i.e., there are two orthonormal matrices $\tilde{U}$, $\tilde{V}$ ($\tilde{U}^T \tilde{U} = I$, $\tilde{V}^T \tilde{V} = I$), which make

$$Z_p^T Z_Q = \tilde{U} \Sigma \tilde{V}^T = (\tilde{U}_1, \tilde{U}_2) \begin{bmatrix} \Sigma_1 & \\ & \Sigma_2 \end{bmatrix} \begin{bmatrix} \tilde{V}_1^T \\ \tilde{V}_2^T \end{bmatrix}$$

3) Let $W = Z_Q \tilde{V}_1 \Sigma_1^{-1/2}$, $V = Z_p \tilde{U}_1 \Sigma_1^{-1/2}$
4) The reduced model is

$$d\frac{z(t)}{dt} = W^T A V z(t) + W^T B u(t),$$
$$\hat{y}(t) = C V z(t). \tag{3.41}$$

Remark

We only need the two matrices W and V to get the reduced model, which only use the inverse of Σ_1. Since Σ_1 only includes those singular values with the largest magnitudes, it is nonsingular and well conditioned.

It can be verified that the SR-method is equivalent to balanced truncation with $T = \Sigma^{-1/2} \tilde{V}^T Z_Q^T$ and $T^{-1} = Z_p \tilde{U} \Sigma^{-1/2}$. It is also easy to see that the matrix Σ in Algorithm 5 is the same as in the balanced truncation.

From Algorithm 5, we see that the balanced truncation also belongs to model reduction based on Petrov-Galerkin projection.

Remark

Since both Gramian matrices are nearly singular matrices, there are also problems with Cholesky factorization in Algorithm 5 because the Cholesky factorization cannot be implemented if the matrix is nearly singular. In practical implementation, we do not compute the Gramian matrices explicitly but use algorithms that return the Cholesky factors directly. Such algorithms can be found in [57–59].

3.8.5
Acceleration Methods of Computing the Gramians

All the above-mentioned Gramian-based methods require the computation of the two Gramians P and Q by solving the two Lyapunov equations in Eq. (3.40).

Direct methods and standard iterative approaches for the solution of Lyapunov equations are restricted to a problem size of $n = O(1000)$ and are therefore of limited use. Using the MATLAB distribution R2007 [60], or newer versions, larger

Lyapunov equations with order up to $n = O(10^5)$ can be solved with direct methods provided that sufficient memory is available. For systems of dimension $n > 1000$, it is recommended to use implementations based on low-rank approximations to the Gramians. This was motivated by the observation that the Gramians of the large-scale systems have a low numerical rank [61–64].

Recently, this problem has attracted the interest of many mathematicians. There are many iterative methods that exploit the low-rank property particularly for the solution of large-scale, sparse Lyapunov equations, see, for example,

- Krylov subspace methods [65–71];
- the factorized sign function iteration [72] and data-sparse implementations of the method [73–75]; and
- methods based on a low-rank alternating direction implicit (ADI) or Smith iterations, for example, the Cholesky Factor-ADI (CF-ADI) algorithm [76–78], cyclic low-rank Smith methods [78, 79], and parallelizations of the low-rank ADI iteration [80].

Most of the methods are shown to be applicable to medium-to-large-scale systems. As some of the iterative methods allow for parallel computing, larger sparse problems with up to $n = O(10^5)$ can be solved on computers with distributed memory architecture [81]. The use of efficient solvers for large-scale Lyapunov equations also improves the implementations of the balanced truncation method, see [16, 74, 82–86].

3.8.6
Extension to More General Systems

During introduction of the Gramian-based methods, we have assumed that the matrix E is the identity matrix. Many methods can be extended to more general systems with $E \neq I$, but E is nonsingular. For the balanced truncation method and the SR-method, the Gramians of the system (Eq. 3.3) with $E \neq I$ can be solved by the general Lyapunov equations

$$APE^T + EPA^T = -BB^T,$$
$$A^T \tilde{Q} E + E^T \tilde{Q} A = -C^T C \tag{3.42}$$

The symmetric Gramian matrix Q can be recovered by $Q = E^T \tilde{Q} E$. The reduced model derived by the SR-metod can be written as

$$W^T EVd \frac{z(t)}{dt} = W^T A V z(t) + W^T B \mathbf{u}(t),$$
$$\hat{y}(t) = CV z(t) \tag{3.43}$$

The same global error bound Eqs. (3.45) and (3.46) also applies to Eq. (3.43) [51, 52].

3.9
Stability, Passivity, and Error Estimation of the Reduced Model

In Section 3.5, we have introduced the concepts of stability and passivity, which are the properties needed to be preserved by the reduced model in order to preserve the stability of the whole system under simulation. We also have introduced the measure of the difference between the outputs of two different systems. In this section, we show how a computable error bound for the reduced model produced by the Gramian-based MOR methods can be derived from this measure.

3.9.1
Stability, Passivity, and Error Bound of Moment-Matching MOR

In general, the moment-matching methods do not preserve stability, passivity of the original system, for example, the moment-matching method based on Petrov-Galerkin projection with $W \neq V$ [25]. For some reduced systems, this can be avoided by using postprocessing techniques [87, 88]. For RLC subcircuits, there exist several approaches based on Galerkin projection where the reduced order models are guaranteed to be stable and passive, see [26, 89–94].

The passivity preservation of the method PRIMA [26] for RLC circuits can be mathematically described as below.

Theorem 5 *If the system matrices E and A satisfy $E^T + E \geq 0$ and $A^T + A \leq 0$, respectively, and if $C = B$, then the reduced model obtained by PRIMA preserves the passivity of the original system in Eq. (3.3).*

The detailed proof can be found in [26]. Benefited from the preservation of passivity, PRIMA, and its similarities, the Band Arnoldi process in Section 3.4 are the widely used MOR methods not only in circuit simulation but also in MEMS simulation.

The moment-matching MOR methods have smaller computational complexity than the Gramian-based methods, but they do not have a global error bound for the reduced model, and therefore, the process of deriving the reduced model cannot be adaptively implemented, which in some sense reduces the efficiency of the methods. Earlier results concerning error bounds for moment-matching methods [95] yield only local bounds for the transfer function. Such bounds can only be used to estimate the accuracy of the transfer function in a certain frequency range but cannot give a global estimation in the whole frequency domain. An expression of the $\mathcal{H}_2$ error of the reduced model obtained by rational interpolation (moments matched up to the first order) is given in [96]. How to estimate the error of the general reduced model by moment-matching still remains an open problem.

3.9.2
Stability, Passivity, and Error Bound of the Gramian-Based MOR

If the two Gramian matrices are computed by the direct methods or standard iterative approaches instead of using the low-rank approximation methods, the balanced truncation as well as the equivalent SR-method compute a reduced model that preserves the stability of the original system. Unfortunately, using the low-rank approximations to the Gramians cannot guarantee that model reduction by balanced truncation still preserves the stability of the original system. However, in practice, this seems to be negligible. In general, the Gramian-based MOR method cannot preserve the passivity of the original system, except for some special types of systems, for example, RLC subcircuit, see [97–99].

One of the most important properties of the Gramian-based methods is the computable global error bound of the reduced model, which makes the generation of the reduced model automatic. For the balanced truncation MOR method in Section 3.8, the global error between the transfer function $H(\cdot)$ of the original system Eq. (3.3) and the transfer function $\hat{H}(\cdot)$ of the reduced model is bounded by

$$\|H - \hat{H}\|_\infty \leq 2(\sigma_{q+1} + \sigma_{q+2} + \cdots + \sigma_n) \tag{3.44}$$

where the nonincreasingly ordered $\sigma_i, i = 1, 2, \ldots, n$ are HSVs. From the measure of the difference between two different systems Eqs. (3.18) and (3.20) in Section 3.5, we see that the error between the output response of the original system and that of the reduced system is bounded by

$$\|y_1 - y_2\|_2 \leq 2\left(\sum_{i=q+1}^{n} \sigma_i\right) \|u\|_2 \tag{3.45}$$

in time domain or by

$$\|Y_1 - Y_2\|_{\mathscr{H}_2} \leq 2\left(\sum_{i=q+1}^{n} \sigma_i\right) \|U\|_{\mathscr{H}_2} \tag{3.46}$$

in frequency domain. Since the HSVs (introduced in Section 3.5.4) are available while the balancing transformation matrix T or the projection matrices W, V in Algorithm 5 are computed, they can be used to control the error of the reduced model on-the-fly. Given a required accuracy of the reduced model *tol*, we can derive a suitable q (Eq. (3.36)) from the above-mentioned global error bound, such that the balanced system can be partitioned according to q Eq. (3.39)). Therefore, the error bound (Eq. 3.44) allows for an adaptive choice of the reduced order q.

The global error bound (Eq. 3.44) is based on the assumption that the transformation matrices for MOR are computed in exact arithmetic. Using approximate low-rank factors, the bound is expected to hold only approximately, see [74, 84]. Moreover, the error bound (3.44) cannot be accurately computed anymore since only a few HSVs were computed by all low-rank methods. The remaining HSVs have to be estimated.

3.10
Dealing with Nonzero Initial Condition

Many results of MOR methods are established based on the assumption $x(0) = 0$, the case of $x(0) \neq 0$ [100] is also of importance and interest. In this section, we address this problem to some extent, although there is still space for research.

When $\mathbf{x}(0) \neq \mathbf{0}$, one can apply the variable transformation $\tilde{\mathbf{x}} = \mathbf{x} - \mathbf{x}_0$ to the original Eq. (3.3) and get

$$E\frac{d\tilde{\mathbf{x}}}{dt} = A\tilde{\mathbf{x}} + \tilde{B}\tilde{\mathbf{u}},$$
$$\tilde{\mathbf{y}}(t) = C\tilde{\mathbf{x}} \qquad (3.47)$$

where $\tilde{B} = [B, A\mathbf{x}_0]$, $\tilde{\mathbf{u}} = [\mathbf{u}(t), 1]^T$. The new Eq. (3.47) has zero initial condition $\tilde{\mathbf{x}}_0 = \mathbf{0}$; therefore, most of the MOR methods can be used. After the solution $\tilde{\mathbf{x}}$ is obtained from the reduced model of Eq. (3.47), $\mathbf{x}$ can be recovered by $\mathbf{x} = \tilde{\mathbf{x}} + \mathbf{x}_0$. The output response of Eq. (3.3) is recovered by $\mathbf{y} = C\mathbf{x} = C\tilde{\mathbf{x}} + C\mathbf{x}_0 = \tilde{\mathbf{y}} + C\mathbf{x}_0$.

By replacing B with $\tilde{B}$ in Eq. (3.17), the transfer function $\tilde{H}(s)$ of Eq. (3.47) has of the same form as $H(s)$

$$H(s) = Y(s)/U(s) = C(sE - A)^{-1}\tilde{B} \qquad (3.48)$$

Unfortunately, the above-mentioned technique is not flexible when one is interested in the system response to varying initial conditions. When $\mathbf{x}_0$ changes to a different value, MOR has to be done once again because $\mathbf{x}_0$ is involved in the input matrix $\tilde{B}$. MOR for systems with varying initial conditions still remains an open question so far.

3.11
MOR for Second-Order, Nonlinear, Parametric systems

Many MOR methods have been developed for more complex systems, such as second-order systems, nonlinear systems, and parametric systems. These models and the corresponding MOR methods are discussed in Part III. Except that more mathematical tools are used, the basic ideas of MOR for the more complex systems are mainly originated from the moment-matching methods or balanced truncation. One exception is the modal superposition MOR method introduced in Chapter 12, Part III which is the earliest developed MOR method, and is different from moment-matching MOR and balanced truncation. Details can be found there.

3.12
Conclusion and Outlook

In this chapter, we have introduced two kinds of very basic model order reduction techniques: moment-matching MOR and Gramian-based MOR. By using model order reduction, a reduced model with much smaller size can be derived. By simulating the reduced model, the transfer function or the output response of the original system can be obtained within acceptable error tolerance. As a result, the reduced model can be used to replace the original system in various analyses, which can be done efficiently at the system level.

Although many issues that involve simulating large-scale dynamical systems can be dealt with by MOR, there remain a lot of open problems. One of the bottlenecks for almost all the MOR methods is the loss of efficiency when solving systems with many parameters, especially when the system is nonlinear. Many methods have been proposed for MOR of parametric systems [101–110]. However, most of them are only suitable for systems including around 10 parameters. Some methods are shown to be efficient for some systems with many parameters, unfortunately they are more or less heuristic [111–113], which need to be further justified mathematically.

Some MOR methods have been applied to design analysis in engineering, and they are proved to be efficient for simple problems, which can be modeled by LTI systems. However, there is still a long way to go for many MOR methods to be really transferred to widely used commercial tools that can solve complex problems, such as nonlinear and/or parametric systems. Even for the LTI systems, there are still challenges for MOR. For example, the big challenge for moment-matching MOR is that the reduced model cannot be generated fully automatically. This is because there is not a general rule on adaptively selecting the appropriate expansion points, the proper number of moments for each expansion point, and the optimal order of the reduced model. In many situations, one has to make several trials in order to get the reduced model within an acceptable order and acceptable accuracy. Although some efforts have been made to attain this goal [3, 96, 114–116], how to fully automatically generate theoretically and numerically reliable reduced models based on moment-matching still remains an open problem.

All in all, more robust numerical algorithms are needed to deal with the numerical issues existing in MOR methods. New mathematical theory is desired to either support the many heuristics used in MOR or propose new MOR methods.

References

1. Rudnyi, E.B. and Korvink, J.G. (2002) Review: automatic model reduction for transient simulation of mems-based devices. *Sensors Update*, **11**, 3–33.
2. Korvink, J.G., Rudnyi, E.B., Greiner, A., and Liu, Z. (2005) MEMS and NEMS simulation, in *MEMS: A Practical Guide to Design, Analysis, and Applications* (eds J.G.Korvink and O.Paul), William Andrew Publishing, Norwich, NY.
3. Bechtold, T., Rudnyi, E.B., and Korvink, J.G. (2005) Error indicators for fully automatic extraction of heat-transfer macromodels for mems. *Journal of Micromechanics and Mountaineering*, **15** (3), 430–440.
4. Bechtold, T., Hauck, T., Voss, L., and Rudnyi, E.B. (2011) Efficient electro-thermal simulation of power semiconductor devices via model order reduction, in *Electronics Cooling a Electrical Magazine*.
5. Moore, B.C. (1981) Principal component analysis in linear systems: controllability, observability, and model reduction. *IEEE Transactions on Automatic Control*, **AC-26**, 17–32.
6. Noor, A.K. and Peters, J.M. (1980) Reduced basis technique for nonlinear analysis of structures. *AIAA Journal*, **18** (4), 455–462.
7. Davison, E.J. (1966) A method for simplifying linear dynamic systems. *IEEE Transactions on Automatic Control*, **AC** (11), 93–101.
8. Marschall, S.A. (1966) An approximate method for reducing the order of a linear system. *Control Engineering*, **10**, 642–648.

9. Craig, R.R. and Bampton, M.C.C. (1968) Coupling of substructures for dynamic analysis. *AIAA Journal*, **6** (7), 1313–1319.
10. Rissanen, J. (1971) Recursive identification of linear systems. *SIAM Journal on Control*, **9** (3), 420–430.
11. Wilson, D.A. (1970) Optimum solution of model-reduction problem. *Proceedings of the Institution of Electrical Engineers (London)*, **117** (6), 1161–1165.
12. Fox, R.L. and Miura, H. (1971) An approximate analysis technique for design calculations. *AIAA Journal*, **9** (1), 177–179.
13. Nickell, R.E. (1976) Nonlinear dynamics by mode superposition. *Computer Methods in Applied Mechanics and Engineering*, **7**, 107–129.
14. Laub, A.J., Heath, M.T., Paige, C.C., and Ward, R.C. (1987) Computation of system balancing transformations and other applications of simultaneous diagonalization algorithms. *IEEE Transactions on Automatic Control*, **34**, 115–122.
15. Antoulas, A.C. (2005) *Approximation of Large-Scale Dynamical Systems*, SIAM Publications, Philadelphia, PA.
16. Benner, P. and Quintana-Ortí, E.S. (2005) Model reduction based on spectral projection methods, in Chapter 1, (pages 5–48) of [117].
17. Patera, A.T. and Rozza, G. (2007) Reduced basis approximation and a posteriori error estimation for parametrized partial differential equations, to appear in (tentative rubric) MIT Pappalardo Graduate Monographs in Mechanical Engineering, Version 1.0, Copyright MIT 2006.
18. Rozza, G., Huynh, D.B.P., and Patera, A.T. (2008) Reduced basis approximation and a posteriori error estimation for affinely parametrized elliptic coercive partial differential equations. *Archives of Computational Methods in Engineering*, **15**, 229–275.
19. Rowley, C.W., Colonius, T., and Murray, R.M. (2004) Model reduction for compressible flow using POD and Galerkin projection. *Physica D*, **189**, (1-2), 115–129.
20. Colonius, T. and Freund, J.B.. POD analysis of sound generation by a turbulent jet. *AIAA Journal*, **2002 – 0072**, 2002.
21. Smith, T.R. (2003) Low-dimensional models of plane Couette flow using the proper orthogonal decomposition. PhD thesis, Princeton Univ.
22. Ravindran, S.S. (2000) A reduced-order approach for optimal control of fluids using proper orthogonal decomposition. *International Journal for Numerical Methods in Fluids*, **34** (5), 425–448.
23. Rathinam, M. and Petzold, L.R. (2002) Dynamic iteration using reduced order models: a method for simulation of large scale modular systems. *SIAM Journal on Numerical Analysis*, **40** (4), 1446–1474.
24. Pillage, L.T. and Rohrer, R.A. (1990) Asymptotic waveform evaluation for timing analysis. *IEEE Transactions on Computer-Aided Design*, **9** (4), 352–366.
25. Feldmann, P. and Freund, R.W. (1995) Efficient linear circuit analysis by Padé approximation via the Lanczos process. *IEEE Transactions on Computer-Aided Design of Integrated Circuits and Systems*, **14** (5), 639–649.
26. Odabasioglu, A., Celik, M., and Pileggi, L.T. (1998) PRIMA: passive reduced-order interconnect macromodeling algorithm. *IEEE Transactions on Computer-Aided Design of Integrated Circuits and Systems*, **17** (8), 645–654.
27. Grimme, E.J. (1997) Krylov projection methods for model reduction. PhD thesis, Univ. Illinois, Urbana-Champaign.
28. Saks, S. (1952) Theory of the integral, Hafner.
29. Chiprout, E. and Nakhla, M.S. (1994) *Asymptotic Waveform Evaluation and Moment Matching for Interconnect Analysis*, Kluwer Academic Publishers, Norwell, MA.
30. Demmel, J.W. (1997) *Applied Numerical Linear Algebra*, SIAM Publications, Philadelphia, PA.
31. Skogestad, S. and Postlethwaite, I. (1997) *Multivariable Feedback Control, Analysis and Design*, John Wiley & Sons, Ltd, Chichester.

32. Saad, Y. (1996) *Iterative Methods for Sparse Linear Systems*, PWS Publishing Company, Boston, MA.
33. Freund, R.W. (2003) Model reduction methods based on Krylov subspaces. *Acta Numerica*, **12**, 267–319.
34. Freund, R.W. (2000) Krylov-subspace methods for reduced-order modeling in circuit simulation. *Journal of Computational and Applied Mathematics*, **123**, (1–2), 395–421.
35. Chu, C.-C., Lai, M., and Feng, W. (2006) MIMO interconnect order reductions by using multiple point adaptive-order rational global Arnoldi algorithm. *IEICE Transactions on Electronics*, **E89-C** (5), 792–802.
36. Chu, C.-C., Lai, M., and Feng, W. (2008) Model-order reductions for MIMO systems using global Krylov subspace methods. *Mathematics and Computers in Simulation*, **79**, 1153–1164.
37. Zadeh, L.A. and Polak, E. (1969) *System Theory*, TaTa McGRAW-Hill Publishing Company Ltd., Bombay, New Delhi.
38. Chen, C.-T. (1999) *Linear System Theory and Desig*, Oxford University Press, New York, Oxford.
39. Oppenheim, A.V., Willsky, A.S., and Nawab, S.H. (1996) *Signals & Systems*, Prentice-Hall, Upper Saddle River, NJ.
40. Freund, R.W. (2000) *Passive Reduced Order Modeling Via Krylov Subspace Methods*, Numerical Analysis Manuscript No. 00-3-02, Bell Laboratories, Murray Hill, NJ.
41. Anderson, B.D.O. and Vongpanitlerd, S. (1973 d) *Network Analysis and Synthesis*, Prentice-Hall, Englewood Cliffs, NJ.
42. Rohrer, R.A. and Nosrati, H. (1981) Passivity considerations in stability studies of numerical integration algorithms. *IEEE Transactions on Circuits and Systems*, **28**, 857–866.
43. Chirlian, P.M. (1967) *Integrated and Active Network Analysis and Synthesis*, Prentice-Hall, Englewood Cliffs, NJ.
44. Gallivan, K., Grimme, E., and Van Dooren, P. (1994) Asymptotic waveform evaluation via a Lanczos method. *Applied Mathematics Letters*, **7** (5), 75–80.
45. Gragg, W.B. and Lindquist, A. (1983) On the partial-realization problem. *Linear Algebra and its Applications*, **50**, (APR), 277–319.
46. Shamash, Y. (1975) Linear system reduction using padé approximation to allow retention of dominant modes. *International Journal of Control*, **21**, 257–272.
47. Bultheel, A. and Van Barel, M. (1986) Padé techniques for model reduction in linear system theory: a survey. *Journal of Computational and Applied Mathematics*, **14**, 401–438.
48. Baker, G.A. (1975) *Essentials of Padé approximations*, Academic Press, New York.
49. Bechtold, T., Rudnyi, E.B., and Korvink, J.G. (2006) *Fast Simulation of Electro-thermal MEMS: Efficient Dynamic Compact Models*, Springer, Berlin.
50. Bai, Z. (2002) Krylov subspace techniques for reduced-order modeling of large-scale dynamical systems. *Applied Numerical Mathematics*, **43**, (1--2), 9–44.
51. Stykel, T. (2004) Gramian-based model reduction for descriptor systems. *Mathematics of Control Signals and Systems*, **16** (4), 297–319.
52. Mehrmann, V. and Stykel, T. (2005) Balanced truncation model reduction for large-scale systems in descriptor form, in Chapter 3, (pages 83–115) of [117].
53. Antoulas, A.C. and Sorensen, D.C. (2001) Approximation of large-scale dynamical systems: an overview. *International Journal of Applied Mathematics and Computer Science*, **11** (5), 1093–1121.
54. Antoulas, A.C., Sorensen, D.C., and Gugercin, S. (2001) A survey of model reduction methods for large-scale systems. *Contemporary Mathematics*, **280**, 193–219.
55. Benner, P. (2009) System-theoretic methods for model reduction of large-scale systems: Simulation, control, and inverse problems, *Proceedings of MathMod 2009*, pp. 126–145.
56. Tombs, M.S. and Postlethwaite, I. (1987) Truncated balanced realization of a stable non-minimal state-space

system. *International Journal of Control*, **46** (4), 1319–1330.

57. Hammarling, S.J. (1982) Numerical solution of the stable, non-negative definite Lypunov equation. *IMA Journal of Numerical Analysis*, **2**, 303–323.
58. Safonov, M.G. and Chiang, R.Y. (1989) A Schur method for balanced truncation model reduction. *IEEE Transactions on Automatic Control*, **34** (7), 729–733.
59. Benner, P., Quintana-Ortí, E.S., and Quintana-Ortí, G. (1999) Solving linear and quadratic matrix equations on distributed memory parallel computers. Proceedings IEEE International Symposium on Computer Aided Control System Design, pp. 46–51.
60. The MathWorks, Inc., http://www.matlab.com. MATLAB.
61. Antoulas, A.C., Sorensen, D.C., and Zhou, Y. (2002) On the decay rate of Hankel singular values and related issues. *Systems & Control Letters*, **46** (5), 323–342.
62. Grasedyck, L. (2004) Existence of a low rank or $\mathcal{H}$-matrix approximant to the solution of a Sylvester equation. *Numerical Linear Algebra with Applications*, **11** (4), 371–389.
63. Penzl, T. (2000) Eigenvalue decay bounds for solutions of Lyapunov equations: the symmetric case. *Systems & Control Letters*, **40** (2), 139–144.
64. Sorensen, D.C. and Zhou, Y. (2002) Bounds on eigenvalue decay rates and sensitivity of solutions to Lyapunov equations. Technical Report TR02-07, Department of Computer Application in Mathematics, Rice University, Houston, TX.
65. Hochbruck, M. and Starke, G. (1995) Preconditioned Krylov subspace methods for Lyapunov matrix equations. *SIAM Journal on Matrix Analysis and Applications*, **16** (1), 156–171.
66. Hodel, A.S., Tenison, B., and Poolla, K.R. (1996) Numerical solution of the Lyapunov equation by approximate power iteration. *Linear Algebra and its Applications*, **236**, 205–230.
67. Hu, D.Y. and Reichel, L. (1992) Krylov-subspace methods for the Sylvester equation. *Linear Algebra and its Applications*, **172**, 283–313.
68. Jaimoukha, I.M. and Kasenally, E.M. (1994) Krylov subspace methods for solving large Lyapunov equations. *SIAM Journal on Numerical Analysis*, **31** (1), 227–251.
69. Jbilou, K. and Riquet, A.J. (2006) Projection methods for large Lyapunov matrix equations. *Linear Algebra and its Applications*, **415** (2-3), 344–358.
70. Saad, Y. (1990) Numerical solution of large Lyapunov equations, in *Signal Processing, Scattering, Operator Theory and Numerical Methods* (eds M.A. Kaashoek, J.H. vanSchuppen, and A.C.M. Ran), Birkhäuser, pp. 503–511.
71. Simoncini, V. (2007) A new iterative method for solving large-scale Lyapunov matrix equations. *SIAM Journal on Scientific Computing*, **29** (3), 1268–1288.
72. Benner, P., Claver, J.M., and Quintana-Ortí, E.S. (1998) Efficient solution of coupled Lyapunov equations via matrix sign function iteration. Proceedings 3rd Portuguese Conference on Automatic Control CONTROLO'98, Coimbra, pp. 205–210.
73. Baur, U. and Benner, P. (2006) Factorized solution of Lyapunov equations based on hierarchical matrix arithmetic. *Computing*, **78** (3), 211–234.
74. Baur, U. and Benner, P. (2008) Gramian-based model reduction for data-sparse systems. *SIAM Journal on Scientific Computing*, **31** (1), 776–798.
75. Grasedyck, L., Hackbusch, W., and Khoromskij, B.N. (2003) Solution of large scale algebraic matrix Riccati equations by use of hierarchical matrices. *Computing*, **70** (2), 121–165.
76. Li, J.-R., Wang, F., and White, J. (1999) An efficient Lyapunov equation-based approach for generating reduced-order models of interconnect. Proceedings Design Automation Conference, pp. 1–6.
77. Li, J.-R. and White, J. (2002) Low rank solution of Lyapunov equations. *SIAM Journal on Matrix Analysis and Applications*, **24** (1), 260–280.
78. Penzl, T. (1999/00) A cyclic low-rank Smith method for large sparse

Lyapunov equations. *SIAM Journal on Scientific Computing*, **21** (4), 1401–1418.
79. Gugercin, S., Sorensen, D.C., and Antoulas, A.C. (2003) A modified low-rank Smith method for large-scale Lyapunov equations. *Numerical Algorithms*, **32** (1), 27–55.
80. Badía, J.M., Benner, P., Mayo, R., and Quintana-Ortí, E.S. (2002) Solving large sparse Lyapunov equations on parallel computers, in *Euro-Par 2002 – Parallel Processing*, Number 2400 in Lecture Notes in Computer Science (eds B.Monien and R.Feldmann), Springer-Verlag, Berlin, Heidelberg, New York, pp. 687–690.
81. Badía, R.M., Benner, P., Mayo, R., Quintana-Ortí, E.S., Quintana-Ortí, G., and Remón, A. (2006) Balanced truncation model reduction of large and sparse generalized linear systems. Technical Report Chemnitz Scientific Computing Preprints 06-04. Fakultät für Mathematik, TU Chemnitz.
82. Benner, P., Quintana-Ortí, E.S., and Quintana-Ortí, G. (2000) Balanced truncation model reduction of large-scale dense systems on parallel computers. *Mathematical and Computer Modelling of Dynamical Systems*, **6** (4), 383–405.
83. Benner, P., Quintana-Ortí, E.S., and Quintana-Ortí, G. (2003) State-space truncation methods for parallel model reduction of large-scale systems. *Parallel Computing*, **29**, 1701–1722.
84. Gugercin, S. and Li, J.-R. (2005) Smith-type methods for balanced truncation of large systems, Chapter 2 (pages 49–82) of [117].
85. Rabiei, P. and Pedram, M. (1999) Model order reduction of large circuits using balanced truncation. Proceedings of Asia and South Pacific Design Automation Conference. pp. 237–240.
86. Van Dooren, P. (2000) Gramian based model reduction of large-scale dynamical systems. Proceedings 18th Dundee Biennial Conference on Numerical Analysis, pp. 231–247.
87. Bai, Z., Feldmann, P., and Freund, R.W. (1998) How to make theoretically passive reduced-order models passive in practice. In Proceedings IEEE Custom Integrated Circuits Conference, pp. 207–210.
88. Bai, Z. and Freund, R.W. (2001) A partial Padé-via-Lanczos method for reduced-order modeling. *Linear Algebra and its Applications*, **332**, 139–164.
89. Bai, Z. and Freund, R.W. (2001) A symmetric band Lanczos process based on coupled recurrences and some applications. *SIAM Journal on Scientific Computing*, **23** (2), 542–562.
90. Freund, R.W. and Feldmann, P. (1996) Reduced-order modeling of large passive linear circuits by means of the SyPVL algorithm. Proceedings IEEE/ACM International Conference on Computer-Aided Design, pp. 280–287.
91. Freund, R.W. and Feldmann, P. (1997) The SyMPVL algorithm and its applications to interconnect simulation. Proceedings International Conference on Simulation of Semiconductor Processes and Devices, pp. 113–116.
92. Freund, R.W. and Feldmann, P. (1998) Reduced-order modeling of large linear passive multi-terminal circuits using matrix-Padé approximation. Proceedings Design, Automation and Test in Europe Conference, pp. 530–537.
93. Kerns, K.J., Wemple, I.L., and Yang, A.T. (1995) Stable and efficient reduction of substrate model networks using congruence transforms. Proceedings IEEE/ACM International Conference on Computer-Aided Design, pp. 207–214.
94. Silveira, L.M., Kamon, M., Elfadel, I., and White, J. (1996) A coordinate-transformed Arnoldi algorithm for generating guaranteed stable reduced order models of arbitrary RLC circuits. Proceedings IEEE/ACM International Conference on Computer-Aided Design, pp. 288–294.
95. Bai, Z., Slone, R.D., and Smith, W.T. (1999) Error bound for reduced system model by Padé approximation via the Lanczos process. *IEEE Transactions on Computer-Aided Design of Integrated Circuits and Systems*, **18** (2), 133–141.

96. Gugercin, S., Antoulas, A.C., and Beattie, C.A. (2008) $\mathcal{H}_2$ model reduction for large-scale linear dynamical systems. *SIAM Journal on Matrix Analysis and Applications*, **30** (2), 609–638.
97. Reis, T. and Stykel, T. (2010) PABTEC: Passivity-preserving balanced truncation for electrical circuits. *IEEE Transactions on Computer-Aided Design of Integrated Circuits and Systems*, **29** (9), 1354–1367.
98. Phillips, J., Daniel, L., and Silveira, L. (2003) Guaranteed passive balancing transformations for model order reduction. *IEEE Transactions on Computer-Aided Design of Integrated Circuits and Systems*, **22** (8), 1027–1041.
99. Yan, B., Tan, S.-D., Liu, P., and McGaughy, B. (2007) Passive interconnect macromodeling via balanced truncation of linear systems in descriptor form. Proceedings Design Automation Conference, pp. 355–360.
100. Feng, L.H., Koziol, D., Rudnyi, E.B., and Korvink, J.G. (2004) Model order reduction for scanning electrochemical microscope: the treatment of nonzero initial condition. Proceedings of IEEE Sensors, pp. 1236–1239.
101. Baur, U. and Benner, P. (2011) Interpolatory projection methods for parameterized model reduction. *SIAM Journal on Scientific Computing*, **33**, 2489–2518.
102. Daniel, L., Siong, O.C., Chay, L.S., Lee, K.H., and White, J. (2004) A multiparameter moment-matching model-reduction approach for generating geometrically parameterized interconnect performance models. *IEEE Transactions on Computer-Aided Design of Integrated Circuits and Systems*, **22** (5), 678–693.
103. Feng, L., Rudnyi, E.B., and Korvink, J.G. (2005) Preserving the film coefficient as a parameter in the compact thermal model for fast electro-thermal simulation. *IEEE Transactions on Computer-Aided Design of Integrated Circuits and Systems*, **24** (12), 1838–1847.
104. Feng, L. and Benner, P. (2007) A robust algorithm for parametric model order reduction. *Proceedings in Applied Mathematics and Mechanics (ICIAM)*, **7** (1), 10215-01–10215-02.
105. Gunupudi, P., Khazaka, R., and Nakhla, M. (2002) Analysis of transmission line circuits using multidimensional model reduction techniques. *IEEE Transactions on Advanced Packaging*, **25** (2), 174–180.
106. Li, Y.-T., Bai, Z., Su, Y., and Zeng, X. (2007) Model order reduction of parameterized interconnect networks via a two-directional arnoldi process. Proceedings IEEE/ACM International Conference on Computer-Aided Design, pp. 868–873.
107. Li, X., Li, P., and Pileggi, L.T. (2005) Parameterized interconnect order reduction with explicit-and-implicit multi-parameter moment matching for inter/intra-die variations. Proceedings IEEE/ACM International Conference on Computer-Aided Design, pp. 806–812.
108. Liu, Y., Pileggi, L.T., and Strojwas, A.J. (1999) Model order reduction of RC(L) interconnect including variational analysis. Proceedings Design Automation Conference, pages 201–206.
109. Phillips, J.R. (2004) Variational interconnect analysis via PMTBR. Proceedings IEEE/ACM International Conference on Computer-Aided Design, pp. 872–879.
110. Weile, D.S., Michielssen, E., Grimme, E., and Gallivan, K.A. (1999) order models of two-parameter linear systems. *Applied Mathematics Letters*, **12** (5), 93–102.
111. Phillips, J. (2004) Variational interconnect analysis via PMTBR. Proceedings IEEE/ACM International Conference on Computer-Aided Design, pp. 872–879.
112. Zhu, Z. and Phillips, J. (2007) Random sampling of moment graph: a stochastic Krylov-reduction algorithm. Proceedings Design, Automation & Test in Europe, pp. 1502–1507.
113. El-Moselhy, T. and Daniel, L. (2010) Variation-aware interconnect extraction using statistical moment preserving model order reduction. Proceedings Design, Automation & Test in Europe, pp. 453–458.

114. Achar, R. and Nakhla, M.S. (2001) Simulation of high-speed interconnects. *Proceedings of the IEEE*, **89** (5), 693–728.
115. Feng, L.H., Korvink, J.G., and Benner, P. (2012) A fully adaptive scheme for model order reduction based on moment-matching, Max-Planck Institute Preprint.
116. Villena, J.F. and Silveira, L.M. (2011) Multi-dimensional automatic sampling schemes for multi-point modeling methodologies. *IEEE Transactions on Computer-Aided Design of Integrated Circuits and Systems*, **30** (8), 1141–1151.
117. Benner, P., Mehrmann, V., and Sorensen, D.C. (eds) (2005) *Dimension Reduction of Large-Scale Systems*, Lecture Notes in Computational Science and Engineering, vol. 45, Springer-Verlag, Berlin/Heidelberg, Germany.

4
Algorithmic Approaches for System-Level Simulation of MEMS and Aspects of Cosimulation

Peter Schneider, Christoph Clauß, Ulrich Donath, Günter Elst, Olaf Enge-Rosenblatt, and Thomas Uhle

4.1
Introduction

Microelectromechanical systems (MEMS), as an important subset of microsystems, are characterized by the interaction of different functional components and various physical effects. Usually, a transducer is combined with analog and digital electrical circuits to realize a certain functionality. In addition, undesirable so-called parasitic effects have to be considered as well. Not only in electrical subsystems, there are components with continuous-time behavior as well as components with discrete-time or discrete-event behavior. Starting with this very rough classification, a general mathematical structure of MEMS models can be derived. From a mathematical point of view, such a generalized model consists of three types of submodels that are described by differential equations, algebraic equations, and Boolean equations.

For the mathematical description of physical effects in the most general form, *partial differential equations* (PDEs) are utilized. Spatially distributed phenomena with an infinite number of degrees of freedom (DoFs) can be described by PDEs. Typical examples are heat conduction, diffusion processes, electromagnetic fields, and mechanical deformations of solids. Beside spatial derivatives, even time derivatives have to be considered in a significant number of applications. For the numerical solution of PDEs, a spatial discretization is carried out. Usually, this results in very large systems of ordinary differential equations. Owing to the specific requirements regarding computational effort and required memory, appropriate methods have to be applied to solve these equations.

Therefore, often a more abstract modeling approach based on a limited number of equations or a composition of basic lumped elements is pursued for system-level simulation. Such systems with a finite number of DoFs and dynamic time behavior are mathematically described by ordinary differential equations (ODEs) [1, 2]. Systems with static time behavior are described by *algebraic equations*. In many practical cases, MEMS or microsystems consist of dynamic as well as static parts. This leads to a mixture of differential and algebraic equations, which is called

System-level Modeling of MEMS, First Edition. Edited by T. Bechtold, G. Schrag, and L. Feng.
© 2013 Wiley-VCH Verlag GmbH & Co. KGaA. Published 2013 by Wiley-VCH Verlag GmbH & Co. KGaA.

differential algebraic equation (DAE) system [3]. The most general representation has an implicit form [4], which is contemplated in the following sections. Although it is possible to divide the complete DAE system into a differential and an algebraic subsystem in some cases at least, the algebraic equations are nonlinear in general and, thus, cannot be solved in a closed form. Typical examples are nonlinear dynamic networks [5], which are mathematically described by implicit DAEs. For a comprehensive discussion of different levels of abstraction for MEMS modeling, see Sections 1.3 and 1.4.

Microsystems also contain parts for digital signal processing and digital controls. The discrete-event behavior of these components can be described at higher levels of abstraction by *Boolean equations* and *finite state machines* (FSMs). FSMs are mostly represented graphically by state diagrams, also called *state charts*. State diagrams describe all valid states as well as transitions between these states and the events at which these transitions are triggered. Usually, actions are executed along with state transitions.

In many cases, the model of a complete system covers more than one of the above-mentioned basic types of mathematical description. This means that the methods for simulating the entire system have to simultaneously solve different systems of equations that are coupled one to another. Common algorithmic approaches for an efficient system-level simulation include methods for nonlinear network simulation, hybrid simulation methods, and cosimulation techniques. These methods will be described in the following sections. For an efficient modeling flow also approaches such as model transformations and model simplifications are needed. However, simplifications and modeling at different levels of abstraction may cause numerical issues. For example, the numerical solution of a continuous-time subsystem with discontinuous input signals or structural changes in its equation system, caused by an event of a discrete-event subsystem, is a challenge for the development of methods and still a subject of present research.

4.2
Mathematical Structure of MEMS Models

In this section, the different types of mathematical representations of MEMS are discussed in more detail. Since PDEs cause high computational effort, they are usually not used in system-level models. However, there are model order reduction methods to transform the spatial discretization of PDEs to a smaller system of ODEs (Chapter 3). That is why ODEs and DAEs, often in combination with Boolean equations and FSMs for the discrete-event parts, are most relevant for system-level modeling.

For system-level modeling, a microsystem is usually divided into several components. These components can be regarded as being connected at conservative and nonconservative connectors. The "data exchange" between models is carried out through these connectors only. In the following, we just use the term *terminal* for a conservative connector as this is the common term in electrical engineering.

4.2.1
Differential and Algebraic Equations

Behavioral modeling of analog MEMS components leads to mathematical representations that consist of nonlinear implicit differential and algebraic equations in general. Each terminal is related to two quantities: a potential quantity and a flow quantity which, therefore, are also called *terminal variables*. In addition, such a DAE system might also depend on component-internal variables needed to mathematically describe the components' behavior. A behavioral model can be mathematically represented in the form of constitutive equations

$$\mathbf{f}(\mathbf{x}_k(t), \dot{\mathbf{x}}_k(t), \mathbf{x}_i(t), \dot{\mathbf{x}}_i(t), t) = 0 \tag{4.1}$$

where $\mathbf{x}_k$ is the vector of all terminal variables, $\dot{\mathbf{x}}_k$ are their time derivatives, $\mathbf{x}_i$ is the vector of all internal variables of the model, $\dot{\mathbf{x}}_i$ are their time derivatives, and t is the instantaneous time. The prevalent dependency of the DAE system (Eq. (4.1)) on parameters is not explicitly mentioned here. Parameters are time-independent quantities that determine the model's functionality and that are assumed to be constant during a particular simulation. The *model behavior* is defined as the *solution of the DAE system* (Eq. (4.1)) on a certain, often physically motivated, region of validity. The term *terminal behavior* is often used for the set of all solutions $\mathbf{x}_k$ [6].

Normally, the number of model equations of a component is less than the overall number of their terminal and internal variables. Solution methods of available simulators require that the number of equations of a system has to match the number of variables. Therefore, open terminals of a component have to be connected to terminals of other models, for example, sources or loads. In electrical systems, this can be voltage or current sources or resistors. In mechanics, sources of deflection or angle as well as load forces or torques are used. Some simulators let set up additional constraints for open terminals to complement the DAE system (Eq. (4.1)).

Usually, behavioral modeling aims at designing the so-called compact models (Sections 2.2.2 and 2.2.3). By dividing the vector of terminal variables $\mathbf{x}_k$ into independent and dependent variables, a more compact representation of the set of equations (Eq. (4.1)) can be achieved in some cases. Likewise, an elimination of algebraic dependencies may lead to a smaller set of equations. The use of modern modeling languages and simulators with features like symbolic preprocessing helps to cut down the number of equations while still retaining the clarity of the behavioral description.

Generally, no universal statements can be made on the number and structure of the DAEs needed for a behavioral model of a component as well as on the choice of internal variables. These aspects strongly depend on the method used to set up the equations. Similar to the formalisms from Lagrange or Hamilton in multibody mechanics, a nodal analysis or a mesh analysis as well as extended methods such as the modified nodal analysis are applied in electrical engineering. In multiphysics modeling, combinations of these methods may also be suitable. Therefore, the mathematical representation for a particular physical system might strongly depend on the utilized modeling approach (e.g., modified nodal analysis or sparse tableau analysis for electrical circuits). However, it is possible to transform different equation systems describing the same physical system one into the other.

Under certain conditions, a prediction for the number of equations can be made. The classical four-pole or two-port representations of a model in electrical engineering, for instance, as well as the generalized multipole or multiport representations assume a mathematical description depending on terminal variables only. Then, the number of terminal variables is twice the number of equations and one half of these terminal variables depends on the other half. These equations are well known as admittance, impedance, hybrid, and chain representations among others. These certain equations only exist if the square submatrix of the Jacobian with respect to the dependent terminal variables has full rank. In general, an implicit representation, where all terminal variables are assumed to be independent, has to be used instead. For linear networks, this implicit representation is well known as the Belevitch representation of the terminal behavior [7].

Common modeling methods for different components of MEMS yield DAE systems with typical structures. These methods and structures of equation systems are explained in the following sections.

4.2.1.1 Multibody Systems

There are different ways of deducing the equations of motion for a multibody system (MBS). The so-called *network method* (read the following section about networks) leading to Newton-Euler equations, on the one hand, and the method using generalized coordinates leading to the *Lagrange equations* [8], on the other hand, are most popular. Using the first method, every body is considered as a component with terminals having a force vector and a velocity vector as terminal variables. Using the second method, Lagrange's motion equations can be derived. Here, the complete MBS may be considered as a single component. The starting point is to distinguish between tree-structured systems and systems with kinematic loops. In tree-structured systems, exactly one path exists from the fixed-in-space basement to every body or its representing node. These paths imply a numbering of the bodies (starting at the basement with zero) and, therefore, the so-called predecessor to every body. Every joint defines a certain DoF for the possible motion of a body with respect to its predecessor. The joint DoF equals the number of generalized coordinates for that joint. Hence, a tree-structured MBS can be described by a vector of n (independent) generalized coordinates $\mathbf{q} = (q_1, \ldots, q_n)^\mathsf{T}$ representing its configuration at the instantaneous time t. Because inertial bodies are connected to the joints, it results from Newton's second law of motion that the number of state variables for every joint is twice as much as the number of generalized coordinates. The dynamic state of an MBS is characterized by the vectors of generalized coordinates $\mathbf{q}$ and generalized velocities $\dot{\mathbf{q}}$. Hence, the equations of motion of a tree-structured system result in a system of nonlinear second-order ODEs. It can be written according to

$$\mathbf{M}(\mathbf{q}(t), t)\,\ddot{\mathbf{q}}(t) = \mathbf{h}(\mathbf{q}(t), \dot{\mathbf{q}}(t), t) \tag{4.2}$$

where **M** is the generalized mass matrix and **h** a quadratic form in $\dot{\mathbf{q}}$ containing generalized forces. Considering generalized forces in more detail, the equations of motion become

$$\mathbf{M}(\mathbf{q}(t),\,t)\,\ddot{\mathbf{q}}(t) = \mathbf{h}_1(\mathbf{q}(t),\,\dot{\mathbf{q}}(t),\,t) + \mathbf{h}_2(\mathbf{q}(t),\,t)\,\dot{\mathbf{q}}(t) + \mathbf{h}_3(\mathbf{q}(t),\,t) \tag{4.3}$$

where $\mathbf{h}_1$ is a quadratic form in $\dot{\mathbf{q}}$ containing the terms of Coriolis forces as well as forces depending on the square of velocity (centrifugal forces, air friction), $\mathbf{h}_2$ represents the matrix of coefficients for velocity-proportional forces, and $\mathbf{h}_3$ denotes a vector of forces depending only on generalized coordinates and/or the instantaneous time.

Often, the topology of an MBS cannot be represented by a tree structure. In such cases, there are kinematic loops that cause additional constraints. Hence, such MBS are also called constraint mechanical systems (CMS). Because of these constraints, the governing equations of a CMS can only be described by a nonlinear DAE system. Generally, it consists of second-order ODEs complemented by algebraic equations (AEs). The AEs describe the constraints of dependent coordinates, often given as a set of equations $\mathbf{g}(\mathbf{q}(t),\,t) = \mathbf{0}$. Hence, the complete DAE system reads

$$\mathbf{M}(\mathbf{q}(t),\,t)\,\ddot{\mathbf{q}}(t) = \mathbf{h}(\mathbf{q}(t),\,\dot{\mathbf{q}}(t),\,t) + \mathbf{G}^{\mathrm{T}}(\mathbf{q}(t),\,t)\,\lambda(t), \quad \mathbf{g}(\mathbf{q}(t),\,t) = \mathbf{0} \tag{4.4}$$

where $\mathbf{G} = \frac{\partial \mathbf{g}}{\partial \mathbf{q}}$ is the Jacobian of the AE system and λ is the vector of Lagrangian multipliers representing constraint forces. This completes the governing equations of an MBS. Generally, higher order DAEs can be reduced to first-order ones by introducing additional variables. The kind of equations depends on the choice of the variables but, nevertheless, they have the same structure as network equation systems (Section 4.4.1).

For more information on MBS, read [9–13] for instance.

4.2.1.2 Fluidic Systems

In MEMS, effects of air compression and air flow (with friction) often play an important role. That is why the following explanations are focused on pneumatic rather than hydraulic systems.

The well-known mathematical model of a gas reads

$$p(t)\,V(t) = m(t)\,R\,T(t) \tag{4.5}$$

where p is the absolute pressure, V is the considered volume, m is the gas mass in the volume, R is the gas constant, and T is the absolute temperature. Assuming that the volume under consideration is determined by some kind of mechanical subsystem (or is regarded as being constant), the equation establishes a relation between pressure, mass, and temperature of a gas in a considered volume. Often, a further simplification can be made by considering the processes to be isothermic. That means that the temperature remains constant and Eq. (4.5) can be reduced to

$$p(t)\,V(\mathbf{x}(t)) = m(t)\,R\,T \tag{4.6}$$

where $\mathbf{x}$ is a vector of mechanical coordinates. The vector $\mathbf{x}$ is determined by the generalized coordinates $\mathbf{q}$.

For a calculation of the gas mass that is present in a volume, the input and output flows have to be considered. Input flows occur if gas particles move from a reservoir with a higher pressure into the volume V. Output flows occur if some gas leaves the volume (into another reservoir or into the atmosphere). Both processes are characterized by some kind of resistance against the gas flow. This resistance highly depends on the geometry of the passing area. Such geometries may be long tall lines, orifices, nozzles, valves, and so on. The physical quantity characterizing the amount of gas passing through the area is the so-called mass flow rate, denoted by $\dot{m}$. It is a highly nonlinear function with respect to the quantities characterizing the gas states of the two reservoirs at the input and the output of the valve: the upstream pressure p_1 and downstream pressure p_2, the upstream temperature T_1, and the temperature at reference conditions T_0. In addition, the mass flow rate depends on the velocity $\vec{v}_g$ of the gas particles when passing the valve. Usually, one distinguishes between subsonic flow and supersonic (or choked) flow. The border between both types is defined by the sonic velocity in the gas. Sometimes, a laminar flow occurs for very low velocities. For all mentioned phenomena [14], the mass flow rate can be expressed by

$$\dot{m}(t) = f(p_1, p_2, T_1, T_0, \vec{v}_g(t)). \tag{4.7}$$

The gas mass in a volume can be calculated by adding all input flows $\dot{m}_{in}$ and output flows $\dot{m}_{out}$. So the mass becomes

$$m(t) = m(t)\Big|_{t=0} + \int_0^t \dot{m}_{in}(\tau)\, d\tau - \int_0^t \dot{m}_{out}(\tau)\, d\tau. \tag{4.8}$$

To sum up the model under constant temperatures, the governing equations are nonlinear first-order ODEs. That is because the inertia of the gas can be neglected. However, if thermodynamic effects have to be considered, the situation becomes much more complicated. In this case, the internal energy

$$U(t) = m(t)\, c_V\, T(t) \tag{4.9}$$

has to be taken into account where c_V is the specific heat capacity at constant volume. Considering a complete model (entering and/or leaving gas, variable volume, heat flow from/to environment), the change of the internal energy is given by the energy due to the entering or leaving gas mass, the mechanical work (if the volume changes), and the heat flow through the boundary of the volume. This leads to a mathematical model of the gas temperature in a volume in the form of a combined differential equation for $\dot{m}$ and $\dot{T}$ (but m and T as well as the pressure p and the time derivative of volume $\dot{V}$ might also occur [14]).

Finally, the governing equations of lumped parameter models for pneumatics are nonlinear first-order ODEs complemented by some additional algebraic equations. Therefore, the complete motion equations are a system of DAEs again, that is, they have the same structure as network equation systems (Eq. (4.15) in Section 4.4.1).

For more information on modeling fluidic systems, we refer to appropriate textbooks [14–17].

4.2.1.3 Networks

Several physical systems, such as electrical, mechanical, thermal, or acoustic systems, can be modeled by networks. A network is an interconnection of various multipoles. The entity of each multipole is described by a nonempty finite set of terminals and a constitutive relation describing the terminal behavior [18–20], which is the relation between the flow quantities through the terminals and the potential quantities across any pair of these terminals. In electrical networks, the flow quantity is called *current* and the potential quantity is a potential difference called *voltage* [21]. In many cases, the constitutive relation approximates the characteristics and properties of a spatially distributed physical system with respect to the terminals, so that the resulting network model is spatially lumped. Such multipoles are often called *lumped elements* and the interconnections of lumped elements are the so-called lumped networks (see Chapter 2 for details on network modeling).

The interconnection of multipoles can be specified by identification of their terminals by means of an equivalence relation. The representatives of its equivalence classes are called *nodes* [21]. Although the term "node" therefore suggests the additional implication of a terminal at which several multipoles are connected, the terms "terminal" and "node" are often used interchangeably. Nodes can be seen as representatives for galvanic connections in electrical circuits, for instance. For the purpose of a simple description of the resulting network equations, we focus on electrical networks in the following, but without loss of generality.

The topology of lumped elements is described by means of a network graph with a nonempty finite set of branches and a finite set of nodes. Two-terminal lumped elements are the simplest and the most common ones. They have only one branch connecting two terminals. In many cases, the constitutive relation of a lumped element can be formulated in terms of the currents i through these branches and the voltages v across these branches. That is why this is also called a current–voltage relation or, shortly, an i–v relation for the branches in electrical networks. Note that the i–v relation of a branch may also depend on currents and voltages of other branches. Examples are controlled sources and coupled inductors. Multipoles are mathematically represented by equations that mostly reveal resistive, inductive, or capacitive behavior, or a mixture of them, that is, ordinary differential and algebraic equations.

The voltages and currents of a network, which are time-dependent functions, have to satisfy Kirchhoff's laws being Kirchhoff's current law (KCL) and Kirchhoff's voltage law (KVL). KCL states that for any node of a lumped electrical network, the algebraic sum of all branch currents leaving the node is zero at any time. KVL states that for any loop in a lumped electrical network, the algebraic sum of the branch voltages around this loop is zero at any time. Thus, terminal voltages of connected terminals have to be equal and, therefore, can also be represented by node voltages, denoted by **u** in the following. The term *node voltages* refers to the voltage differences between the nodes of a network and one arbitrarily picked node called the *reference node*, which is supposed to have zero voltage. Hence,

$$\mathbf{v}(t) - \mathbf{A}^\mathrm{T}\mathbf{u}(t) = \mathbf{0} \tag{4.10}$$

where t is the value of the instantaneous time again and $\mathbf{A}$ is the reduced incidence matrix of the network graph.

The solution of a network is defined as the set of currents and voltages that satisfies Kirchhoff's laws as well as its i–v relation. There are different methods to get to a reduced equation system. The most favorite one is probably to set up an equation system with respect to the rules of a modified nodal analysis [6, 22], which is exemplarily shown in the following. With KCL we have the equilibrium equations

$$\mathbf{A}\,\mathbf{i}(t) = \mathbf{A}\begin{pmatrix} \mathbf{i}_1(t) \\ \mathbf{i}_2(t) \end{pmatrix} = \mathbf{0} \tag{4.11}$$

where $\mathbf{i}_1$ is the vector of currents of those branches with an i–v relation in admittance representation and $\mathbf{i}_2$ is the vector of all the other branch currents. With respect to $\mathbf{i}_1$ and $\mathbf{i}_2$, the i–v relation of an electrical network can generally be written in the form of a nonlinear, first-order DAE system

$$\begin{aligned} \mathbf{i}_1(t) - \mathbf{g}\big(\mathbf{v}(t),\,\dot{\mathbf{v}}(t),\,\mathbf{i}_2(t),\,\dot{\mathbf{i}}_2(t),\,t\big) &= \mathbf{0} \\ \mathbf{h}\big(\mathbf{v}(t),\,\dot{\mathbf{v}}(t),\,\mathbf{i}_2(t),\,\dot{\mathbf{i}}_2(t),\,t\big) &= \mathbf{0} \end{aligned} \tag{4.12}$$

with $\dot{\mathbf{v}} = \dfrac{d\mathbf{v}}{dt}$ and $\dot{\mathbf{i}}_2 = \dfrac{d\mathbf{i}_2}{dt}$.

With Eqs. (4.10)–(4.12), we get the resulting implicit DAE system

$$\mathbf{A}\begin{pmatrix} \mathbf{g}\big(\mathbf{A}^\mathsf{T}\mathbf{u}(t),\,\mathbf{A}^\mathsf{T}\dot{\mathbf{u}}(t),\,\mathbf{i}_2(t),\,\dot{\mathbf{i}}_2(t),\,t\big) \\ \mathbf{i}_2(t) \\ \mathbf{h}\big(\mathbf{A}^\mathsf{T}\mathbf{u}(t),\,\mathbf{A}^\mathsf{T}\dot{\mathbf{u}}(t),\,\mathbf{i}_2(t),\,\dot{\mathbf{i}}_2(t),\,t\big) = \mathbf{0} \end{pmatrix} = \mathbf{0} \tag{4.13}$$

which shortly can also be written as

$$\mathbf{f}\big(\mathbf{x}(t),\,\dot{\mathbf{x}}(t),\,t\big) = \mathbf{0} \tag{4.14}$$

with the vector $\mathbf{x} = (\mathbf{u},\,\mathbf{i}_2)^\mathsf{T}$ containing the system variables.

For more information on this topic, we recommend well-known textbooks, for example, [7, 23–31].

4.2.2
Boolean Equations and Finite State Machines

For digital controls, Boolean equations and deterministic FSMs are common mathematical representations. *State diagrams* (synonym: *state charts*) are graphical representations of FSMs. They describe all possible *states* of objects in which *events* are expected and the *transitions* between these states. The state transitions are triggered by dedicated events. Along with the state transitions, state entries, and/or state exits, related *actions* are executed.

Classically, a deterministic FSM is described by a 6-tuple $(Z, X, Y, f, g,\,\text{and } z_0)$ [32] with

- *state set* Z: a nonempty finite set of vertices, graphically represented by circles and labeled with unique designator symbols inside them,

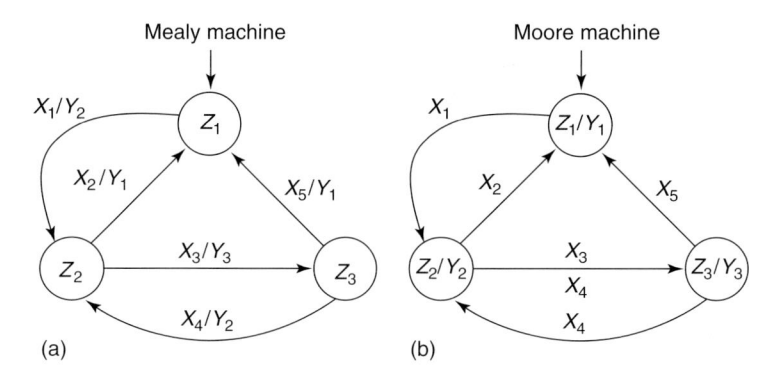

Figure 4.1 (a,b) Finite state machines.

- *input alphabet* X: a nonempty finite set of input symbols,
- *output alphabet* Y: a nonempty finite set of output symbols,
- *transition function* f: a mapping $f: Z \times X \to Z$ describing the state transitions with respect to the input symbols (each transition is graphically represented by an arrow directed from the present state to the next state and labeled by the corresponding input symbol),
- *output function* g: a mapping $g: Z \times X \to Y$ determining the output symbols with respect to the state transition,
- *initial state* z_0: $z_0 \in Z$, graphically represented by an arrow with a dot as origin or no origin pointing to the initial state.

Representatives of classical FSMs are *Mealy machines* and *Moore machines* (Figure 4.1). Both are triggered by inputs denoted on each arrow. For Mealy machines, the output depends on the inputs and the states; for Moore machines, the output depends on the states only. This means that the output function g can be reduced to $g: Z \to Y$ in Moore machines. Therefore, they also differ in the labeling of outputs in the graphical representation. In Mealy machines, the denotation of outputs follows the inputs, separated by a slash "/" from the input symbol. In Moore machines, outputs are denoted in the states, separated by a slash "/" from the state label.

An extension of classical state diagrams are *Harel state charts* and *UML behavioral state machines* (synonym: *UML state charts*) [33–35], which are an object-based variant of Harel state charts. They have the characteristics of both Mealy machines and Moore machines. This is achieved by the denotation of actions at transition arrows as well as in states (Figure 4.2). Beside computing outputs, the actions

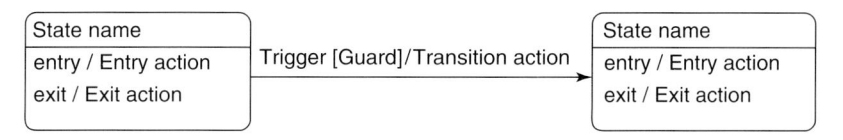

Figure 4.2 State chart annotations.

may also comprise assignments to variables, execution of simple algorithms, function calls, and generation of events. The actions are nonpreemptable and do not consume time. Especially for digital controls, the following kinds of triggers for state transitions can be used: signal triggers, change triggers, and time triggers. Signal triggers give notice of the receipts of signal instances or messages. Change triggers fire when Boolean-valued conditions become true. Time triggers cause state transitions after a virtual timer has expired. For fine-grained controls of the firing of transitions, guards can be set subsequently on triggers. A guard is a Boolean expression that is evaluated whenever the corresponding trigger fires. Only if the guard is true, the state transition proceeds. Otherwise, it is disabled. At a state transition, the actions are executed in the following order: the exit action of the current state at first, then its transition action, and after that the entry action of the next state.

A new concept in Harel state charts and UML state charts are *hierarchically nested states*. The state hierarchy is built of composite states that contain one or more regions, each with a distinct set of states which, in turn, are simple states or composite states itself. For a graphical representation of control flow, state charts may also contain additional pseudostates such as forks, joins, junctions, and choices.

4.3
General Approaches for System-Level Model Description

As seen in the previous section, the application of DAEs is most common in describing MEMS. In this section, approaches of description methods are presented.

From a historical point of view, electrical circuits were investigated at first. Nonelectrical networks were mapped to analogous electrical networks, and macromodels were designed for more complex components. This approach is specific to SPICE, which is presented in the following section. Later on, the fixed set of built-in models in SPICE was extended with the introduction of model description languages such as MAST, VHDL-AMS, Verilog-AMS, SystemC AMS, and Modelica. They allow a domain-specific behavioral description next to the usage of analog and digital signal sets. Now designers could write their own models while powerful domain-specific model libraries were also available.

4.3.1
SPICE and Macromodeling

Berkeley SPICE (Simulation Program with Integrated Circuit Emphasis) [36] is a general-purpose circuit simulator with built-in models (basic elements such as resistors, capacitors, inductors, controlled and independent sources, and semiconductor devices such as diodes, MOSFETs, and BJTs) (Figure 4.3). Instead of implementing additional models, designers can customize the predefined built-in models by a large variety of parameters. The set of predefined SPICE models has become a quasi standard in circuit simulation.

4.3 General Approaches for System-Level Model Description

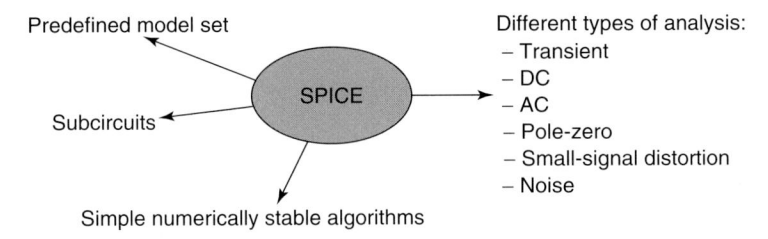

```
Predefined model set              Different types of analysis:
                                   – Transient
                                   – DC
                  SPICE            – AC
Subcircuits                        – Pole-zero
                                   – Small-signal distortion
                                   – Noise
     Simple numerically stable algorithms
```

Figure 4.3 Essential features of SPICE.

The behavior of more complex components (amplifiers for instance) is typically described by macromodels that are composed of existing built-in models or other macromodels as well. Then, these macromodels, called *subcircuits* in SPICE, can be used like any other SPICE model.

Schematics or netlists contain the instances of utilized models, their parameter settings, and how they are interconnected. SPICE netlists are available for a large amount of electrical and electronic devices.

SPICE sets up a system of DAEs to be solved during simulation. The performance of SPICE is high because model equations and solution methods are tightly coupled. Values of element-related parameters are directly written into the Jacobian, for instance.

By describing nonelectrical networks (thermal, mechanical, fluidic, and magnetic) as electrical ones using common analogies [21, 37] in combination with macromodeling, a variety of MEMS components can be simulated. This advantage turns out to be also a disadvantage if systems become too complex for macromodeling and the transformation to electrical terms is no longer intuitive.

4.3.2
Model Description Languages

The fixed set of predefined SPICE models has the advantage that solution methods in SPICE simulators can be tuned to become highly efficient. On the contrary, a disadvantage for the designer is the missing possibility to design behavioral models. This has been overcome by the introduction of model description languages. They provide the designer with a high flexibility to describe model behavior. One of the first such languages was MAST [38], a proprietary language for the Saber simulator. Compared to the SPICE approach, Saber is a representative of next generation circuit simulators and has been developed with the aim to completely separate models from the simulator's internals. However, the language MAST has not become a standard for modeling and it is only used with the Saber simulator.

Besides, other model description languages such as VHDL-AMS, Verilog-AMS, SystemC AMS, and Modelica were developed, which are supported by several simulators and became standards. An overview of the development of modeling languages is given in Figure 4.4. Starting with domain-specific languages in the

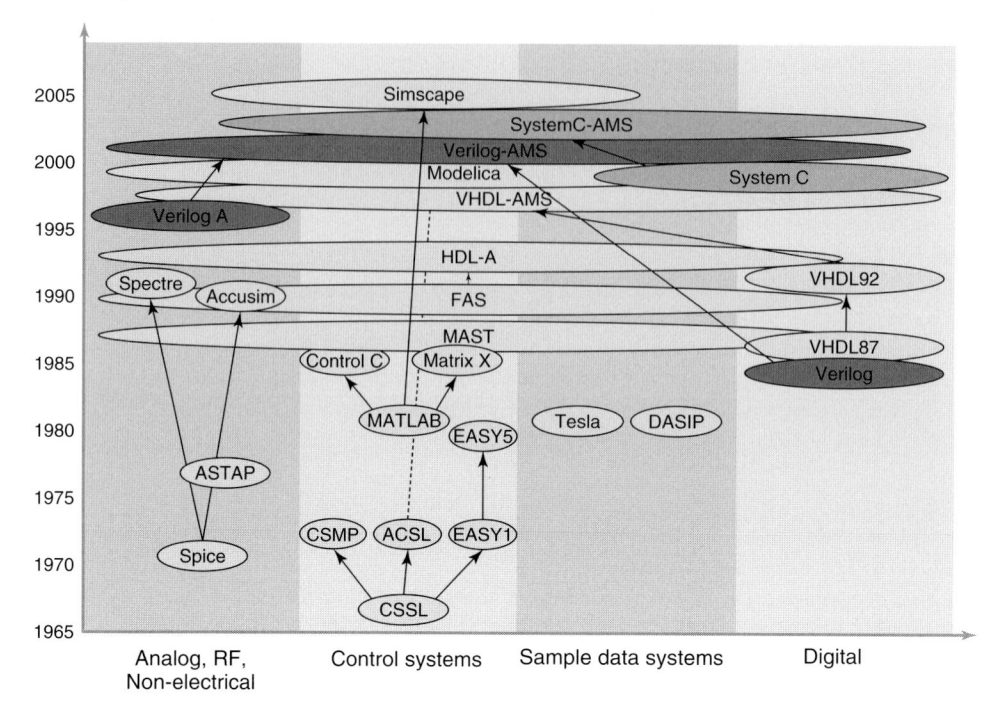

Figure 4.4 Modeling languages.

mid-1960s, languages evolved to comprehensive description means. VHDL-AMS as a representative of advanced model description languages (well suited for MEMS modeling) is described in more detail here.

VHDL-AMS (very high speed integrated circuit hardware description language– analog and mixed-signal extensions) [39] is a powerful and well-established language for the description and simulation of analog, digital, and mixed-signal systems. The language was standardized by IEEE in the VHDL-1076.1 standard. It is an extension to VHDL that is widely used for the design of digital electronics. Analog electrical and nonelectrical systems can be described as well as control systems and sampled data systems among others.

VHDL-AMS is based on the concept of separating the model interface (ENTITY) and the model implementation (ARCHITECTURE). The entity containing the model name describes the connectors of the model (PORT) as well as model parameters (GENERIC). For one entity, different implementations of the model behavior can be specified, which is put into separate architectures. When instantiating a model in a netlist, one of the architectures has to be selected (CONFIGURATION). Models can be combined in packages. Nearly every type of equations (linear, nonlinear, ODE, DAE, and conditional equations (IF ⋯ USE)) can be formulated using quantities and signal values. They are accessed via the ports or are additionally defined as *internal variables* (QUANTITY).

Another advanced widespread model description language is Verilog-AMS [40], which is as powerful as VHDL-AMS. Verilog-AMS is based on the Verilog standard (IEEE 1364). It is suitable for analog, mixed-signal, and integrated circuit design and enables behavioral descriptions as well as structural descriptions of systems and components. It also supports both electrical and nonelectrical system descriptions.

SystemC AMS, a similar powerful model description language with a particular focus on system-level modeling, is an extension to the SystemC standard (IEEE 1666) and separately described in Chapter 15.

The object-oriented language Modelica [41] has been designed for models in any physical domain being most powerful for analog system models, electrical as well as nonelectrical ones.

The above-mentioned description languages are used as input languages for a lot of simulators. A brief overview on simulators suitable for MEMS simulation is given in [42, 43].

4.4
Numerical Methods for System-Level Simulation

4.4.1
Solution of Nonlinear DAEs

The mathematical task of simulating MEMS is generally solving DAEs. Since nonlinear DAEs cannot be solved analytically in most cases, the DAEs are usually solved numerically according to the following basic solution method. It was first utilized in SPICE, and, later on, investigated and implemented with many modifications. An initial value problem is given by a DAE system and an initial condition

$$\mathbf{f}(\mathbf{x}(t), \dot{\mathbf{x}}(t), t) = \mathbf{0}, \quad \mathbf{x}(t_0) = \mathbf{x}_0 \qquad (4.15)$$

which is to be solved for the time-dependent function $\mathbf{x}$ on an interval $[t_0, t_N]$. This interval is a countable union of disjoint intervals $(t_{\ell-1}, t_\ell]$ with $1 \leq \ell \leq N; \ell, N \in \mathbb{N}$. The interval lengths $h_\ell = t_\ell - t_{\ell-1}$ are called *step size* of the ℓth time step. The derivatives $\dot{\mathbf{x}}$ are approximated using an integration formula, the backward Euler formula $\dot{\mathbf{x}}(t_\ell) \approx \frac{1}{h_\ell}(\mathbf{x}_\ell - \mathbf{x}_{\ell-1})$ for instance.[1] This way, only a nonlinear equation system

$$\mathbf{f}\left(\mathbf{x}_\ell, \frac{1}{h_\ell}(\mathbf{x}_\ell - \mathbf{x}_{\ell-1}), t_\ell\right) = \mathbf{0} \qquad (4.16)$$

has to be solved at each point in time $t_\ell, 1 \leq \ell \leq N$. The error $\|\mathbf{x}_\ell - \mathbf{x}(t_\ell)\|$ is called *local truncation error*. The step sizes h_ℓ are determined in such a way that estimates of the local truncation error are smaller than a given tolerance ε_{lte}. For solving Eq. (4.16), methods of the Newton-Raphson type [44] calculating a series $\{\mathbf{x}_\ell^k\}_k$ can be

---

1) Note that due to this Euler discretization, only a numerical approximation $\mathbf{x}_\ell$ of the real solution $\mathbf{x}(t_\ell)$ can be calculated.

applied starting with an initial guess x_ℓ^0, which is usually predicted by an extrapolation formula, for example, the forward Euler formula $x_\ell^0 = x_{\ell-1} + \frac{h_\ell}{h_{\ell-1}}(x_{\ell-1} - x_{\ell-2})$. This method is iterated until a sufficiently accurate value x_ℓ^k is reached. That is if $\left\|x_\ell^k - x_\ell^{k-1}\right\| < \varepsilon_{\text{ite}1}$ and $\left\|f\left(x_\ell^k, \frac{1}{h_\ell}(x_\ell^k - x_{\ell-1}), t_\ell\right)\right\| < \varepsilon_{\text{ite}2}$ hold. If the linear equation system in each Newton-Raphson iteration is solved by Gaussian elimination, sparse matrix techniques can be utilized because the Jacobians for the equation systems of most models are sparse.

Many investigations have been done during the past decades to improve performance, stability, and scalability of this basic method. Important issues were the following.

- **Choice of tolerances**: The numerical errors have to be smaller than the corresponding tolerances ε_{lte}, $\varepsilon_{\text{ite}1}$, and $\varepsilon_{\text{ite}2}$. The number of Newton-Raphson iterations and the number of step size rejections, on the one hand, as well as the accuracy of the solution, on the other hand, depend on these tolerances. Practically, simulations are started with very small tolerances calculating nearly exact solutions. In repeated simulations with larger tolerances, the tolerances are adapted until the solution differs significantly from the nearly exact solution.
- **High-dimensional equation systems**: Especially if electrical circuits are involved, the number of DAEs is often large and the corresponding Jacobian is sparse. Then, sparse matrix techniques, where pivoting is not only based on decreasing the numerical errors but also based on reducing the fill-in, are applied. The fill-in of the Jacobian denotes those additional nonzeros produced by the LU decomposition during Gaussian elimination. It is again a trade-off between accuracy and performance because less fill-in provides a better performance. Another approach for dealing with high-dimensional equation systems are methods called *model order reduction* (see Chapter 3). They calculate an approximation of the system that is smaller by orders of magnitudes. Also, symbolic preprocessing [45, 46] may reduce the number of equations.
- **Stiffness and multiscale systems**: Although it has been proven that it is difficult to formulate a precise definition of stiffness, generally, a DAE system is *stiff* if components with different time constants (e.g., from slow thermal and fast electrical processes) contribute to the same DAE system. The problem is that the fastest process limits the step size of the integration even if rapidly changing parts of the solution have already decayed. Specific implicit integration formulas such as in multistep methods are said to be A-stable. Similar problems are a matter of investigation for multiscale systems, where scale-different granularity is simulated together with regard to both time and length.
- **Homotopy**: The solution of a nonlinear equation system is sometimes hard to find because the Newton-Raphson method only converges if the initial guess is sufficiently close to the solution. This is especially problematic when determining valid initial values while suitable initial conditions are missing. The idea of homotopy methods is to start with a simplified equation and change it gradually into the equation of interest using the known solution of the simpler equation as a starting point for the more complex equation. This way, complicated nonlinear

equations can be solved in many cases. A lot of simulators utilize homotopy methods. In Modelica for instance, the designer can formulate appropriate equations by means of a homotopy operator [47].

- **Index reduction**: An essential mathematical property of the mathematical DAE system (not of the physical system) is the index [3]. The index is, roughly spoken, a measure of how "far away" the DAE is from an explicit ODE where every variable is differentiated. The following simple DAE system $\dot{x}_1(t) + x_2(t) = 0$, $x_1(t) = t$ has index two, for instance. There is exactly one solution: $x_1(t) = t$, $x_2(t) = -1$. The initial values cannot be chosen independently.

There are different index definitions that coincide in the case of linear DAEs with constant coefficients. For nonlinear DAEs, the linearization at a given time instant defines an index that depends on time in general. There are other index definitions as well, for example, the number of differentiations of the given DAE system (or parts of it) necessary to transform the equation into an ODE system. Higher index systems (higher than index one) typically suffer from bad numerical properties [48] because the numerical solution can slide into partial spaces where no solution trajectory is defined. Therefore, it is useful to know the DAE index and to avoid higher index systems as often as possible. There are simulation tools that determine the numerical index. Symbolic methods can be applied if the DAE system is available in a symbolic way. Then beside the calculation of index, the transformation into a lower index DAE system is also possible (e.g., Pandelides' algorithm [49]).

4.4.2
Mixed-Signal Simulation Cycle

In many cases, DAEs with only analog behavior cannot sufficiently describe MEMS. In general, parts with discrete-event behavior are also needed. Values of digital signals in those parts only change at discrete points in time, the so-called events. Digital signals can be switching variables (on, off) as well as real-valued signals of digital controls. However, digital signals may also originate from inequality conditions like $x_1(t) > x_2(t)$. The signal value becomes true at the event when x_1 "starts" to exceed x_2. It becomes false at the event when x_2 starts to exceed x_1. In this example, the value is a Boolean value and it changes only at discrete points in time.

The challenge is the combination of an analog and a digital simulation algorithm if digital signals are connected to analog models. This is covered by simulators such as Saber, AMS Designer, Questa ADMS (formerly known as AdvanceMS), or SystemC AMS. Sometimes, the mixed-signal simulation cycle is defined by the standard of the model description language to make sure that the designer understands the interaction of analog and digital model components. The mixed-signal simulation cycle of VHDL-AMS is exemplarily shown in Figure 4.5. Similar mixed-signal simulation cycles are defined for other modeling languages.

VHDL, which provides a discrete-event simulation cycle, was extended by means for analog behavioral description. Therefore, the discrete-event simulation cycle

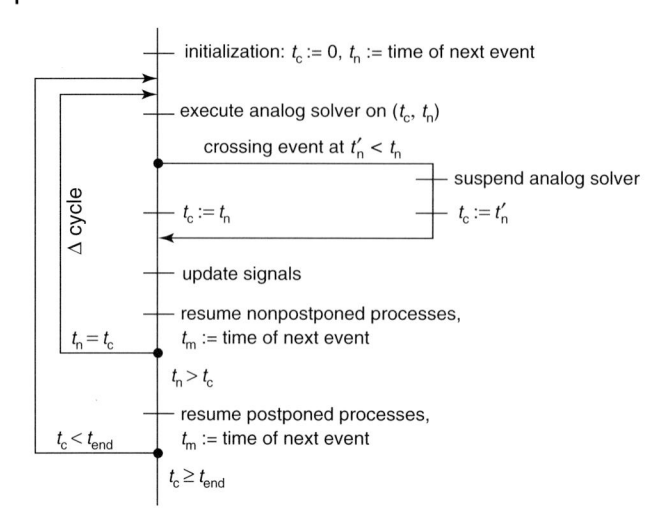

```
                    ├── initialization: t_c := 0, t_n := time of next event
                    │
            ┌──────►├── execute analog solver on (t_c, t_n)
            │       │   crossing event at t'_n < t_n
            │       ●─────────────────────────────┬── suspend analog solver
   Δ cycle  │       ├── t_c := t_n                ├── t_c := t'_n
            │       │◄────────────────────────────┘
            │       ├── update signals
            │       │
            │       ├── resume nonpostponed processes,
      t_n=t_c│      │   t_m := time of next event
            │       │
            │       │ t_n > t_c
            │       ├── resume postponed processes,
      t_c<t_end│    │   t_m := time of next event
            │       │
            │       │ t_c ≥ t_end
```

Figure 4.5 VHDL-AMS mixed-signal simulation cycle.

was extended to an algorithm incorporating the analog solver. In a purely digital event-driven simulation, events occur at discrete points in time that are multiples of a minimal resolution time (MRT, usually 1 ps or 1 fs). If the current time t_c has been reached, the earliest next event occurs at time t_n. The value of t_n is assigned to t_c in the next simulation cycle unless an earlier event is scheduled during the current execution of the processes.

After reaching t_c and determining t_n, the analog solver is invoked in the mixed-signal simulation cycle to calculate the values of the analog signals on $(t_c, t_n]$. A threshold crossing might happen in this interval during simulation, which results in an additional crossing event at $t'_n < t_n$. In such cases, the analog solver stops at t'_n and the scheduler of the digital simulator proceeds at t'_n, which then becomes the value of t_c. This algorithm ensures that the analog solver does not need to reject valid time steps.

Only within so-called delta cycles, the analog solver has to cope with a "zero time" integration on $(t_c, t_c]$. Then, changes of the instantaneous values of the functions forcing analog quantities result in discontinuities that are a challenge for every simulator (Section 4.5.1).

4.4.3
Cosimulation

Since different physical domains closely interact in MEMS, models of different domains have to be coupled. One successful way is to model one system using one simulator and a common multidomain description language. Often, special simulators are highly adapted to the respective physical domain and their very detailed models might be of high value for system simulation. If existing approved models developed for dedicated simulators shall be reused, it might be reasonable

4.4 Numerical Methods for System-Level Simulation

and less time consuming to couple different simulators. Therefore, this section deals with simulator coupling, which is cosimulation with at least two simulators that work independently but communicate during simulation for synchronization and exchange of instantaneous values.

4.4.3.1 Coupling Algorithms

A simulator that can be coupled to other simulators receives data to be considered during its own simulation from other simulators and sends data to other simulators in return. Therefore, a simulator can be regarded as a function V that works on received data $u(t_{com,\ell})$ and calculates data $y(t_{com,\ell}) = V(u(t_{com,\ell}))$ to be returned. The time instant $t_{com,\ell}$ denotes the ℓth communication instant. Within the intervals $(t_{com,\ell}, t_{com,\ell+1})$, that is, between communication instants, normally no communication occurs between simulators.

If two simulators are coupled, the following task

$$y_1 = V_1(y_2)$$
$$y_2 = V_2(y_1) \quad (4.17)$$

has to be solved at every communication instant.

For the coupled problem, different solution methods can be used. Direct analytical solutions are not feasible in general since V_k usually describes complicated, often nonlinear equations. Iterative methods are used instead. The simplest approaches are the Gauss-Jacobi iteration

$$y_1^{k+1} = V_1(y_2^k)$$
$$y_2^{k+1} = V_2(y_1^k) \quad (4.18)$$

and the Gauss-Seidel iteration

$$y_1^{k+1} = V_1(y_2^k)$$
$$y_2^{k+1} = V_2(y_1^{k+1}) \quad (4.19)$$

with $0 \leq k \leq N$; $k, N \in \mathbb{N}$. If y_1 and y_2 are scalar values, the Gauss-Seidel iteration can be illustrated as shown in Figure 4.6(a). Both methods only converge to

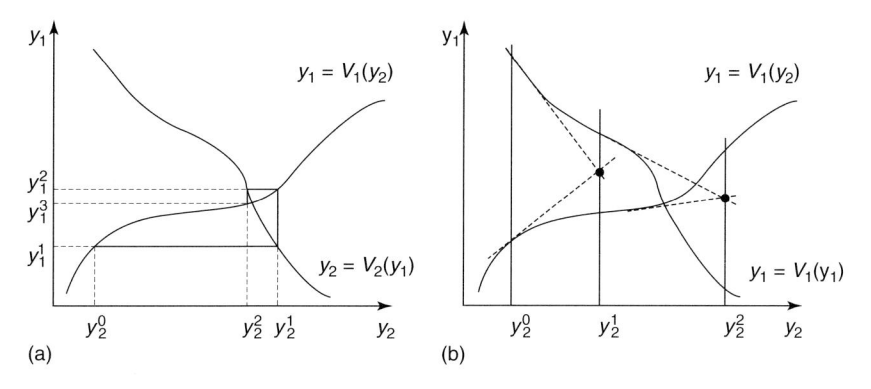

Figure 4.6 Illustrations of (a) Gauss-Seidel iteration and (b) Newton iteration.

a fixed-point solution if the mapping given by the simulators is contracting. If these simple fixed-point iterations do not converge or the convergence is poor, Newton-type methods can be used. These methods are more complex since linearizations are needed for each Newton step. If y_1 and y_2 are again scalar values, the Newton iteration can be illustrated as shown in Figure 4.6(b).

Since the above-mentioned solution methods use the simulators but work on top of them, these algorithms are called *master algorithms*, whereas the simulators are called *slaves*. They are controlled by the master.

For using a simulator in a cosimulation approach, it must have the following properties at least:

- The simulator time generally increases within the interval $[t_0, t_N]$.
- The simulator accepts times $t_{\text{com},\ell} \in [t_0, t_N]$ and can interrupt at $t_{\text{com},\ell}$.
- During interrupted simulation, the simulator can both receive values $u(t_{\text{com},\ell})$ and send values $y(t_{\text{com},\ell})$, and it can receive a new time $t_{\text{com},\ell+1}$.

If the simulator involved in cosimulation is able to reject intermediate solutions to repeatedly simulate the same communication interval, amended master algorithms could be utilized. For better master algorithms, it could be reasonable to exchange the instantaneous values of the time derivatives also.

4.4.3.2 Cosimulation Interface

Simulators are usually prepared differently for cosimulation. Data exchange is often possible via a proprietary interface. Algorithmic aspects (instantiating, initializing, interrupting, and error reporting) are handled individually. Hence, simulator coupling is managed by special task-dependent individual solutions. A standardized functional mock-up interface (FMI) for the coupling of at least two simulators in a cosimulation environment was developed for Modelica to overcome these drawbacks [50]. Once a simulator supports the FMI for cosimulation standard, it can be involved in cosimulation easily. Moreover, master algorithms can be implemented as separate tools (simulation backplane) only having to provide the basic communication interface for involved simulators [51].

The FMI for cosimulation defines both a small set of functions and a slave simulator specific XML file. These functions cover

- the creation and destruction of cosimulation slaves,
- the exchange of data between master and slaves,
- functions for controlling the simulation progress, and
- functions for retrieving status information of the slaves.

To prepare a simulator for cosimulation, this set of functions must be provided.

The slave-specific XML files contain the relevant information like the number of inputs and outputs and information about the slave simulator, which have to be evaluated once before the simulation starts. Capability flags specify the properties of the slave simulator, the ability to reject time steps for instance. These flags enable the master to adapt its algorithm to the cosimulation task by utilizing the task's structure as well as the properties of the slave.

4.5
Emerging Problems and Advanced Simulation Techniques

Models of MEMS are usually developed by partitioning the system and describing the submodels at different levels of abstraction. This approach aims to maintain simulation performance and to enable a thorough and efficient overall system-level simulation of complex application scenarios. As these submodels may belong to the digital or the analog domain, these systems are often called *hybrid* or *mixed-signal systems*.

As already discussed in Sections 4.4.2 and 4.4.3, simulation of mixed-signal systems, that is, the combination of an analog and a digital simulator, is normally done by coupling the various parts of such a system via directed signals. Some of these signals may be considered as input signals with respect to the analog parts, the others as output signals. Value changes of these input signals at discrete points in time, known as *events*, may cause discontinuous behavior. This comprises discontinuous forcing functions as prescribed quantities of independent sources in the analog parts as well as structural changes in its DAE systems, for instance. Some examples of hybrid systems from mechanical and electrical engineering are

- Mechanics: clutches, collision of masses, Coulomb friction, "maximum distance" phenomena (Figure 4.7),
- Electronics: driver stages with pulse width modulation, DC power supplies, circuits in general with switches and relays as well as diodes, thyristors, and transistors (if considered as ideal switches).

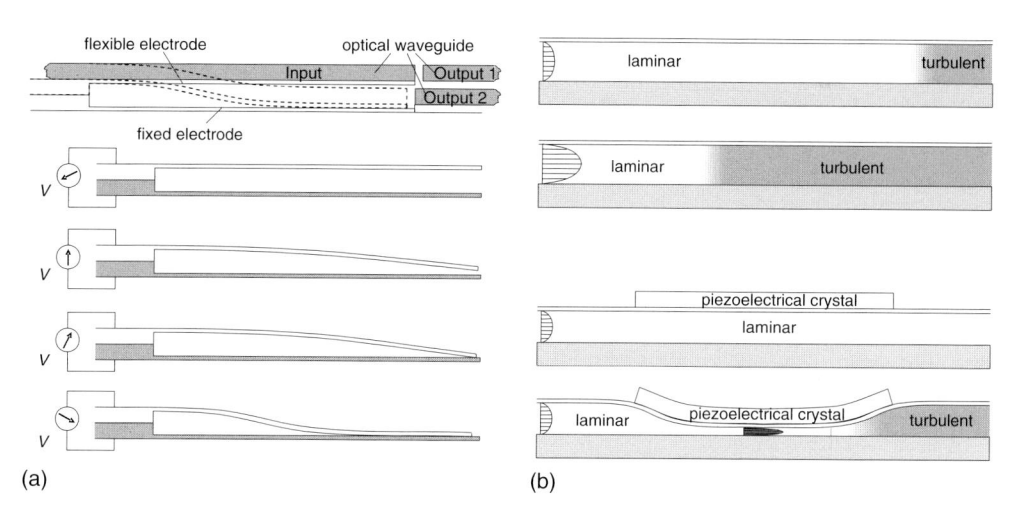

Figure 4.7 Examples for possible changes in model equations in MEMS: (a) pull-in in optical switches and (b) change of valid model paradigm due to flow velocity or geometry variations.

In the following two sections, methods for handling discontinuous forcing functions and structural changes of DAE systems are described separately.

4.5.1
Discontinuous Forcing Functions

Owing to the nature of digital signals, the waveforms of prescribed quantities caused by these signals are expected to be discontinuous. Usually, a common approach of most simulators is to replace the steps in the stimulus functions by steep finite slopes. Then, the resulting waveforms of the stimuli are continuous again. These steep slopes are really challenging for simulation methods. Huge values in the derivatives of the system variables may occur and so may harm both the simulation results and the simulation performance. That is mainly due to the local truncation error, which is increasing and, in turn, is decreasing the step size of the applied integration method.

In addition, iteration methods as described in Section 4.4.3 often cannot be applied as digital signals cannot be reset to past states in most digital simulators. Hence, a time reset mechanism is not feasible. Moreover, during the concept phase of a design process, rise and fall times of the logic gates' output voltages, for instance, are not known. In these cases, discontinuities in input signals cannot be replaced by finite slopes and, thus, are indispensable for a modeling methodology at higher levels of abstraction.

Although the following method can be applied to a wider range of use cases, the probably best way to present the following ideas might be by means of a simple example circuit as shown in Figure 4.8. The DAE system for its network model due to the modified nodal analysis (Section 4.2.1) may be

$$\frac{v_C}{R} + i_C - I_{\text{sat}}\left(e^{\frac{v_S - v_C}{V_T}} - 1\right) = 0 \tag{4.20}$$

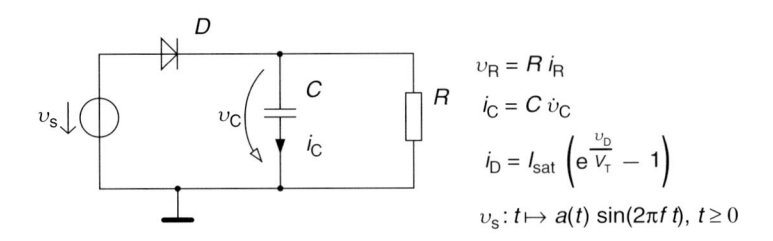

$$C\dot{v}_C - i_C = 0 \tag{4.21}$$

$$v_R = R\, i_R$$
$$i_C = C\, \dot{v}_C$$
$$i_D = I_{\text{sat}}\left(e^{\frac{v_D}{V_T}} - 1\right)$$
$$v_s : t \mapsto a(t)\sin(2\pi f\, t),\ t \geq 0$$

Figure 4.8 Network of a half-wave rectifier with the corresponding constitutive equations.

4.5 Emerging Problems and Advanced Simulation Techniques

where v_C and i_C are the system variables and v_s is the prescribed voltage of the voltage source, a sinusoid with the frequency $f = 50$ Hz and the amplitude a,

$$a(t) = \begin{cases} 10\,\text{V} & \text{if } t \leq 105\,\text{ms} \\ 3\,\text{V} & \text{if } t > 105\,\text{ms} \end{cases}. \tag{4.22}$$

Surely, the system variable i_C can be eliminated by adding Eqs. (4.20) and (4.21), but, as it turns out, it is necessary to keep the differential Eq. (4.21) in addition to Eq. (4.20).

To answer the question which system variables should be treated continuously across the discontinuity in a at $t_s = 105$ ms, let us substitute the time t for a new independent variable τ such that t is itself a function of τ (Figure 4.9), which remains constant on a defined interval $[\tau_{s1}, \tau_{s2}]$, that is,

$$t(\tau) = \begin{cases} \tau - \tau_{s1} + t_s & \text{if } \tau \leq \tau_{s1} \\ t_s & \text{if } \tau_{s1} < \tau \leq \tau_{s2} \\ \tau - \tau_{s2} + t_s & \text{if } \tau > \tau_{s2} \end{cases}. \tag{4.23}$$

So the DAE system (Eqs. (4.20) and (4.21)) becomes

$$\frac{\bar{v}_C}{R} + \bar{i}_C - I_{\text{sat}} \left(e^{\frac{\bar{v}_s - \bar{v}_C}{V_T}} - 1 \right) = 0 \tag{4.24}$$

$$C \bar{v}'_C - \bar{i}_C\, t' = 0 \tag{4.25}$$

where $\bar{v}_C(\tau) = v_C(t(\tau))$, $\bar{v}'_C(\tau) = \dot{v}_C(t(\tau))\, t'(\tau)$, $\bar{i}_C(\tau) = i_C(t(\tau))$, and $\bar{v}_s(\tau) = v_s(t(\tau))$. The waveforms of the sinusoid's amplitudes a (before the substitution) and $\bar{a}$ (after the substitution) are shown in Figure 4.10. Notice that $\bar{a}(\tau)$ cannot be defined by $a(t(\tau))$ on (τ_{s1}, τ_{s2}), but we already know that

$$\bar{a}(\tau) = \begin{cases} 10\,\text{V} & \text{if } \tau \leq \tau_{s1} \\ 10\,\text{V} - \lambda(\tau)\, 7\,\text{V} & \text{if } \tau_{s1} < \tau \leq \tau_{s2} \\ 3\,\text{V} & \text{if } \tau > \tau_{s2} \end{cases} \tag{4.26}$$

with $\lambda : [\tau_{s1}, \tau_{s2}] \to [0, 1]$.

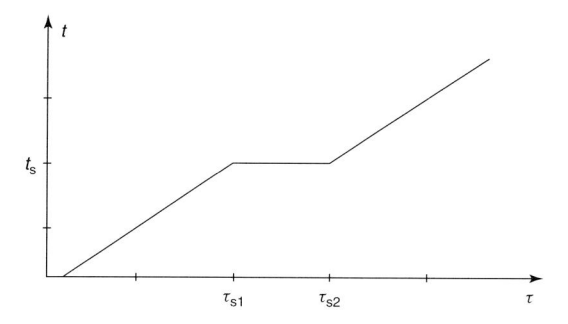

Figure 4.9 Time t as a function of τ in the neighborhood of $t(\tau) = t_s$ (note that $\frac{dt}{d\tau} = 1$ if $\tau \notin [\tau_{s1}, \tau_{s2}]$).

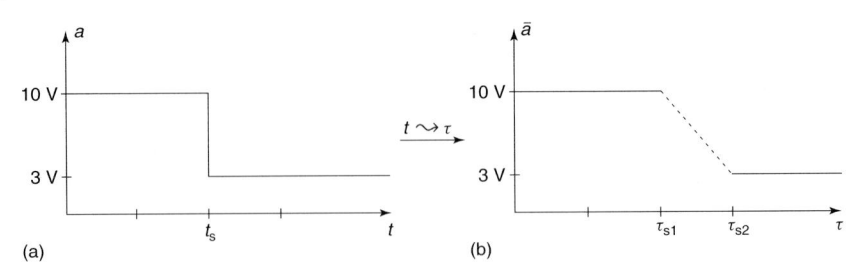

Figure 4.10 Amplitude of the sinusoidal voltage v_s as a function of t (a) and as a function of τ (b).

It is important to realize that charge conservation is preserved at the event of a discontinuity due to Eq. (4.25) because with $t' = 0$ we have $C\bar{v}'_C = \bar{q}' = 0$,[2] and, hence, $\bar{v}'_C(\tau) = 0$ on (τ_{s1}, τ_{s2}). That is why $\bar{v}_C$ stays constant and, thus, v_C is continuous across a discontinuity in v_s, whereas i_C may be discontinuous. Notice that $\bar{i}_C$ is determined by Eq. (4.24) only.

Similar to the method of *source stepping*, assume $\lambda(\tau) := \frac{\tau - \tau_{s1}}{\tau_{s2} - \tau_{s1}}$ across the discontinuity at $t(\tau) = t_s$. Hence, the waveform of $\bar{a}$ is linear on $[\tau_{s1}, \tau_{s2}]$ like it is illustrated by the dotted line in Figure 4.10. So in contrast to v_s, $\bar{v}_s$ is continuous, and even smooth on (τ_{s1}, τ_{s2}). Now the DAE system (Eqs. (4.24) and (4.25)) with $t' = 1$ is solved on $(0, \tau_{s1}]$ using a common integration method (Section 4.4). After that, the DAE system (Eqs. (4.24) and (4.25)) with $t' = 0$ is solved on $(\tau_{s1}, \tau_{s2}]$ using the same integration method beginning with the initial value $(\bar{v}_C(\tau_{s1}), \bar{i}_C(\tau_{s1}))$ until $\lambda = 1$.[3] Its step size control ensures convergence of the applied integration method. So we get $(\bar{v}_C(\tau_{s2}), \bar{i}_C(\tau_{s2}))$ with $\bar{v}_C(\tau_{s2}) = \bar{v}_C(\tau_{s1})$, which becomes the initial value for the integration after the discontinuity when $t' = 1$ again. This procedure is analogously repeated for any other discontinuity.

The results of the complete simulation of the example circuit with an additional discontinuity at $t_s = 305$ ms are shown in Figure 4.11. The calculation of the solution on the complete domain $(0, 350$ ms$]$ is sequentially done like it is described above (resulting waveforms are dash-dotted if $t' = 1$ and solid black if $t' = 0$). There is no difference to solving any other initial value problem because the left-hand limits of the derivatives of the system variables do not equal their right-hand limits. Only the instantaneous values of the system variables are taken as the initial values for the next integration.

4.5.2
Structural Changes in Model Equations

Structural changes in equation systems during simulation are still a subject of research. Such changes are caused either by significant changes in the structure of

---

2) Note that for $t' = 0$ the equation system (Eqs. (4.24) and (4.25)) corresponds to a resistive equivalent network where the capacitor is replaced by an independent voltage source with the prescribed voltage $v_C(t_s)$.
3) Generally, λ is not linear w.r.t. τ, so that λ also needs to be a system variable and a continuation method can be applied to solve the equation system [18, 52].

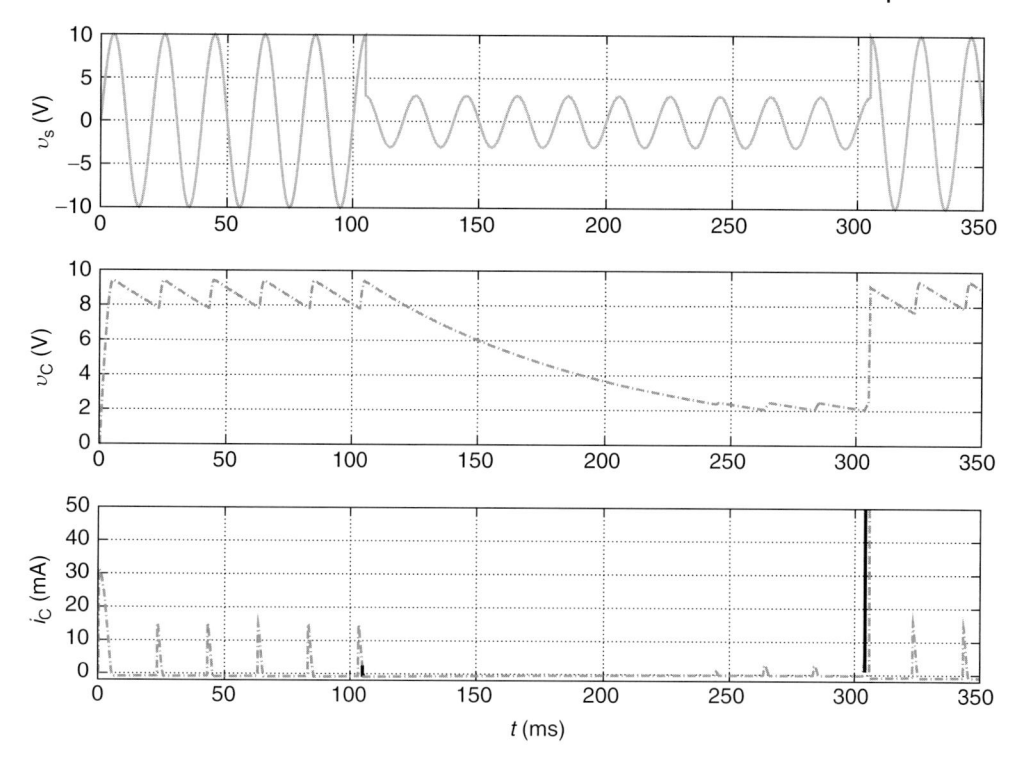

Figure 4.11 Simulation results: prescribed voltage v_s drawn solid, calculated quantities v_C and i_C drawn dash-dotted if $t'=1$ and solid black if $t'=0$.

a system (e.g., breaking of a rigid body being part of a mechanical system) or by the utilization of dedicated modeling techniques (e.g., replacing a spatially lumped model with a spatially distributed model and vice versa).

Models of MEMS at higher levels of abstraction may sometimes cause unilateral constraints when certain physical phenomena are simplified. These unilateral constraints mostly occur in the analog parts, either in the multibody subsystems or in the electrical subsystems. In rare situations, such a phenomenon affects both physical domains [53, 54].

Anyway, unilateral constraints can always be described by complementary conditions. A simple example from mechanics is a 1D body contact problem. The complementary condition can be found between the contact force f (in normal direction) and the distance g between the two bodies. They are called *complementary variables*. The condition can be formulated as $0 \leq f \perp g \geq 0$ or $f \geq 0$, $g \geq 0$, $fg = 0$, which is actually the same. A simple example from electronics is an ideal diode. Here, the complementary variables are the voltage difference v between anode and cathode and the current i through the diode. Analogously, the complementary condition can be formulated as $0 \leq v \perp i \geq 0$ or $v \geq 0$, $i \geq 0$, $vi = 0$.

An indicator for a possible structural change during a simulation is the violation of the currently valid condition. Considering one of the present examples, a zero crossing of the variable not currently diminishing should be used as an indicator. With respect to the contact phenomenon, either the distance g (if both bodies are currently separated) or the contact force f (if both bodies are currently in contact) has to be monitored staying nonnegative. For the ideal diode, the case is quite similar. Either the voltage v across the diode (being currently in blocking mode) or the current i through it (being currently in conducting mode) has to be monitored staying nonnegative.

The event of a zero crossing in an indicator function implies the activation or deactivation of a unilateral constraint. That is why the transient simulation, that is, the numerical integration of the present DAE system, stops since the structure of the system is no longer valid. Now a new valid DAE system has to be set up, which may not be an easy task [55]. In some cases, a-priori knowledge can be used [56]. Updating a state machine repeatedly until all signal values have settled may help in other cases. Another way is to set up and solve a so-called complementarity problem, a linear complementarity problem (LCP) for most tasks [57–60] and a nonlinear complementarity problem (NCP) for spatial contact/friction tasks [61–63]. However, these approaches are only suitable if events are sparse and occasionally. If there is a huge number of events (such as in digital circuits), these approaches tremendoulsy slow down the simulation performance.

A lot of research has still to be done on this subject. For more information, we exemplarily refer to [53, 55, 62, 64, 65] as well as to very good textbooks [60, 66].

4.6
Conclusion

System-level simulation of MEMS based on mathematical descriptions with DAEs is an efficient method in MEMS design. Owing to the heterogeneous structure of MEMS comprising transducers, analog circuits, as well as digital electronics for controls and signal processing, a combination of DAE solvers and discrete-event simulation algorithms is usually needed. Simulator coupling could be an alternative possibility that reduces the effort for a unified modeling approach but may cause additional numerical issues because of synchronization and convergence tests. A comprehensive consideration of physical effects and their modeling on an appropriate level of abstraction yields new challenges for simulation methods. Especially, discontinuities in forcing functions as well as structural changes in the model equations require new approaches for accurate and efficient system-level simulation.

References

1. Braun, M. (1975) *Differentialgleichungen und ihre Anwendungen*, Springer, Berlin.

2. Hartman, P. (1964) *Ordinary Differential Equations*, John Wiley & Sons, Inc., New York.

3. Brenan, K.E., Campbell, S.L., and Petzold, L.R. (1989) *Numerical Solution of Initial-Value Problems in Differential-Algebraic Equations*, North-Holland, New York.
4. Roos, H.-G. and Schwetlik, H. (1999) *Numerische Mathematik: Das Grundwissen für jedermann*, Teubner, Stuttgart.
5. Lenk, A. (1973) *Elektromechanische Systeme*, vol. 3, Verlag Technik, Berlin.
6. Reibiger, A. (2008) Auxiliary Branch Method and Modified Nodal Voltage Equations. *Adv. Radio Sci.*, **6**, 157–163.
7. Belevitch, V. (1968) *Classical Network Theory*, Holden-Day, San Francisco, CA.
8. Maißer, P. (1991) A Differential-Geometric Approach to the Multibody System Dynamics, in *ZAMM – J. Appl. Math. Mech.*, **71** (4), T116–T119.
9. García de Jalón, J. and Bayo, E. (1993) *Kinematic and Dynamic Simulation of Multibody Systems*, Springer, New York.
10. Haug, E.J. (1989) *Computer Aided Kinematics and Dynamics of Mechanical Systems*, Allyn and Bacon, Boston, MA.
11. Nikravesh, P.E. (1988) *Computer Aided Analysis of Mechanical Systems*, Prentice-Hall, Englewood Cliffs, NJ.
12. Roberson, R.E. and Schwertassek, R. (1988) *Dynamics of Multibody Systems*, Springer, New York.
13. Shabana, A.A. (2001) *Dynamics of Multibody Systems*, 2nd edn, John Wiley & Sons, Inc., New York.
14. Beater, P. (2007) *Pneumatic Drives*, Springer, Berlin.
15. Andersen, B.W. (1967) *The Analysis and Design of Pneumatic Systems*, John Wiley & Sons, Inc., New York.
16. Dixon, S.L. (1978) *Fluid Mechanics, Thermodynamics and Turbomachinery*, 3rd edn, Pergamon Press, Oxford.
17. Çengel, A. (1997) *Introduction to Thermodynamics and Heat Transfer*, Irwin/McGraw-Hill, Boston, MA.
18. Haase, J. (1983) *Verfahren zur Beschreibung und Berechnung des Klemmenverhaltens resistiver Netzwerke*, PhD (Dr.-Ing.) thesis, Dresden University of Technology.
19. Reibiger, A. (1985) On the Terminal Behaviour of Networks, in *Proceedings of European Conference on Circuit Theory and Design (ECCTD'85)*, Prague, pp. 224–228.
20. Reibiger, A. (2003) Terminal behaviour of networks, multipoles and multiports, in *Proceedings of 4th Vienna International Conference on Mathematical Modelling (MATHMOD, 2003)*, Vienna.
21. Reibiger, A. (2009) Foundations of network theory, in *Proceedings of International Symposium on Theoretical Electrical Engineering (ISTET'09)*, Lübeck.
22. Ho, C.-W., Ruehli, A.E., and Brennan, P.A. (1975) The Modified Nodal Approach to Network Analysis, in *IEEE Trans. Circ. Syst.*, **22** (6), 504–509.
23. Chua, L.O., Desoer, C.A., and Kuh, E.S. (1987) *Linear and Nonlinear Circuits*, McGraw-Hill, New York.
24. Desoer, C.A. and Kuh, E.S. (1969) *Basic Circuit Theory*, McGraw-Hill, New York.
25. Dorf, R.C. (ed.) (1997) *The Electrical Engineering Handbook*, 3rd edn, CRC Press, Boca Raton, FL.
26. Irwin, J.D. (2011) *Basic Engineering Circuit Analysis*, 10th edn, John Wiley & Sons, Inc., New York.
27. Philippow, E. (1992) *Grundlagen der Elektrotechnik*, 9th edn, Verlag Technik, Berlin.
28. Smith, R.J. and Dorf, R.C. (1991) *Circuits, Devices and Systems – A First Course in Electrical Engineering*, 5th edn, John Wiley & Sons, Inc., New Work.
29. Seshu, S. and Reed, M.B. (1961) *Linear Graphs and Electrical Networks*, Addison-Wesley, Reading, MA.
30. Sudhakar, A. (2006) *Circuits and Networks – Analysis and Synthesis*, McGraw-Hill, New York.
31. Vlach, J. and Singhal, K. (1993) *Computer Methods for Circuit Analysis and Design*, 2nd edn, Springer, New York.
32. Gill, A. (1962) *Introduction to the Theory of Finite-State Machines*, McGraw-Hill, New York.
33. Harel, D. and Politi, M. (1998) *Modeling Reactive Systems with Statecharts, the STATEMATE Approach*, McGraw-Hill, New York.
34. Samek, M. (2008) *Practical UML Statecharts in C/C++, Event-Driven Programming for Embedded Systems*, 2nd edn, Newnes, Burlington, MA.

35. Object Management Group (2009) OMG Unified Modeling Language (OMG UML), Superstructure Specification, version 2.2, http://www.omg.org/spec/UML/2.2/Superstructure. (release date February 2009).
36. Nagel, L.W. and Pederson, D.O. (1975) *Simulation Program with Integrated Circuit Emphasis (SPICE)*, ERL-M520, University of California, Berkeley.
37. Reinschke, K. and Schwarz, P. (1976) *Verfahren zur rechnergestützten Analyse linearer Netzwerke*, Akademie-Verlag, Berlin.
38. Vlach, M. (1990) Modeling and Simulation with Saber, in *Proceedings of 3rd Annual IEEE ASIC Seminar and Exhibition*, Rochester, pp. T11.1–T11.9.
39. Christen, E. and Bakalar, K. (1999) VHDL-AMS – A Hardware Description Language for Analog and Mixed-Signal Applications, in *IEEE Trans. Circ. Syst.-II*, **46** (10), 1263–1272.
40. Kundert, K.S. and Zinke, O. (2004) *The Designer's Guide to Verilog-AMS*, 1st edn, Kluwer Academic Publishers, London.
41. Modelica Association (2012) Modelica – A Unified Object-Oriented Language for Physical Systems Modeling. Language Specification, version 3.2, https://www.modelica.org/documents/. (last update February 2012).
42. Schneider, P. (2010) *Modellierungsmethodik für heterogene Systeme der Mikrosystemtechnik und Mechatronik*, PhD (Dr.-Ing.) thesis. Dresden University of Technology, TUD press, Dresden.
43. Mosterman, P.J. (1999) An overview of hybrid simulation phenomena and their support by simulation packages, in *Hybrid Systems: Computation and Control*, Lectures Notes in Computer Science, **vol. 1569**, (eds F.W. Vaandrager and J.H.V. Schuppen), Springer, Berlin.
44. Ortega, J.M. and Rheinboldt, W.C. (1970) *Iterative Solution of Nonlinear Equations in Several Variables*, Academic Press, New York.
45. Sommer, R., Hennig, E., Dröge, G., and Horneber, E.-H. (1993) Equation-based Symbolic Approximation by Matrix Reduction with Quantitative Error Prediction. *Alta Frequenza - Rivista di Elettronica, 6/93*. **5** (6), 29–37.
46. Broz, J., Clauß, C., Halfmann, T., Lang, P., Martin, R., and Schwarz, P. (2006) Automated symbolic reduction for mechatronical systems, in *Proceedings of the Conference on Computer Aided Control System Design*, Munich, pp. 408–415.
47. Sielemann, M., Casella, F., Otter, M., Clauß, C., Eborn, J., Mattsson, S.E., and Olsson, H. (2011) Robust initialization of differential-algebraic equations using homotopy. *Proceedings of 8th International Modelica Conference*, Dresden.
48. März, R. (1991) Numerical Methods for Differential-Algebraic Equations, in *Acta Numerica*, **1**, 141–198.
49. Pantelides, C.C. (1988) The Consistent Initialization of Differential-Algebraic Systems. in *SIAM J. Sci. Stat. Comput.*, **9**, 213–231.
50. MODELISAR Consortium (2010) Functional Mock-up Interface for Co-Simulation, version 1.0, http://www.functional-mockup-interface.org/. (last update 12 October 2012).
51. Bastian, J., Clauß, C., Wolf, S., and Schneider, P. (2011) Master for Co-simulation using FMI, in *Proceedings of 8th International Modelica Conference*, Dresden.
52. Allgower, E.L. and Georg, K. (1990) *Numerical Continuation Methods: An Introduction*, Series in Computational Mathematics, vol. 13, Springer, Berlin.
53. Enge, O. and Maißer, P. (2005) Modelling Electromechanical Systems with Electrical Switching Components Using the Linear Complementarity Problem, in *J. Multibody Syst. Dyn.*, **13** (4), 421–445.
54. Enge, O. (2005) *Analyse und Synthese elektromechanischer Systeme*, PhD (Dr.-Ing.) thesis. Chemnitz University of Technology, Shaker, Aachen.
55. Enge-Rosenblatt, O., Bastian, J., Clauß, C., and Schwarz, P. (2007) Numerical simulation of continuous systems with structural dynamics, in *Proceedings of 6th EUROSIM Congress on Modelling and Simulation (EUROSIM 2007)*, Ljubljana.
56. Pfeiffer, F. (1991) Dynamical Systems with Time-varying or Unsteady Structure, in *ZAMM – J. Appl. Math. Mech.*, **71** (4), T6–T22.

57. Cottle, R.W., Pang, J.-S., and Stone, R.E. (1992) *The Linear Complementarity Problem*, Academic Press, Boston, MA.
58. Glocker, C. and Pfeiffer, F. (1993) Complementarity Problems in Multibody Systems with Planar Friction, in *Arch. Appl. Mech.*, **63**, 452–463.
59. Murty, K.G. (1988) *Linear Complementarity, Linear and Nonlinear Programming*, Sigma Series in Applied Mathematics, vol. 3, Heldermann, Berlin.
60. Pfeifer, F. and Glocker, C. (1996) *Multibody Dynamics with Unilateral Contacts*, John Wiley & Sons, Inc., New York.
61. Kwak, B.M. (1991) Complementarity Problem Formulation for Three-Dimensional Frictional Contact, in *ASME J. Appl. Mech.*, **58**, 134–140.
62. Glocker, C. (1999) Formulation of Spatial Contact Situations in Rigid Multibody Systems, in *Comput. Methods Appl. Mech. Eng.*, **177** (3– 4), 199–214.
63. Glocker, C. (2001) Spatial Friction as Standard NLCP, in *ZAMM – J. Appl. Math. Mech.*, **81** (S3), 665–666.
64. Lötstedt, P. (1982) Mechanical Systems of Rigid Bodies Subject to Unilateral Constraints, in *SIAM J. Appl. Mech.*, **42** (2), 281–296.
65. van der Schaft, A.J. and Schumacher, J.M. (1998) Complementarity Modelling of Hybrid Systems, in *IEEE Trans. Automat. Control*, **43** (4), 483–490.
66. Acary, V. and Brogliato, B. (2008) *Numerical Methods for Nonsmooth Dynamical Systems – Applications in Mechanics and Electronics*, Lecture Notes in Applied and Computational Mechanics, vol. 35, Springer, Berlin.

Part II
Lumped Element Modeling Method for MEMS Devices

5
System-Level Modeling of Surface Micromachined Beamlike Electrothermal Microactuators

Ren-Gang Li and Qing-An Huang

5.1
Introduction

Microactuators are enabling devices that perform physical functions and interact with the environments by altering their geometries to convert energy into force and motion. For some applications with large mechanical output, low input voltage, and limited size, electrothermal microactuators are the most suitable choice. Thermal actuation in microelectromechanical system (MEMS) has been demonstrated as a compact high-force actuation technique that compliments electrostatic actuation. Joule heating is typically used to power thermal actuators, which generally operate at lower and more desirable voltages than electrostatic ones. Additional advantage of these actuators is their relatively simple fabrication process.

Thermal expansion cannot be directly used to drive other devices because of its small magnitude. In order to operate as an actuator, thermoelastic stresses and strains of compliant structures, such as beams, were used in practice. Guckel *et al.* [1] first adopted the German acronym for Lithographie, Galvansformung, Abformung (LIGA) technology to produce a nickel-based U-shaped flexure actuator that deforms to one side of the structure because of different thermal expansions of each leg of "U." Soon the patterns, processes, and actuation principles of the U-shaped actuator were modified to increase the actuation displacements [2–8] and evoked the birth of a long-short-beam actuator [5], the operation principle of which is similar. Another typical principle of thermal actuation discovered by Que *et al.* [9, 10] was the bent-beam actuation, where the thermal expansions of both beams contribute to the same direction of actuation almost vertical to the direction of expansion. This principle has been used very successfully for strain sensing [11]. The combination of both principles produced actuators with movement in more directions, mostly out-of-plane actuators [12–17]. Each principle is useful in real application. The U-shaped principle is more suitable for building parts that bend flexibly, such as hinges, legs, grippers [18–22]. The bent-beam actuators show excellent performance in producing large and straight forces that push or drag things [23–27], although proper designs may also give them different usages [28–33].

The success of these applications have driven people to model them accurately so as to find out the maximum extent of the force and displacement a microactuator can reach within limited layout area and operation temperature. However, the modeling of a thermal actuator is actually more complicated than it seems to be, as it is a coupled field problem. Each parameter in the mechanical, thermal, and electrical domain has to be included for simulation as well as each type of thermal conduction mechanism. Finite element analysis (FEA) is the only tool at the early stage [34–37]. But it is undoubtedly time consuming, as the same model may be meshed three times in each energy domain (in the worst case where the direct coupling FEA is not applicable) to produce the final results. Indeed, it is not difficult to see that nearly all the existing thermal actuators are made up of one element, the beam. If the classical beam theory is rightly used, the problem would be given an easier and a faster solution. This has motivated Huang et al. [38–41] to set up a pure analytical model for simulation of a surface-micromachined polysilicon thermal actuator. Zhang et al. [42] built a macromodel for cascaded bent-beam actuators. To make the model more useful, the critical material properties were measured in the following efforts [43–45]. These works finally may have inspired a more valuable thinking that thermal actuators are reproducible, that is, they are actually a system consisting of reproducible parts such as beam, anchor, and air gap. Actually, it is also correct to regard all MEMS structures and even the MEMS package in the same way [46] (as the other chapters have revealed). But in the thermal actuator field, the character is more visible – there are only beams. As a result, Li et al. [47, 48] introduced the nodal analysis method for the thermal actuator to start a system-level simulation. The reproducible parts of any actuator are modeled into a package with nodal interfaces and assembled together to form the object required. Since the models are nodal, the simulation method is compatible with that of the conventional integrated circuits (ICs). At last, the only thing left to do is to include hybrid but the same form of (nodal) models of thermal actuators and ICs in a system-level simulator and run the whole simulation in one step. It is clear that this system level of simulation is more promising than the conventional tool, not only in time and memory but also in its hierarchic structure. The following sections begin with such a concept.

This chapter first provides a brief overview and classification on surface micromachined polysilicon electrothermal microactuators. The specific device geometries of several types of commonly used electrothermal microactuators are then summarized. Material properties generally affect the performance of the electrothermal microactuators. Therefore, some basic material properties that are used for the modeling of the electrothermal microactuators are then provided with emphasis on the actuators made of polysilicon. On the basis of a nodal analysis method, the system-level simulation for the electrothermal microactuators is finally presented. Using this method, the electrothermal model could be simulated with the mechanical model as well as electrical components.

5.2
Classification and Problem Description

This section first describes two categories of electrothermal microactuators according to actuator motion type: in-plane actuators and out-of-plane actuators. Their structures and operating principles are discussed individually. The material properties of these actuators are also listed.

5.2.1
In-Plane

5.2.1.1 U-Shaped Actuator

The U-shaped [1] in-plane actuator is sometimes referenced to as a *pseudo bimorph electrothermal actuator* or *basic electrothermal actuator*. A basic layout of the U-shaped actuator is given in Figure 5.1. It is composed of three serialized beams: one hot beam, one flex beam, one cold beam, and two anchors. As voltage is applied across the two anchors of the actuator, current passes through the series of beams, more heat is generated in the narrow beam than that in the wide beam. Since the flexure beam is much shorter and the cold arm beam much wider, the resistances and generated Joule heat are much lower in these beams. As a result, the temperature and thermal expansion in the hot arm beam are larger than those in the cold arm and flexure beams. Since the beams are connected at the free end, they move the free end in an arc trial.

An additional hot arm is added to the U-shaped actuator [6] as shown in Figure 5.2. This additional hot arm provides a new current return path while removing the unnecessary Joule heating in the cold beam and flexure beam. A bidirectional U-shaped electrothermal microactuator can also be designed to provide in-plane bidirectional displacement. Two hot beams are placed at each side of the cold beam. When voltage is applied between the anchor connected with one hot beam and that

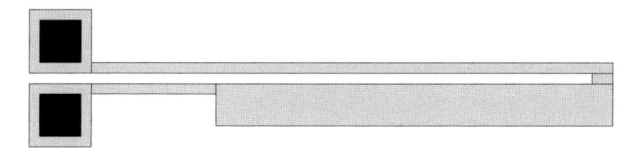

Figure 5.1 Schematic top view of a U-shaped polysilicon thermal microactuator.

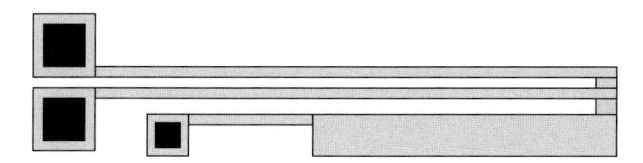

Figure 5.2 Schematic top view of a double hot beam U-shaped polysilicon thermal microactuator.

adjacent to the flexure beam, this hot beam is heated and the beam moves to the opposite direction.

5.2.1.2 Bent-Beam Actuators

Bent-beam or chevron type actuators are different compared to U-shaped actuators, in that they do not rely on segments with different thermal expansion. Instead, the bent-beam actuator operates on local thermal expansion of a beam situated between two anchor points, as shown in Figure 5.3. As both ends of the beam are fixed, compressive stress exerts within the beam and, given an initially defined chevron shape, the apex moves in plane. This design offers rectilinear displacements that are normal to the direction connected the two anchors, an improvement to the curved response of the U-shaped actuator.

An array of bent-beam actuators were designed with a common work coupling beam [32], which efficiently multiplies the actuator force while conserving the displacement, illustrated in Figure 5.4. While studying actuation versus angle in this array, it was noted that out-of-plane buckling may occur at low angles.

5.2.1.3 Long-Short Beam Actuator

Similar to U-shaped thermal actuator, this category of actuator uses the asymmetry of the thermal expansion [5]. As illustrated in Figure 5.5, current passes through the long and the short beams when voltage is applied across the two fixed anchors

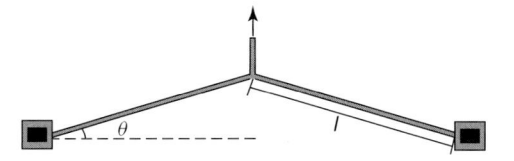

Figure 5.3 Schematic top view of a bent-beam thermal actuator.

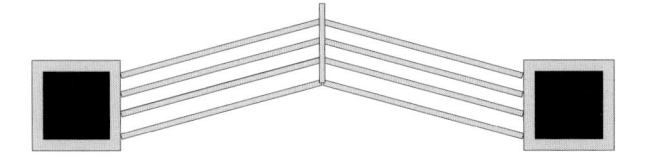

Figure 5.4 Schematic top view of an arrayed bent-beam thermal actuator.

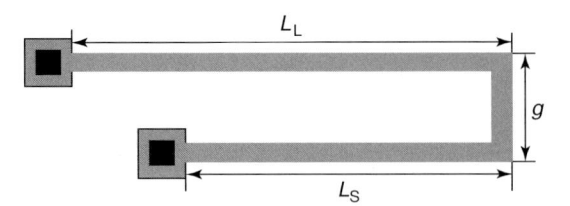

Figure 5.5 Schematic top view of a long-short-beam thermal microactuator.

of the actuator. Owing to the same intersectional area and the same material properties, the long beam and the short beam generate identical heat per unit length and therefore the same thermal expansion. However, for the whole beam, the long one expands more than the short one, which pulls the tip of the actuator move approximately normal to the two beam.

5.2.2
Out of Plane

An out-of-plane actuator [16, 17] is composed of several beams in two different polysilicon layers as shown in Figure 5.6. Anchor 3 and the thin arm are in the lower layer, poly 1. Anchors 1 and 2, the flexure arm, the wide arm, and the connecting arm are in the upper layer, poly 2. The actuator has two symmetric parts along the thin arm to guarantee that there is only a displacement perpendicular to the substrate. When a voltage difference is applied across anchor 1 and anchor 3 while keeping anchor 1 and anchor 2 at the same voltage, electric current passes through the flexure arm, wide arm, connecting arm, and thin arm and produces Joule heat. The thin arm produces more heat than the other arms because of its high resistance. So it expands more than the wide arm, which produces an additional moment introduced by the different-layer connection. It is this moment that actuates the actuator moving out of plane. The flexure arm is designed to reduce its rotary inertia, which intends to improve the stroke of the tip of this actuator.

5.2.3
Material Properties

Some basic material properties that are used for the modeling and analysis of thermal actuators are provided, with emphasis on actuators made of polysilicon.

5.2.3.1 Thermal Conductivity
Thermal conductivity is a critical material property of thermal devices. In most cases, crystalline materials exhibit anisotropy and are of higher conductivity than amorphous materials [49]. The effect of temperature on crystalline materials can

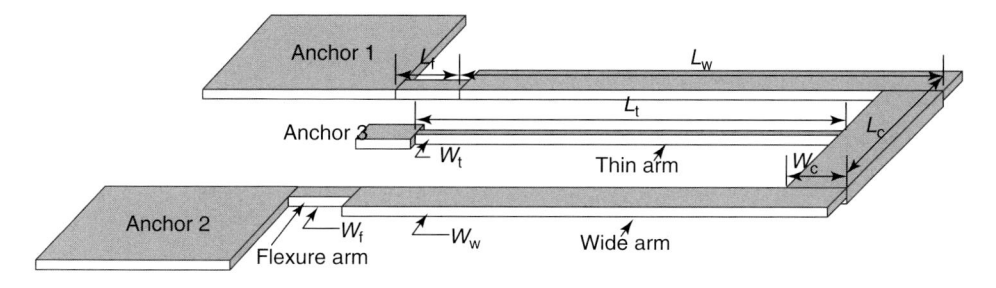

Figure 5.6 Schematic view of a two-layer out-of-plane electrothermal microactuator.

be quite significant. The temperature also affects the properties of polysilicon, but the fabrication process links to the variation of thermal conductivity more closely. There can be a considerable range in these properties because of the variation in grain size for different fabrication processes [50]. Studies have found that constant thermal conductivity between 29 and 34 W (mK^{-1}) could be used for many polysilicon models [18]. A value of 150 W (mK^{-1}) is often used as a constant value of thermal conductivity of single crystalline silicon.

Most electrothermal actuators operate in an environment of air or similar fluid where thermal conductivity has a large influence on the performance of the devices. Thermal conductivity of air is well understood and can be found in many articles [49, 51]. When temperature ranges from room temperature to 1100 K, a linear fit to tabular values in [49] provides a reasonable approximation.

5.2.3.2 Specific Heat

For transient thermal analysis, the specific heat is required. It is reasonable to consider the specific heat of polysilicon as similar to that of single crystal silicon, which has been experimentally verified at room temperature [52]. Isolated measurement of polysilicon silicon has suggested that a practical error bound of 5% on the silicon specific heat is acceptable for polysilicon [53]. A constant value 705 J kg^{-1} K^{-1} can be used for specific heat of polysilicon as an approximation.

5.2.3.3 Resistivity

The resistivity of polysilicon depends on its dopant material and concentrations. For electrothermal devices in polysilicon, the resistivity at room temperature is well studied in the IC industry. However, resistivity is less understood at higher temperatures.

The resistivity (ρ) is often modeled as a linear function of material temperature using a temperature coefficient of resistance (ξ) at relatively lower temperature. This relationship can be written as

$$\rho(T) = \rho_0 [1 + \xi (T - T_0)] \tag{5.1}$$

where T is the temperature; T_0, the reference temperature and ρ_0, the resistivity at T_0. This equation has been used by a number of researchers to model the temperature dependence of resistivity of electrothermal actuators. At higher temperature, the grain size of the polysilicon may vary and rediffusion may occur, which affects the resistivity.

5.2.3.4 Coefficient of Thermal Expansion

The coefficient of thermal expansion is a critical parameter for the coupled thermomechanical analysis of thermal actuators. This coefficient is a function of temperature. A curve fit of empirical data for silicon is given by Okada and Tokumaru [54]

$$\alpha(T) = 3.725 \times 10^{-6} \left[1 - e^{-5.88 \times 10^{-3}(T-125)} + 5.548 \times 10^{-4} T \right] \tag{5.2}$$

where T is the absolute temperature that ranges from 120 to 1500 K. This expression is for high-purity silicon, but has been used by many researchers to model the coefficient of thermal expansion of polysilicon [55]. A near room temperature value of $\alpha = 2.7 \times 10^{-6}(K^{-1})$ is sometimes approximately used as a constant value.

By far, the structures, operation mechanisms, material properties of typical thermal actuators have been described explicitly. For accurate simulation of each actuator, at least two coupled field models should be established, that is, the electrothermal and thermomechanical models, and the thermal conduction mechanisms that play the dominant role need to be considered properly. Fortunately, for actuators of beamlike structures, the modeling can start just with a simple beam.

5.3
Modeling

The electrothermal microactuator is actuated by the thermal expansion caused by Ohmic heating, which is produced by the electric current passing through each beam. This section first gives the heat-transfer equation followed by the electrothermal model and finally the thermomechanical model. In order to perform a system-level simulation for the thermal actuator, a high-level simulation method should be used. As described earlier, since the beam element is simple and fundamental among the thermal actuators, a nodal analysis method is an optimal choice. In the following analysis, only the nodal quantities of a beam are used to set up the models so that they are highly reproducible for constituting complex (structure) system by interconnection through nodes.

5.3.1
Electrothermal Model of a Beam

The behavior of heat transfer may be governed by the Fourier equation

$$\rho c \frac{\partial T(\vec{r}, t)}{\partial t} - \nabla \left[\kappa \nabla T(\vec{r}, t) \right] = g(\vec{r}, t) \ \vec{r} \in R, t > 0 \tag{5.3}$$

where $T(\vec{r}, t)$ is the temperature distribution, $g(\vec{r}, t)$ is the heat generation per unit volume at position $\vec{r}$ and time t, ρ is the density, c is the specific heat, κ is the thermal conductivity, and R is the region of the MEMS structure. The boundary conditions are applied at the boundary of the region, R.

A model in nodal analysis should be in the format of ordinary differential equations (ODEs) or arithmetic equations, while Eq. (5.3) is a partial differential equation (PDE). In order to model the beam as shown in Figure 5.7 using the nodal analysis method, Eq. (5.3) should be simplified using the following assumptions.

The heat loss by the radiation is negligible when the electrothermal microactuators operate normally. It shows that less than 1% of the heat is lost to the surroundings by radiation even at high temperature (1000 °C) [38].

The temperature distributes in one dimension in the beam elements. It also means that the temperature at any location along the cross-section of the beam is uniform. This is proper if the Biot number, B_i, is less than 0.1. The Biot number is identified as $B_i = (\kappa_{air}/\kappa)(w/t_{air})$, where w and κ are the width and the thermal conductivity of the beam, respectively. κ_{air} is the thermal conductivity of the air layer between the beam and the silicon substrate, and t_{air} is the gap height between the beam and the substrate. For a beam with $\kappa_{air} = 0.026 \text{W m}^{-1}\text{K}^{-1}$ $\kappa = 131 \text{W m}^{-1}\text{K}^{-1}$ $w = 2\,\mu\text{m}$, and $t_{air} = 2\,\mu\text{m}$, the calculated Biot number is 0.0002. Therefore, the temperature gradients along the cross-section are very small in the most part except for the joint.

The ambient temperature, which is identical with the temperature of the substrate, is assumed to be constant. When the temperature is not too high along the actuator, this assumption is reasonable.

The anchor of the actuator is assumed to be the ideal heat sink. If the power applied to the anchor is in a proper range, the shift of the temperature of the anchor is not remarkable and could be ignored. This assumption proven by the finite element method (FEM) analysis and experiments is commonly taken in most of the published analysis.

On the basis of these assumptions, the governing equation in the beam element could be simplified significantly as a 1D equation by taking the coordinates shown in Figure 5.7. The dimensions of the beam are labeled in Figure 5.7.

$$\rho c w b \frac{\partial T_b(x,t)}{\partial t} - wb \frac{\partial}{\partial x}\left[\kappa \frac{\partial T_b(x,t)}{\partial x}\right] = -\left(hw + \frac{wS}{R_T}\right)[T_b(x,t) - T_\infty] + \frac{i^2(t)\rho_e}{wb} \tag{5.4}$$

where $T_b(x,t)$ is the temperature distribution of the beam, T_∞ is the temperature of the substrate, κ is the thermal conductivity of the beam, h is the convection coefficients, ρ_e is the electrical resistivity of the material of the beam, $i(t)$ is the electric current through the beam, and S is the shape factor that accounts for the impact of the shape of the element on heat conduction to the substrate. R_T would be the thermal resistance between the beam and the substrate if the beam is wide enough. For example, in the PolyMUMPs process, R_T is calculated as in Eq. (5.5) by assuming that the substrate is kept at the ambient temperature and ignoring

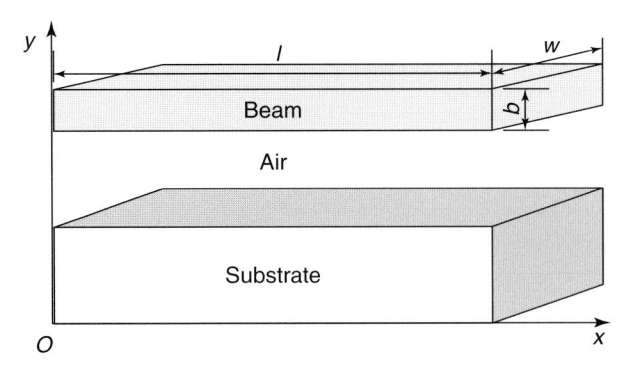

Figure 5.7 Schematic view of a beam.

the heat resistance of the layer t_{air} of silicon nitride because the thermal resistance of the air layer is much larger than that of the silicon nitride.

$$R_T = \frac{t_{air}}{\kappa_{air}} \tag{5.5}$$

where κ_{air} and t_{air} are the thermal conductivity and thickness of the air gap, respectively. The following equation gives the shape factor for heat conduction [38].

$$S = \frac{b}{w}\left(\frac{2t_{air}}{b}+1\right)+1 \tag{5.6}$$

In order to simplify the analysis, a relative temperature distribution $T(x,t)$ is used.

$$T(x,t) = T_b(x,t) - T_\infty \tag{5.7}$$

To consider the dependence of the material properties on the temperature, the thermal conductivity, heat convection coefficient, and electrical resistivity are assumed to be linear functions of the temperature, as shown in Eq. (5.8). And the thermal conductivity of the air is also assumed as a linear function of temperature. This is a trade-off between the accuracy and the complexity of the model.

$$\kappa = \kappa_0 + c_\kappa T, \ h = h_0 + c_h T, \ \rho_e = \rho_{e0}(1+\zeta T), \ \kappa_{air} = \kappa_{air0} + c_{\kappa\,air} T \tag{5.8}$$

The governing equation could be written as

$$\rho c w b \frac{\partial T(x,t)}{\partial t} - w b \frac{\partial}{\partial x}\left(\kappa \frac{\partial T(x,t)}{\partial x}\right) = \\ -\left[hw + \left(2+\frac{b+w}{t_{air}}\right)\kappa_{air} - i^2\frac{\rho_{e0}\zeta}{wb}\right]T(x,t) + i^2\frac{\rho_{e0}}{wb} \tag{5.9}$$

The boundary conditions of this equation are

$$\begin{cases} T|_{x=0} = T_1(t), & \kappa w b \dfrac{\partial T}{\partial x}\bigg|_{x=0} = q_1(t) = \kappa w b \eta_1(t) \\ T|_{x=l} = T_2(t), & -\kappa w b \dfrac{\partial T}{\partial x}\bigg|_{x=l} = q_2(t) = -\kappa w b \eta_2(t) \end{cases} \tag{5.10}$$

where $q_1(t)$ and $q_2(t)$ are the heat flux out of each end of the beam and $\eta_1(t)$ and $\eta_2(t)$ are the temperature gradients.

A trial solution is raised, which can meet the above boundary conditions in order to solve the equation. In the electrothermal microactuator, the temperature of most part changes smoothly because of the uniform heat source distribution and relatively high thermal conductivity. In other words, the temperatures of adjacent parts in the actuator correlate closely. Taking advantage of the close correlation of the temperature along the beam, the computing scale of the model could be significantly reduced. After performing second-order Hermit polynomial fitting with the temperature of the two nodes at each end of the beam and their derivatives, the temperature along the beam could be expressed as below by offsetting the temperature residual using Fourier transformation with proper orders. The order is determined as 3, which is optimized with numerical simulations.

$$T = N\phi \tag{5.11}$$

where

$$\begin{cases} N = \left[\dfrac{2x^3}{l^3} - \dfrac{3x^2}{l^2} + 1, -\dfrac{2x^3}{l^3} + \dfrac{3x^2}{l^2}, \dfrac{x^3}{l^2} - \dfrac{2x^2}{l} + x, \dfrac{x^3}{l^2} - \dfrac{x^2}{l}, 1 - \cos\left(\dfrac{2\pi x}{l}\right)\right] \\ \phi = [T_1\ T_2\ \eta_1\ \eta_2\ \beta]^T \end{cases}$$

(5.12)

β is a function of time to be determined. Substituting Eq. (5.12) in Eq. (5.9), and using the weighted residual method and the Galerkin method, Eq. (5.9) may be converted into a series of ODEs by which the nodal analysis model may be built, as shown in Eq (5.13).

$$\overline{M}\phi' + P(\phi) + \overline{K}\phi - i^2 G\phi + Q(\phi) = \overline{F} \qquad (5.13)$$

where

$$\begin{cases} \overline{M} = \rho cwb \displaystyle\int_0^l N^T N dx & \text{(a)} \\ P(\phi) = \kappa_1 wb \displaystyle\int_0^l \left(N_x^T(T) N_x \phi\right) dx + h_1 w \displaystyle\int_0^l \left(N^T(T) N\phi\right) dx & \text{(b)} \\ \overline{K} = \kappa_0 wb \displaystyle\int_0^l \left(N_x^T N_x\right) dx + h_0 w \displaystyle\int_0^l N^T N dx & \text{(c)} \\ \overline{F} = i^2 \dfrac{\rho_{e0}}{wb} \displaystyle\int_0^l N^T dx - q_1 N^T|_{x=0} - q_2 N^T|_{x=l} & \text{(d)} \\ Q(\phi) = \displaystyle\int_0^l N^T\left[\left(2 + \dfrac{b+w}{t_{air}}\right)\kappa_{air}\right] N\phi dx & \text{(e)} \\ G = \dfrac{\rho_{e0}\zeta}{wb} \displaystyle\int_0^l N^T N dx & \text{(f)} \end{cases}$$

(5.14)

Both t_{air} and κ_{air} are the functions of the coordinates along the beam, so $Q(\phi)$ is difficult to be written in an explicit formulation. However, it is necessary to make all the formulation in explicit format to build the nodal analysis model. So t_{air} takes the mean value along the beam in this analysis. The accuracy of this assumption is reasonable and verified by the available data. Equation (5.13) is a group of ODEs. The PDE in Eq. (5.3) is transformed into ODEs, which may be utilized in the nodal analysis method and simulated with any nonlinear ODE solver. The electrical resistance is temperature dependent due to the nonconstant electrical resistivity. It could be expressed as

$$R = \int_0^l \dfrac{(1+\zeta T)\rho_{e0}}{wb} dx = \dfrac{\rho_{e0} l}{wb} + \dfrac{\zeta \rho_{e0} l}{wb}\left[\dfrac{T_1+T_2}{2} + \dfrac{l(\eta_1-\eta_2)}{12} + \beta\right] \qquad (5.15)$$

5.3.2
Thermomechanical Model of the Beam

The effect of the thermal expansion along the axis in the beam may be equivalent to a pair of counterbalance along the axis applied at the two nodes of the beam under the assumption of small deflection [56].

$$F_{eqv} = \frac{Ewb}{l}\left(\int_0^l \varepsilon_t dx - \Delta l\right) \quad (5.16)$$

where E is the Young modulus and ε_t is the thermal strain in the x-direction. It's a function of the temperature-dependent thermal expansion coefficient. And the thermal expansion coefficient is generally a function of the temperature. In this analysis, the thermal expansion coefficient is assumed to be a function of the temperature as shown in Eq. (5.17).

$$\varepsilon_t = \alpha_m T + \beta_m T^2 \quad (5.17)$$

Δl is the change of the effective length of the beam when it is bent. Under small deflection assumption, the change in the beam length because of the deflection is traditionally ignored. But in the electrothermal microactuator, the thermal strain may be released by this change, which could significantly influence the behavior of the actuator.

$$\Delta l = l_{bent} - l_{original} \approx \int_0^l \frac{1}{2}\left[\left(\frac{dv}{dx}\right)^2 + \left(\frac{dw}{dx}\right)^2\right] dx \quad (5.18)$$

where w and v are the deflections parallel and perpendicular to the substrate, respectively.

The axial stretching force N_s could be expressed as

$$N_s = \frac{Ewb}{l}\left(\Delta x + \Delta l - \int_0^l \varepsilon_t dx\right) \quad (5.19)$$

where Δx is the relative axial displacement of the two nodes

The mean gap shift used in the electrothermal model mentioned earlier is a function of the nodal displacement, as shown in Eq. (5.20). It is an integration of the shape function of the beam.

$$\Delta g_{mean} = \frac{1}{2}(w_1 + w_2) - \frac{l}{12}(\varphi_{y1} + \varphi_{y2}) \quad (5.20)$$

where w_1 and w_2 are the deflection perpendicular to the substrate of the two nodes of the beam, respectively. φ_{y1} and φ_{y2} are the rotary angle about y-axis of the two nodes, respectively. The symbol t_{air} could be expressed as follows:

$$t_{air} = t_{air,org} - \Delta g_{mean} \quad (5.21)$$

where $t_{air,org}$ is the original thickness of the air layer, or the lift of the polysilicon beam.

In the out-of-plane electrothermal microactuator, the actuating motion comes out of the different-layer joint of beams, which introduces an additional moment from axial forces. So this kind of joint plays a key role in these actuators. Here, this joint is abstracted out as an independent component by ignoring the overlap of the beams to model it individually. So there is no mass of this component. The relation between the forces and moments applied at this joint, as shown in Figure 5.8, is

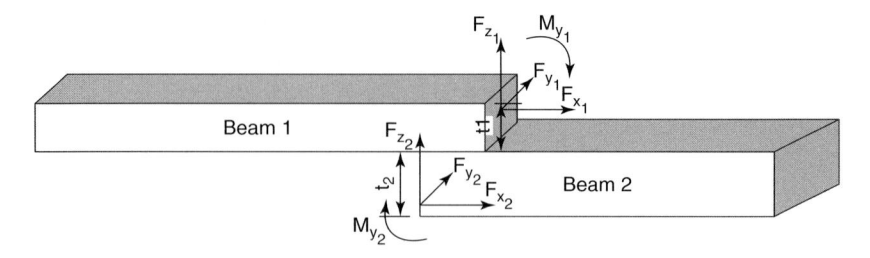

Figure 5.8 Cross-sectional view of a different-layer joint.

listed below. Although the overlapped part is ignored, its accuracy is acceptable in the simulation because the overlapped area is small compared to the length of the beam.

$$\begin{cases} F_{x_1} + F_{x_2} = 0 \\ F_{y_1} + F_{y_2} = 0 \\ F_{z_1} + F_{z_2} = 0 \\ M_{y_1} + M_{y_2} + F_{x_1}\left(\dfrac{t_1 + t_2}{2}\right) = 0 \end{cases} \quad (5.22)$$

The transformation from the global coordinate system (die coordinate system) to the local coordinate system (beam coordinate system) is also important because it may couple the pure bending and the torsion of different beams in the actuator. For the displacement vector, $\begin{bmatrix} u & v & w & \varphi_x & \varphi_y & \varphi_z \end{bmatrix}^T$, including three translations and three rotations, and the force vector, $\begin{bmatrix} F_x & F_y & F_z & M_x & M_y & M_z \end{bmatrix}^T$, including three forces and three moments of a node, the transformation matrix is written as follows:

$$\begin{bmatrix} \cos\varphi & \sin\varphi & 0 & 0 & 0 & 0 \\ -\sin\varphi & \cos\varphi & 0 & 0 & 0 & 0 \\ 0 & 0 & 1 & 0 & 0 & 0 \\ 0 & 0 & 0 & \cos\varphi & \sin\varphi & 0 \\ 0 & 0 & 0 & -\sin\varphi & \cos\varphi & 0 \\ 0 & 0 & 0 & 0 & 0 & 1 \end{bmatrix} \quad (5.23)$$

where φ is the rotation of the local coordinate system relative to the global one in the area parallel to the substrate.

5.4 Solving

The next step is to properly solve the nodal models. Since the nodal models are explicitly analytical, it is quite nature to regard them as circuit models, which is also explicitly analytical and nodal, and uses a circuit simulator as the solver. This concept is well known as the *equivalent circuit method* that solves the noncircuit problem in a circuital way.

To take it more clearly, for each degree of freedom in one energy domain, there is usually and only a pair of nodal quantities at one node, the "across" quantity and the "through" quantity (see also Chapter 2). The former quantity is mostly featured as something that can be set across an element, such as the voltage in electrical domain, translational displacement in mechanical domain, and temperature in thermal domain. The latter quantity is featured as something that flows through an element, such as the current in electrical domain, force in mechanical domain, and heat flux in thermal domain. If we are speaking a noncircuit nodal model as an equivalent circuit model, the across quantities at a node are regarded as equivalent electrical currents and the through quantities all as the equivalent voltages at the circuit terminals. From such a perspective, Kirchhoff's current law (KCL) and Kirchhoff's voltage law (KVL) that dominate the circuit problem also work well here, that is, all the through quantities at one node sum up to zero, all the across quantities along one loop sum up to zero, and of course, the across quantities of different elements at one interconnected node value the same. Generally, for the electro-thermo-mechanical beam model, there are totally 7 pairs, 14 nodal quantities at one node, where 6 pairs correspond to the displacements and forces in the mechanical domain and 1 pair corresponds to temperature and heat flux in the thermal domain. Thus, the analytical models in the thermal and mechanical domains are transformed into a circuit network that contains seven circuit nodes. Solving the network in a circuit simulator works out the whole problem. However, since there are couplings and unreasonable nonlinearities in the parameters, the real problem is always more complex than this conceptual model. This point is discussed further in the following section.

5.4.1
Equivalent Circuit of a Coupled Electrothermal Model

SPICE is often used as circuit simulator. The equivalent circuit of a coupled electrothermal model for the beam is derived from Eq. (5.13). The method of the conversion from the linear ODE to the equivalent circuit representation is well studied in the past [57]. In this equation, all the linear items are represented with passive circuit components, such as capacitors and resistors. The nonlinear items in the equation are represented by some nonlinear controlled sources in SPICE.

The equivalent circuit of the coupled electrothermal model for the beam is illustrated in Figure 5.9. Although the concept of the equivalent circuit method is clear, one may still find it difficult to visualize the correlation of Eq. (5.13) to this figure. In fact, to interpret a time-variant differential equation in a circuital way, more manipulation of the equation is still needed. One can refer to literatures for details [57]. The resistance of a resistor and the capacitance of a capacitor refer to the corresponding item in matrix $\overline{K}$ and $\overline{M}$ in Eq. (5.14), respectively. Other matrices and functions, both linear and nonlinear ones, are all represented with nonlinear controlled sources. The controllers of this controlled source are elements in the vector ϕ and the electric currents through the beam. Function $Q(\phi)$ is

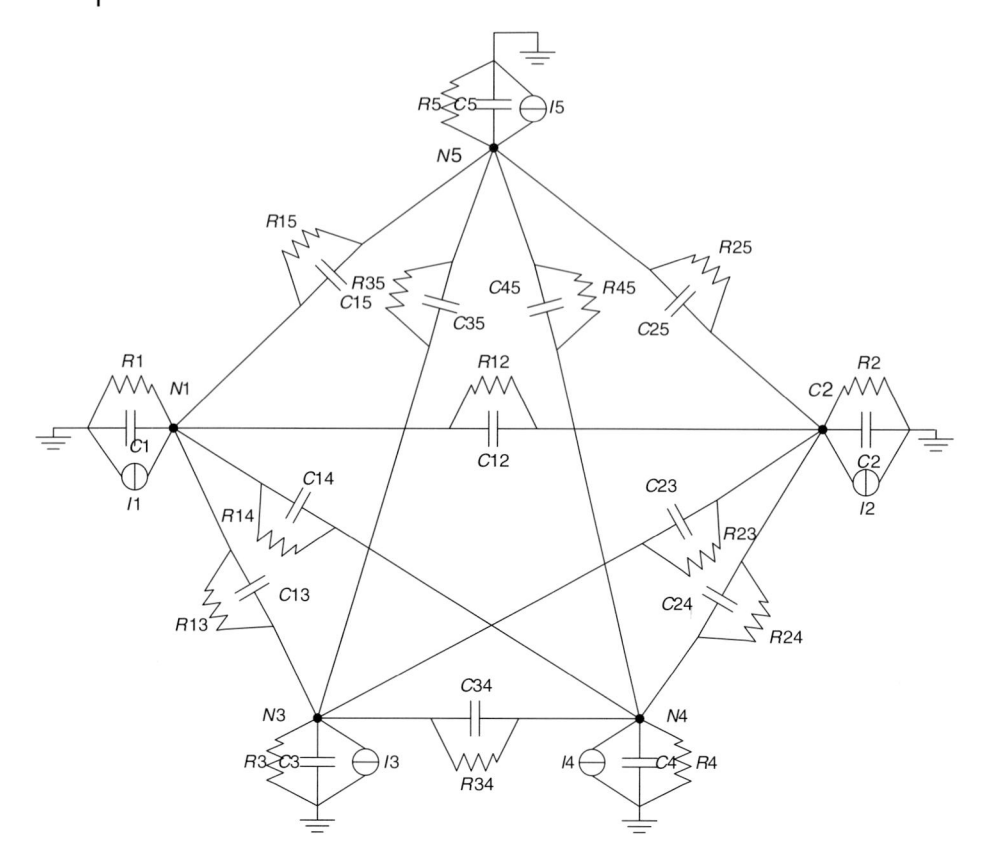

Figure 5.9 Equivalent circuit of the coupled electrothermal model of a beam.

also controlled by the thickness of the air gap, which is obtained by solving the thermomechanical model of the beam described in the following section.

5.4.2
Equivalent Circuit of the Thermomechanical Model of a Beam

As mentioned, the effect of the thermal expansion is equivalent to a pair of counterbalance applied at the two ends of the beam. So the model could be partitioned into two parts naturally, one is the equivalent force and the other the traditional 3D mechanical beam model. The 3D mechanical beam model may also be partitioned into a core part and peripheral part. The core part constructs the main frame of the mechanical properties of the beam, and the peripheral part adds some nonlinear or unideal effects into the model to improve its accuracy. In the equivalent circuit, there are also these partitions. The equivalent circuit of the core of the beam is represented with some passive components using the method introduced in the analysis [57]. The equivalent circuit of the unideal effects, such as the change in the effective axial length of the beam and the large axial-stress effect, is modeled with some nonlinear controlled sources, which are controlled by the displacement of the beam nodes. In the out-of-plane electrothermal microactuator, there is a cross-influence between the thermal field and the mechanical field. The effect of thermal expansion is the forward influence, which is represented with controlled sources acting in the rules of Eq. (5.16). The back influence is the change of the air gap while actuating, which is represented with the controlled sources whose controller is the nodal displacement of the beam.

The equivalent circuit of the different-layer joint is also represented with some controlled sources. KVL and KCL are also directly used in the conversion. There are six nodes in this model. Except for the moment of y-axis, other nodes are linked with some resistors with zero resistance, which connects the across and through quantities directly. The two nodes of rotation of y-axis are connected with the controlled sources, which follow the last item in Eq. (5.22).

5.5 Case Study

An elaborately designed out-of-plane electrothermal microactuator fabricated in MUMPs is taken as an example to verify the nodal analysis model of the electrothermal microactuator [58]. This actuator is composed of several beams in two different polysilicon layers, as shown in Figure 5.6.

This actuator is then represented in nodal analysis models as presented in Figure 5.10. This model is composed of seven beams, three anchors, and a different-layer joint. There are totally six free nodes in this model. In order to confirm the nodal analysis method, the connecting arm is divided into two beams by the different-layer joint.

The dimensions of this nodal analysis model refer to the paper [58] and then offset the linking overlap as listed in Table 5.1.

The material properties of the actuator refer to the paper [58] and the data supplied by MUMPs. In order to simplify the model, some data are obtained by linear fitting of the original reference data. They are shown in Table 5.2.

In the simulations, the ambient temperature as well as the temperature of the given substrate is taken as 320 K and kept constant. All the simulation results of

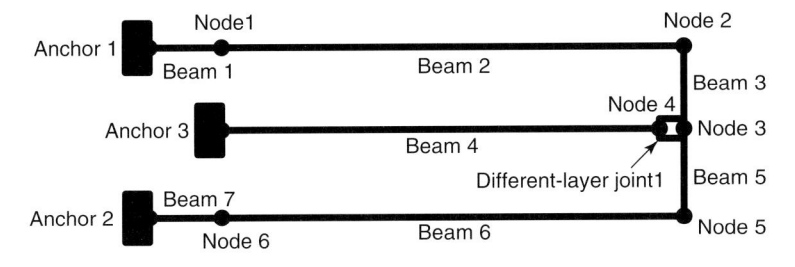

Figure 5.10 Nodal representation of an out-of-plane electrothermal microactuator.

Table 5.1 The dimensions of the elements in the simulated electrothermal microactuator.

Dimensions	Length (μm)	Width (μm)	Thickness (μm)	Lift (μm)
Beam 1, 7	30	3	1.5	4.75
Beam 2, 6	134	8	1.5	4.75
Beam 3, 5	24	20	3.5	4.75
Beam 4	136	2	2	2

Table 5.2 Material properties used in the simulation.

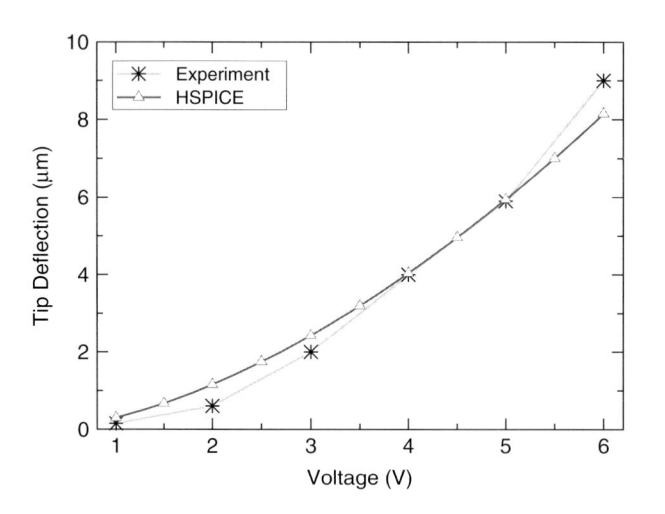

Material properties	Value	Unit
ρ	2330	kg m^{-3}
c	700	$\text{J Kg}^{-1}\text{K}^{-1}$
κ_0	48.095	$\text{W m}^{-1}\text{K}^{-1}$
κ_1	−0.05554	$\text{W m}^{-1}\text{K}^{-2}$
h_0	517.45	$\text{W m}^{-1}\text{K}^{-1}$
h_1	1.388	$\text{W m}^{-1}\text{K}^{-2}$
ρ_{e0}	2e−5	$\Omega\text{ m}$
ζ	1.25e−3	K^{-1}
κ_{air0}	0.0251	$\text{W m}^{-1}\text{K}^{-1}$
κ_{air1}	6.8e−5	$\text{W m}^{-1}\text{K}^{-2}$
α_{m0}	2.9976e−6	K^{-1}
α_{m1}	2.125e−9	K^{-2}
E	1.69e11	Pa

Figure 5.11 Tip deflection of the actuator versus the applied voltage.

temperature are relative to this temperature. The same voltage is applied to anchor 1 and anchor 2 in all the simulations below.

The applied voltage across anchor 1 and anchor 3 sweeps from 1 to 8 V with a step size of 0.1 V. The tip deflection of the actuator, the z-axial displacement of node 3 in the nodal analysis model, is obtained and compared with the published data, as shown in Figure 5.11 [58]. The simulation results agree well with the experimental data with the maximum error less than 5%.

The transient response is an important characteristic of the actuators. Especially, the response time on step stimulus is widely concerned. In this simulation, a pulse stimulus with an amplitude of 5 V, a pulse width of 0.8 ms, and a period of 1.6 ms is applied. The transient response of the nodal temperature and deflection of nodes 1, 2, and 3 are illustrated in Figure 5.12. The deflections of nodes 1 and 2 are almost identical.

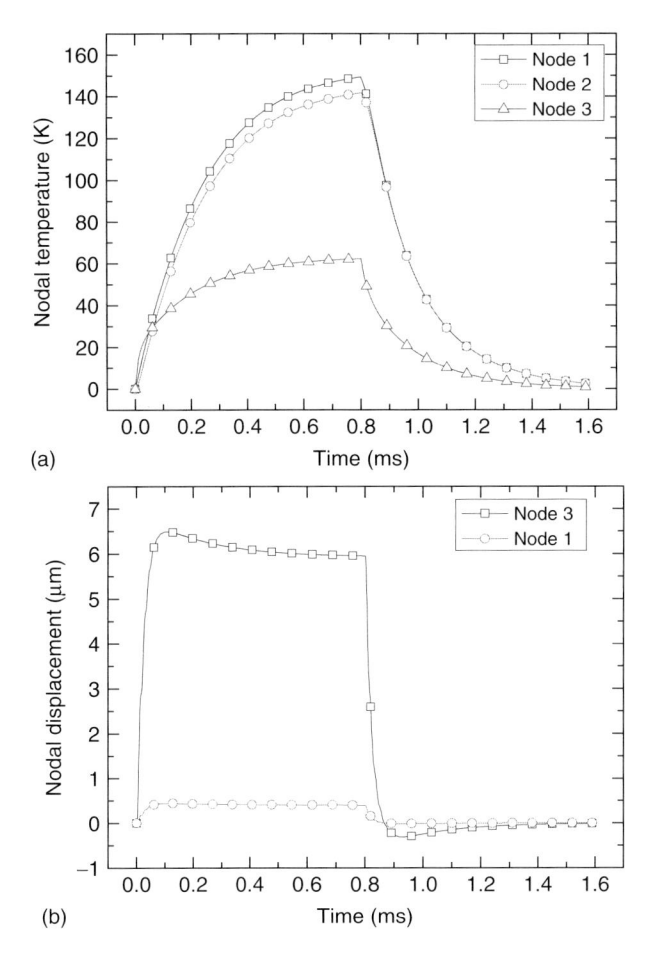

Figure 5.12 (a) and (b) Transient response of an out-of-plane electrothermal microactuator.

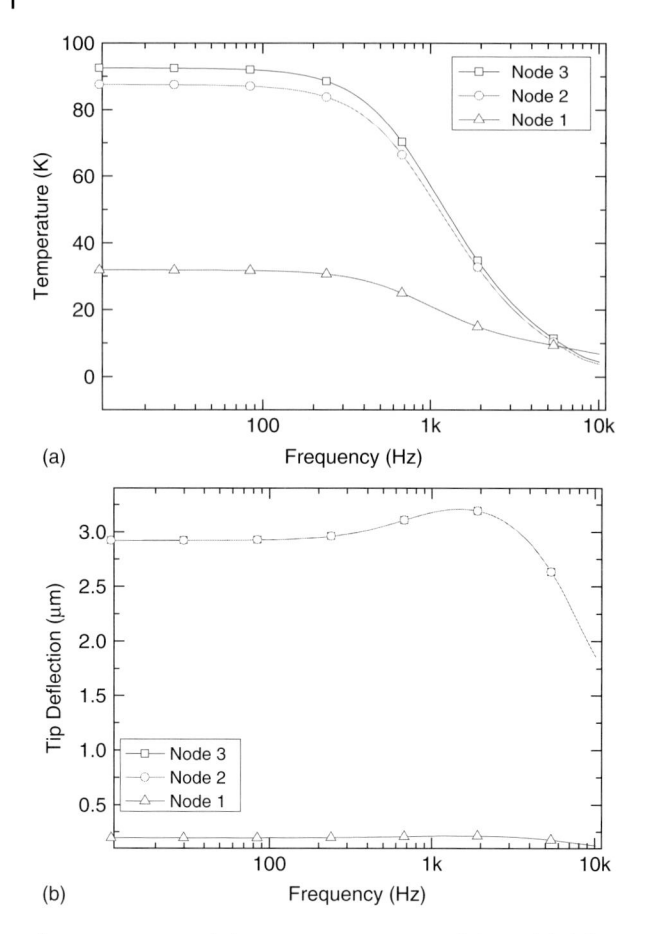

Figure 5.13 (a) and (b) Frequency response of the nodal deflection and temperature.

The behavior of the actuator as a function of the frequency of the applied voltage is also concerned. Here, the frequency response of the actuator is simulated to give this characteristic. The frequency of the applied voltage ranges from 1 Hz to 10 kHz with an amplitude of 3 V and a DC bias of 2 V. As shown in Figure 5.13, the nodal temperature and deflection are characterized by a low-pass filter behavior. The peak about 1.5 kHz in the frequency response of tip deflection is caused by the mechanical resonance.

5.6
Conclusion and Outlook

Electrothermal microactuators are reviewed. A nodal analysis model is presented in detail. On the basis of the heat transfer equation and the traditional mechanical nodal analysis model, a coupled electro-thermo-mechanical model is developed, which takes into account some nonideal effects. The model is in a form of an array of ODEs, which may be solved by a number of ODE solvers. A robust and commercially widely used solver, SPICE, is selected to simulate this model. For SPICE is a circuit simulation tool, all models built in this chapter should be in a form of circuit. So representation of equivalent circuit of the nodal analysis model is provided. Some representative actuators are studied following the introduction of representation of equivalent circuit of the thermal actuator. All the nodal analysis models have been simulated in comparison with FEA model to verify the results.

By far, the nodal analysis of an arbitrary microactuator constructed by pure beam elements has been demonstrated. It clearly shows the feasibility and advantages of modeling a MEMS device with elements and nodes. With more complex microstructures to be simulated in a nodal way, we should continue to set up the models of other fundamental element, such as plates, anchors, or even the impact of the package or the substrate. Chapter 6 will provide us some interesting research of this kind in fields beyond the microactuator.

References

1. Guckel, H., Klein, J., Christenson, T., Skrobis, K., Laudon, M., and Lovell, E.G. (1992) Thermo-magnetic metal flexure actuators. Proceedings of IEEE Solid-State Sensor and Actuator Workshop, pp. 73–76.
2. Nguyen, N., Ho, S., and Low, C. (2004) *Journal of Micromechanics and Microengineering*, **14**, 969–974.
3. Chronis, N. and Lee, L. (2004) Polymer MEMS-based microgripper for single cell manipulation. Technical Digest of IEEE Conference on Micro Electro Mechanical Systems, pp. 17–20.
4. Comtois, J.H., Bright, V.M., and M. Phipps (1995) Thermal microactuators for surface. micromachining process. Proceedings of SPIE 2642, pp. 10–21.
5. Pan, C.S. and Hsu, W. (1997) *Journal of Micromechanics and Microengineering*, **7**, 7–13.
6. Burns, D.M. and Bright, V.M. (1997) Design and performance of a double hot arm polysilicon thermal actuators. Proceedings of SPIE 3224, pp. 296–306.
7. Kolesar, E., Odom, W., Jayachadran, J., Ruff, M., Ko, S., Howard, J., Allen, P., Wilken, J., Boydston, N., Bosch, J., Wilks, R., and McAllister, J. (2004) *Thin Solid Films*, **447–448**, 481–488.
8. Kolesar, E., Ruff, M., Odom, W., Jayachadran, J., McAllister, J., Ko, S., Howard, J., Allen, P., Wilken, J., Boydston, N., Bosch, J., and Wilks, R. (2002) *Thin Solid Films*, **420–421**, 530–538.
9. Que, L., Park, J.-S., and Gianchandani, Y.B. (2002) Bent-beam electro-thermal actuators for high force applications. Proceedings of IEEE Conference on Micro Electro Mechanical Systems, pp. 31–34.
10. Que, L., Park, J.-S., and Gianchandani, Y.B. (2001) *Journal of Microelectromechanical Systems*, **10**, 247–254.
11. Gianchandani, Y.B. and Najafi, K. (1996) *Journal of Microelectromechanical Systems*, **5**, 52–58.
12. Cao, A., Kim, J., Tsao, T., and Lin, L. (2004) A bi-directional electrothermal electromagnetic actuator. Technical Digest of IEEE Conference on Micro Electro Mechanical Systems, pp. 450–453.
13. Jonsmann, J., Sigmund, O., and Bouwstra, S. (1999) Compliant electro-thermal microactuators. Proceedings of IEEE Conference on Micro Electro Mechanical Systems, pp. 588–511.
14. Comtois, J.H. and Bright, V.M. (1997) *Sensors and Actuators A*, **58**, 19–25.

15. Yan, D., Khajepour, A., and Mansour, R. (2004) *Journal of Micromechanics and Microengineering*, **14**, 841–850.
16. Chen, W., Chu, C., Hsieh, J., and Fang, W. (2003) *Sensors and Actuators A*, **103**, 48–58.
17. Chen, W.-C., Hsieh, J., and Fang, W. (2002) A novel single-layer bi-directional out-of-plane electrothermal microactuator. Proceedings of IEEE Conference on Micro Electro Mechanical Systems; 2002, pp. 693–697.
18. Comtois, J.H. and Bright, V.M. (1997) *Sensors and Actuators A*, **58**, 19–25.
19. Reid, J., Bright, V., and Comtois, J. (1996) A surface micromachined rotating micro-mirror normal to the substrate. Proceedings of Summer Topical Meetings of Advanced Applications of Lasers in Materials Processing, pp. 39–40.
20. Lee, C., Lin, Y., Lai, Y., Tasi, M., Chen, C., and Wu, C. (2004) *IEEE Photonics Technology Letters*, **16**, 1044–1046.
21. Chiou, J. and Lin, W. (2004) *Optics Communications*, **237**, 341–350.
22. Lerch, P., Silimane, C., Romanowicz, B., and Renaud, P. (1996) *Journal of Micromechanics and Microengineering*, **6**, 134–137.
23. Park, J.-S., Chu, L.L., Oliver, A.D., and Gianchandani, Y.B. (2001) *Journal of Microelectromechanical Systems*, **10**, 255–262.
24. Saini, R., Geisberger, A., Tsui, K., Nistorica, C., Ellis, M., and Skidmore, G. (2003) Assembled MEMS VOA. Proceedings of International Conference of Optical MEMS, pp. 139–140.
25. Chu, L. and Gianchandani, Y. (2003) *Journal of Micromechanics and Microengineering*, **13**, 279–285.
26. Lu, S., Dikin, D., Zhang, S., Fisher, T., Lee, J., and Ruoff, R. (2004) *Review of Scientific Instruments*, **25**, 2154–2162.
27. Syms, R., Zou, H., and Stagg, J. (2004) *Journal of Micromechanics and Microengineering*, **14**, 667–674.
28. Oh, Y., Lee, W., and Skidmore, G. (2003) Design, optimization, and experiments of compliant microgripper. Proceedings of International Mechanical Engineering Congress and Exposition, pp. 345–360.
29. Sharma, M. and Udeshi, T. (2003) Extensions of spring model approach for continuum-based topology optimization of compliant mechanisms. Technical Proceedings of Nanotechnology Conference and Trade Show, pp. 440–443.
30. Sigmund, O. (2001) *Structural and Multidisciplinary Optimization*, **21**, 120–127.
31. Harsh, K., Su, B., Zhang, W., Bright, V., and Lee, Y. (2000) *Sensors and Actuators A*, **80**, 108–118.
32. Sinclair, M. (2002) A high frequency resonant scanner using thermal actuation. IEEE International Conference on Micro Electro Mechanical Systems, pp. 698–701.
33. Kim, C., Lee, M., and Jun, C. (2004) *IEEE Photonics Technology Letters*, **16**, 1894–1896.
34. Mankame, N. and Ananthasuresh, G. (2000) The effects of thermal boundary conditions and scaling on electro-thermal-compliant micro devices. Technical Proceedings of the 2000 International Conference on Modeling and Simulation of Microsystems, 3, pp. 609–612.
35. Mankame, N.D. and Ananthasuresh, G.K. (2001) *Journal of Micromechanics and Microengineering*, **11**, 452–462.
36. Atre, A. and Boedo, S. (2004) Effect of thermophysical property variations on surface micromachined polysilicon beam flexure actuators. Technical Proceedings of Nanotechnology Conference and Trade Show, 2, pp. 263–266.
37. Yang, Y.-J. and Yu, C.-C. (2004) *Journal of Micromechanics and Microengineering*, **14**, 587–596.
38. Huang, Q.A. and Lee, N.K.S. (1999) *Journal of Micromechanics and Microengineering*, **9**, 64–70.
39. Huang, Q.A. and Lee, N.K.S. (1999) *Microsystem Technologies*, **5**, 133–137.
40. Huang, Q.A. and Lee, N.K.S. (2000) *Sensors and Actuators A*, **80**, 267–272.
41. Kuang, Y., Huang, Q.A., and Lee, N.K.S. (2002) *Microsystem Technologies*, **8**, 17–21.
42. Zhang, Y.X., Huang, Q.A., and Li, R.G. (2006) *Sensors and Actuators A*, **128**, 165–175.

43. Xu, G.B., Li, Y., and Huang, Q.A. (2007) *Sensors and Actuators A*, **136**, 249–254.
44. Xu, G.B. and Huang, Q.A. (2006) *IEEE Sensors Journal*, **6**, 428–433.
45. Huang, Q.A., Xu, G.B., and Qi, L. (2006) *Journal of Micromechanics and Microengineering*, **16**, 981–985.
46. Fedder, G.K. and Jing, Q. (1999) *IEEE Transactions on Circuits and Systems II*, **46**, 1309–1315.
47. Li, R.G., Huang, Q.A., and Li, W.H. (2008) *Microsystem Technologies*, **14**, 119–129.
48. Li, R.G., Huang, Q.A., and Li, W.H. (2009) *Microsystem Technologies*, **15**, 217–225.
49. Incropera, F.P. and DeWitt, D.P. (1996) *Introduction to Heat Transfer*, John Wiley & Sons Inc., New York.
50. Slack, G.A. (1964) *Journal of Applied Physics*, **35**, 3460–3466.
51. Holman, J.P. (1997) *Heat Transfer*, McGraw-Hill.
52. Manginell, R.P. (1997) Physical properties of polysilicon. PhD Thesis. University of New Mexico.
53. Manginell, R.P. (1997) Polycrystalline-silicon microbridge combustible gas sensor. Ph.D. Dissertation. University of New Mexico.
54. Okada, Y. and Tokumaru, Y. (1984) *Journal of Applied Physics*, **56**, 314–320.
55. Butler, J.T. and Bright, V.M. (1998) *ASME DSC-MEMS*, **66**, 571–576.
56. Jing, Q. (2003) Model and simulation for design of suspended MEMS. PhD Dissertation. Carnegie Mellon University.
57. Hsu, J.T. and Quoc, L.V. (1996) *IEEE Transactions on Circuits and Systems I*, **43**, 721–732.
58. Atre, A. (2006) *Journal of Micromechanics and Microengineering*, **16**, 205–313.

6
System-Level Modeling of Packaging Effects of MEMS Devices

Jing Song and Qing-An Huang

6.1
Introduction

In microelectromechanical system (MEMS) devices, the deformation of MEMS structure due to packaging-induced stress is of great concern as it directly affects the performance of the device. With the development of advanced package technologies such as system in package (SiP) and wafer level package (WLP), MEMS device will interact with the package more closely. Package will become a functional part of the system rather than a separate carrier of the device. Therefore, understanding the influence of the packaging on MEMS device performance is essential to a successful package-device codesign.

Packaging effects were observed as soon as MEMS devices were invented. At the early stage, the packaging effects were simply compensated using calibrations. However, with the development of MEMS devices with high performance and reliability, designers start to call for a more fundamental study of the packaging effects so that the major part of effects could be eliminated at the beginning of the design and the compensations become easier and faster. The major studies on packaging effects are summarized as follows with respect to various MEMS applications.

Although finite element method (FEM) can be used to simulate any packaged MEMS devices, the simulation time and computational cost are often problems. With the increasing complexity of MEMS components and micromodules that contain MEMS parts, this problem becomes more severe. So we turn to another way to settle it. It has been demonstrated clearly in Chapter 5 that MEMS thermal actuators can be modeled as a system consisting of beam elements. Here the system-level modeling concept goes further. A packaging MEMS device is also regarded as a system made up of different fundamental parts. But unlike the only beam element in the thermal actuator case, there are more such fundamental elements here. One of the key elements is the package substrate, whose deformation may significantly affect the performance of MEMS devices on it. If we could construct a packaged MEMS device using these reproducible elements in a system way, the nodal analysis method described earlier can also be applied here. The

System-level Modeling of MEMS, First Edition. Edited by T. Bechtold, G. Schrag, and L. Feng.
© 2013 Wiley-VCH Verlag GmbH & Co. KGaA. Published 2013 by Wiley-VCH Verlag GmbH & Co. KGaA.

obvious advantages of the method are high simulation speed, reproducibility of the element model, compatibility with the system-level simulation tool, and easy access for new MEMS designers.

6.2
Packaging Effects of MEMS and Their Impact on Typical MEMS Devices

Basic structures include microcantilevers and microbridges. They are also called *one-end fixed beams* and *doubly fixed beams*, respectively. They contain the most fundamental features of all MEMS structures. Study on packaging effect of the basic structures will provide a clear and general knowledge that applies to the design of other structures and the package. The microcantilevers are free of axial stresses, but susceptible to stiction induced by narrowed gap. The microbridges are greatly affected by built-in stresses. Hence packaging-induced stresses and deformations, either mechanical or thermomechanical, will affect both structures in different ways.

Die attach is one of the most critical processes that cause packaging effects [1]. The thermal cure required for die attach leads to thermal mismatch between MEMS chip and package substrate that results in chip warpage and surface strain distribution. Warpage changes the gap heights, and surface strain produces built-in stresses for microbridges. Hence, relative shifts over 10% of resonant frequencies [2], pull-in voltage [3], and radio-frequency (RF) performances [4] have been observed in packaged microbridges. Surface strain has proven to vary across the chip so that the packaging effect will behave differently according to the location of the MEMS structures [5]. It will also contribute significantly to the temperature coefficient of the structure performance [6]. Besides the materials of chip and package substrate, undesirable adhesive materials, dimensions, and anchor type also account for the packaging effects [3].

Other important packaging processes that contribute to MEMS packaging effect include molding and capping. Here capping mostly refers to precapping of MEMS devices using the front-end MEMS process instead of back-end packaging process. Only capped MEMS device can be encapsulated using low-cost molding processes. To improve the production, the caps are preferred to be fabricated in the wafer level that uses thin film deposition and sacrificial etching. However, it was demonstrated that thin film cap may crack under the molding pressure [7].

6.2.1
Accelerometers

As one of the most mature devices in MEMS industry, the packaging effect of MEMS accelerometers was investigated adequately. A packaged accelerometer may experience output signal shift (offset) due to thermomechanical stresses caused by the packaging assembly processes and external loads applied during application [8–12]. Using a reduced-order MEMS sensor and package model extracted by

CoventorWare™, the device offset can be predicted accurately in an acceptable simulation time, so that feedback design and optimization analysis is possible to improve the device performance by at least five times [12]. Besides the static die attach, effects of seal adhesives on dynamic performances were also investigated. Seal adhesives with small Young's moduli greatly affect the mode shapes of a packaged accelerometer, which results in strong distortion of output signal of accelerometer [10].

6.2.2
Gyros

Packaging affects MEMS gyroscope structure because package deformation or package-induced stress results in a frequency shift of the gyroscope structures. Deformation behavior of a MEMS gyroscope package induced by temperature changes was investigated using Moiré interferometry technique [13]. Analysis shows that global bending occurs because of the thermal mismatch between the chip, the molding compound, and the printed circuit board. Using a modified design of the suspended legs of the gyro, the frequency shift can be greatly reduced [13].

6.2.3
Pressure Sensors

MEMS pressure sensors can be bulk micromachined or surface micromachined. Its realization as silicon piezoresistive pressure sensor is one of the most mature technology. Besides the common effect of die attach/bonding and molding [14–20], the packaged performance of such a pressure sensor is also greatly influenced by the silicone gel material that is used to protect the die surface [20]. Simulation results show that the different geometries of the protection gel (convex or concave) will influence pressure sensitivity significantly. In addition, as pressure sensors are relatively large, the design of sensor structure itself will affect the packaging effect considerably [18].

6.2.4
Thermal Actuators

Packaging stress affects the stiffness of thermal microactuators and therefore shifts their displacements in operation. The problem was investigated at varying stress levels using a four-point bending stage [21]. The thermal microactuators were fabricated using a surface micromachining process and the microactuator displacement measured at varying stress levels. Increasing the tensile stress on the test die decreases the initial displacement and decreases the displacement per driving current ratio. The maximum displacement can be reduced by more than 60% at an applied tensile stress of around 200 MPa.

6.2.5
Hall Sensors

The performance of packaged Hall sensors is strongly dependent on temperature changes because the thermomechanical stresses induced by packaging will have a great effect through piezo-Hall effect and the piezoresistive effect. Temperature effects during the critical die attach and molding processes were investigated [22]. To enable an accurate simulation, the viscoelastic, viscoplastic and shrinking behaviors of the adhesives and molding materials were taken into account. According to the simulation and test results, offset voltage can change by up to 80% of the full-scale, and the sensitivity changes by 4% after die attach and molding process.

6.3
System-Level Modeling

Previous efforts have shown that coupling between the device and package causes stresses and deformations, lowers the performance, and compromises the reliability of a MEMS product. Moreover, with the development of advanced package technologies, such as SiP and WLP [23], the package and device will interact more closely. Package will become a functional part of the system rather than a separate carrier of the device. Hence, besides the conventional simulation hierarchy of MEMS, it is beneficial to develop a new design tool that integrates the package design and device design into the same level of simulation to promote the efficiency of design.

FEM is the most common tool to simulate the packaging effect of MEMS. Solid modeling and fine meshing of both MEMS and package structures were implemented to find out the packaging effect as accurately as possible. The method is proper for most bulk-micromachined structures such as MEMS pressure sensors and microphones, since the sizes of MEMS and package are similar. However, for most surface-micromachined MEMS devices, there is the large difference in the size scale between the device and its package. Any combined device-package FEM with resolution to accurately model the device will exceed practical computational resources. Hence the reported simulations usually separate the FEM models for the package and the device analysis, and use the package simulation results as boundary conditions to the devices. This simulation technique has proven to work reasonably in most cases, and has been integrated into package design module of the commercial simulation software of MEMS such as CoventorWare (see also Chapter 17) and Intellisuite™ (see also Chapter 20). To further increase the simulation efficiency, parametric behavioral package models can be numerically established according to the simulation results of the package [24, 25]. However, the separate simulation approaches are clearly not the final solution, and cost-effective simulation technique still requires to be developed to ensure a real compact package-device codesign. The system-level nodal method (NM) described in the next section tries to provide such a solution.

As explained in Chapter 5, the NM sets up the behavioral model of a specific element (such as a beam or a transistor) using only the nodal quantities based on physical analysis. If the element is electrical, the model is just the circuit model. If not, the model can be regarded as a generalized Kirchhoffian network-based model, that is, an equivalent circuit model by regarding all the "across" quantities as the equivalent voltages, and all the "through" quantities as the equivalent electrical currents (see also Chapter 2). Thus the model is solvable in any circuit simulator. In contrast with FEM that partitions a solid model into general elements with simple shape, NM partitions a model into specific elements (like a beam or a transistor) that are individually modeled by a specific theory. Since the theory is specific, the NM simulation is often fast and efficient.

However, the NM normally uses lumped nodes to interconnect elements. In other words, the nodal quantities of the model are all lumped parameters. These lumped nodes are not reasonable to describe distributed interconnections. To describe a distributed interconnection using lumped nodes, one would usually partition a body into several elements, but then these elements no longer neatly correspond to the design elements. Considering these individual elements in such a partition makes NM difficult to use. To address the problem, this chapter presents a distributed nodal method (DNM) for system-level modeling of packaged MEMS components. A typical microsystem is partitioned into function elements that are interconnected by distributed interfaces. Nodes with distributed quantities are introduced to describe their interconnections. The details of the method are described in the following text.

6.3.1
System Partitioning

A model depicted in Figure 6.1 is used as an example to describe DNM. It comprises a surface-micromachined microbridge fabricated on a chip. The chip is attached

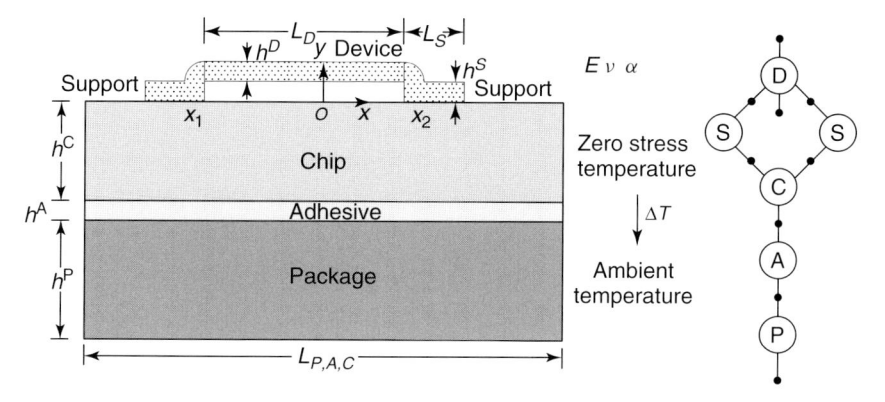

Figure 6.1 Cross-sectional view of a typically die-bonded MEMS device and its topology.

to a package substrate via an adhesive. After the curing of the adhesive, the whole assembly returns to ambient temperature. To simplify the modeling, a 2D model is used. It has been shown that 2D and 3D analyses predict almost the same warpage in a square chip package [26].

The model is divided into multiple elements including the device (D), support (S), chip substrate (C), adhesive (A), and package substrate (P). Each filled pattern in the model corresponds to an individual material with its Young's modulus E, Poisson ratio v, and coefficient of thermal expansion (CTE) α. The fixture point O of the model is set at the surface of the chip. x_1 and x_2 identify the horizontal coordinate of both supports, respectively. The vertical and horizontal dimension of each element is h_X and L_X, with subscript X indicating the name of the element, for example, L_S denotes the length of the support element. In the topology of the model, each circle stands for an element, and each dot stands for an interconnection between the elements. Either lumped or distributed nodes can be defined considering the nature of the interconnection. Using the distributed nodes, the topology of model is compact and easy to understand.

6.3.2
Behavioral Modeling of Single Substructures

Behavioral models of each substructures are set up according to own physics. Here three major substructures are focused on, the microbeam, the bulk substrate, and the support. Their modeling is very different.

6.3.2.1 Microbeam Model

Modeling of a microbeam has been well studied. The model usually contains two lumped nodes at both ends of the beam. Nodal quantities are shown in Figure 6.2. $\{F_{x_i}, F_{y_i}, M_i\}$ are defined as the nodal horizontal force, vertical force, and bending moment and $\{u_i, v_i, \theta_i\}$ the nodal horizontal displacement, vertical displacement, and angle of rotation, respectively. Subscript i denotes the nodal number. The element model can be established using conventional nodal analysis method. For the static analysis, the final model can be represented as a stiffness matrix as follows [27], where t, L, E, and I are the thickness, length, Young's modulus, and rotary moment of inertia of the beam, respectively.

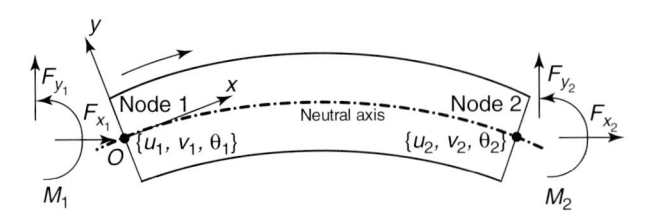

Figure 6.2 Nodal model of microbeam element.

6.3 System-Level Modeling

$$\begin{bmatrix} F_{x_1} \\ F_{y_2} \\ M_1 \\ F_{x_2} \\ F_{y_2} \\ M_2 \end{bmatrix} = \frac{EI}{L^3} \begin{bmatrix} \frac{tL^2}{I} & 0 & 0 & -\frac{tL^2}{I} & 0 & 0 \\ 0 & 12 & 6L & 0 & -12 & 6L \\ 0 & 6L & 4L^2 & 0 & -6L & 2L^2 \\ -\frac{tL^2}{I} & 0 & 0 & \frac{tL^2}{I} & 0 & 0 \\ 0 & -12 & -6L & 0 & 12 & -6L \\ 0 & 6L & 2L^2 & 0 & -6L & 4L^2 \end{bmatrix} \begin{bmatrix} u_{x_1} \\ u_{y_1} \\ \theta_1 \\ u_{x_2} \\ u_{y_2} \\ \theta_2 \end{bmatrix}$$

$$+ N \begin{bmatrix} 0 & 0 & 0 & 0 & 0 & 0 \\ 0 & \frac{6}{5L} & \frac{1}{10} & 0 & -\frac{6}{5L} & \frac{1}{10} \\ 0 & \frac{1}{10} & \frac{2L}{15} & 0 & -\frac{1}{10} & -\frac{L}{30} \\ 0 & 0 & 0 & 0 & 0 & 0 \\ 0 & -\frac{6}{5L} & -\frac{1}{10} & 0 & \frac{6}{5L} & -\frac{1}{10} \\ 0 & \frac{1}{10} & -\frac{L}{30} & 0 & -\frac{1}{10} & \frac{2L}{15} \end{bmatrix} \begin{bmatrix} u_{x_1} \\ u_{y_1} \\ \theta_1 \\ u_{x_2} \\ u_{y_2} \\ \theta_2 \end{bmatrix} \quad (6.1)$$

6.3.2.2 Chip/Adhesive/Package Model

In this example, the chip, adhesive, and package all looks similar to a beam. They can all be modeled as a typical beam element as shown in Figure 6.3. However, this beam model is different from the previous microbeam model. There are four nodes here, including two lumped nodes, 1 and 2, at the ends of the neutral axis, and two distributed nodes, 3 and 4, at the top and bottom interfaces. The definitions of the lumped nodes are the same as in the microbeam model. For the distributed node, nodal stress functions are defined including shear stress $\tau_i(x)$ and peeling stress $q_i(x)$, and nodal displacement functions are defined including horizontal displacement $u_i(x)$ and vertical displacement $v_i(x)$. The two shear functions actually describe the shear stress and y-directional normal stress along the bottom edge of the beam. $u(x)$ and $v(x)$ are defined as the horizontal and vertical displacement function along the neutral axis of the undeflected beam, respectively.

To establish the model, a local Cartesian coordinate system is selected to provide a homogenous boundary condition at the left end.

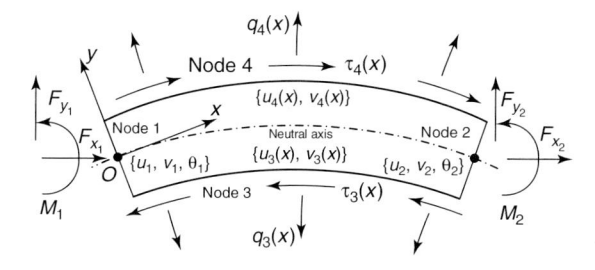

$$u(0) = v(0) = v'(0) = 0 \quad (6.2)$$

Figure 6.3 Compatible nodal model of elements for chip, adhesive, and package substrate.

Equilibrium of the nodal forces gives

$$F_{x_1} + F_{x_2} + F_{x\tau}(0) = 0$$
$$F_{y_1} + F_{y_2} + F_{yq}(0) = 0$$
$$M_1 + M_2 + M_\tau(0) + M_q(0) = 0 \tag{6.3}$$

where

$$F_{x\tau}(x) = \int_x^L [\tau_4(t) - \tau_3(t)]dt, \; F_{yq}(x) = \int_x^L [q_4(t) - q_3(t)]dt, \; M_\tau(x)$$
$$= -\frac{h}{2}\int_L^x [\tau_3(t) + \tau_4(t)]dt,$$

and

$$M_q(x) = \int_L^x \int_L^s [q_3(t) - q_4(t)]dt \, ds$$

are the distributed horizontal force, vertical force, the moment from horizontal force, and the moment from vertical force, respectively. It can be seen from Eq. (6.3) that the lumped forces at both ends are correlated with each other given the distributed stresses. Hence the set of $\{F_{x_2}, F_{y_2}, M_2\}$ at the right end is enough to represent all lumped forces. For simplicity this set will be denoted as $\{F_x, F_y, M\}$ in the following text.

In Figure 6.3, the governing equation of the microbridge under the nodal loads can be derived according to the Euler–Bernoulli beam theory

$$EI\frac{d^4v(x)}{dx^4} - \frac{d}{dx}\left\{[F_x + F_{x\tau}(x)]\frac{dv(x)}{dx}\right\}$$
$$= q_3(x) + q_4(x) - \frac{h}{2}\frac{d[\tau_3(x) - \tau_4(x)]}{dx} \tag{6.4}$$

and the equilibrium of force and moment at node 2 is given by

$$F_y = -EI\frac{d^3v(x)}{dx^3}\bigg|_{x=L} + [F_x + F_{x\tau}(L)]\frac{dv(x)}{dx}\bigg|_{x=L} - \frac{h}{2}[\tau_3(L) - \tau_4(L)] \tag{6.5}$$

$$M = -EI\frac{d^2v(x)}{dx^2}\bigg|_{x=L} \tag{6.6}$$

where E, α, L, h, and I are the effective Young's Modulus, CTE, length, height, and second moment of area of the beam, respectively. For the 2D case, the effective Young's Modulus is denoted by $E = E_0/(1 - v^2)$ under the assumption of zero plane strain, where E_0 and v are the Young's Modulus and Poisson ratio of the material. $F_{x\tau}(x)$ is the line force of the beam from the distributed loads. For simplicity to solve, it is approximated as a constant, $F_{x\tau} = \int_0^L F_{x\tau}/(x)dxL$. Integrating both sides of Eq. (6.4) from L to x and taking into account the boundary conditions yields

$$EI\frac{d^2v(x)}{dx^2} - (F_x + F_{x\tau})v(x) = -F_y(x - L)$$
$$- [M + (F_x + F_{\lambda x})v(L)] - M_q(x) - M_\tau(x) \tag{6.7}$$

According to the integral equation theory, the integral of both sides of the Eq. (6.7) from 0 to L in combination with Eq. (6.2) yields

$$v(x) = f(x) + \lambda \int_0^x (x-t) v(t) dt \tag{6.8}$$

with

$$\lambda = \frac{(F_x + F_{x\tau})}{(EI)}$$

$$f(x) = -\frac{x^3 - 3x^2 L}{6EI} F_{y2} - \frac{x^2}{2EI} \left[-M + (F_x + F_{x\tau}) v(L) \right] - \int_0^x \int_0^s \frac{1}{EI} M_q(t) dt ds$$

$$- \int_0^x \int_0^s \frac{1}{EI} M_\tau(t) dt ds$$

The solution to Eq. (6.8) is

$$v(x) = \begin{cases} f(x) + \sqrt{\lambda} \int_0^x \sin\left[\sqrt{\lambda}(x-t)\right] f(t) dt & \lambda \geq 0 \\ f(x) - \sqrt{-\lambda} \int_0^x \sin\left[\sqrt{-\lambda}(x-t)\right] f(t) dt & \lambda < 0 \end{cases} \tag{6.9}$$

The explicit form of $v(x)$ is obtained after substituting x by L at both sides.

Considering the axial thermal expansion and the stress-stiffening effect of the beam element, the horizontal strain along the neutral axis of the undeflected beam in Figure 6.3 is

$$\frac{du(x)}{dx} = \frac{F_x + F_{x\tau}(x)}{Eh} - \frac{1}{2} \left[\frac{dv(x)}{dx} \right]^2 + \alpha T(x) \tag{6.10}$$

Integrating both sides from 0 to x in combination with Eq. (6.2) yields

$$u(x) = \frac{1}{Eh} \int_0^x \left[F_x + F_{x\tau}(t) \right] dt - \frac{1}{2} \int_0^x \left[\frac{dv(t)}{dt} \right]^2 dt + \int_0^x \alpha T(t) dt \tag{6.11}$$

From Eqs. (6.9) and (6.10), the expression of nodal displacement can be derived using the theory of interfacial compliance [28], which describes how the layered beam interfaces can be locally deformed by shearing and peeling stresses.

$$\begin{cases} \Delta u = u_2 - u_1 = u(L), & \Delta v = v_2 - v_1 - \theta_1 L = v(L), & \Delta \theta = \theta_2 - \theta_1 = v'(L) \\ u_3(x) = u(x) + v'(x) h/2 + k_u \tau_3(x), & v_3(x) = v(x) + k_v \left[q_1(x) - q_1(0) \right] \\ u_4(x) = u(x) - v'(x) h/2 + k_u \tau_4(x), & v_4(x) = v(x) + k_v \left[q_4(x) - q_4(0) \right] \end{cases} \tag{6.12}$$

where k_u and k_v are the horizontal and vertical interfacial compliances, respectively.

6.3.2.3 Support Model

The effect of support compliance on the performance of MEMS devices has proven to be significant. The support can be modeled as a nodal matrix using lumped forces and displacements at the joints with devices [29]. Figure 6.4 shows the 2D model of a surface-micromachined step-up support. Node 1 stands for the support bottom. Node 2 stands for the support joint. Both nodes are lumped. $\{F_{x_i}, F_{y_i}, M_i\}$ are defined as nodal forces, $\{u_i, v_i, \theta i\}$ are defined as nodal displacements. Subscript i denotes the nodal number. $F_{x_i}, F_{y_i}, M_i, u_i, v_i, \theta_i$ are the horizontal force, vertical

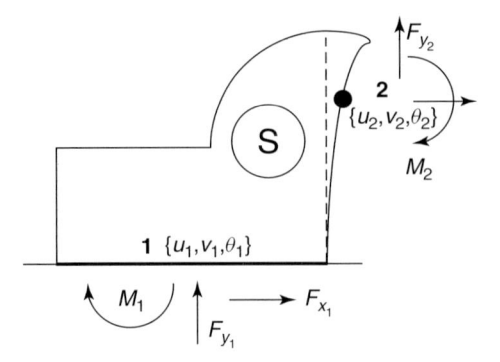

Figure 6.4 Nodal model of a step-up support element (left side).

force, bending moment, horizontal displacement, vertical displacement, and angle of rotation, respectively. To set the support model, the following quantities are also defined

$$\Delta u = u_2 - u_1 - \theta_1 \left(h^S + h^D/2\right), \Delta v = v_2 - v_1, \Delta \theta = \theta_2 - \theta_1$$
$$F_x = F_{x_1} = -F_{x_2}, F_y = F_{y_1} = -F_{y_2}, M = M_1 = -M_2 \tag{6.13}$$

where h^S and h^D are the thickness (height) of the support and device, respectively. Then, the nodal matrix is

$$\begin{bmatrix} \Delta u \\ \Delta v \\ \Delta \theta \end{bmatrix} = \begin{bmatrix} S_{11} & S_{12} & S_{13} \\ S_{21} & S_{22} & S_{23} \\ S_{31} & S_{32} & S_{33} \end{bmatrix} \begin{bmatrix} F_x \\ F_y \\ M \end{bmatrix} \tag{6.14}$$

where S_{ij} are the coefficients obtained from numerical fitting or analytical modeling.

6.3.3
Element Integration

The elements are then combined into a complete system through the continuity conditions of the nodes. For the lumped node, the continuity is interpreted by the equilibrium of forces and equality of displacements at the joint node of different elements. For the distributed node, the distributed stress functions, $q(x)$ and $\tau(x)$, are first expanded as an N-term series of Legendre polynomials $P_k(x)$ $(k = 1, 2, \ldots, n)$ that are orthogonal and complete in the space of continuous functions. $2N$ undetermined coefficients C_1 to C_{2N} are obtained in Pascal.

$$q(x) = \sum_{i=1}^{N} C_i P_{k(i)}(x)$$

$$\tau(x) = \sum_{i=N+1}^{2N} C_i P_{k(i)}(x) \tag{6.15}$$

where subscript $k(_i)$ indicates the sequence number of Legendre polynomials in accordance with the coefficient C_i. Substituting Eq. (6.12) into an element model yields the distributed displacement functions $u(x)$ and $v(x)$ in terms of C_1 to C_{2N}. Next, N equispaced points x_1 to x_n are selected along the distributed node. $u(x)$ and $v(x)$ are evaluated at these points to provide two N-dimensional vectors that also represent the displacement distribution.

$$u(x) \rightarrow \left[u(x_1), u(x_2) \cdots u(x_N)\right]^T$$
$$v(x) \rightarrow \left[v(x_1), v(x_2) \cdots v(x_N)\right]^T \tag{6.16}$$

The continuity of the distributed functions are thus interpreted as the equality of the finite quantities at the joint nodes, that is, all the horizontal and vertical displacements at $x_1 - x_n$ are equal, and all the coefficient of the stress functions at the interconnection interface are equal for both of the interconnected elements. Combined with the boundary conditions, $2N$ equations are set up with $2N$ undetermined coefficients at each distributed node. The whole model is thus solvable. Increasing N improves the precision but slows down the simulation. Here $N = 6$ was selected as a trade-off between simulation speed and precision. At present, the DNM model can be easily solved using mathematical tool such as Matlab software. The equivalent circuit interpretation of the model is also possible since the model is explicitly analytical. It is seen that the difference between the DNM model and the normal nodal model is that the distributed nodal quantities are introduced. In order to apply the Kirchhoffian rules to the distributed quantities, they must be first interpreted using new lumped quantities. Here the polynomial decomposition and the displacement vector are used to create such lumped quantities. Other discretization method may also be applicable.

6.3.4
FEM and Experimental Validation

We used a die-bonded MEMS microbridge to experimentally validate the system-level model. The die warpage and natural frequency shift of surface-micromachined microbridges were investigated in the experiments. Each parameter was validated in an individual test.

Two bare chips (B1, B2) and three chips with test structures (T1, T2, T3) were prepared for experimental validation, as shown in Figure 6.5. In chip T1–T3, microbridges and microrotation strain gauges were surface-micromachined using standard polysilicon surface micromachining process. The strain gauge measured the residual in-plane strain for model simulation. The microbridges and gauges were fabricated on distributed positions on a chip. In total, 5 groups of 44 microbridges were mounted on a test chip. The lengths of the microbridges varied from 450 to 850 µm, and their widths varied from 10 to 40 µm. The microbridges were oriented along the horizontal x-axis. The spans of the gauges varied from 550 to 850 µm, and their widths were 4 µm. They were also oriented along the x-axis at different locations of the chip. After surface micromachining, chips with a size of 0.97 mm × 0.97 mm × 540 µm were diced. The test structures were then released. Each chip was attached to a package substrate using the epoxy CB603. The epoxy was cured at 120 °C for 120 s and returned to the ambient temperature of 25 °C. The

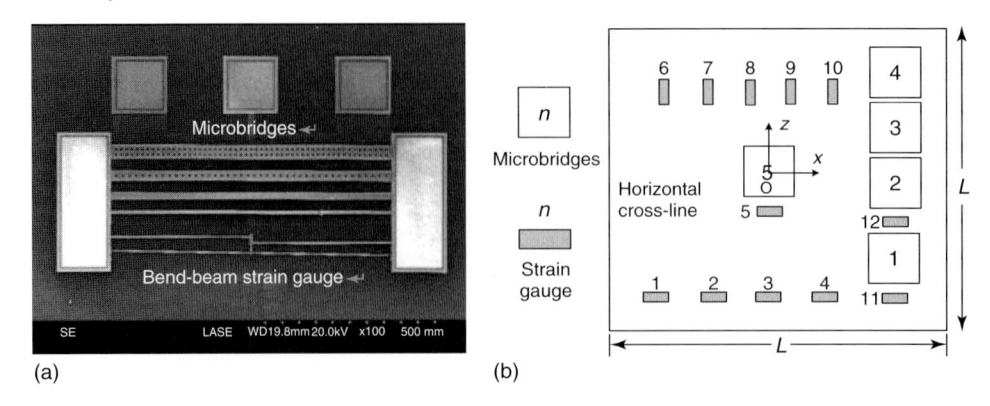

Figure 6.5 View of the test chips T1–T3. (a) Test structures including microbridges and a strain gauge. (b) Schematic layout of the test structures on the test chip.

material of the package substrate of all test chips was a flame retardant epoxy (FR4) except chip B2. B2 was attached to an alumina ceramic substrate. Table 6.1 lists the geometrical dimensions and material properties of the packages and devices.

Bare chips of B1 and B2 were used to validate the die warpage results. The digital image correlation (DIC) method was used to measure the warpage of the die-attached bare chips. A Peltier effect semiconductor heater and a FLUKE17B thermometer were used to control the environmental temperature from 0 to 120 °C. Figure 6.6a shows the simulated and measured warpage of B1 at different temperatures after die bonding. The DNM model results agree well with FEM and approach the measured results. Figure 6.6b shows the warpage comparison of B1 and B2 at 80 °C. Warpage of B2 was much smaller because the ceramic substrate is 20 times stiffer than the FR4 substrate, and the CTE mismatch between the device

Table 6.1 Material properties and geometries in the DNM, FEM simulation, and the experiment.

	Length L (μm)	Height h (μm)	Young's Modulus E (GPa)	Poisson ratio ν	CTE α (1× 10^{-6} per °C)
Beam (D)	450 ± 0.5 ~840 ± 0.5	2 ± 0.05	170 ± 5	0.28	2.6 ± 0.1
Support (S)/Gap (G)	10 ± 0.5	2 ± 0.05	170 ± 5	0.28	2.6 ± 0.1
Chip (C)	5000 ± 50[a] 9700 ± 50[b]	350 ± 5[a] 540 ± 5[b]	170 ± 5	0.28	2.6 ± 0.1
Adhesive (A)	7000 ± 50[a] 9700 ± 50[b]	25 ± 5[a] 25 ± 5[b]	6 ± 4	0.3	28 ± 2
Package (P)	7000 ± 50[a] 16 600 ± 50[b]	750 ± 5[a] 600 ± 5[b]	16 ± 2	0.28	16 ± 2

[a] DIC test.
[b] LDV test.

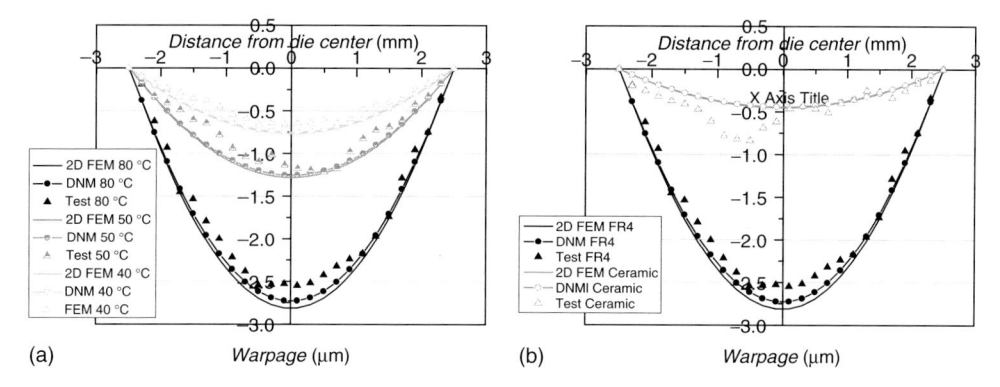

Figure 6.6 Results of the DIC test after die bonding. (a) Comparison of the warpage and its temperature dependence of B1 among the DNM, 2D FEM model, and test results. (b) Comparison of the warpage between the B1 (FR4 substrate) and B2 (ceramic substrate) at 80 °C.

and the substrate is three times smaller. According to the results, FR4 substrate was selected in the next test to induce larger stresses in the test structures.

T1, T2, and T3 were used to investigate the packaging effect of MEMS structures. A laser Doppler vibrometer (LDV) system was used to measure the natural frequency of the microbridges. Figure 6.7a shows the frequency shifts of the microbridges with different lengths but similar axial strains before and after die bonding. Their strains are considered to be similar because they were located at the same horizontal coordinate, so that the residual strain from the fabrication and packaging are similar. As the simulation has predicted, die bonding caused a relative shift range of 12–26% for the first natural frequency and 19–26% for the third one. In this test configuration, die bonding caused a tension strain along the surface of the chip that compensates part of the compressive strain from the fabrication process. Figure 6.7b shows the frequency shifts of the microbridges with the same length but different horizontal coordinates on the chip surface before

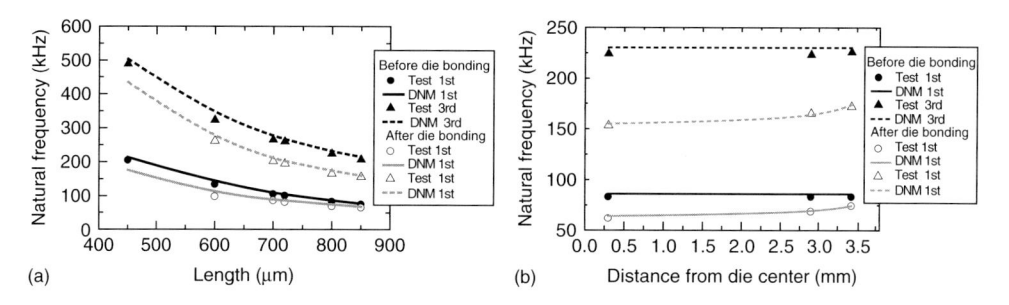

Figure 6.7 Results of the LDV test before and after die bonding. (a) Frequency shifts of microbridges with an identical [x] coordinate among all lengths. The microbridges were located at inner side of the row of groups 1–4. (b) Frequency shifts of microbridges at distributed positions but the same lengths. A total of three microbridges were included here. One is from group 5 and the others are from group 3.

and after die bonding. Since the frequency shifts are varying along the chip surface, it is clear that the device location has considerable effect on the packaged behavior. On the basis of the results, it is concluded that the distributed package–device interaction is highly responsible for the observed distributed effect of MEMS. The results of the DNM model matched the test data with a maximum relative error of 4%.

6.4
Conclusion and Outlook

As revealed in the validation tests, the packaged behavior of a MEMS microbridge may deviate greatly from the design. However, it is predictable using a compact system-level model. A relative error of less than 10% can be reached for different dimensions of the MEMS microbridges. The system-level modeling significantly saves the computational cost and simplifies the simulation hierarchy. After the elements and nodes are classified and defined, one can simply combine the elements together in the right topology to simulate the behavior of the packaged devices fastly and accurately. The implementation of the partition concept may lower the knowledge requirement for MEMS designers, as is the case in the integrated circuit (IC) industry.

The next work left to do is to establish more models of fundamental elements of MEMS devices and its packages. It is fortunate that MEMS and the packages are mostly fabricated using that simple planar process, so most of the fundamental elements are regular, such as beams, rectangular plates, and cuboid masses. The regularity of these elements geometry may greatly simplify the modeling and makes nodal analysis method a real system-level tool for simulating complex microsystems that contain MEMS and three-dimensional mechanical parts. What's more, the modern packaging technology, which uses more and more three-dimensional interconnections of regular packaging parts, such as vias, bumps, and thermal sinks, may also benefit greatly from the same simulation tool.

References

1. Dickerson, T. and Ward, M. (1997) Low deformation and stress packaging of micromachined devices. IEE Colloquium on Assembly and Connections in Microsystems, pp. 7-1–7-3.
2. Li, M., Huang, Q.A., Song, J., Tang, J., and Chen, F. (2006) Theoretical and experimental study on the thermally induced packaging effect in COB structures. International Conference on Electronic Packaging Technology, pp. 198–201.
3. Lishchynska, M., O'Mahony, C., Slattery, O., Wittler, O., and Walter, H. (2007) IEEE Transactions on Advanced Packaging, 30, 629–635.
4. Yang, L., Liao, X.P., and Song, J. (2008) Effect of bonding on the packaged RF MEMS switch. International Conference on Electronic Packaging Technology, pp. 327–330.

5. Song, J., Huang, Q.A., Li, M., and Tang, J.Y. (2009) *Journal of Microelectromechanical Systems*, **18**, 274–286.
6. Song, J., Huang, Q.A., Li, M., and Tang, J.Y. (2009) *Microsystem Technologies*, **15**, 925–932.
7. De Anna, R. and Roy, S. (1999) *Ansys Solutions*, **1**, 22–24.
8. Li, G. and Tseng, A.A. (2001) *IEEE Transactions on Electronics Packaging and Manufacturing*, **24**, 18–25.
9. Huang, W., Cai, X., Xu, B., Luo, L., Li, X., and Cheng, Z. (2003) *Sensors and Actuators A*, **102**, 268–278.
10. Zhang, X.R., Tee, T.Y., and Luan, J.E. (2005) Comprehensive warpage analysis of stacked die MEMS package in accelerometer application. International Conference on Electronic Packaging Technology, pp. 581–586.
11. Zhang, X., Park, S.B., and Judy, M.W. (2007) *Journal of Microelectromechanical Systems*, **16**, 639–649.
12. Joo, J.W. and Choa, S.H. (2007) *IEEE Transactions on Components and Packaging Technologies*, **30**, 346–354.
13. Chiou, J.A. (2003) *IEEE Transactions on Advanced Packaging*, **26**, 327–333.
14. Meyyappan, K., McClusky, P., and Chen, L.Y. (2003) *IEEE Transactions on Device and Materials Reliability*, **3**, 152–158.
15. Krondorfer, R., Kim, Y.K., Kim, J., Gustafson, C.-G., and Lommasson, T.C. (2004) *Microelectronic Reliability*, **44**, 1995–2002.
16. Peng, C.-T., Lin, J.-C., Lin, C.-T., and Chiang, K.-N. (2005) *Sensors and Actuators A*, **119**, 28–37.
17. Lee, C.C., Peng, C.T., and Chiang, K.N. (2006) *Sensors and Actuators A*, **126**, 48–55.
18. Krondorfer, R.H. and Kim, Y.K. (2007) *IEEE Transactions on Components and Packaging Technologies*, 285–293.
19. Chou, T.-L., Chu, C.-H., Lin, C.-T., and Chiang, K.-N. (2009) *Sensors and Actuators A*, **152**, 29–38.
20. Phinney, L.M., Spletzer, M.A., Baker, M.S., and Serrano, J.R. (2010) *Journal of Micromechanics and Microengineering*, **20**, 095011.
21. Fischer, S. and Wilde, J. (2008) *IEEE Transactions on Advanced Packaging*, **31**, 594–603.
22. Miettinen, J., Mäntysalo, M., Kajia, K., and Ristolainen, E.O. (2004) System design issues for 3D system-in-package. IEEE Electronic Components and Technology Conference, pp. 610–615.
23. Rabinovich, V.L., van Kujik, J., Zhang, S., Bart, S.F., and Gilbert, J.R. (1999) Extraction of compact models for MEMS/MOEMS package-device co-design. Symposium on Design, Test, and Microfabrication of MEMS and MOEMS (SPIE 3680), pp. 114–119.
24. Bart, S.F., Zhang, S., Rabinovich, V.L., and Cunningham, S. (2000) *Microelectronic Reliability*, **40**, 1235–1241.
25. Tsai, M.Y., Lin, Y.C., Huang, C.Y., and Wu, J.D. (2005) *IEEE Transactions on Electronic Packaging and Manufacturing*, **28**, 328–337.
26. Vandemeer, J.E. (1998) Nodal Design of Actuators and Sensors (NODAS). Technical Report, Carnegie Mellon Universit.
27. Suhir, E. (1989) *Journal of Applied Mechanics*, **56**, 596–600.
28. Kobrinsky, M.J., Deutsch, E.R., and Senturia, S.D. (2000) *Journal of Microelectromechanical Systems*, **9**, 361–369.
29. Zaal, J.J.M., van Driel, W.D., Bendida, S., Li, Q., van Beek, J.T.M., and Zhang, G.Q. (2008) *Microelectronics Reliability*, **48**, 1567–1571.

7
Mixed-Level Approach for the Modeling of Distributed Effects in Microsystems

Martin Niessner and Gabriele Schrag

7.1
General Concept of Finite Networks and Mixed-Level Models

Generalized Kirchhoffian network theory (Chapter 2) offers an intuitive and efficient framework for the lumped element model based simulation of microeletromechanical devices including electrical components such as control and readout units (Chapters 5, 6, 8, and 17). However, physics-based or order-reduction-based lumped element models are not always available or obtainable. This is especially the case for distributed effects that inherently require spatially discretized models, for example, the nonuniform thermal heating of microstructures [1] or the fluidic damping in microdevices [2–4].

Commonly, the finite element method (FEM) is employed to model distributed effects of this kind. However, several complications arise when FEM-based models are directly employed in the simulation of the full microsystem including also the electric circuitry. Electric circuits are often very complex and may consist of a large number of elements. As they are usually modeled on the Kirchhoffian network level, they cannot be easily and efficiently integrated in FEM simulation software [1, 5]. Consequently, the FEM and the circuit simulation tools need to be coupled externally. The external coupling prohibits, in turn, simultaneous solving of the different model parts and may require a highly sophisticated numerical coupling algorithm, depending on the kind of simulation performed and interaction between the model parts (Chapter 4). Moreover, the electrical components within the circuits are modeled with computationally efficient lumped element models in order to enable the fast simulation of the transient behavior of the circuit for a large number of cycles, whereas FEM simulations of complex geometries require a lot of computational resources and are very time consuming. This results in relatively long simulation times when the transient behavior of the full microsystem has to be simulated.

An alternative for the co-simulation of distributed effects, electric circuitry and other lumped element models is the finite network (FN) approach – the equivalent

System-level Modeling of MEMS, First Edition. Edited by T. Bechtold, G. Schrag, and L. Feng.
© 2013 Wiley-VCH Verlag GmbH & Co. KGaA. Published 2013 by Wiley-VCH Verlag GmbH & Co. KGaA.

of the FEM approach, but on the level of Kirchhoffian networks. When applying the FN approach, the governing partial differential equations (PDEs) are first spatially discretized, as in the FEM approach. Second, Kirchhoffian network elements are formulated on the basis of the governing PDEs. These network elements are used to connect the respective mesh nodes obtained from the discretization, turning the spatial mesh into a Kirchhoffian network. This leads to a large system of flux-conserving coupled ordinary differential equations (ODEs) governed by across and through variables (the concept of across and through variables is explained in Chapter 2). This system of ODEs can be compiled and solved in standard analog and mixed-signal circuit simulation tools such as CADENCE Spectre, SYNOPSYS Saber, or SPICE. If the system of ODEs is linear, a netlist composed of simple RLC equivalent circuits can be used to represent the ODEs. If the system of ODEs is nonlinear, parametric behavioral models coded in hardware description languages such as VHDL-AMS or Verilog-AMS have to be used and instantiated in a netlist.

Using FNs for the modeling and simulation of distributed effects in microsystems has several advantages. First, FNs allow for the cosimulation of the microdevice with its electric circuitry within one single software environment: a standard circuit simulator. As the system of coupled equations can be solved simultaneously, and special numerical features of circuit simulators [6], such as their ability to deal with stiff problems [7], can be exploited, faster convergence of strongly coupled and nonlinear problems can be achieved than in staggered simulation schemes [1, 8].

Second, FNs allow for an intuitive coupling with lumped element models either through the nodal through-variables or across-variables. This enables the compilation of the so-called mixed-level models (MLMs), that is, models that combine both, discretized FNs and lumped element models.

Third, the number of degrees of freedom (DOFs) of FNs, and especially of MLMs, can be reduced significantly when compared to FEM models. This is because the FEM approach aims at an accurate calculation of the potential, whereas the FN approach is primarily aimed at the accurate conservation of fluxes. This is manifested in the requirements of the mesh for FEM and FN simulations. FEM simulations need a highly resolved mesh where the gradient of the potential is high. By contrast, FN simulations might not need a mesh that is as highly resolved at the respective positions, for example, if only the flux variable is of interest. Moreover, computationally expensive features of the analyzed geometry can be described by lumped element models with only a few DOFs (Section 7.4.1). This enables a dedicated tailoring of the model complexity according to the required accuracy and simulation time.

The following sections of this chapter explain the workflow for the generation of mixed-level models using squeeze film damping (SQFD) in microelectromechanical systems (MEMS) as example. The resulting mixed-level model will be benchmarked with respect to FEM simulations, experimental data, and alternative modeling approaches.

7.2
Approaches for the Modeling of Squeeze Film Damping in MEMS

Damping determines the dynamic performance of micromechanical systems. Several devices such as a certain type of micromachined accelerometers [9] even require a specific amount of damping for a proper operation. Consequently, the estimation of damping is one of the most important steps in the development of MEMS.

The dissipative mechanisms responsible for the damping in microsystems are numerous [10, 11]: support loss, internal friction, thermoelastic, electronic, and viscous air damping. However, at pressures higher than medium vacuum pressure, viscous air damping, and especially SQFD, dominates over other losses [11]. Hence, the reliable prediction of SQFD is one of the main concerns in MEMS research since several years [2, 4, 12–22].

Briefly speaking, the term SQFD denotes the situation when a suspended microstructure is moving against a thin film of fluid that is enclosed in a small gap between this moving structure and a fixed plate (Figure 7.1). The process of squeezing the fluid (air in the case considered here) out of the small gap causes the major part of the viscous damping force. For the evaluation of SQFD, one has, firstly, to calculate the pressure profile underneath the moving plate. Secondly, the pressure profile has to be integrated across the plate in order to obtain the damping force.

Figure 7.2 shows different kinds of pressure profiles underneath moving microplates. The profile of a nonperforated plate is parabolic with the maximum located in the center of the plate (Figure 7.2a). This is because the air can only leave and enter across the boundary. When the plate is slightly perforated, the profile is still parabolic, but the maximum pressure is decreased as air can now also leave and enter through the holes (Figure 7.2b). Near the boundary, most of the fluid still leaves across the boundary. When the plate is heavily perforated, the air leaves and enters mainly through the holes changing the profile from a parabolic one to a "flat" one, that is, the pressure profile is determined by the perforation holes, except at the boundary (Figure 7.2c).

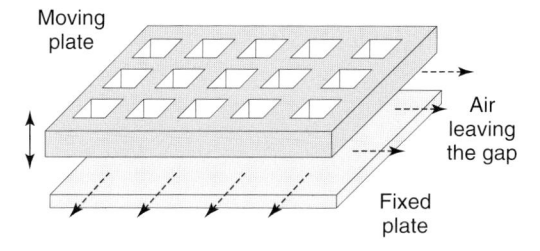

Figure 7.1 Illustration of an SQFD-dominated micromechanical structure: a suspended perforated plate is vibrating above a fixed plate. Owing to this motion, the air in the gap between the plates enters and leaves not only across the boundary, as shown by the arrows in the figure, but also through the perforation holes.

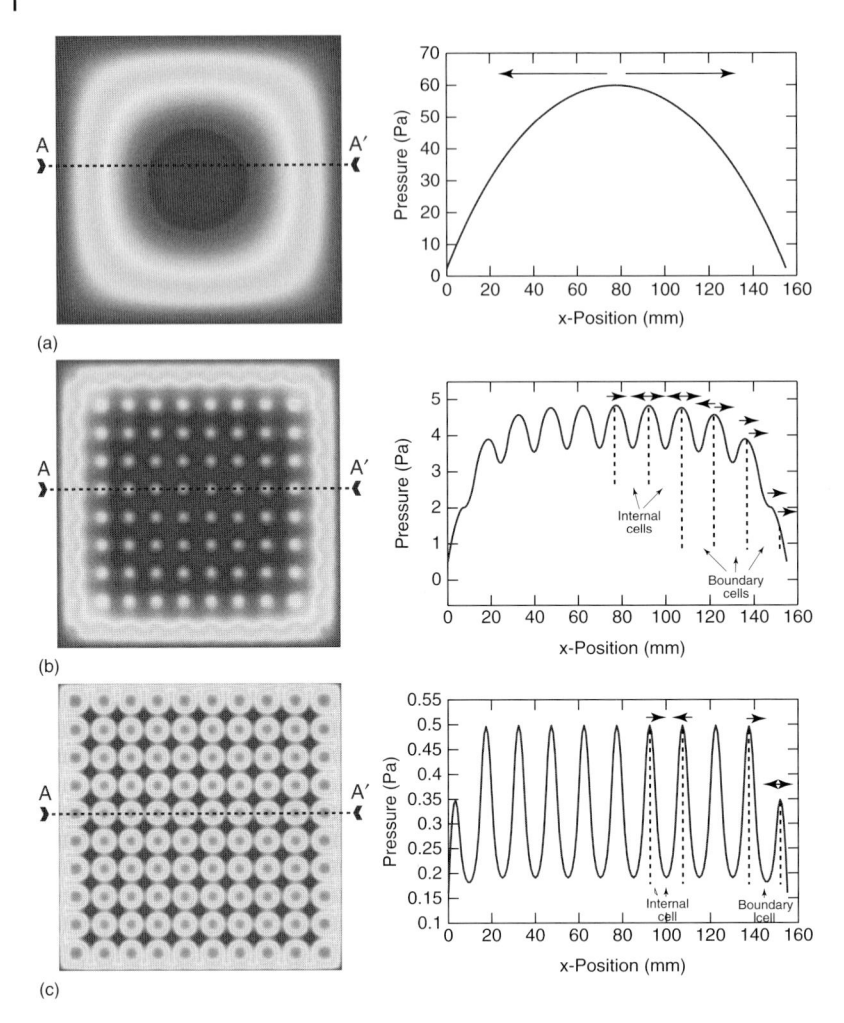

Figure 7.2 Pressure profiles underneath moving plates calculated using the mixed-mode FEM model presented in Ref. [19]. (a) Left: Pressure profile underneath a nonperforated plate. Right: Cut along the horizontal line A–A'. The air leaves across the boundary of the plate. (b) Left: Profile underneath a plate with perforation ratio 10% and thickness 15 µm. Right: Cut along the horizontal line A–A'. In the inner area of the plate, the air leaves through the holes. Toward the boundary of the plate, the air leaves increasingly across the boundary. (c) Left: Profile underneath a plate with perforation ratio 42% and thickness 15 µm. Right: Cut along the horizontal line A–A'. The air leaves through the holes of the plate except for the cells along the boundary.

7.2.1
Reynolds Equation-Based Modeling Strategies

The Navier-Stokes equations are the most general set of equations that can be employed for the calculation of the pressure profile underneath those moving plates, but an evaluation on the basis of the Navier-Stokes equations would be of high computational expense. For that reason, the Reynolds equation is commonly employed for the modeling of SQFD in MEMS, which is a simplified form of the Navier-Stokes equations, but applicable for most microdevices (the Reynolds equation is discussed in detail in Section 7.3.1). However, the Reynolds equation is originally derived assuming two nonperforated and infinitely extended parallel plates [23], whereas the moving components in MEMS are usually both perforated and of finite size (Figure 7.1).

As a consequence, different strategies for including the flow of air through the perforation holes and the finite size effects along the boundary have evolved and developed into several Reynolds equation-based SQFD modeling approaches. These approaches can be classified into the following four types:

- **Analytical approach based on a modified Reynolds equation:** In this approach, an additional term accounting for the flow of air through the perforation holes is introduced into the Reynolds equation. The Reynolds equation thus modified is then integrated over the area of the moving MEMS component. This way, the influence of the perforations is distributed over the whole area of integration and an analytical expression for the damping force is obtained. An example is the model presented by Bao *et al.* [14].
- **Analytical approach based on decomposition into cells:** In this approach, the perforated structure is decomposed into an array of uniformly distributed perforation cells. A perforation cell can be perceived as a commonly circular cell with a hole in its center. The total damping force on the moving structure is evaluated by calculating first the force on a single cell and then using a homogenization approach to extend to the total number of cells. An example is the model presented by Veijola [18].
- **Mixed-mode approach based on FEM:** In this approach, the total damping force is calculated using the FEM. The Reynolds equation in its original form is solved underneath the nonperforated part of the MEMS structure, whereas a modified Reynolds equation that accounts for the flow-through perforations is solved at the locations of holes. A pressure boundary condition accounts for the finite size effects at the boundary of the MEMS structure. The FEM simulator iterates until the pressure is continuously solved for. The integration of the pressure profile yields the damping force on the MEMS structure. An example is the model derived by Veijola and Raback [19].
- **Mixed-level approach based on GKN theory:** In this approach, the total damping force is calculated using a mixed-level model [13, 24] (Section 7.1). The Reynolds equation is solved underneath the nonperforated part of the structure using an FN. Physics-based lumped element models are attached to respective nodes of the FN to account for the flow-through holes and for the effects along the boundary. The total damping force is obtained by summing up the nodal damping forces. In contrast to the FEM-based mixed-mode model, the approach primarily aims at an accurate conservation of the fluxes. Section 7.3 describes this type of model in detail.

7.2.2
Motivation for Using Mixed-Level Modeling

None of the approaches described in Section 7.2.1 has established as "the" standard or reference for the simulation of SQFD in MEMS. Commonly, analytical models based on the decomposition approach (type 2 in Section 7.2.1) are very popular with MEMS designers because these models usually offer a closed form analytical expression for the calculation of SQFD. However, studies by Veijola et al. [20] and De Pasquale et al. [21] showed that the damping calculated by an analytical model based on decomposition into cells [18] may deviate up to 63% from the experimental value at normal pressure.

These significant deviations are due to assumptions made in this approach that do not hold true for all types of perforated microstructures. Figure 7.3 illustrates how a perforated microplate is decomposed into cells and which assumptions are made for the flow of air within the cell according to this approach: the damping of a single cell is calculated assuming that no air leaves across the outer boundary of the cell but that all air leaves through the hole in the center of the cell. This assumption is very reasonable for highly perforated structures with flat pressure profiles, as illustrated in the right panel of Figure 7.2. For structures with nonflat pressure profiles, as is the case for the structure in Figure 7.2b, this assumption is more than arguable. The cut along the profile (Figure 7.2b-right) shows that the pressure varies considerably and decreases with the distance to the boundary of the plate, that is, there is a pressure gradient across the cells near the boundary resulting into a flow of air toward and across the boundary. If this flow is neglected, damping is overestimated even though homogenization methods are used, as observable in [20, 21].

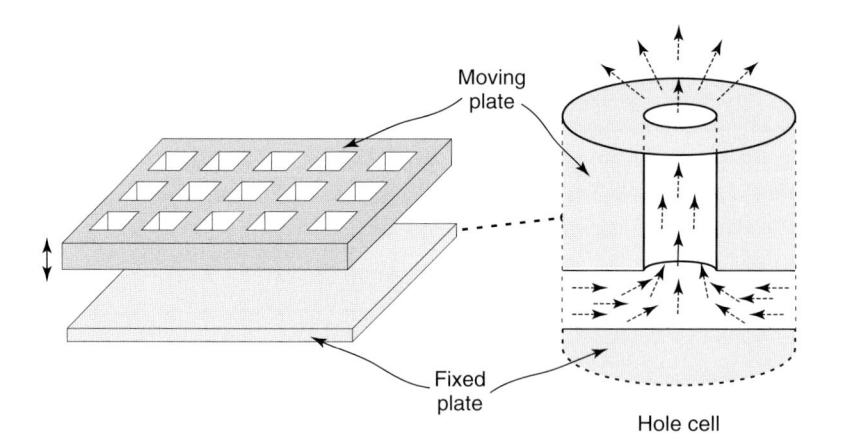

Figure 7.3 Conceptual idea of the approaches based on decomposition into cells: the perforated microstructure (left) is decomposed into circular cells with effective cell and hole radius (right). Within a cell, it is assumed that all air (symbolized by arrows) leaves through the hole of the cell and that no air leaves across the boundary of the cell.

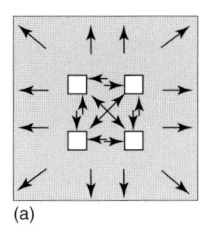

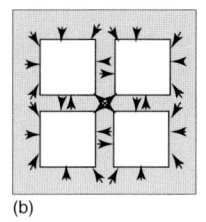

Figure 7.4 (a) Geometry where most of the flow leaves across the boundary. (b) Geometry where most of the flow leaves through the perforation holes.

For this reason, Mohite et al. [16, 17] and Pandey and Pratap [2] proposed to divide the array of cells into internal cells, located in the interior of the plate, and different types of boundary cells, located along the boundary of the plate. The boundary cells allow for a flow of air across their boundaries. With further modifications, Pandey and Pratap [2] were able to achieve very good agreement with the experimental data of a single, specific MEMS structure.

As can be inferred from that, the initially intuitive idea of decomposing a perforated MEMS structure into single cells gets more and more complicated. This complication is due to the distributed nature of SQFD. The air underneath a moving perforated microplate will always try to leave along the path with the minimum fluidic resistance. Depending on the respective geometry, the air will consequently tend to leave mostly through the perforation holes (Figure 7.4b), or mostly across the boundary (Figure 7.4a), or there will be a mixture of both flow patterns.

Pandey and Pratap [2] and Veijola et al. [20] also benchmarked analytical models based on a modified Reynolds equation (approach type 1 in Section 7.2.1) and showed that these models do not perform better than analytical models based on the decomposition approach.

In summary, it is very difficult for MEMS designers, who do not have a deep understanding of microfluidics, to anticipate the flow pattern underneath a newly designed microstructure and to choose an appropriate analytical model, if necessary, also with a reasonable separation into boundary and internal cells. Thus, it is advisable to use models that are inherently able to reflect the distributed nature of SQFD and to account for different flow patterns at the same time: mixed-mode and mixed-level models (approaches types 3 and 4 in Section 7.2.1). Indeed, mixed-mode [21] and, especially, mixed-level models [22] showed to deliver more accurate results for various MEMS geometries.

7.3
Mixed-Level Modeling of Squeeze Film Damping in MEMS

This section derives a mixed-level model for the calculation of SQFD in arbitrarily shaped MEMS structures. The general idea of the MLM is to use an FN for evaluating the nonlinear Reynolds equation where applicable and to employ physics-based lumped element models to account for the flow-through perforation holes and effects along the boundary.

7.3.1
Finite Network-Based Evaluation of the Reynolds Equation

In the most general and rigorous physical approach, SQFD should be described by the complex and nonlinear Navier-Stokes equations. However, the well-known and much simpler Reynolds equation [23] can be employed in the situation of SQFD (Figure 7.1), if the following assumptions are met.

- The Reynolds number Re is small (Re $\ll$ 1), that is, friction forces dominate over inertial effects and the flow within the microdevice is laminar. This assumption restricts the validity of Reynolds equation in terms of the frequency f of device operation:

$$f \ll \frac{\eta}{2\pi \cdot \rho \cdot g_{min}^2 \cdot \Upsilon_{Reynolds}} \tag{7.1}$$

Here, η, ρ, g_{min}, and $\Upsilon_{Reynolds}$ denote the viscosity and density of the air, the minimum gap width of the structure, and a correction for gas rarefaction effects (Section 7.3.3).

- The thickness of the air film between the two moving plates is small compared to the lateral dimensions of the plates. As a result, the vertical variation of the pressure underneath the plate is negligible when compared to the lateral one.

These assumptions are met for a variety of microdevices. The Reynolds equation accounts in its most general form for the situation where both plates are free to move in horizontal as well as in vertical direction. If one plate is assumed to be fixed and only vertical movement is considered, the Reynolds equation reads

$$\underbrace{\nabla \left(\frac{\rho g^3}{12\eta} \nabla p \right)}_{\text{Poiseuille flow}} = \underbrace{\rho \frac{\partial g}{\partial t}}_{\text{generation}} + \underbrace{g \frac{\partial \rho}{\partial t}}_{\text{compressibility}} \tag{7.2}$$

Here, g denotes the local gap width. In its physical interpretation, the term on the left-hand side is the Poiseuille flow, which is responsible for the dissipative losses. The first term on the right-hand side models the fluid flow generated by the time-varying gap height due to the moving plate, and the second term on the right-hand side represents the compression of the air inside the gap. As the air in MEMS can be assumed to be at constant temperature, we can use the ideal gas equation to obtain a proportional relation between density and pressure

$$\rho = \rho_0 \cdot \frac{p}{P_0} \tag{7.3}$$

Here, ρ_0 is the reference density at a reference pressure P_0, which is generally the pressure of the surrounding air. Equation (7.2) can thus be simplified to

$$\nabla \left(\frac{p}{P_0} \frac{g^3}{12\eta} \nabla p \right) = \frac{p}{P_0} \frac{\partial g}{\partial t} + \frac{g}{P_0} \frac{\partial p}{\partial t} \tag{7.4}$$

Following the FN idea (Section 7.1), the next steps are to find an interpretation of Eq. (7.4) in terms of the Kirchhoffian network formalism, to choose a discretization method, and to derive expressions for the lumped network elements.

As Kirchhoffian networks are primarily flux conserving, it is natural to start with considering the continuity equation in its local form (Chapter 2)

$$\nabla \vec{j}_x = -\dot{\rho}_x + \Pi_x \tag{7.5}$$

Here, $\vec{j}_x$ denotes the flow density of a thermodynamically extensive quantity x (a through-quantity) and ρ_x and Π_x denote the density and the generation rate of the extensive quantity. The continuity equation states that a local change of the density over time ($\dot{\rho}_x$) of an extensive quantity is due to either an in-/outflow ($\nabla \vec{j}_x$) or a generation (Π_x) of this quantity. According to Onsager [25] (Chapter 2), we can derive a relation between the flow density $\vec{j}_x$ and the gradient of the corresponding thermodynamically intensive quantity ϕ (an across-quantity)

$$\vec{j}_x = \lambda(-\nabla\phi) = -\lambda \nabla \phi \tag{7.6}$$

Here, λ denotes a transport coefficient. For instance, the continuity equation for the electric charge in the electric domain reads

$$\nabla \vec{j}_e = -c\dot{\varphi} + \Pi_e \quad \text{with} \quad \vec{j}_e = -\sigma \nabla \varphi \tag{7.7}$$

Here, $\vec{j}_e$ denotes the electric current density, σ the electric conductivity, φ the electric potential, c a local capacity, and Π_e a local electric source. When neither capacities ($c = 0$) nor sources ($\Pi_e = 0$) are present, we arrive by integration of Eq. (7.7) at the well-known Kirchhoff's current law (KCL): $\sum I_i = 0$. When $c \neq 0$ and $\Pi_e \neq 0$, these terms need to be taken into account on the Kirchhoffian network level. Figure 7.5 shows the three types of Kirchhoffian network elements that correspond to the three terms of the continuity equation: resistors are used to model dissipative flows, capacitors to model accumulation, and sources to model generation.

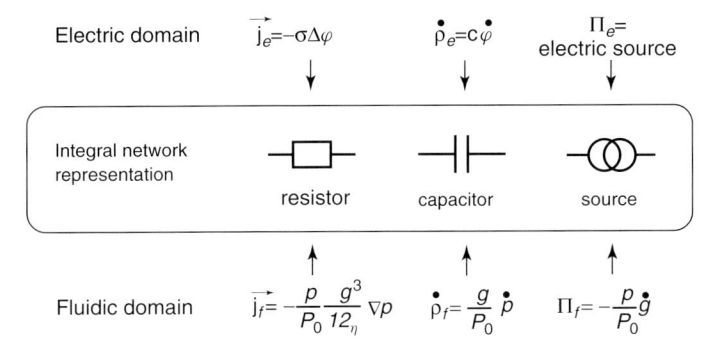

Electric domain	$\vec{j}_e = -\sigma \Delta \varphi$	$\dot{\rho}_e = c\dot{\varphi}$	$\Pi_e =$ electric source
Integral network representation	resistor	capacitor	source
Fluidic domain	$\vec{j}_f = -\dfrac{p}{P_0}\dfrac{g^3}{12\eta}\nabla p$	$\dot{\rho}_f = \dfrac{g}{P_0}\dot{p}$	$\Pi_f = -\dfrac{p}{P_0}\dot{g}$

Figure 7.5 Illustration of the correspondence between the different terms in the continuity equations in the electrical domain and the respective lumped Kirchhoffian network elements. Also, the Reynolds equation (Eq. (7.4)) is interpreted in this context.

Furthermore, Figure 7.5 interprets the Reynolds equation (Eq. (7.4)) in the context of the Kirchhoffian network formalism. Here, already a major difference compared to the electric domain can be identified: the electric network elements depend on electric across and through quantities only, whereas the fluidic network elements resulting from Eq. (7.4) will not only depend on fluidic quantities but will also be strongly coupled with the mechanical domain through the gap width g. The resistance of the fluidic resistor will depend nonlinearly, the capacitor linearly on the gap width, and the source will depend on its change over time.

Equation (7.4) is a local equation. In order to evaluate Eq. (7.4) for a real microstructure, we have to discretize its geometry and to solve Eq. (7.4) on the resulting mesh. This procedure will be illustrated using the finite box discretization method [26]. Figure 7.6a shows the mesh obtained from applying this method and allowing only rectangular finite boxes. The network elements can now be directly derived from integration of the different terms in Eq. (7.4) within this mesh (Chapter 2).

In order to describe the flow $Q_{S,k0}$ generated by the displacement of the moving plate (see source in Figure 7.6c) at a node k, the generation rate $\dot{g}$ is multiplied by the nodal area A_k

$$Q_{S,k0} = A_k \cdot \frac{P_{k0}}{P_0} \dot{g}_k \tag{7.8}$$

Here, P_{k0} denotes the pressure difference between the node k and the reference pressure P_0 and g_k is the local gap width at the node k. The nodal area A_k (Figure 7.6b) can be extracted from the Voronoi graph of the discretization. The flow $Q_{C,k0}$ due to the change in density (see capacitor in Figure 7.6c) at a node k is given in the same way by

$$Q_{C,k0} = A_k \cdot g_k \cdot \frac{\dot{P}_{k0}}{P_0} \tag{7.9}$$

The Poiseuille flow Q_{kl} between two adjacent nodes k and l (see resistor in Figure 7.6c) is calculated by multiplying the Poiseuille flow term in Eq. (7.2) with

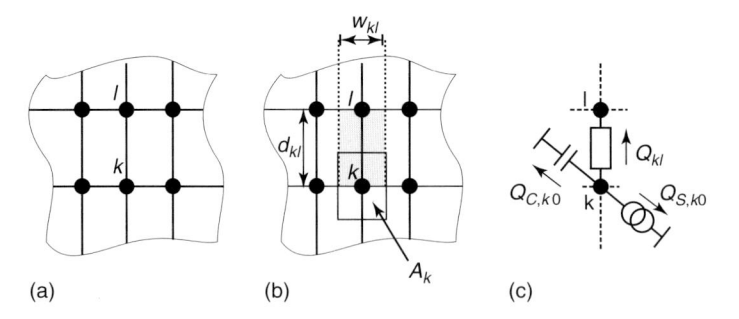

Figure 7.6 (a) Rectangular mesh resulting from a discretization of the plate using the finite box method [26]. (b) Geometry parameters obtainable from the discretized geometry. (c) Kirchhoffian network elements that result according to the FN concept for a node k and its connection to the node l.

the ratio between the width w_{kl} of the two-dimensional Poiseuille flow "channel" from node k to l and the length of this channel, that is, the distance d_{kl} between the nodes k and l (Figure 7.6b).

$$Q_{kl} = \frac{w_{kl} \cdot g_{kl}^3}{12\eta \cdot d_{kl}} \cdot \frac{\tilde{P}_{kl}}{P_0} \cdot P_{kl} \cdot \Upsilon_{\text{Reynolds}} \tag{7.10}$$

Here, g_{kl} denotes the average gap width of the nodes, $\tilde{P}_{kl}$ the averaged pressure of the two nodes, and P_{kl} the pressure difference between the nodes. $\Upsilon_{\text{Reynolds}}$ is a correction factor, which accounts for gas rarefaction, that is, for regimes, where the mean free path of the air molecules is large or of the same order as the device dimensions, and the validity of continuum theory becomes doubtful. This is explained in Section 7.3.3.

This procedure is repeated for each node, so that an FN consisting of nonlinear fluidic resistors, sources, and capacitors is obtained (Figure 7.7). The resulting Kirchhoffian network is governed by the pressure differences P_{kl} between nodes and the flow rates Q_{kl} along network edges as "across"- and "through"-variables.

7.3.2
Physics-Based Lumped Element Models

By its nature, the Reynolds equation-based FN model shown in Figure 7.7 does not yet account for the flow-through perforation holes and the effects along the boundary. These effects will be included using physics-based lumped element models that will be attached to the respective network nodes.

7.3.2.1 Boundary Model
An air flow leaving and entering the gap across the boundary can be considered similar to an air flow passing through an elliptical orifice [27]. Consequently, a pressure drop exists between the air along the boundary and the ambient atmosphere, whereas Eq. (7.4) assumes ambient pressure directly at the structure's boundaries. This pressure difference is modeled by fluidic resistors attached to the nodes of the FN that are located along the boundary (Figure 7.8). Sattler [27] derived the respective boundary resistance $R_{B,k}$ of one resistor analytically by calculating

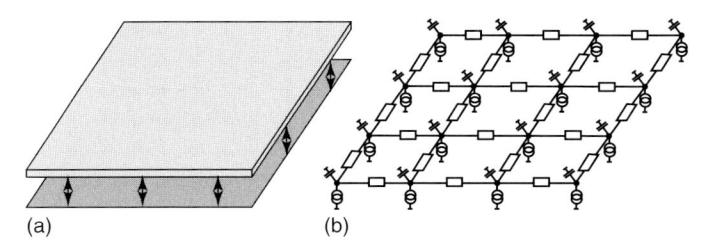

(a)   (b)

Figure 7.7 Rectangular plate moving toward a fixed plate (a) and the corresponding finite network model (b) consisting of fluidic resistors, sources, and capacitors.

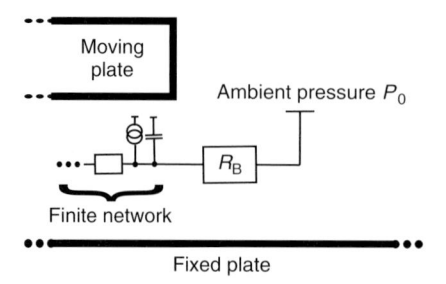

Figure 7.8 Illustration of how boundary effects are modeled. A boundary resistor R_B is attached to each node of the finite network that is situated along the boundary of the plate.

the flow through an orifice of equivalent dimensions

$$R_{B,k} = \frac{P_{k0}}{Q_{k0}} = \tau \cdot \frac{3\pi\eta}{g_k^2 \cdot b_k} \cdot \Upsilon_{BT}^{-1} \tag{7.11}$$

Here, $\tau = 0.84$ denotes a parameter fitted by FE simulations that is constant throughout all geometrical variations and b_k the length of the plate boundary assigned to the kth node. Υ_{BT} is a correction factor for gas rarefaction, which is explained in Section 7.3.3.

7.3.2.2 Hole Model

Equivalently, perforation holes are modeled by three fluidic resistors connected in series [27, 28] (Figure 7.9). The first resistor $R_{T,r}$ models the transition of the flow from the gap underneath the moving plate into the rth hole. As this situation is

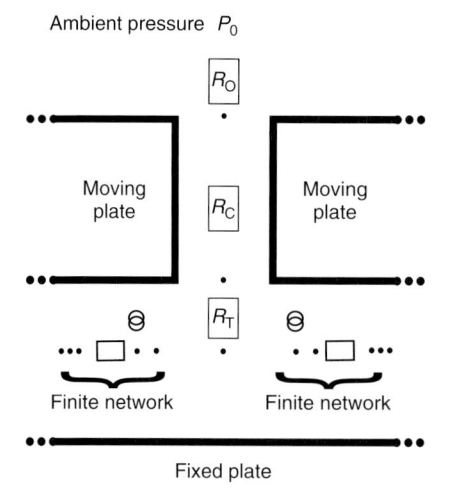

Figure 7.9 Illustration of how holes are modeled and embedded within the finite network. All nodes of the finite network that are located next to the hole are connected to the resistor R_T modeling the transition region when the fluid enters the channel. The resistor R_C models the channel resistance. The resistor R_O models the orifice flow.

similar to the situation along the boundary, the expression is equivalent to Eq. (7.11). The second resistor $R_{C,r}$ models the fluidic resistance of the "channel" through the plate. The third resistor $R_{O,r}$ describes the contribution due to the orifice, that is, when the air leaves or enters the channel. Sattler [27] derived these expression for square holes

$$R_{T,r} = \tau \cdot \frac{3\pi\eta}{g_r^2 b_r} \cdot \Upsilon_{BT}^{-1} \tag{7.12}$$

$$R_{C,r} = \frac{12\eta L_r}{0.42 s_r^4} \cdot \Upsilon_C^{-1} \tag{7.13}$$

$$R_{O,r} = \frac{21\eta}{s_r^3} \cdot \Upsilon_O^{-1} \tag{7.14}$$

Here, g_r denotes the averaged gap width at the perimeter of the rth hole, s_r the side length of the hole, $b_r = 4s_r$ the perimeter of the hole, and L_r the length of the channel, that is, the thickness of the plate. Υ_C and Υ_O are correction factors for the rarefied gas regime (Section 7.3.3).

The total pressure drop P_{r0} between the perimeter of the hole and the ambient atmosphere results in

$$P_{r0} = (R_{C,r} + R_{O,r} + R_{T,r}) \cdot Q_{r0,\text{rey}} + (R_{C,r} + R_{O,r}) \cdot Q_{r0,\text{rel}} \tag{7.15}$$

Here, $Q_{r0,\text{rey}}$ denotes the flow originating from the FN, that is, from underneath the nonperforated part of the moving plate (left and right from the hole in Figure 7.9). $Q_{r0,\text{rel}}$ accounts for the air that is situated directly underneath a perforation hole and that is also squeezed through the hole as the plate moves. This "relative" flow is calculated by multiplying the area of the hole times the velocity of the moving plate: $Q_{r0,\text{rel}} = s_r^2 \dot{g}_r$.

7.3.3
Gas Rarefaction Effects

The small air gaps present in micromachined structures often require taking the effect of gas rarefaction into account [27], that is, when the mean free path of the gas molecules is larger than or of the same order as the dimensions of the device. In this case, continuum theory can no longer be applied [27]. Instead, molecular dynamics (MD) models and simulations are imperative when the effect of rarefaction becomes pronounced.

Typically, in order to avoid the field of MD, the so-called "correction factors" are employed. These correction factors are obtained by fitting Navier-Stokes-based FE simulations of selected geometries to respective data from MD simulations. These correction factors are then used to extend the fluidic models, derived by means of continuum theory, into the MD regime where continuum theory is no longer applicable. More precisely, the non-physics-based factors reduce the fluidic resistances of the MLM with falling pressure and/or air gap widths. The factors Υ_{BT}, Υ_C, and Υ_O in the above-derived fluidic submodes are taken from Sattler [27], who did a thorough and systematic review of data and publications by Beskok

[15, 29], Sharipov and Seleznev [30–32], and Veijola [12], [30–35]. The factors $\Upsilon_{Reynolds}$, Υ_{BT}, and Υ_O are directly taken from or based on publications by Veijola [12, 34, 35], and the factor Υ_C is taken from Beskok [29]. The expressions for the correction factors read

$$\Upsilon_{Reynolds} = 1 + 9.638 Kn^{1.159} \tag{7.16}$$

$$\Upsilon_{BT} \approx \Upsilon_{Reynolds} \cdot \frac{1 + 0.5 D^{-0.5} \times 30^{-0.238}}{1 + 2.471 D^{-0.659}} \tag{7.17}$$

$$\Upsilon_C = \left(1 + 1.085 Kn \cdot \arctan(8\sqrt{Kn})\right) \left(1 + \frac{6 Kn}{1 + Kn}\right) \tag{7.18}$$

$$\Upsilon_O = \left(1 + \frac{6.703 Kn(1.577 + Kn)}{2.326 + Kn}\right) \cdot \left(\frac{1 + 0.688 D^{-0.858} \Lambda_0^{-0.125}}{1 + 1.7 D^{-0.858}}\right) \tag{7.19}$$

Here, $Kn = \lambda_f/d_f$ denotes the Knudsen number, where λ_f is the mean free path, which changes with the ambient pressure P_0, and d_f denotes the characteristic length of the flow. The characteristic length is different for each correction factor (Table 7.1). Furthermore, $D = \sqrt{\pi}/(2Kn)$ represents the parameter of rarefaction and $\Lambda_0 = L_r/(s_r/\sqrt{\pi})$ the ratio of the channel length L_r and the side length of the hole s_r. The presented correction factors are valid for square holes only. The given expression for Υ_{BT} is an approximation that is adequate in case of the investigated structure. A more general expression for Υ_{BT} and correction factors for circular holes can be found in [27].

Note that the critical issue when using correction factors is that the factors are not yet systematically validated by experiments. The factors are supposed to have a validity up to Knudsen numbers of $Kn \approx 880$ [12, 27], but only with respect to the data from MD simulations used for the fits. Consequently, applying these correction factors for Knudsen numbers so high is very doubtful as this would mean that the device operates deeply within the MD regime. Moreover, the correction factors were derived for simple, basic geometries only, for example, for a single channel but not for complex structures such as perforated microplates.

Table 7.1 Correspondence of correction factors and characteristic lengths.

Correction factor	d_f
$\Upsilon_{Reynolds}$ (Eq. 7.16)	g_k
Υ_{BT} (Eq. 7.17)	g_k
Υ_C (Eq. 7.18)	s_r
Υ_O (Eq. 7.19)	$s_r/2$

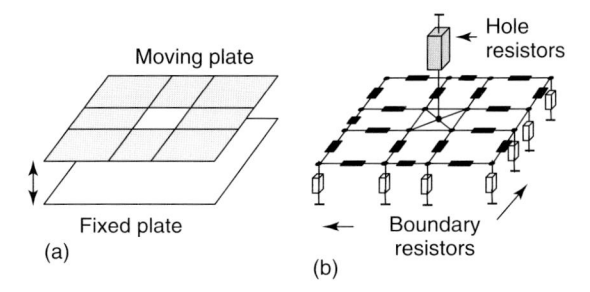

Figure 7.10 (a) Schematic diagram of the full mixed-level model using the example of a membrane with only one perforation. (b) Black 2D-boxes symbolize the finite network and 3D-boxes the lumped resistors along the boundary and at perforation holes.

7.3.4
Calculation of the Total Damping Force

Figure 7.10 illustrates the resulting mixed-level model for calculating the SQFD on perforated microstructures: the FN for the evaluation of the Reynolds equation is combined with physics-based lumped element models that account for the perforations and the pressure drop at the boundary.

The nodal damping forces $F_{\text{rey},k}$ are calculated in the MLM by multiplying the nodal pressures P_{k0} and the nodal areas A_k. The total damping force on the plate $F_{\text{rey, total}}$ is obtained by summing up all nodal contributions

$$F_{\text{rey,total}} = \sum_k F_{\text{rey},k} = \sum_k P_{k0} \cdot A_k \qquad (7.20)$$

In addition to the nodal damping forces exerted by squeezing the air out of the gap underneath the membrane ($F_{\text{rey},k}$), also the shear forces $F_{\text{shear},r}$ along the channel walls of the perforations have to be accounted for. Using the analytic expression derived by Sattler [27], the shear forces $F_{\text{shear},r}$ read

$$F_{\text{shear,total}} = \sum_r F_{\text{shear},r} = \sum_r s_r^2 \cdot R_{C,r} \cdot \left(Q_{r0,\text{rey}} + Q_{r0,\text{rel}}\right) \qquad (7.21)$$

7.3.5
Coupling with Mechanical Models

The fluidic mixed-level model is bidirectionally coupled with the mechanical model of the plate through the local gap widths $g_k(t)$ and through the damping forces in Eqs. (7.20 and 7.21). In the MLM, the gap width is allowed to vary from node to node. This enables to simulate deformable structures with nonuniform gap widths also [36].

Furthermore, the mixed-level model allows to tailor the mechanical model of the perforated plate according to the needs of the designer who is free to choose between rigid body motion [13, 24], torsional motion [37], the modal superposition technique [36] (Chapter 12), or any other. In the case of modal superposition, the local gap widths read

$$g_k(\underline{q}, t) = g_{0,k} + \sum_{i=1}^{m} \Phi_{i,k} \cdot q_i(t) \quad (7.22)$$

Here, $g_{0,k}$ denotes the initial gap width at the kth node, $\Phi_{i,k}$ denotes the value of the discretized eigenmode shape of the ith eigenmode at the kth fluidic network node, q_i denotes the amplitude of the ith eigenmode, and $\underline{q}$ denotes the vector of modal amplitudes. In order to obtain the modal moment for the coupling to the ith eigenmode, the nodal damping and the hole shear forces have to be multiplied with the discretized eigenmode shape function (Chapter 3.2)

$$M_{\text{Reynolds},i} = \sum_{k} \Phi_{i,k} \cdot F_{\text{rey},k} \quad (7.23)$$

$$M_{\text{shear},i} = \sum_{r} \Phi_{i,r} \cdot F_{\text{shear},k} \quad (7.24)$$

7.3.6
Automated Model Generation

As the models described in Sections 7.3.1 and 7.3.2 are rigorously based on geometry and material parameters, the presented mixed-level model is suitable

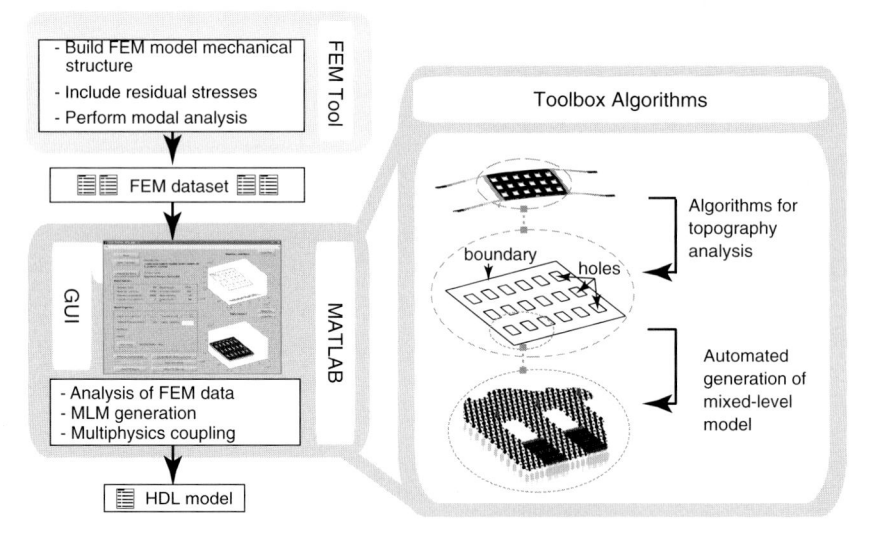

Figure 7.11 Workflow of the automated generation of MLMs using a MATLAB-based toolbox. First, the perforated plate has to be discretized and analyzed in a FEM tool, for example, ANSYS or COMSOL Multiphysics. Second, a FEM dataset comprising geometry and mechanical information has to be passed to the toolbox. Using a GUI and especially developed algorithms, the user can easily generate a fluidic MLM for calculating SQFD and couple this model to the other energy domains, for example, the mechanical or the electrostatic domain.

for automated model generation. To this purpose, a toolbox was developed in MATLAB [38]. Starting from the modal mechanical information and the discretized geometry of the perforated plate, the toolbox allows for the automatic compilation of mixed-level models. The resulting models are coded in an hardware description language (see workflow in Figure 7.11). An essential element of this toolbox are algorithms that analyze the geometry and automatically attach the boundary and hole models at the respective positions.

7.4 Evaluation

In this section, the mixed-level model is benchmarked with respect to Navier-Stokes-based FEM simulations, experimental data, and models representing the alternative approaches listed in Section 7.2.1.

7.4.1 Numerical Evaluation

For enabling a numerical assessment of the mixed-level model described in Section 7.3, the damping forces from MLM simulation are compared to results from Navier-Stokes-based FEM simulations performed with COMSOL Multiphysics. Following the assumptions made in Section 7.3.1, inertial effects of air were not considered in the FEM models. In order to enable an evaluation of the quality of the models used in the MLM, rarefaction effects (Section 7.3.3) were neglected, that is, no-slip conditions were used along walls in the FEM simulation and all correction factors in MLM were set to one.

A quadratic plate with dimensions 25 µm × 25 µm × 5 µm (width × length × thickness) with a gap width of 2 µm was considered for this study. The evaluation was performed step by step starting from a nonperforated plate with zero-pressure boundary conditions, going to a nonperforated plate with real boundary conditions, and finishing with a nonregularly perforated plate with real boundary conditions (Figure 7.12).

In order to give a first idea of the accuracy of the FN and the lumped element models, nonlinearities arising from a varying gap width and compressibility were neglected in both the MLM and the FEM simulations shown in Figure 7.13. A sinusoidal vertical displacement with an amplitude of 20 nm was enforced on the quadratic plate in the simulations. Figure 7.13a shows the results when zero-pressure boundary conditions directly at the boundaries of the plate and when real boundary conditions are assumed. In the case of the zero-pressure boundary conditions, the FEM- and MLM-simulated damping forces show excellent agreement with a maximum error of only 3.4%. This proves that FNs can really perform equivalently to FEM models.

In the case of the real boundary conditions, the FEM simulation also calculates the flow in the volume next to the boundary of the plate, whereas the MLM

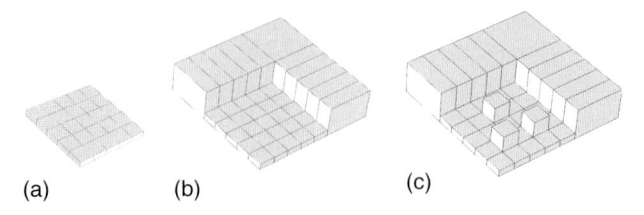

Figure 7.12 View of the FEM-discretized fluid domains that were simulated in COMSOL Multiphysics. In model (a), only the fluid in the gap was simulated assuming zero-pressure boundary conditions. In model (b), real boundary conditions were simulated using additional volumes attached to the plate. For the sake of clarity, some of the boundary volumes were omitted in the figure. In model (c), a plate perforated with three holes was simulated using real boundary conditions. Some of the boundary volumes were also omitted in this figure. In order to reduce the number of DOFs, outlet conditions directly at the ends of the channels were assumed. This was also taken into account in the MLM by reducing R_O accordingly.

simulation uses the lumped element models presented in Section 7.3.2. The agreement between the FEM and MLM results are still very good. The maximum error is only 6.1%. This proves that the approach of using lumped element models for taking boundary effects into account is viable. Please note the difference in the amplitudes of the damping forces: ignoring the real (and thus correct) boundary conditions results in a considerable error of 32%.

Finally, Figure 7.13b compares the results for a quadratic plate with three perforation holes and real boundary conditions. The two models still have an acceptable agreement with a maximum error of 11.7%.

Figure 7.14 shows damping forces calculated when the amplitude of the plate oscillation is increased and both varying gap widths and compressibility are taken into account. In this case, the damping forces are no longer in phase and have no longer a sinusoidal form. The MLM is able to reproduce this behavior. For the nonperforated plate with zero-pressure boundary conditions, the maximum amplitude error is 2.2% (Figure 7.14a). For the quadratic plate with three perforation holes and real boundary conditions, the maximum amplitude error is 10.9% (Figure 7.14b).

In summary, the error with respect to FEM simulation can slightly exceed 10%, but this is definitely acceptable when simulation time is also important. The MLM needs only a fraction of the simulation time and computational resources which FEM simulations need (Table 7.2). On the other hand, good agreement with FEM simulations is without doubt crucial, but the agreement with experimental data is more important and finally decides whether a model is worth using it or not.

7.4.2
Experimental Evaluation

The mixed-level model was evaluated with respect to the experimental data of three perforated microstructures: a large and a small silicon resonator (abbreviated as

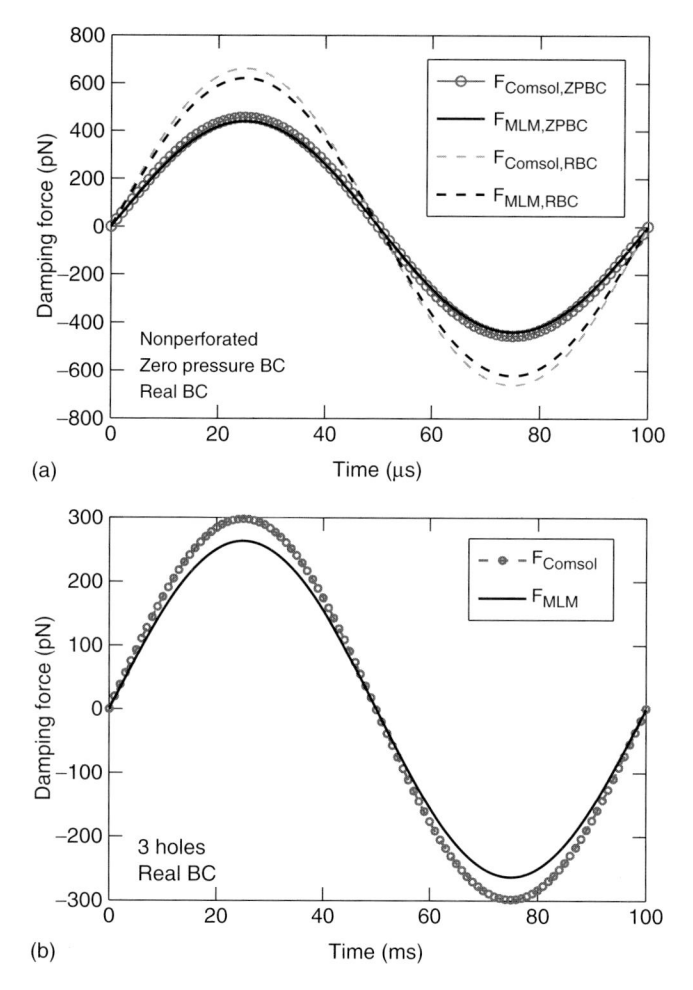

Figure 7.13 Comparison of the damping forces from FEM and MLM simulation neglecting compressibility and nonlinear effects. (a) Forces calculated for the nonperforated plate using zero-pressure ($F_{Comsol, ZPBC}$, $F_{MLM, ZPBC}$) and real boundary conditions ($F_{Comsol, RBC}$, $F_{MLM, RBC}$). (b) Forces calculated for the perforated plate using real boundary conditions.

"Resonator A" and "Resonator B"; see Figure 7.15) fabricated using SOI technology and one RF-MEMS switch (abbreviated as "RFS"; see Figure 7.15) fabricated using a surface micromachining process based on gold [39]. The technical data of the three devices is summarized in Table 7.3. A laser vibrometer equipped with a pressure-controlled chamber [36] was used for extracting the quality factors of the devices at different pressure levels. The electrostatically controlled structures were excited with a low-voltage white noise signal. The amplitudes of the resulting out-of-plane vibrations were in the magnitude of only ∼20 nm, thus ensuring the linearity of the devices concerning electromechanical phenomena. The quality

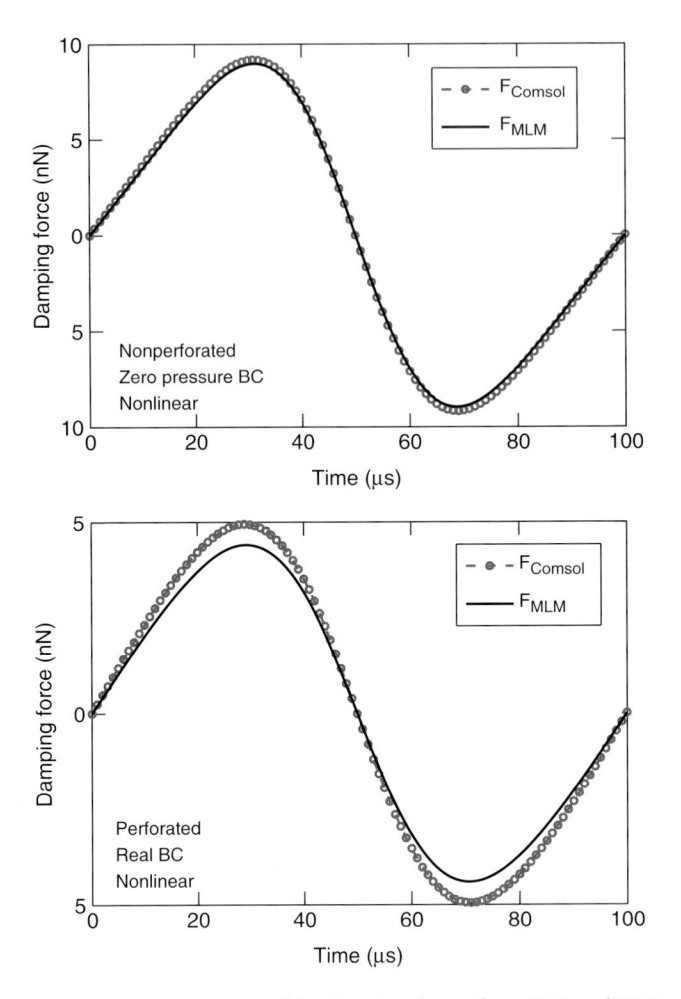

Figure 7.14 Comparison of the damping forces from FEM and MLM simulation taking compressibility and nonlinear effects into account and increased displacement amplitude. (a) Forces calculated for the nonperforated plate using zero-pressure boundary conditions. (b) Forces calculated for the perforated plate using real boundary conditions.

Table 7.2 Computational resources of the MLM and FEM models for the nonlinear simulations shown in Figure 7.14.

Geometry	MLM model			FEM model		
	DOFs	CPUs	Time	DOFs	CPUs	Time
Plate with zero-pressure BC	2604	1	10 s	$8.2 \cdot 10^5$	16	4 h
Plate with three holes and real BC	2245	1	9 s	$1.8 \cdot 10^6$	24	2 h 11 min

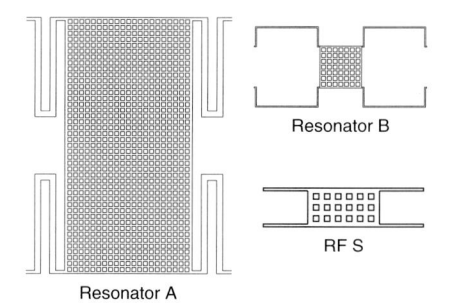

Figure 7.15 Schematic view of the investigated MEMS devices. All structures are electrostatically controlled. A folded beam design is used for the suspensions of the silicon resonators ("Resonator A" and "Resonator B") in order to reduce the resonance frequency of these devices. The RF-MEMS switch ("RFS") uses a straight beam design for the suspensions. Table 7.3 lists the technical data of the devices.

Table 7.3 Technical data of the characterized devices.

	Resonator A	Resonator B	RFS2
Material	Silicon	Silicon	Gold
Membrane width (µm)	425	133	140
Membrane length (µm)	850	127.6	260
Membrane thickness (µm)	15.65	15.65	5.2
Average gap width (µm)	2.2	2	3.1
Hole side length (µm)	13.3	13.3	20
Spacing between holes (µm)	6.3	5.7	20
Boundary frame width (µm)	(avg.) 8.45	(avg.) 3.5	20
Number of holes	21 × 43	7 × 7	3 × 6
Perforation level (%)	44.2	46.9	23
Resonance frequency (kHz)	30	44	14
Q_{Meas} at 960 mbar	17.35	37.47	13.58
Q_{Meas} around 0.1 mbar	553.43	404.5	280.8
Q_{Limit} (estimated)	700	500	350

As the gap width of the "RFS" varies between 2 and 3.7 µm, an averaged gap width is given.

factors were calculated by applying the half-power bandwidth method to the measured frequency spectra.

Figure 7.16 shows the measured quality factors of the three devices versus pressure. For all devices, we can observe that the quality factors increase with falling pressure but saturate at low pressures. This behavior can either be due to other dissipative mechanisms than SQFD (see effects listed in introduction to Section 7.2) that limit the quality factors of the devices at low pressures or the quality factor extraction method. In order to take these effects into account, a total

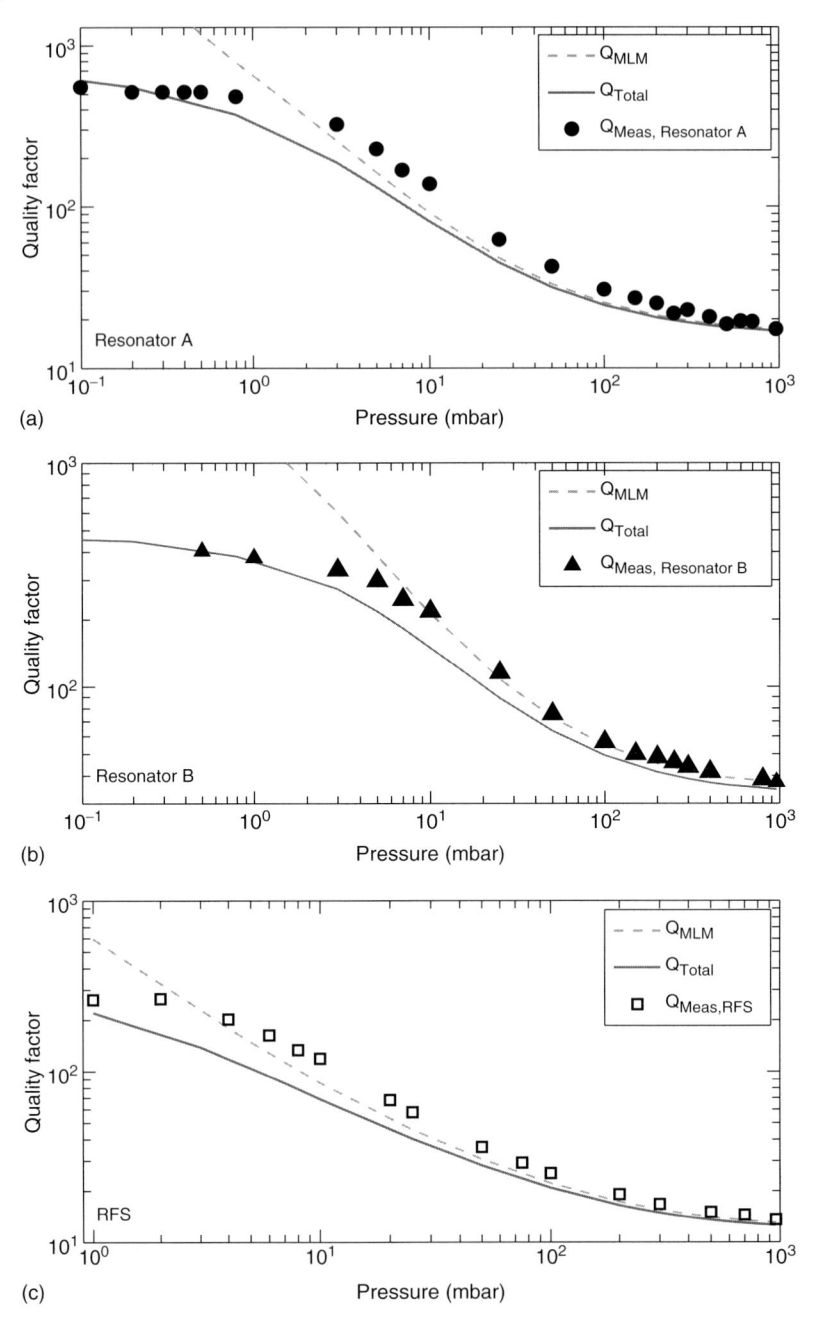

Figure 7.16 (a–c) Measured and simulated quality factors of the three investigated devices versus pressure. Q_{Meas} denotes the quality factor extracted from measurement, Q_{MLM} the quality factor obtained from MLM simulation, and Q_{Total} the quality factor calculated using Eq. (7.25.)

quality factor Q_{Total} is calculated

$$\frac{1}{Q_{Total}} = \frac{1}{Q_{MLM}} + \frac{1}{Q_{Limit}} \quad (7.25)$$

Here, Q_{Limit} denotes the quality factor due to the limiting damping mechanisms that is extracted for each device from measurements performed at low pressure (Table 7.3) and Q_{MLM} the quality factor obtained from MLM simulation. A mechanical model based on modal superposition was used to account for the dynamic deformation of the perforated structures and their suspensions.

The agreement of the measured and simulated quality factors is remarkably good (Figure 7.16). At normal pressure, the maximum error with respect to the measured quality factors is 7.2% for Q_{Total} and 3.7% for Q_{MLM}, that is, without taking Q_{Limit} into account (Table 7.4 for details)

The maximum error of Q_{Total} occurs for pressure levels in the transition regime, that is, between the slip flow and MD regime. Q_{MLM} is more accurate than Q_{Total} in this regime, but consequently overestimates the quality factor once the other mechanisms dominate the damping characteristics. In general, Q_{Total} and Q_{MLM} do not exceed an error threshold of 15% for pressures higher than 100 mbar.

It is very difficult to clearly identify the main source of error at pressures lower than 100 mbar. Possible sources of error are, for example, the lumped modeling of the limiting damping mechanisms (Q_{Limit}) and the not yet validated correction factors accounting rarefied gas effects (Section 7.3.3) that dominate the resistors of the MLM at these pressures. Only additional pressure-dependent experimental data of more and other types of microstructures would enable a clear identification.

7.4.3
Comparison with Alternative Damping Models

In order to benchmark the MLM, the damping model by Bao et al. [14] (analytical approach type 1 described in Section 7.2.1), the model by Veijola [18] (analytical approach type 2) and the mixed-mode model by Veijola et al. [19] (numerical approach type 3) were also evaluated with respect to the measured data. As the objective is to really assess the quality of the different approaches and not the quality

Table 7.4 Comparison of quality factors at normal pressure (960 mbar).

	Resonator A	Resonator B	RFS
Q_{Meas}	17.35	37.47	13.58
Q_{MLM} (relative error)	17.36 (−0.1%)	37.39 (0.2%)	13.08 (3.7%)
Q_{Total} (relative error)	16.94 (2.3%)	34.79 (7.2%)	12.61 (7.2%)

The relative error is put in brackets next to the absolute value. A positive error value means that damping is overestimated. A negative value means that damping is underestimated.

Table 7.5 Comparison of quality factors predicted by the alternative modeling approaches.

	Resonator A	Resonator B	RFS
Q_{Meas}	17.35	37.47	13.58
Q_{Bao} [14] (relative error)	16.18 (+6.7%)	24.96 (+33.4%)	6.98 (+48.6%)
Q^*_{Bao} [14] (relative error)	15.81 (+8.9%)	23.77 (+36.6%)	6.84 (+49.6%)
$Q_{\text{Veijola, A}}$ [18] (relative error)	15.25 (+12.1%)	23.36 (+37.7%)	18.04 (−32.8%)
$Q^*_{\text{Veijola,A}}$ [18] (relative error)	14.92 (+14%)	22.32 (+40.4%)	17.16 (−26.3%)
$Q_{\text{Veijola, N}}$ [19] (relative error)	21.43 (−23.5%)	44.05 (−17.6%)	19.87 (−48.3%)
$Q^*_{\text{Veijola,N}}$ [19] (relative error)	20.79 (−19.8%)	40.48 (−8%)	18.8 (−38.5%)

[a] The evaluation is performed at normal pressure (960 mbar). The relative error is put in brackets next to the absolute value. A positive error value means that damping is overestimated. A negative value means that damping is underestimated. $Q_{\text{Veijola, A}}$ denotes the quality factor calculated from the analytical model of Veijola [18], and $Q_{\text{Veijola, N}}$ denotes the quality factor calculated from the numerical mixed-mode model of Veijola and Raback [19].

of the factors accounting for rarefied gas effects, the evaluation was performed at normal pressure only. The procedures in [20, 21] were followed to calculate the quality factors from the different models.

The obtained quality factors are summarized in Table 7.5. They are calculated with and without taking the limiting damping mechanisms into account, respectively. If Q_{Limit} was taken into account using Eq. (7.25), the quality factor is marked with an asterisk.

In general, all analytical models tend to overestimate damping, especially in the case of the small resonator "Resonator B." The absolute error of the analytical models is in the range from 8.9% to 49.6%, when Q_{Limit} is taken into account. Please note that the error of the analytical models would change if also the losses due to the fluidic damping on the suspensions and the boundary frame would be taken into account [21].

The numerical mixed-mode model of Veijola shows an error in the range between 8% and 38%. Possible reasons for the high deviation in the case of the RF-MEMS switch are discussed in [22].

In summary, the MLM model performs very good, when compared to alternative models and has proved to be more reliable, when geometrical dimensions are varied and predictive extrapolation is required.

7.5
Conclusion

The mixed-level approach, which brings together both discretized and lumped modeling, was successfully applied to the fluid–structure interaction problem of SQFD in MEMS. The automatically generated mixed-level models showed good agreement with FEM-based Navier-Stokes simulations, excellent agreement with measured data, and furthermore proved to be more reliable and generally more accurate than three other state-of-the-art SQFD models. This demonstrates the power of the mixed-level modeling approach and, especially, its predictiveness.

An inherent advantage of this approach is that it allows for "tailored modeling." By interlinking models with different levels of abstraction across multiple energy domains, the designer can tailor the accuracy and computation time of the mixed-level model according to his or her needs. For instance, Voigt [1] uses a fully coupled three-dimensional thermoelectromechanical FN to simulate the self-test of an MEMS accelerometer. In the work presented in [36], the dynamic pull-in of an electrostatically actuated ohmic contact-type RF-MEMS switch is simulated using a mechanical model based on modal superposition with eigenmodes extracted by FEM simulation, a lumped electrostatic model, and a mixed-level model for the calculation of SQFD. The only limitation concerning the models is that they have to conform with the framework of generalized Kirchhoffian network theory.

The employed models should be, of course, as physics based as possible in order to make the resulting mixed-level model as predictive as possible. If no physics-based model is available or obtainable for a specific effect, a "black box" model extracted from FEM models using mathematical order reduction (Chapter 3) can also be integrated in this approach.

Concerning the design of microsystems, FNs and mixed-level models can be employed in the initial, conceptual design phase, for example, when no accurate analytical model is available for the effect of interest, or in the more detailed component design phase, for example, when FEM simulations are too time consuming or too difficult to integrate in the full system simulation.

The key factors for enabling the routine and successful use of the mixed-level approach at a company or an institute are the availability of models for the problem of interest, that is, a respective library of calibrated models, and an easy-to-use software tool for automatic model generation. This demands, of course, a considerable investment before first results are obtained from this approach. However, as was demonstrated in Section 7.4 and many other works [1, 13, 24, 36], this investment pays off.

References

1. Voigt, P. (2003) *Compact modeling of microsystems*, Selected Topics of Electronics and Micromechatronics, vol. 7, Shaker, Aachen, Germany.
2. Pandey, A. and Pratap, R. (2008) *J. Microfluid Nanofluid*, **4**, 205–218.
3. Del Tin, L., Iannacci, J., Gaddi, R., Gnudi, A., Rudnyi, E., Greiner, A., and Korvink, J. (2007) Non linear compact modeling of RF-MEMS switches by means of model order reduction. Technical Digest of the 14th International Conference on Solid-State Sensors, Actuators and Microsystems (Transducers'07), France, pp. 635–638.
4. Veijola, T. (1999) Finite-difference large-displacement gas-film model. Technical Digest of the 10th International Conference on Solid-State

Sensors, Actuators and Microsystems (Transducers'99), Japan, pp. 1152–1155.
5. Schroth, A., Blochwitz, T., and Gerlach, G. (1996) *Sens. Actuators, A*, **54**, 632–635.
6. Ngoya, E., Rousset, J., and Obregon, J. (1997) *IEEE Trans. Comput. Aided Des. Integr. Circ. Syst.*, **16** (6), 638–644.
7. Kundert, K. (1995) *The Designer's Guide to SPICE and SPECTRE*, Kluwer, Boston, MA.
8. Schrag, G., Voigt, P., Sieber, E., Wiest, U., Hoppe, R., and Wachutka, G. (1997) Device- and system-level models for micropump simulation. Proceedings of the Conference on Micromaterials (Micro Mat 1997), Germany, pp. 941–944.
9. Khalilyulin, R., Steinhuber, T., Schrag, G., and Wachutka, G. (2010) Hardware/software co-simulation for the rapid prototyping of an acceleration sensor system with force-feedback control. Proceedings of the 40th European Solid-State Device Research Conference (ESSDERC 2010), Spain, pp. 186–189.
10. Weinberg, M., Candler, R., Chandorkar, S., Varsanik, J., Kenny, T., and Duwel, A. (2009) Energy loss in MEMS resonators and the impact on inertial and RF devices. Technical Digest of the 15th International Conference on Solid-State Sensors, Actuators and Microsystems (Transducers'09), USA, pp. 688–695.
11. Hosaka, H., Itao, K., and Kuroda, S. (1995) *Sens. Actuators, A*, **49**, 87–95.
12. Veijola, T., Kuisma, H., and Lahdenperä, J. (1998) *Sens. Actuators, A*, **66**, 83–92.
13. Schrag, G. and Wachutka, G. (2002) *Sens. Actuators, A*, **97-98**, 193–200.
14. Bao, M., Yang, H., Sun, Y., and French, P. (2003) *J. Micromech. Microeng.*, **13**, 795–800.
15. Bahukudumbi, P. and Beskok, A. (2003) *J. Micromech. Microeng.*, **13**, 873–884.
16. Mohite, S., Kesari, H., Sonti, V., and Pratap, R. (2005) *J. Micromech. Microeng.*, **15**, 2083–2092.
17. Mohite, S., Sonti, V., and Pratap, R. (2006) Analytical model for squeeze film effects in perforated MEMS structures including open border effects. Proceedings of EUROSENSORS XX, Sweden, pp. 154–155.
18. Veijola, T. (2006) *J. Microfluid Nanofluid*, **2**, 249–260.
19. Veijola, T. and Raback, P. (2007) *J. Sensors*, **7**, 1069–1090.
20. Veijola, T., De Pasquale, G., and Somà, A. (2009) *J. Microsyst. Technol.*, **15**, 1121–1128.
21. De Pasquale, G., Veijola, T., and Somà, A. (2010) *J. Micromech. Microeng.*, **20** (12), 015010.
22. Niessner, M., Schrag, G., Iannacci, J., and Wachutka, G. (2011) Mixed-level modeling of squeeze film damping in MEMS: simulation and pressure-dependent experimental validation. Technical Digest of the 16th International Conference on Solid-State Sensors, Actuators and Microsystems (Transducers'11), China, pp. 1693–1698.
23. Hamrock, B. (1994) *Fundamentals of Fluid Film Lubrication*, McGraw-Hill, Singapore.
24. Schrag, G. and Wachutka, G. (2004) *Sens. Actuators, A*, **111**, 222–228.
25. Onsager, L. (1931) *Phys. Rev.*, **37**, 405–426.
26. Banks, R., Rose, D., and Fichtner, W. (1983) *IEEE Trans. Electron. Devices*, **ED-30** (9), 1031–1041.
27. Sattler, R. (2007) *Physikalisch basierte Mixed-Level Modellierung von gedaempften elektromechanischen Mikrosystemem*, Selected Topics of Electronics and Micromechatronics, vol. 28, Shaker, Aachen, Germany.
28. Sattler, R. and Wachutka, G. (2004) Analytical compact models for squeeze-film damping. Proceedings of Design, Test, Integration and Packaging of MEMS/MOEMS Conference (DTIP 2004), Switzerland, pp. 377–382.
29. Karniadakis, G. and Beskok, A. (2002) *Micro Flows: Fundamentals and Simulation*, Springer, New York.
30. Sharipov, F. and Seleznev, V. (1998) *J. Phys. Chem. Ref. Data*, **27**, 657–706.
31. Sharipov, F. (1999) *J. Vac. Sci. Technol. A*, **17**, 3062–3066.
32. Sharipov, F. (2000) Data on the slip coefficients. Proceedings of the 3rd International Conference on Modeling and Simulation of Microsystems (MSM'00), USA, pp. 570–573.

33. Veijola, T., Kuisma, H., Lahdenperä, J., and Ryhänen, T. (1995) *Sens. Actuators, A*, **48**, 239–248.
34. Veijola, T. (2002) End effects of rare gas flow in short channels and in squeeze film dampers. Proceedings of the 5th International Conference on Modeling and Simulation of Microsystems (MSM'02), USA, pp. 104–107.
35. Veijola, T., Pursula, A., and Raback, P. (2004) Extending the validity of existing squeezed-film damper models with elongations of surface dimensions. Proceedings of the 7th International Conference on Modeling and Simulation of Microsystems (MSM'04), USA, pp. 235–238.
36. Niessner, M., Schrag, G., Iannacci, J., and Wachutka, G. (2011) Macromodel-Based Simulation and Measurement of the Dynamic Pull-in of Viscously Damped RF-MEMS Switches. *In: Sensors and Actuators: A. Physical*, **172** (1), pp. 269–279.
37. Sattler, R., Plötz, F., Fattinger, G., and Wachutka, G. (2002) *Sens. Actuators, A*, **97-98**, 337–346.
38. Niessner, M., Schrag, G., and Wachutka, G. (2008) Reduced-order modeling and coupled multi-energy domain simulation of damped highly perforated microstructures. Proceedings of the 8th Congress on Computational Mechanics and the 5th European Congress on Computational Mechanics in Applied Sciences and Engineering (WCCM8/ECCOMAS 2008), Italy,
39. Rangra, K., Giacomozzi, F., Margesin, B., Lorenzelli, L., Mulloni, V., Collini, C., Marcelli, R., and Soncini, G. (2004) Micromachined low actuation voltage RF MEMS capacitive switches, technology and characterization. Proceedings of the IEEE 2004 International Semiconductor Conference, Romania, pp. 165–168.

8
Compact Modeling of RF-MEMS Devices
Jacopo Iannacci

8.1
Introduction

RF-MEMS, that is, microelectromechanical systems for radio frequency applications, have been reported in literature since more than one decade, highlighting their significant performance and characteristics, concerning basic passive components, such as variable capacitors (i.e., varactors) [1, 2], inductors [3, 4], and switches [5–8], and complex networks, such as phase shifters [9], impedance matching networks [10], and switching matrices [11–15]. Basic passive components in RF-MEMS technology present outstanding performance compared to their counterparts in standard semiconductor technology, such as high Q-factor, high linearity, low losses, and good isolation. By replacing standard passive components with their implementations in RF-MEMS technology within transceivers and circuits for telecommunication platforms, the performance of the whole systems can be boosted. Moreover, realizations of complex networks based on RF-MEMS components can replace entire subblocks of standard RF circuits (e.g., phase shifters, switching matrices, and so on), extending the reconfigurability and operability of the whole device, such as telecommunication platforms, satellites, and radar systems. Given these considerations, it is clear that RF-MEMS need to be properly modeled and simulated, as is typically done when dealing with standard semiconductor devices and circuits. However, MEMS technology, with its multiphysical nature that always implies the coupling of different physical domains with the mechanical properties of materials, makes the availability of proper simulation tools more difficult. Moreover, the integration of RF-MEMS devices with standard CMOS circuitry incorporated in the same system demands for the possibility of predicting the characteristics of a new hybrid RF-MEMS/CMOS block within a unique simulation environment. For all the above-mentioned reasons, a MEMS compact model library was developed by Iannacci *et al.* [16], implemented in the VerilogA$^{©}$ programming language [17], and used within the Cadence^{TM} IC (Integrated Circuit) development framework, exploiting the Spectre$^{©}$ simulator machine.

System-level Modeling of MEMS, First Edition. Edited by T. Bechtold, G. Schrag, and L. Feng.
© 2013 Wiley-VCH Verlag GmbH & Co. KGaA. Published 2013 by Wiley-VCH Verlag GmbH & Co. KGaA.

In this chapter, the MEMS model library is exploited to build up in Cadence the schematic of a complex MEMS-based network, namely, a multistate RF power attenuator, whose coupled electromechanical and electromagnetic behavior is simulated in Spectre and validated against experimental data. Very fast simulations are possible thanks to the above-mentioned MEMS model library, despite the rather pronounced complexity of the RF-MEMS network geometry. The working principle of the RF power attenuator simulated in this chapter has been already presented and discussed by Iannacci *et al.* in [18]. Briefly, the network features resistors of different value, implemented in a boron-doped polysilicon layer, loading the RF line. Such resistors can be selectively shorted by actuating series ohmic (cantilever type or clamped–clamped) MEMS switches, consequently changing the whole resistive load on the RF line and, in turn, the attenuation level realized by the network.

The whole network is composed of two sections, namely, a parallel and a series one. In the first one, the RF line is split into two parallel branches, each of which is loaded by three polysilicon resistors. Moreover, each RF branch can be selected or not, and when both are activated, two identical resistive loads are connected in parallel, halving the whole resistance and, in turn, doubling the number of different configurations the section can realize (refer to [18] for any detail). The switching stages of the parallel section are 4, realizing 16 different configurations, and the MEMS switches are series ohmic cantilevertype ones. On the other hand, the series section features just one RF line with three polysilicon resistors that can be selected or deselected by three clamped–clamped series ohmic MEMS switches.

The chapter is arranged as follows. Section 8.2 briefly introduces the simulation approach and the features of the developed software tool, exploited throughout the chapter for the modeling of the RF-MEMS network chosen as case study. Section 8.3 reports the measured and simulated electromechanical and electromagnetic behavior of the RF power attenuator parallel section, while Section 8.4 focuses on the series section. Moreover, the whole network formed by the parallel and series sections is simulated in Section 8.5. Conclusions and the list of references complete the chapter.

8.2
Brief Description of the MEMS Compact Modeling Approach

The MEMS compact model library exploited in this chapter for the simulation of an RF-MEMS-based complex network has been already presented in [16]. On the basis of the compact modeling approach, a MEMS device to be simulated can be subdivided into elementary components with certain characteristics. For example, a typical MEMS/RF-MEMS switch with a central suspended movable plate and flexible suspensions can be considered as an electromechanical plate transducer connected to straight elastic beams. The VerilogA software library includes a rigid plate element featuring a parallel plate transducer mathematical model and a flexible straight beam model based on the structural analysis theory. The elements feature a limited number of interconnection points (i.e., the two ends for the slender beam and the four vertices for the rigid plate) through which they can be joined together, defining the whole structure to be simulated (e.g., a MEMS switch). The library also includes boundary condition elements, such as anchors defining the mechanical constraints of the structure. An example of exploitation of the MEMS software library is reported in [19] and shown graphically in Figure 8.1. The top image reports the experimental 3D view of an RF-MEMS shunt variable capacitor (i.e., varactor), obtained with an optical profiling system. The bottom image shows the Cadence schematic of the same device composed with the elementary components available in the VerilogA library. The central rigid plate is connected to flexible straight beams suitably composed together in order to reproduce the serpentine-like suspensions of the physical device.

The solution of the coupled electromechanical problem is managed in Spectre according to Kirchhoff's laws, as the magnitudes at each node are described in terms

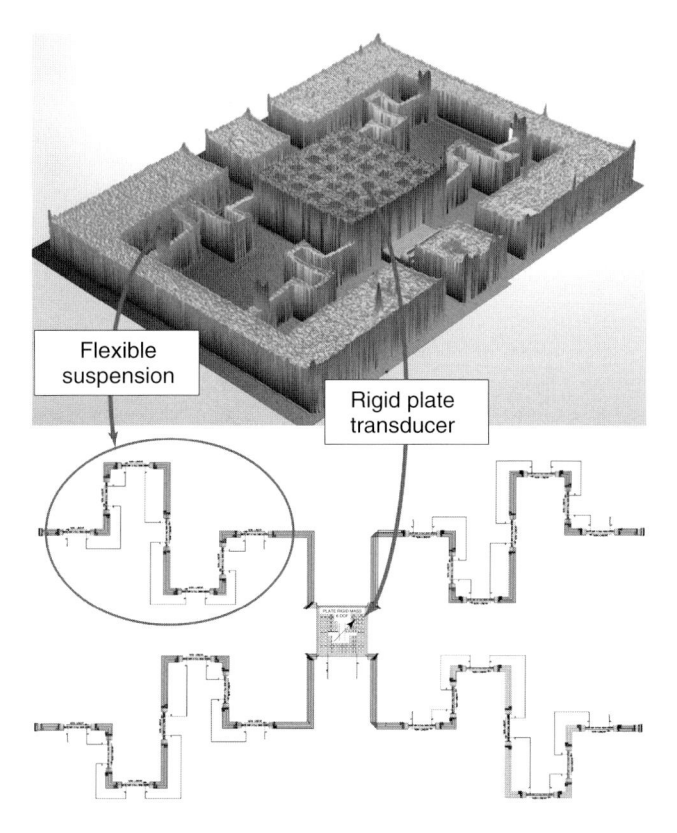

Figure 8.1 Top image: Measured 3D profile of an RF-MEMS variable capacitor (i.e., varactor) acquired with an optical profiling system. Bottom image: Cadence schematic of the RF-MEMS varactor composed of the elementary models included in the VerilogA software library.

of through and across variables [17], that is, current and voltage in the electrical domain and force and displacement in the mechanical domain, respectively. The availability of the discussed MEMS compact model library enables the fast and accurate simulation of complete devices and components based on RF-MEMS technology as well as the coupling, within the same simulation environment, of the electrical, mechanical, and electromagnetic physical domains. Such a tool is particularly useful in the preliminary design phase of a new device concept, as it enables varying several geometry features and observing in a short time their influence on the electromechanical and electromagnetic characteristics of the analyzed structure. The predictive accuracy of the VerilogA model library is discussed in detail in [19] and is extensively reported in the following sections of this chapter.

8.3
RF-MEMS Multistate Attenuator Parallel Section

This section, focused on the RF attenuator parallel section, starts with the description and simulation of the MEMS series switch employed in network architecture and on its coupled electromechanical characteristics. The top image in Figure 8.2 shows a microphotograph of the cantilever type series ohmic switch.

The left gold area is the anchored end of the membrane, while the two fingers in the right-hand image are the free end of the membrane, and they short the polysilicon resistive load (gray rectangle in figure) when the switch is pulled-in. The central plate is the actual electromechanical transducer suspended above the fixed polysilicon actuation electrode. The darker areas of the switch are due to a double-layer electrodeposition of gold (i.e., thicker and more stiff metal). Details of the technology can be found in [14]. The bottom image in Figure 8.2 shows the Cadence schematic of the switch assembled with the elementary MEMS components of the VerilogA library previously mentioned. The flexible parts of the switch, namely, the suspending part linking the anchored end to the transducer and the contact fingers (free membrane end), are implemented with the elastic straight beam model available in the library [20]. On the other hand, the electromechanical transducer is described by the suspended rigid plate [21]. Anchoring points complete the schematic, defining the mechanical constraints of the membrane-hinged end. The bias is applied in the schematic by a voltage source available in the standard component libraries of Cadence. The in-plane features of the switch are about 250 μm in length and 90 μm in width, while the gold thickness is around 2 and 5 μm, depending on whether one or both the metalizations are performed. For more details on the geometry, refer to [18]. Figure 8.3 compares the measured pull-in/pull-out characteristic of the cantilever, observed with a dynamic optical profilometer based on interferometry (WYKO 1100 DMEMS by Veeco [22]), with the Spectre DC simulation of the schematic in the bottom image of Figure 8.2.

The measured curve shows a double pull-in characteristic because of the increasingly uneven position of the suspended membrane with respect to the underlying

8.3 RF-MEMS Multistate Attenuator Parallel Section | 195

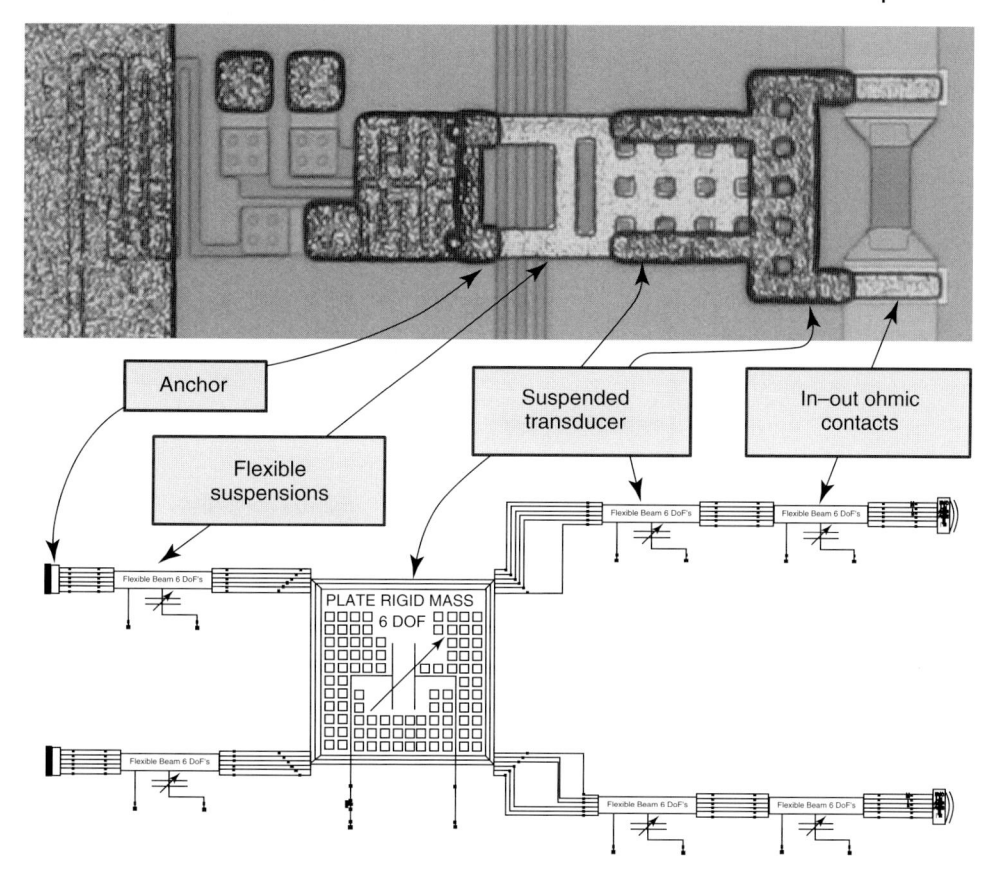

Figure 8.2 Top image: Microphotograph of the cantilever series ohmic MEMS switch present in the reconfigurable RF power attenuator (parallel section). Bottom image: Cadence schematic of the cantilever switch composed with the basic elements available in the VerilogA library. The anchoring points are visible in the left of the photo/schematic. The deformable parts of the switch are implemented with flexible straight beams, while the electromechanical transducer is described by the rigid suspended plate model.

surface, as the controlling bias voltage ramps up. Consequently, the first collapsing part is the tip of the two contacting fingers (the top image in Figure 8.2), changing (i.e., increasing) the elastic constant of the still suspended part, and indeed shifting the pull-in of the suspended transducer plate (second pull-in) higher. This characteristic is accurately predicted by the Spectre simulation (snap-down voltages at 25 and 30 V). The quantitative difference in the measured and simulated vertical displacement, around the double transition, is due to the fact that the interferometer averages the detected vertical position on a large portion of the suspended membrane, while the Spectre output refers to a node of the schematic. The pull-out voltage (15 V) is also predicted by Spectre with a reasonable accuracy.

After discussing the electromechanical simulation of the cantilever switch, the RF behavior of the whole parallel section is now treated. The top left image of

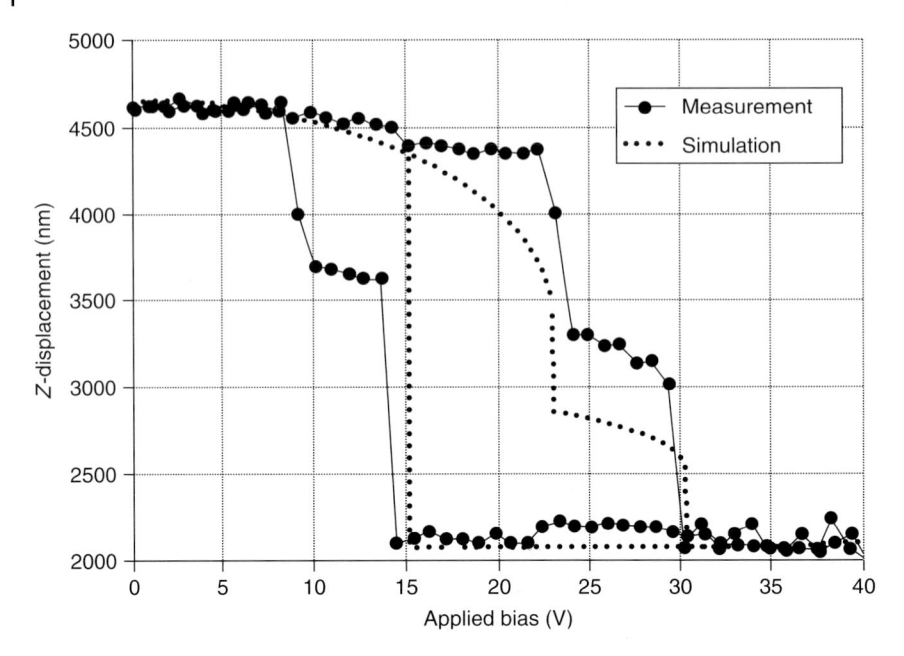

Figure 8.3 Comparison of the measured (dynamic optical profilometer) and simulated (DC simulation in Spectre) pull-in/pull-out characteristic of the cantilever MEMS switch in Figure 8.2. The double pull-in, due to the uneven position of the cantilever when biased and occurring at 24 and 30 V, is accurately predicted by the Spectre simulation. Also, the measured and simulated pull-out levels (around 15 V) show a good match.

Figure 8.4 shows the microphotograph of the entire RF attenuator parallel section, highlighting the CPW (coplanar waveguide) input and output (Port 1 and Port 2, respectively). The top right image of Figure 8.4 shows the close-up of the intrinsic MEMS part of the network, where the two parallel RF lines are visible, as well as the polysilicon resistive loads and the eight cantilever switches reconfiguring the state of the network. The switches of the stages labeled b, c, and d are actuated in pairs. This means that the three resistors $R1$, $R2$, and $R3$ (25, 100, and 500 Ω, respectively) are both shorted or are loading the RF line. Alternatively, the switches a1 and a2 can be controlled independently, selecting one or both the RF line branches (i.e., series or parallel). The Spectre schematic, based on the VerilogA compact models, corresponding to the parallel network in the top left image of Figure 8.4, is depicted in the bottom image of Figure 8.4. The schematic of each of the eight switches is the same reported in the bottom image of Figure 8.2. The vertical displacement pin of each switch is connected to a two-state resistor, also available in the VerilogA library, highlighted in the a1 switch and shown in the magnified part in Figure 8.5. A minimum and maximum resistance value is specified in the device parameters, and the Rs resistor realizes the first one when the cantilever is actuated and the second one when it is not pulled-in. The values of $R1$, $R2$, and $R3$ are adjusted in order to account for the measured contact resistance

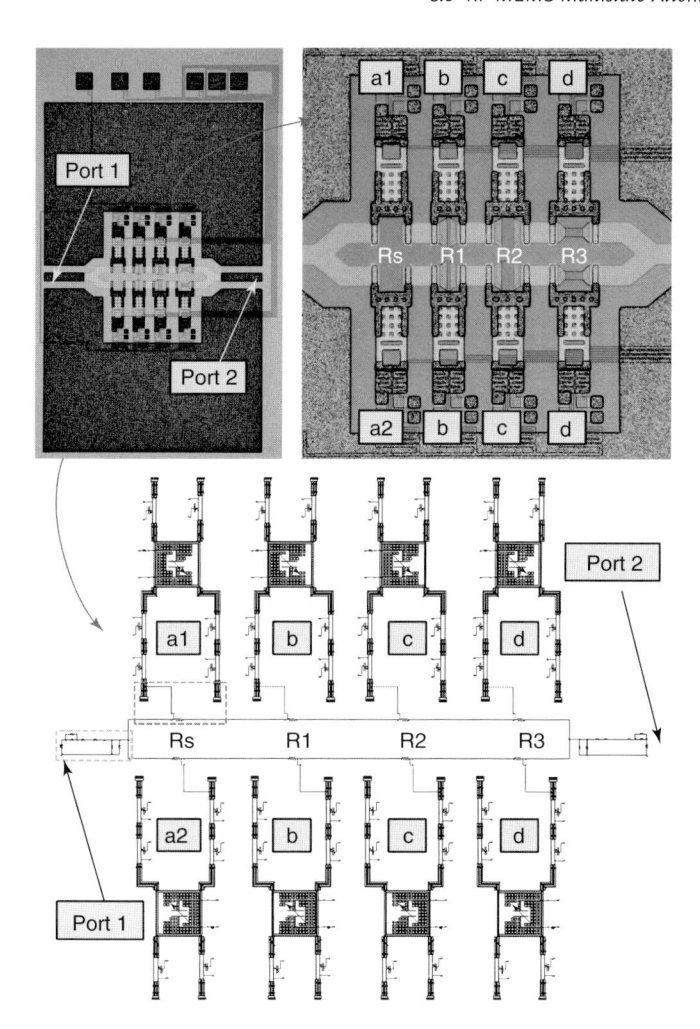

Figure 8.4 Top left image: Microphotograph of the RF attenuator parallel section, comprising four switch sections. Top right image: Magnified image of the switch sections, where the double RF line is visible. The switches a1 and a2, independently controllable, select the upper and lower RF lines when actuated, respectively, while the switch pairs b, c, and d, when actuated, short the load resistors $R1$, $R2$, and $R3$. Bottom image: Cadence schematic of the RF attenuator parallel section. The schematic of each of the eight switches is the same as shown in the bottom panel of Figure 8.2. The two-state resistor (Rs) and the input lumped element network (close to Port 1), highlighted in the schematic, are detailed in Figures 8.5 and 8.6, respectively.

of the transition between the RF line (multimetal) and the polysilicon resistor [18]. The RF ports for the Spectre S-parameter simulation (Port 1 and Port 2) are highlighted in the schematic (bottom image of Figure 8.4), and a lumped element network, accounting for the distributed parasitic effects, is visible close to them. In particular, the network close to Port 1 is highlighted. Its topology is detailed in

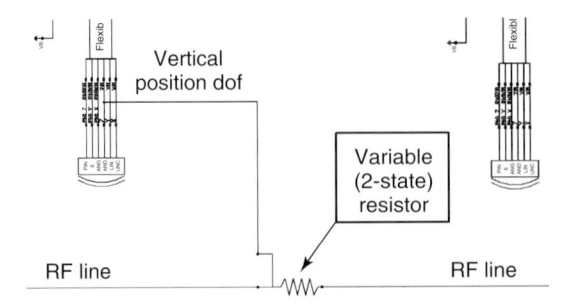

Figure 8.5 Close-up of the schematic in bottom image of Figure 8.4 in correspondence with section a1, highlighting the two-state resistor symbol loading the RF line. When the switch is not actuated, the resistance of section a1 is maximum (i.e., open circuit), while when it pulls-in, the resistance switches to the minimum value (i.e., short circuit).

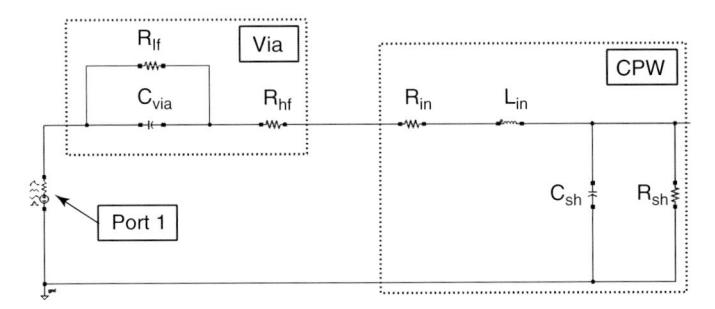

Figure 8.6 Lumped element network describing the nonidealities of the RF power attenuator. The section labeled as "Via" includes the series parasitic capacitance (C_{via}) due to the residual oxide layer, and two resistive contributions, namely, R_{lf}, acting in the low-frequency range, and R_{hf}, acting in the whole frequency range. The section labeled as CPW models the distributed effects of the input/output waveguide sections of the RF power attenuator (R_{in}, L_{in}, C_{sh}, and R_{sh}). The network is included in the schematic in the bottom image of Figure 8.4, while the values of the lumped elements are listed in Table 8.1.

Figure 8.6 and defined according to a well-known approach presented by Dambrine et al. [23, 24], already applied by the author for the electromagnetic modeling of RF-MEMS [19].

The lumped element network in Figure 8.6 is composed of two sections, highlighted and labeled as Via and CPW. The first one models the nonidealities of the gold to multimetal transition (for details concerning the technology refer to [18]). A residual thin titanium oxide layer (due to a technology issue) introduces a large parasitic series capacitance on the RF line, represented in the lumped network by C_{via}, and resistive losses acting both in parallel to the parasitic capacitance (R_{lf}) in the low-frequency regime and in series, through the whole analyzed spectrum (R_{rf}). This network topology, modeling the nonideal via, was already reported by the author in [16]. The values of these lumped elements are reported in Table 8.1 and obtained by experimental measurements performed on test structures belonging to the same fabrication batch of the measured RF power attenuator. On the other

Table 8.1 Values of the lumped elements extracted in the network accounting for the non-idealities of the gold to multimetal transitions and input/output CPWs, reported in Figure 8.6 and included in the schematic in the bottom image of Figure 8.4 both at the input (Port 1) and at the output (Port 2).

C_{via}	R_{lf}	R_{hf}	R_{in}	L_{in}	C_{sh}	R_{sh}
10 pF	15 Ω	15 Ω	1 Ω	300 pH	20 fF	1 GΩ

hand, the section of the lumped element network in Figure 8.6, labeled as CPW, accounts for the distributed parasitic effects due to the input/output portions of CPW, bringing the RF signal from the GSG (ground-signal-ground) probes to the intrinsic RF-MEMS network (top left image of Figure 8.4 and top right image of Figure 8.4). The topology of the CPW is well known in literature [25, 26], and the values of R_{in}, L_{in}, C_{sh}, and R_{sh}, reported in Table 8.1 have been extracted according to the approaches already discussed in Refs [16, 19]. The just described lumped element network, included at Port 1 and Port 2 of the Cadence schematic in the bottom image of Figure 8.4, is capable of reproducing the behavior of the measured S-parameters of the analyzed network. Such a network founds both its topology and the values of the components it features on physical assumptions. The influence of the parasitic effects on the electromagnetic characteristic of the real device is discussed in the following text.

Figure 8.7 shows the comparison between a few measured and simulated attenuation levels (S21 parameter) of the RF power attenuator parallel section, up to 30 GHz. The minimum and maximum attenuation levels are plotted, and, referring to the labels in Figure 8.4, they correspond to the switches a1-b-c-d and a1 actuated, respectively. Other three configurations presenting an increasing level of attenuation are also measured and plotted, corresponding to the switches a1-d, a1-a2, and a1-c actuated, respectively. The S21 characteristic is rather flat over the whole frequency range, especially for the lower attenuation levels, while for the larger ones it presents at most a decrease of 4 dB from DC up to 30 GHz. The characteristic of the lower attenuation levels also highlights more clearly the bend because of the parasitic series capacitance of the gold to multi-metal transitions (discussed earlier and modeled in Figure 8.6). The capacitance C_{via} lets the parallel parasitic resistance (R_{lf}) dominate at low frequency (up to 4–5 GHz), leading to larger losses that tend to decrease the more the parasitic capacitance gets closer to a short circuit. The S21 characteristics simulated in Spectre (S-parameter simulation) for the same network configurations are superimposed on the measured curves in the plot in Figure 8.7.

Starting from the curve corresponding to the switches a1-b-c-d actuated (i.e., minimum attenuation level) and looking also to the next two attenuation levels (switches a1-d and a1-a2 actuated), the Spectre simulation is in good agreement with the experimental characteristic. In particular, the curve bending in the low-frequency range (up to about 5 GHz), due to the parasitic effects of the gold to multimetal transitions, is qualitatively well predicted. This implies that the topology

200 | *8 Compact Modeling of RF-MEMS Devices*

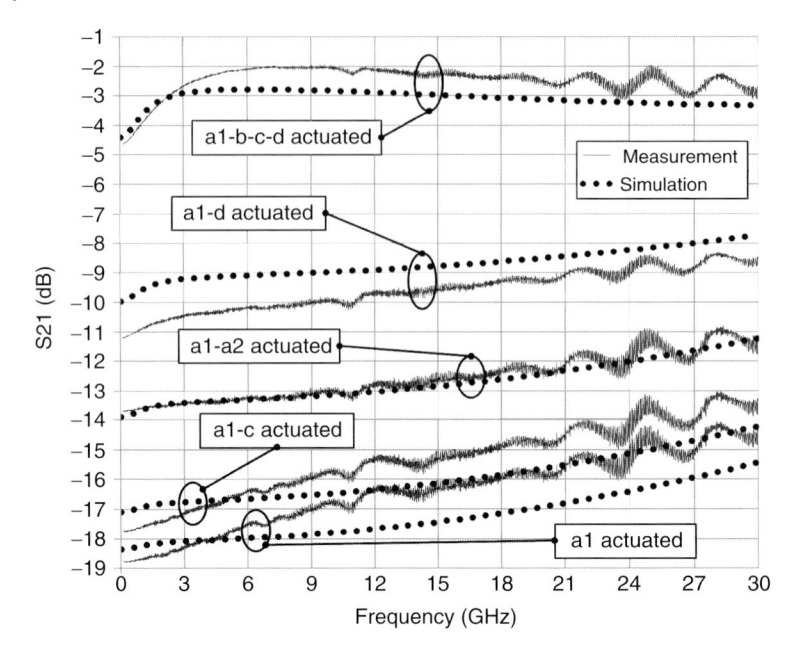

Figure 8.7 Comparison of the measured and simulated (S-parameter in Spectre) S21 parameter for a few configurations of the network in the bottom image of Figure 8.4. The quantitative behavior of the simulated curves with respect to the experimental ones is satisfactory, although in a few cases the qualitative trend does not fully match the measurements. However, the resistive loads are set to the nominal value, and possible technology spreads of the layer resistivity are not accounted for. In any case, the shift between simulations and measurements is never larger than 1 dB.

of the network modeling the via parasitics (Via section in Figure 8.6) is correct on a physical basis, and it is not simply a best fit case. Moreover, the resistance value for R_{lf} and R_{hf} is not freely chosen but is experimentally extracted from test structures, purposely included in the same fabrication batch, to determine the process characteristics. Looking at the other two S21 characteristics (switches a1-c and a1 actuated), the Spectre qualitative prediction of the curve behavior is definitely coarser if compared to the other attenuation levels. Neglecting the influence of the via's parasitic effects that mainly dominate the lower attenuation levels in the low-frequency range, the larger the resistive load on the RF line, the more pronounced is the slope of the attenuation level that decreases as the frequency rises. This trend is due to the resistive load that does not behave just as a resistor at high frequency; while in the Spectre schematic, it is instanced as an ideal component. This consideration is also valid for the CPW sections (with a buried multimetal signal line) between one switch section and the adjacent one (top left image of Figure 8.4) that are not accounted for at all in the Spectre schematic in the bottom image of Figure 8.4. Eventually, the multimetal to polysilicon transition nonidealities are accounted for in the Spectre simulation by just adding the measured contact resistance (7 Ω per each transition). However, such a transition can also be more accurately modeled relying on a lumped element network that features at least one capacitor (in series on the RF line), in order to better describe the frequency response of a resistor [27]. It is reasonable to assume that the slope of the experimental curves with the switches a1-c and a1 actuated (i.e., the larger attenuation levels in Figure 8.7) is mainly due to the reactive parasitic effects of the multimetal to polysilicon resistors, and it tends to increase with the number of resistors loading the RF line. To this purpose, it should be noted that in this network, when a resistor is shorted by the actuated MEMS switch, the two multimetal to polysilicon transitions (before and after the load) are also shorted. Despite all the effects just described and not accounted for in the Spectre simulation, the superimposition of the simulated curves with the experimental ones is rather satisfactory in all the analyzed cases. Even when the qualitative trend of the simulated versus measured characteristic shows a certain disagreement, the quantitative comparison highlights a difference that is never larger than 1 dB. This is a good achievement, especially when keeping in mind that all the resistance nominal values set in Spectre are averaged quantities collected by the batch testing, and consequently do not take into account possible local spreads in different areas of the same processed silicon wafer.

The philosophy underlying the discussed example is that the predictive accuracy refinement of the Spectre simulation can be defined by the RF-MEMS engineer, depending on his or her needs. This means that the intrinsic MEMS network, based on the compact models in VerilogA, can be enriched with lumped elements within the Cadence schematic, in order to achieve the desired level of accuracy. This approach is the same followed in the modeling of high-frequency (Microwave) transistors [23, 24], in which the complexity of the equivalent lumped element network increases as additional parasitic and second-order effects are to be accounted for. The Cadence simulation environment, as well as other commercial IC development frameworks, is particularly suitable for a modular and iterative approach aimed at the best fit of simulations against experimental data. Moreover, it is also appropriate for the optimization of new designs, bearing in mind a set of specifications that must be complied. This is possible thanks to two significant features that such a typology of commercial tools offers. On one hand, they enable an easy and fast assembly of the device of interest, entirely managed with libraries of symbols at schematic level, that also exploit the concept of hierarchy to nest device parts and network portions within simplified symbols to be used at higher level. For example, in the schematic in the bottom image of Figure 8.4, adding more lumped elements to account for second-order effects, like the ones discussed above, would require a very limited effort for the RF-MEMS engineer. On the other hand, the ease in modifying and tuning the network topology, on the basis of the needs of the RF-MEMS engineer, is sustained by the very limited amount of time typically necessary to run a simulation within Cadence. For instance, each of the Spectre S-parameter simulations of the network in the bottom image of Figure 8.4, in which the state of each switching section is determined applying a DC bias larger than the pull-in level (in this case 40 V DC) to the switches meant to be actuated (mixed electromechanical and electromagnetic simulation), takes about 20 s for its complete execution on a standard desktop PC.

8.4
RF-MEMS Multistate Attenuator Series Section

This section focuses on the series section of the MEMS-based RF power attenuator. Following the same approach as in the previous section, the first aspect to be discussed is the electromechanical simulation of the switching unit exploited in the network. The top image in Figure 8.8 shows a microphotograph of the RF-MEMS switch.

Similar to the parallel section, the switches are also series ohmic ones, although they are now based on a clamped–clamped configuration. As the microphotograph highlights, two cantilever membranes, similar to the one in the top image of Figure 8.2, are mirrored with respect to the central RF line and joined together by the two contact fingers visible in the central part in the top image in Figure 8.8. Consequently, in this case, there are two biasing electrodes, acting on the two opposite sides of the central line. The bottom image in Figure 8.8 shows the Spectre schematic of the clamped–clamped switch, and all the correspondences between the physical device and the schematic, such as the anchors, flexible suspensions, and electromechanical transducers, are highlighted. The simulated and measured pull-in characteristic of the clamped–clamped series ohmic RF-MEMS switch discussed in this section is shown in Figure 8.9.

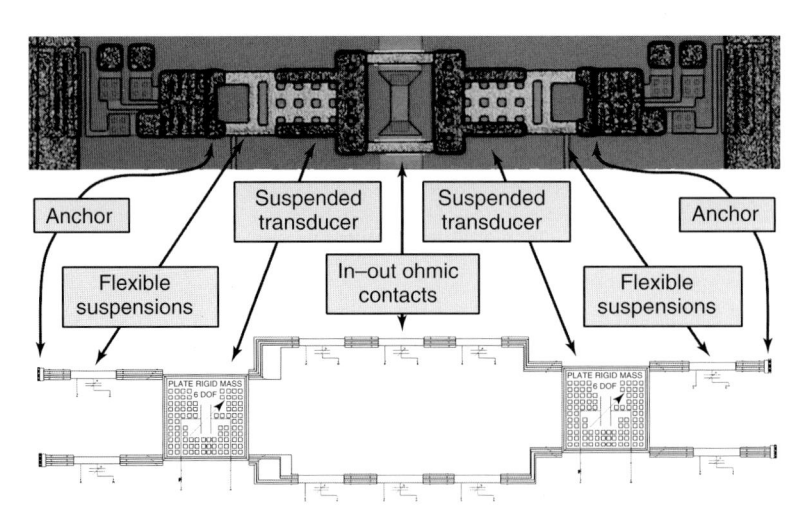

Figure 8.8 Top image: Microphotograph of the clamped–clamped series ohmic MEMS switch present in the reconfigurable RF power attenuator (series section). Bottom image: Cadence schematic of the clamped–clamped switch, composed of the basic elements available in the VerilogA library.

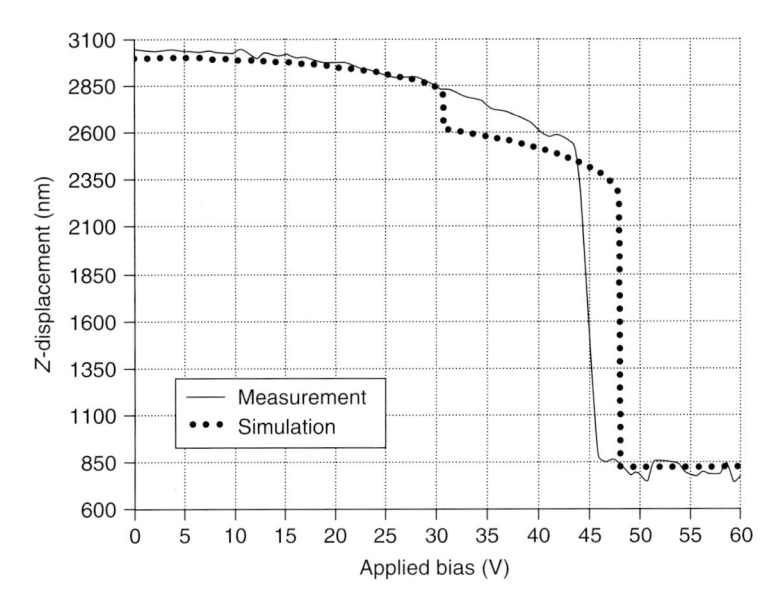

Figure 8.9 Comparison of the measured (dynamic optical profilometer) and simulated (DC simulation in Spectre) pull-in characteristic of the clamped–clamped MEMS switch in Figure 8.8. The measured actuation (46 V) is overestimated by about 2 V by Spectre, while the measurement of the pull-out is not available.

The simulated curve (Spectre DC simulation) presents a double pull-in, as it was for the cantilever switch discussed in Section 8.3, even though in this case the same characteristic is not visible in the measured vertical displacement versus the applied bias (WYKO 1100 DMEMS by Veeco [22]). The simulated displacement is detected at a node of the flexible suspensions close to the anchoring area of the MEMS structure, while the measured vertical displacement characteristic is averaged in the central part of the suspended membrane. Because of this reason, the double pull-in is not visible in the experimental characteristic. Beside the qualitative mismatch, the pull-in voltage is predicted by the Spectre simulation with a difference of 2 V (4% error) with respect to the measurement (46 V, Spectre; 48 V, measurement). In this case, experimental data concerning the pull-out of the RF-MEMS switch are not available for the validation of the simulated results. After the validation of the electromechanical behavior of the MEMS switch, the attenuator series section RF behavior is now discussed.

The microphotograph of the RF power attenuator series section is reported in the left panel of Figure 8.10, highlighting Port 1 and Port 2, as well as the e-f-g switching sections, composed of the clamped–clamped switching units just discussed (Figure 8.8). The resistive loads are characterized by the same values previously reported for the network parallel section, namely, 25, 100, and 500 Ω for the e-f-g switching sections, respectively. The right panel of Figure 8.10 shows the Cadence schematic of the attenuator series section implemented with the compact models available in the VerilogA library.

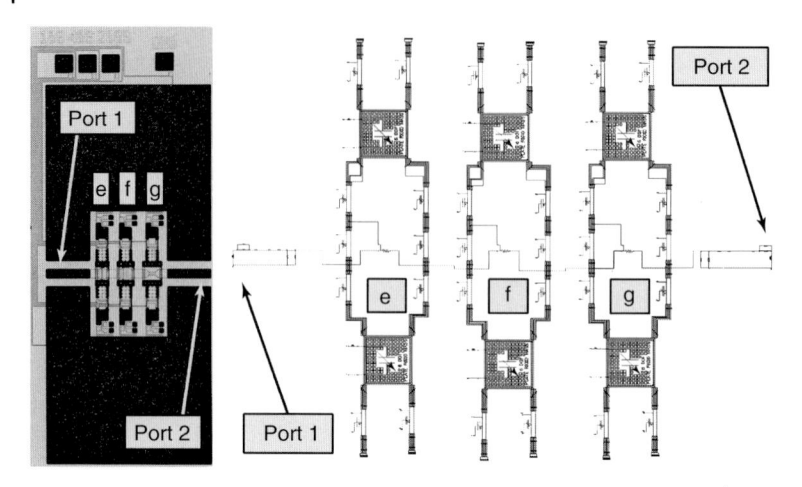

Figure 8.10 Left image: Microphotograph of the RF attenuator series section, comprising three switch units. The switch pairs e, f, and g, when actuated, short the load resistors R1, R2, and R3 on the RF line. Right image: Cadence schematic of the RF attenuator series section. Each of the three switches schematic is the same as reported in the bottom image of Figure 8.8.

The lumped element networks accounting for the connecting CPW sections and for the nonidealities of vias are visible nearby Port 1 and Port 2. The topology of such a network is the same as employed in the parallel section, detailed in Figure 8.6, and the values of the comprised elements are listed in Table 8.1. Moreover, the ohmic contact of each switch with the underlying RF line is modeled with the same two-state resistor, already exploited in the parallel section, and highlighted in Figure 8.5. Figure 8.11 depicts a comparison of the measured and simulated maximum and minimum attenuation level (S21 parameter) of the network series section. Unlike in the parallel section (Figure 8.7), the S-parameter characteristic is measured only up to 13.5 GHz with a less performing VNA (vector network analyzer) [28]. In the maximum attenuation level, none of the three switching sections is biased, and the three resistances load the RF line. On the other hand, to reach the minimum attenuation level, a 60 V DC bias is applied to the three switching units, both in the measurement and in the Spectre S-parameter simulation.

The minimum attenuation level of the series section presents a rather high value (around 8.5 dB) if compared to the 2–3 dB level of the parallel section. This is due to a nonperfect physical contact of the clamped–clamped switch suspended fingers with the underlying input–output RF ohmic contacts (top image in Figure 8.8). The contact fingers tend to be arched when the switch is actuated, reducing drastically the contact area and making worse, in turn, the quality of the ohmic contact. Such an undesired effect, ignored during the design of the RF power attenuator network, has been observed by means of 3D optical inspections (with a white light optical interferometer), and the collected data are not reported here for the sake of brevity. An additional resistance in series to each of the three resistive loads accounts

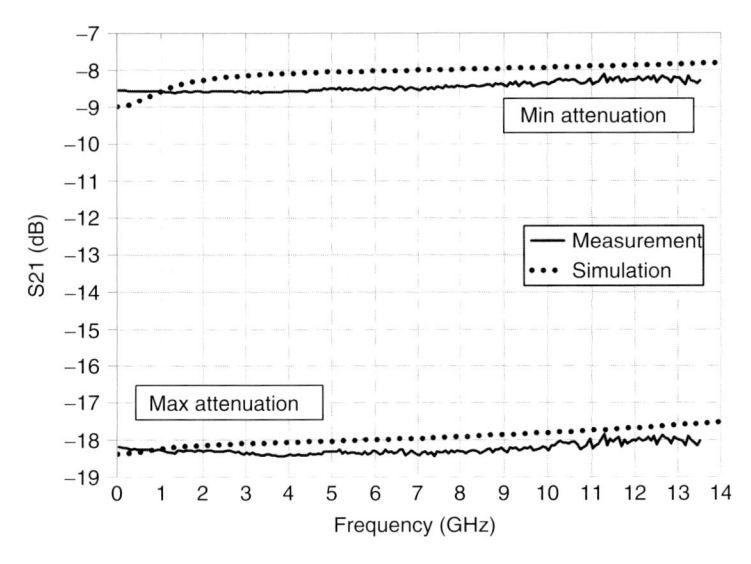

Figure 8.11 Comparison of the measured and simulated (S-parameter in Spectre) S21 parameter for the minimum and maximum attenuation levels of the network in the left panel of Figure 8.10 up to 13.5 GHz.

for the via parasitic resistance. This issue does not affect the parallel section as it employs cantilever-type switches. Consequently, when the switches reach the actuated state, the contact fingers tilt over the ohmic contacts below, and the contact area is sufficient to ensure a good ohmic contact (i.e., a low contact resistance). On the other hand, in the clamped–clamped switch, the contact fingers move downward till reaching the contact with the pads below. At this stage, the contact fingers are hinged on the two edges of the ohmic contact area, and when the two lateral movable electrodes move further down, attracted by the electrostatic force, the central part of the contact fingers arches upward because of a leverage-like behavior, increasing the contact resistance.

8.5
Whole RF-MEMS Multistate Attenuator Network

In this section that concludes the chapter, the parallel and series parts of the network (discussed in Sections 8.3 and 8.4, respectively) are considered together to analyze the whole RF power multistate network. As measurements of the entire network were not available at the time of writing, only Spectre simulations are reported. However, the composition of the whole network schematic is based on the results of the two constitutive subnetworks that were tailored to match the available experimental evidences. The schematic of the whole RF-MEMS reconfigurable power attenuator is shown in Figure 8.12. In this figure, the schematics of the parallel and series sections, previously depicted in the bottom image in Figure 8.4

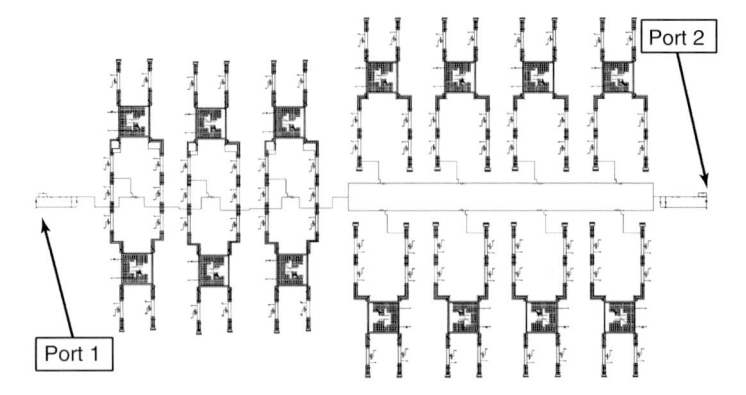

Figure 8.12 Spectre schematic of the whole network, comprising both the series and parallel sections. The network, not tested yet, realizes $2^7 = 128$ different attenuation levels, and the input/output lumped element networks are the same as previously reported in Figure 8.6.

and the right panel in Figure 8.10, are visible. The topology of the input and output lumped element network is the same as that employed in the parallel section and reported in detail in Figure 8.6. The values of the component elements are reported in Table 8.1. Figure 8.13 shows several of the 128 attenuation levels (S21 parameters) implemented by the reconfigurable network, occurring between the minimum and maximum ones, and simulated in Spectre (S-parameter simulation) up to 40 GHz. The switches of the parallel section (cantilever type) are actuated in Spectre by applying a DC bias of 40 V, while the ones of the series section

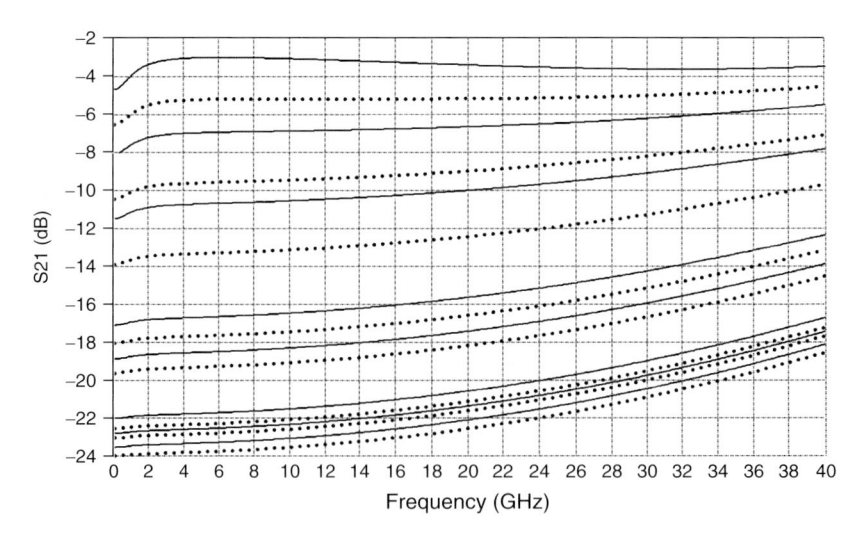

Figure 8.13 Spectre simulation (S-parameter from DC up to 40 GHz) for a few of the 128 attenuation levels implemented by the RF reconfigurable power attenuator schematic reported in Figure 8.12.

(clamped–clamped) are controlled with 60 V DC. In the simulation of the whole network, the nonideality of the ohmic contact referred to in the case of the clamped–clamped switches is neglected (Section 8.4).

8.6 Conclusions

In this chapter, the exploitation of a VerilogA-based MEMS model library within the Cadence simulation environment was presented and discussed. The focus was on tool exploitation for the accurate mixed-domain (electromechanical and electromagnetic) simulation of a complex RF-MEMS-based network, namely, a reconfigurable multistate RF power attenuator.

The network features two main sections, a parallel and a series one. The first type features two branches of resistors loading the RF line connected in parallel, while the second one comprises a unique line with resistors. The coupled electromechanical behavior of the ohmic switches (a cantilever type and a clamped–clamped one) of which the two subnetworks are composed, was analyzed. The switches geometry (at schematic level) was composed within Cadence by connecting the elemental components of the VerilogA library, and the pull-in/pull-out characteristic was simulated in Spectre (DC simulation). The comparison of simulations and measurements showed a rather accurate prediction of the pull-in and pull-out voltages of the switches in Spectre. All the simulations were performed within reduced computation times (in the range of several seconds on a desktop PC).

The electromagnetic behavior of the RF attenuator subsection was then analyzed. Each of the two parts features multiple replicas of the RF-MEMS ohmic switches previously mentioned. They are then included in the schematic of the network, and several elements are added in order to model the RF behavior as well as nonidealities and distributed parasitic effects. Suitable lumped element networks are included in the device schematic, in order to model additional effects. All the values of the parasitic effects have been extracted by conducting batch measurements on dedicated test structures. Both concerning the series and parallel sections, the comparison of S-parameter measurements and the Spectre (S-parameter) simulations show a good agreement. This proves the correctness and conceptual soundness of the proposed modeling approach. Finally, the whole network, featuring both the series and parallel networks, was simulated in Spectre, referring to several of the 128 states it realizes.

In conclusion, this contribution reported on the practical exploitation of a MEMS model library, based on the VerilogA programming language, for the simulation of real RF-MEMS devices and networks within the Cadence IC development framework, by means of the Spectre circuit simulator. The parasitic effects have been accounted for, relying on physical considerations around their causes as well as on experimental data collected from ad hoc test structures. The availability of the MEMS model library, along with techniques devoted to the inclusion of the parasitic effects linked to a certain technology platform, enables the fast and accurate mixed-domain simulation (electromechanical and electromagnetic) of RF-MEMS devices and complex networks.

References

1. Mahameed, R. and Rebeiz, G.M. (2010) Electrostatic RF MEMS tunable capacitors with analog tunability and low temperature sensitivity. Microwave Symposium Digest (MTT), 2010 May IEEE MTT-S International, pp. 1254–1257.
2. McFeetors, G. and Okoniewski, M. (2006) Custom fabricated high-Q analog dual-gap RF MEMS varactors. International Conference on Microwaves, Radar Wireless Communications, 2006. MIKON 2006 May, pp. 155–158.
3. Dong-Ming, F., Quan, Y., Xiu-Han, L., Hai-Xia, Z., Yong, Z., and Xiao-Lin, Z. (2010) High performance MEMS spiral inductors. 2010 5th IEEE International Conference on Nano/Micro Engineered and Molecular Systems (NEMS), Jan, pp. 1033–1035.
4. El Gmati, I., Fulcrand, R., Calmon, P., Boukabache, A., Pons, P., Boussetta, H., Kallala, A., and Besbes, K. (2010) RF MEMS fluidic variable inductor. 2010 5th International Conference on Design and Technology of Integrated Systems in Nanoscale Era (DTIS), March, pp. 1–3.
5. Yong-Seok, L., Yun-Ho, J., Jung-Mu, K., and Yong-Kweon, K. (2010) A 50-110 GHz ohmic contact RF MEMS silicon switch with high isolation. 2010 IEEE 23rd International Conference on Micro Electro Mechanical Systems (MEMS), Jan, pp. 759–762.
6. Reines, I., Pillans, B., and Rebeiz, G.M. (2010) J. Microelectromech. Syst., **99**, 1–11.
7. Solazzi, F., Tazzoli, A., Farinelli, P., Faes, A., Mulloni, V., Meneghesso, G., and Margesin, B. (2010) Active recovering mechanism for high performance RF MEMS redundancy switches. Microwave Conference (EuMC), 2010 European, Sept, pp. 93–96.
8. Solazzi, F., Palego, C., Halder, S., Hwang, J.C.M., Faes, A., Mulloni, V., Margesin, B., Farinelli, P., and Sorrentino, R. (2010) Electro-thermal analysis of RF MEM capacitive switches for high-power applications. Solid-State Device Research Conference (ESSDERC), 2010 Proceedings of the European, Sept, pp. 468–471.
9. Buck, T. and Kasper, E. (2010) RF MEMS phase shifters for 24 and 77 GHz on high resistivity silicon. 2010 Topical Meeting on Silicon Monolithic Integrated Circuits in RF Systems (SiRF), Jan, pp. 224–227.
10. Domingue, F., Fouladi, S., and Mansour, R.R. (2010) A reconfigurable impedance matching network using dual-beam MEMS switches for an extended operating frequency range. Microwave Symposium Digest (MTT), 2010 IEEE MTT-S International, May, pp. 1552–1555.
11. Fomani, A.A. and Mansour, R.R. (2009) Miniature RF MEMS switch matrices. Microwave Symposium Digest, 2009. MTT '09. IEEE MTT-S International, June, pp. 1221–1224.
12. Fomani, A.A. and Mansour, R.R. (2009) IEEE Trans. Microw. Theory Tech., **57** (12), 3434–3441.
13. Ocera, A., Farinelli, P., Cherubini, F., Mezzanotte, P., Sorrentino, R., Margesin, B., and Giacomozzi, F. (2007) A MEMS-reconfigurable power divider on high resistivity silicon substrate. Microwave Symposium, 2007. IEEE/MTT-S International, June, pp. 501–504.
14. Iannacci, J., Giacomozzi, F., Colpo, S., Margesin, B., and Bartek, M. (2009) A general purpose reconfigurable MEMS-based attenuator for radio frequency and microwave applications. EUROCON 2009, EUROCON '09. IEEE, May, pp. 1197–1205.
15. Xiaoguang, L., Katehi, L.P.B., Chappell, W.J., and Peroulis, D. (2010) J. Microelectromech. Syst., **19** (4), 774–784.

16. Iannacci, J., Gaddi, R., and Gnudi, A. (2010) *J. Microelectromech. Syst.*, **19** (3), 526–537.
17. Zimmer, T., Milet-Lewis, N., Fakhfakh, A., Ardouin, B., Levi, H., Duluc, J.B., and Fouillat, P. (1999) Hierarchical analogue design and behavioural modelling. IEEE International Conference on Microelectronic Systems Education, 1999. MSE '99, pp. 59–60.
18. Iannacci, J., Faes, A., Mastri, F., Masotti, D., and Rizzoli, V. (2010) A MEMS-based wide-band multi-state power attenuator for radio frequency and microwave applications. Proceedings of TechConnect World, NSTI Nanotech 2010, Jun, vol. 2, pp. 328–331.
19. Iannacci, J. (2010) in *Advanced Microwave Circuits and Systems* (ed. V. Zhurbenko), Advanced Microwave Circuits and Systems, INTECH, pp. 313–338.
20. Iannacci, J. and Gaddi, R. (2010) *Mixed-Domain Simulation and Wafer-Level Packaging of RF-MEMS Devices*, Lambert Academic Publishing - LAP.
21. Iannacci, J., Del Tin, L., Gaddi, R., Gnudi, A., and Rangra, K.J. (2005) *Compact modeling of a MEMS toggle-switch based on modified nodal analysis*. Proceedings of Symposium on Design, Test, Integration and Packaging of MEMS/MOEMS (DTIP 2005), June, pp. 411–416.
22. Novak, E., Der-Shen, W., Unruh, P., and Schurig, M. (2003) MEMS metrology using a strobed interferometric system. Proceedings of the XVII IMEKO World Congress, Jun, pp. 178–182.
23. Dambrine, G., Cappy, A., Heliodore, F., and Playez, E. (1988) *IEEE Trans. Microw. Theory Tech.*, **36** (7), 1151–1159.
24. Danneville, F., Fan, S., Tamen, B., Dambrine, G., and Cappy, A. (1998) A new two-temperature noise model for FET mixers suitable for CAD. ARFTG Conference Digest, 1998. Computer-Aided Design and Test for High-Speed Electronics. 52nd, pp. 59–66.
25. Pozar, D.M. (2005) *Microwave Engineering*, John Wiley and Sons, Inc.
26. Bonagnide, G., Sherman, J., and Dunleavy, L. (1998) A lumped-element modeling approach for waveguide of arbitrary geometry. Southeastcon '98. Proceedings. IEEE, Apr, pp. 190–193.
27. Besser, L. and Gilmore, R. (2003) *Practical RF Circuit Design for Modern Wireless Systems, Volume I: Passive Circuits and Systems*, Artech House.
28. Zhao, H., Tang, A.-Y., Sobis, P., Bryllert, T., Yhland, K., Stenarson, J., and Stake, J. (2010) VNA-calibration and S-parameter characterization of sub-millimeter wave integrated membrane circuits. 2010 35th International Conference on Infrared Millimeter and Terahertz Waves (IRMMW-THz), Sept, pp. 12.

Part III
Mathematical Model Order Reduction for MEMS Devices

System-level Modeling of MEMS, First Edition. Edited by T. Bechtold, G. Schrag, and L. Feng.
© 2013 Wiley-VCH Verlag GmbH & Co. KGaA. Published 2013 by Wiley-VCH Verlag GmbH & Co. KGaA.

9
Moment-Matching-Based Linear Model Order Reduction for Nonparametric and Parametric Electrothermal MEMS Models

Tamara Bechtold, Dennis Hohlfeld, Evgenii B. Rudnyi, and Jan G. Korvink

9.1
Introduction

In this chapter, we present the application of moment-matching-based model order reduction (MOR) to electrothermal MEMS models. We use the Block–Arnoldi algorithm with deflation [1], described in Chapter 3 (Algorithm 4). The resulting reduced models are easily convertible into hardware description language (HDL) form and can be directly used for system-level simulation, as the speed up in transient simulation is of several orders of magnitude. We further construct parametric reduced-order models with multivariate moment-matching approach, which are described later in this chapter. Parametric-reduced models are employed within an optimization loop to efficiently determine thermal parameters of thin films. We use the software tool "MOR for ANSYS" to automatically create reduced-order models directly from ANSYS finite element thermal models. "MOR for ANSYS" is described in more details in Chapter 18.

Section 9.2 summarizes the general methodology for applying MOR for automatic generation of dynamic compact thermal models for electrothermal MEMS, presented in [2]. Section 9.3 describes a silicon-based microhotplate, which has been used as a case study in Sections 9.4 and 9.5. Section 9.4 demonstrates the application of reduced-order model for cosimulation with control circuitry and parameterization of the controller. In Section 9.5, a methodology for the efficient determination of material properties via parametric model order reduction (pMOR) and subsequent automatic parameter fitting is presented. Section 9.6 concludes the chapter and gives an outlook to future developments in the area.

9.2
Methodology for Applying Model Order Reduction to Electrothermal MEMS Models: Review of Achieved Results and Open Issues

In [2], a methodology for applying MOR for automatic generation of dynamic compact thermal models for several hotplate-based MEMS devices was presented. Microhotplates are structures composed of a membrane, which is fabricated from a thin-film material and carries heating and sensing elements. Microhotplates, are essential components in gas sensors, microfilaments, and other thermally tunable microsystems. Our first example is a thermally tunable optical filter [3] (Figure 9.1). Here, temperature is used to control the wavelength of light passing perpendicular through the membrane. Second example uses a gas sensor [4] (Figure 9.2) in which the gas sensing material is deposited on top of the membrane. Elevated temperature is essential for high detection sensitivity and responsivity. Last example details an array of gas sensors [5] (Figure 9.3). Each sensor includes transistors for heating and dedicated structures for improving temperature homogeneity.

Figure 9.1 Thermally tunable optical filter [3].

Figure 9.2 Gas sensor [4]. (Picture courtesy of J. Wöllenstein (Fraunhofer IPM, Germany).)

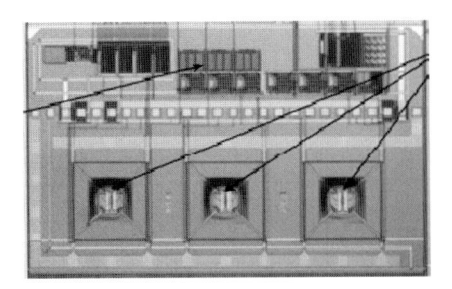

Figure 9.3 Gas sensor array [5]. (Picture courtesy of M. Graf (ETH Zurich, Switzerland).)

The operation of the above devices requires the control of the membrane temperature and, hence, thermal modeling is a crucial issue. The heat transfer within a hotplate is described through the following equations (which apply to Joule heating phenomena in general):

$$\nabla \cdot (\kappa \nabla T) + Q - \rho c_p \frac{\partial T}{\partial t} = 0 \tag{9.1}$$

$$Q = j^2 R$$

where $\kappa(r)$ is the thermal conductivity in Watts per meter Kelvin at the position r, $c_p(r)$ is the specific heat capacity in Joule per kilogram Kelvin, $\rho(r)$ is the mass density in kilogram per cubic meter, $T(r,t)$ is the temperature distribution, and $Q(r,t)$ is the heat generation rate per unit volume in Watts per cubic meter. We solve Eq. (9.1) with the initial condition $T_0 = 0$, the Dirichlet boundary condition $T = 0$ at the bottom of computational domain.

We assume that the heat generation is uniformly distributed within the heater and that the system matrices can be considered as temperature independent at the operation temperature. Finite element method (FEM)-based spatial discretization of Eq. (9.1) leads to a large linear ordinary differential equations (ODEs) system with temperature-dependent input:

$$C \cdot \dot{T} + K \cdot T = F \cdot \frac{U^2(t)}{R(T)} \tag{9.2}$$

$$y = E \cdot T$$

The form (Eq. (9.2)) is also called *state-space representation*, where $C, K \in R^{n \times n}$ are the global heat capacity and heat conductivity matrices, $T(t), B, E \in R^n$ are the state (temperature) vector, the load, and the output vector, respectively, and n is the dimension of the system, that is, the number of degrees of freedom (DOF) in finite element model. Note that F and E are matrices in cases when there are more than one source and more than one output defined (this is noted as multiple-input–multiple-output – MIMO system in Chapter 3). We further take into account the temperature dependence of the heater's resistivity that causes the weak nonlinearity of the model (Eq. (9.3)), that is, the dependence of the input function on the state vector, given by

$$R(T) = R_0(1 + \alpha \Delta T) \tag{9.3}$$

where R_0 is the resistance at $\Delta T = 0$ K and α is the linear temperature coefficient of the material. Equation (9.2) is a starting point for MOR, which leads to a system of the same form but with much smaller dimension, as schematically shown in Figure 9.4.

As already mentioned in Chapter 3, the most often used MOR approach in engineering problems is a Krylov-subspace-based moment-matching approach [1], which is based on expanding the transfer function of Eq. (9.2):

$$H(s) = E(sC + K)^{-1} F \tag{9.4}$$

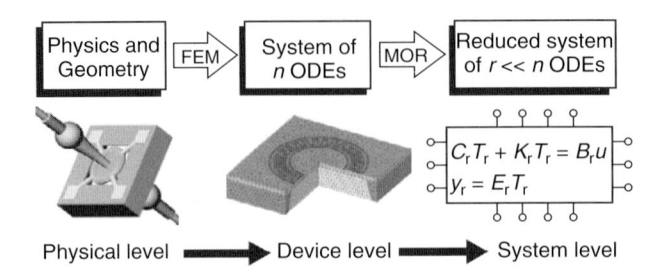

Figure 9.4 Model order reduction is a part of conversion process from physical to compact model.

in the Taylor series around some value of the Laplace variable $s = s_0$:

$$H(s) = \sum_{i=0}^{\infty} m_i(s_0)(s - s_0)^i \qquad (9.5)$$

where $m_i(s_0) = E(-(K + s_0 C)^{-1} C)^i \cdot (K + s_0 C)^{-1} F$ is called the i-th moment around s_0, and then finding a much lower order system (system composed of $r \ll n$ linear ODEs of the form (Eq. (9.2)) whose transfer function $H_r(s)$ has the same moments as $H(s)$. There are two main algorithms belonging to Krylov-subspace-based moment-matching approach, Arnoldi and Lanczos. Arnoldi algorithm matches r moments and Lanczos algorithm $2r$ moments (provided the order of the reduced system is r), as described in more details in Chapter 3 and its references. The moment-matching property is achieved by constructing an orthonormal projection matrix $V \in R^{n \times r}$, according to Algorithm 4 in Chapter 3, and then projecting the system (Eq. (9.2)) as follows:

$$V^T C V \times \dot{T}_r + V^T K V \times T_r = V^T F \times \frac{U^2(t)}{R(T)} \qquad (9.6)$$

$$y_r(t) = EV \times T_r$$

The matrices of the new, smaller system are $C_r = V^T C V$, $K_r = V^T K V$, $F_r = V^T F$, $E_r = EV$ and it can be proved [1] that for Eqs. (9.6) and (9.2) the moment-matching property holds, that is, the first r moments of both system transfer functions are the same. The very name "Krylov-based" emerges from the fact that V is computed as an orthogonal basis for the Krylov subspace of dimension r, induced by $A = K^{-1}C$ and $b = K^{-1}F$:

$$K(A, b) = \text{colspan}\{b, Ab, \ldots, A^{r-1}b\} \qquad (9.7)$$

A main drawback of the described approach is that the new state vector T_r in Eq. (9.6) is not any more composed of temperatures in finite element nodes, but rather of generalized coordinates, that is, the physical meaning has been lost through MOR.

Figure 9.5 shows the excellent results of MOR based on Block–Arnoldi algorithm with deflation, for the optical filter device. Frequency domain is best approximated around the zero frequency, as the expansion point for Taylor series of the transfer function was chosen to be $s_0 = 0$.

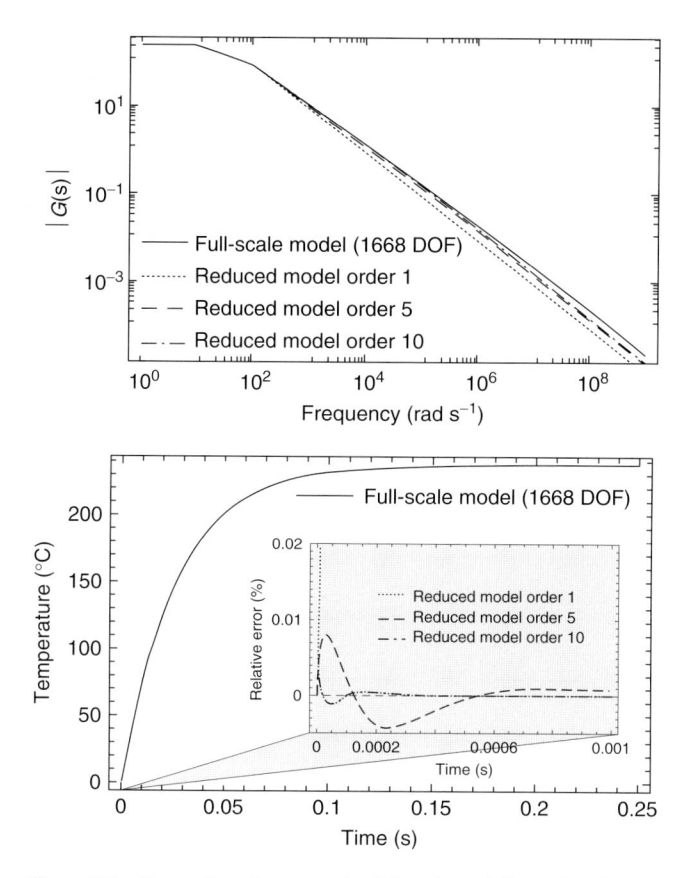

Figure 9.5 Comparison between the full-scale and the reduced-order models of the optical filter device from Ref. [3] in a single defined output node for the constant heating power of 1 mW. (a) Frequency response and (b) step response (outer plot) and step response errors (inner plot).

The results in Figure 9.5b represent the temperature increase in a single output node, which is defined by a row-vector C in Eq. (9.2). As there is a single heat source (voltage applied to the heating resistor), B in Eq. (9.2) is a column vector. We refer to the single-input–single-output (SISO) setup of the system (Eq. (9.2)), which can be easily reduced by, for example, Algorithm 3 from Chapter 3. In more general cases, the temperature response in several or even in all finite element nodes might be required. In [2], it was demonstrated that using the Block–Arnoldi algorithm with deflation from [1], it is possible to approximate not only a single output response but also the transient thermal response in all finite element nodes of the device. This is due to the fact that Block–Arnoldi algorithm with deflation does not take into account the output vector (matrix) C when constructing the basis V of the single projection subspace (see Chapter 3 for explanation of the projection subspace); Figure 9.6 shows the mean approximation error in all finite element nodes of the optical filter device, as a function of the reduced system's order.

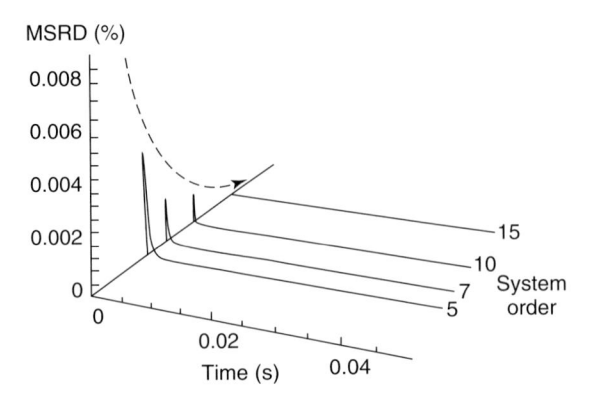

Figure 9.6 Mean square relative difference (defined in [2]) for all of the 1668 finite element nodes of the optical filter [3] during the initial 0.05 s of heating.

It is further possible to transfer the nonlinearities (Eq. (9.3)) of the input function into the reduced system. This is implemented by considering the arithmetic mean of the maximum and minimum temperature values of the heater section as the lumped heater's temperature, that is, by defining additional outputs in Eq. (9.2) (for more details see Section 5.5.1 in [2]). Figure 9.7 shows the step response and the step response error between the full-scale and the reduced-order 10 model of the gas sensor device from [4] when the temperature-dependent heating power is taken into account.

To apply the MOR based on Block–Arnoldi algorithm with deflation, an MEMS designer has to provide a discretized model (e.g., a finite element model) of the device and to specify which frequency band should be well approximated by the compact model. This is done by choosing one or more expansion points in the frequency domain, as explained in Chapter 3. The next important step is

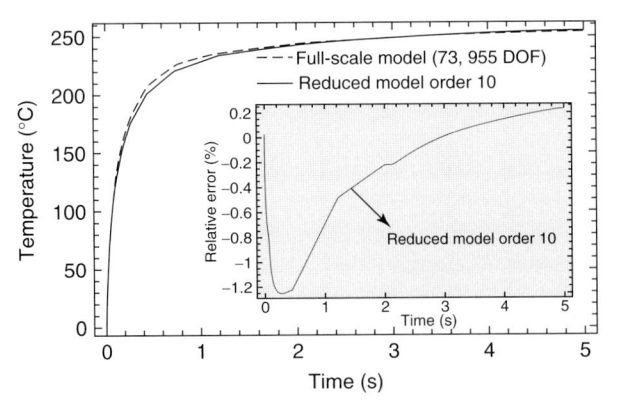

Figure 9.7 Step response (outer plot) and step response error (inner plot) of the gas sensor [4] in a single output node for temperature-dependent heating power according to Eq. (9.4) with $\alpha = 1.469 \times 10^{-3} K^{-1}$ for a platinum heater.

to specify the proper order of the target reduced system, such as to achieve a desired accuracy. To automate the MOR process completely, one should be able to estimate the error between the full-scale and the reduced-order model as a function of the reduced model's dimension. As already mentioned in Chapter 3, an effective error estimate for the Krylov-subspace methods is still an open research question. In [2], three heuristic methods to estimate the error of reduction based on Block–Arnoldi algorithm with deflation were suggested. The first method is based on the convergence of relative error between two successive reduced-order models of the order r and $r+1$, the second is based on the convergence of Hankel singular values (defined in Chapter 3) of the reduced-order models, and the third method is based on the sequential application of Block–Arnoldi algorithm with deflation and mathematically superior Grammian-based methods (described in Chapter 3). Figure 9.8 shows the first type of error estimator for the gas sensor device from [4].

As MEMS are often composed of interconnected subsystems, for example, array structures, it is desirable to reduce each subsystem on its own and then to couple all reduced subsystems, following, for example, the principle of Kirchoffian networks (Chapter 2). The main problem thereby is that the thermal flow is not a lumped physical phenomenon as, for example, the electrical flow is along metallic wire interconnects. The latter is the case since the ratio of electrical conductivity of metals and that of insulators is of the order of 10^8. Hence, the electrical current flow takes place almost solely in metal paths. This is not the case with heat flow because the ratio of thermal conductivities in microtechnology is only of the order of 10^2. Therefore, it is unclear how to lump the thermal fluxes at shared surfaces between two finite element models to form the thermal ports that would serve to couple together several compact models. Note that, if one would keep all the surface nodes as ports, that is, inputs/outputs in Eq. (9.3), the dimension of the reduced model when using Block–Arnoldi algorithm with deflation would become prohibitively large. Hence, it is only efficient to apply Block–Arnoldi algorithm with deflation to the complete coupled system, as suggested in [2], as long as the dimension of the array model stays moderate. Figure 9.9 compares the step

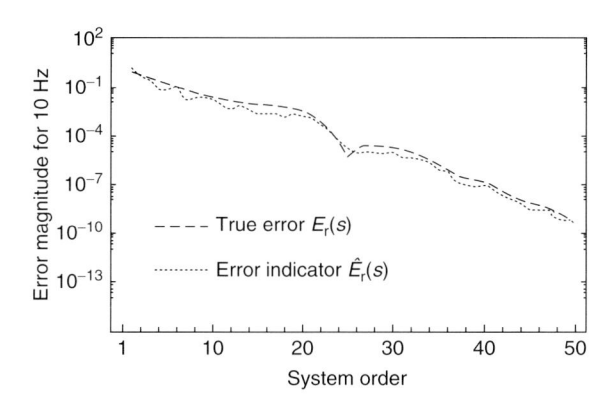

Figure 9.8 Error indicator from Ref. [2] for the gas sensor model [4] with 73 955 DOF.

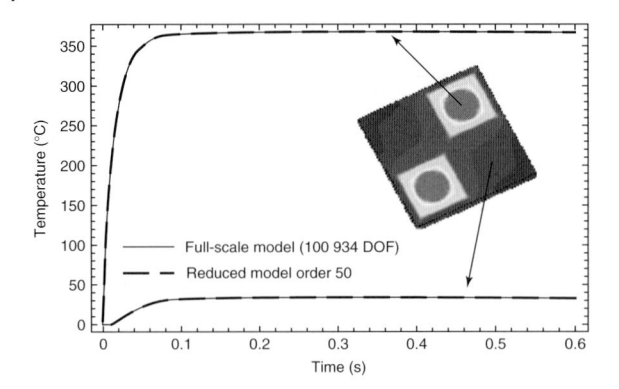

Figure 9.9 Step response of the full-scale and reduced-order models for the microhotplate array based on a gas sensor device from Ref. [5]. Reduction is done by Block–Arnoldi algorithm with deflation from Ref. [1], in case when two heat sources of 40 mW each are switched on.

responses of the full-scale and reduced-order model for the microhotplate array based on a gas sensor device from [5].

General technique of coupling two reduced thermal models gained by Krylov-subspace-based projection is discussed in [2]. In this respect, the structure-preserving model reduction techniques [6] should be further investigated. For now, the coupling of reduced-order models, as possible solution to reducing models of MEMS array structures, remains a further open question.

9.3
MEMS Case Study – Silicon-Based Microhotplate

In the following, we consider the test structure shown in Figure 9.10.

The microstructured hotplate consists of a thin-film membrane composed of silicon nitride suspended over a silicon frame. Thin-film metal resistors are fabricated on top of the membrane for heating and temperature sensing. To achieve

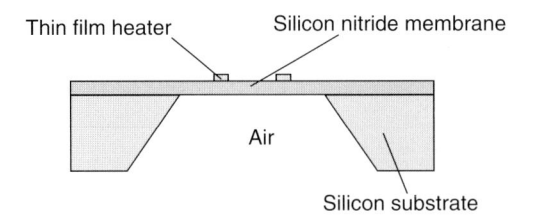

Figure 9.10 A silicon nitride membrane with integrated heater and sensing element was fabricated by low-frequency plasma enhanced chemical vapor deposition. The square membrane is 500 nm thick with a side length of 550 μm, and the thin-film heater is made from 150 nm platinum with a 50 nm titanium adhesion layer.

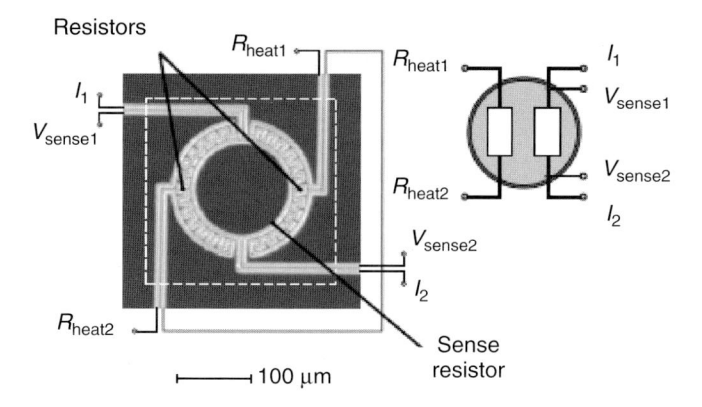

Figure 9.11 Schematic view of the thin-film resistors for heating and temperature sensing. The heating resistor is operated at constant voltage. The sensing resistor is configured for four-point measurement.

a preferably circular symmetric and homogeneous temperature distribution at the center of the square membrane, both resistors are arranged as shown in Figure 9.11.

Characterization of the static and transient thermal properties of the membrane is performed on a temperature-controlled mount. For measuring the transient temperature changes, a constant current of 100 µA is passed through the sensing resistor, while the voltage across the inner terminals is measured using an oscilloscope. To use the sensing resistor as a temperature sensor, the linear temperature coefficient of the material's resistivity has to be known. The temperature coefficient is measured by acquiring the sensor's resistance at various temperatures, which are precisely set by the Peltier mount. The electrical resistance depends linearly on the temperature over the investigated temperature range and is modeled by Eq. (9.3). A temperature coefficient $\alpha = 2.293 \times 10^{-4}$ K^{-1} is obtained for a metallization of 150 nm platinum with 50 nm titanium. The transient thermal response of the membrane is characterized by applying a rectangular voltage signal to the heating resistor using a function generator. The signal output is configured as a voltage source with a fixed output impedance of 50 Ω. The thermal response over a whole period is presented in Figure 9.12. One can recognize the drop in the heating power shortly after the voltage is applied. This is due to the fact that the heating resistor also depends on temperature (nonlinearity of the right-hand side in Eq. (9.2)). An increase in temperature causes the heating resistance to grow, which leads to slightly smaller heating power. After applying the heating power, the membrane's temperature increases until a maximum value is reached. This temperature is defined as the steady-state value. After setting the power to zero, the heat stored in the membrane's volume is dissipated to the surrounding media by conduction and free convection. Thus, the temperature drops down to its initial value.

A three-dimensional finite element model of the test structure has been implemented in ANSYS, as displayed in Figure 9.13.

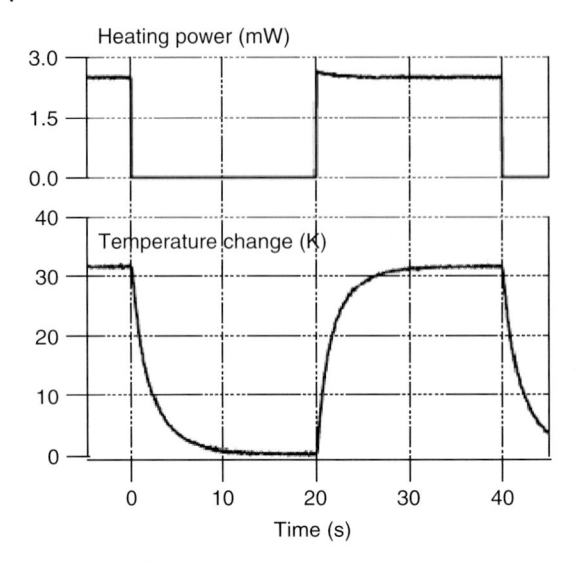

Figure 9.12 Temperature modulation of a silicon/nitride membrane. A square wave heating power with a frequency of 25 Hz was applied to the membrane.

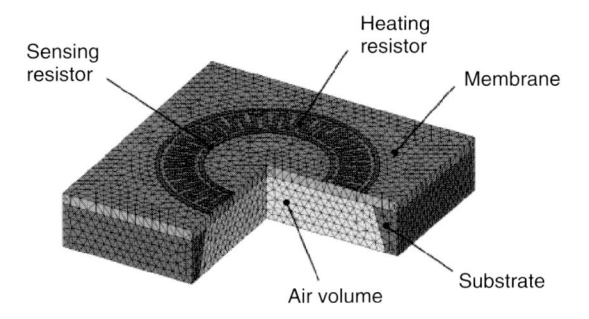

Figure 9.13 FEM mesh of the three-dimensional model with 66.000 nodes.

The FEM model incorporates 66.000 DOF, which corresponds to 66.000 ODEs of the form (Eq. (9.2)). It considers the heat conduction through the solid material and the air beneath the membrane as well as convection to the air above the membrane. The latter is considered in the form of convection boundary condition:

$$q = h(T - T_{\text{air}}) \tag{9.8}$$

where h is the heat transfer coefficient between the membrane and the ambient air in Watts per meter square Kelvin. Radiation mechanism has not been considered with respect to MOR, but it could be included via defining an additional temperature dependent. Figure 9.14 shows the considered heat loss mechanisms in more detail.

For the presented case study, MOR based on Block–Arnoldi algorithm with deflation and implemented within the software tool MOR for ANSYS (described in more details in Chapter 18) has been applied. The output of MOR for ANSYS

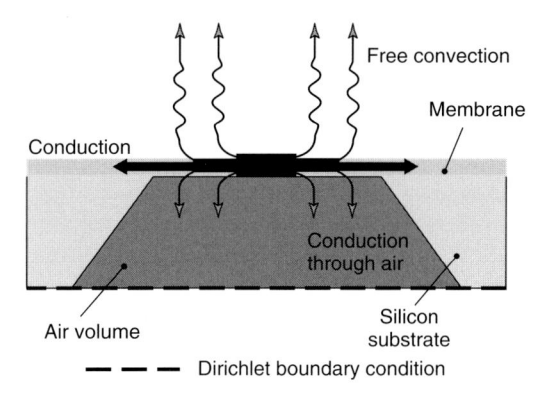

Figure 9.14 Heat loss mechanisms taken into account by the FEM model. Dirichlet boundary condition T = 0 K is set at the bottom edge of the simulation domain.

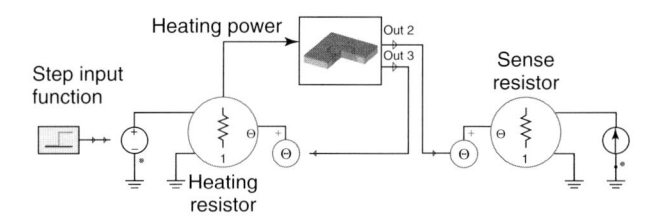

Figure 9.15 System-level model of microstructure and electrical components. Resistors are modeled as temperature dependent. A step input function is applied to a controlled voltage source that drives the heating resistor.

is a reduced model in the form (Eq. (9.2)) but with only 30 DOF (the size of the reduced model is based on our experience). Time integration of the reduced model is performed with system-level simulator SIMPLORER® as shown in Figure 9.15. Reduced model can be imported in SIMPLORER in either state space form or as VHDL-AMS code.

Figure 9.16 shows a good match between the full-scale model and the reduced-order model, which now can be used for system-level simulation, control, or design optimization. Two different applications are presented in the following sections. Comparison with the measurements is presented in Figure 9.23.

9.4
Application of the Reduced-Order Model for the Parameterization of the Controller

After having derived the reduced-order model of the MEMS device, we are able to simulate it within a system-level simulator, for example, together with driving and/or control circuitry. On the basis of the reduced model, it is possible to efficiently determine the control parameters while preserving the accuracy of the

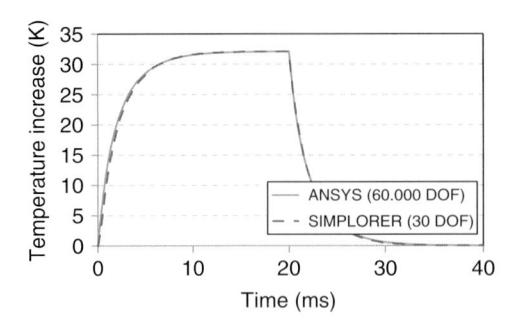

Figure 9.16 Simulation results of full-scale FE model of silicon nitride membrane from Figure 9.10 in ANSYS and the reduced-order model in SIMPLORER (measurements are displayed in Figure 9.23).

device simulation. As already mentioned, the operation of microhotplate-based microsystems requires the control of the membrane's temperature. The absolute temperature of the membrane should be independent of variations in the surrounding temperature and changes in the convection boundary conditions. Furthermore, adjustments to new temperature set points should be performed by the thermal microsystem as quick as possible.

In [7] two application scenarios for operation of the microhotplate case study under temperature control were investigated, constant-value (set-point) temperature control and tracking control. In the first case, a fast thermal response, which leads to the prescribed temperature value, is desired with minimum overshoot. The second scenario requires the temperature profile to track a prescribed function of time. In both cases, the resistance values of the heating and sensing resistor depend on their respective temperatures according to Eq. (9.3).

Figure 9.17 shows the implementation of the temperature control scheme for the microhotplate in the system-level simulator. With suitable values for the proportional and integral gain parameters of the control unit, this loop can efficiently eliminate the effects of ambient temperature variations and will also

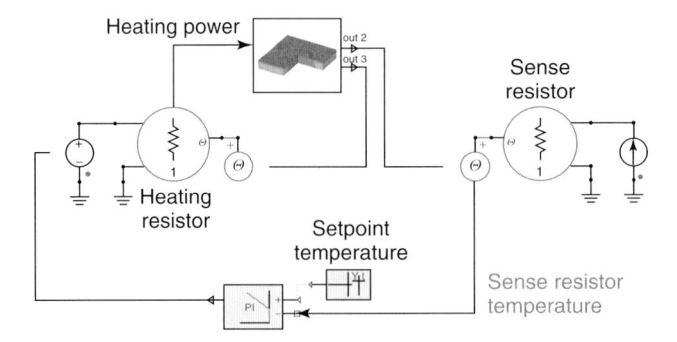

Figure 9.17 System-level model of microhotplate and control loop. A PI controller is used in the control loop.

allow the membrane temperature to follow time-varying set-point temperatures. The heating resistor acts as an actuator onto the process, by transferring the control signal to the membrane in form of a heat generation rate, which in turn changes its temperature. The sensing resistor supplies the temperature information that is compared to an external set value. The resulting difference is passed to the controller whose output is used to set a voltage source that drives the temperature-dependent heating resistor according to the nonlinear input function in Eq. (9.2).

Often the parameterization of the controller is performed manually, based on the designer's expertise and heuristic approaches [8, 9]. No assurance can be given to the overall quality of the adjusted parameter set. However, in cases where the control should fulfill specific goals, standard procedures are not applicable and further adjustments of the control parameters are required. In this case, an optimization strategy, for the entire system comprising the microhotplate and the controller, using the reduced-order model of the microstructure, can be applied. In the system simulator SIMPLORER®, one can define suitable cost functions and use optimization algorithms (available are quasi-Newton algorithm, genetic algorithms, etc.) to identify control parameters, which yield an optimum system performance. Several goal functions with different weighting factors can be defined. For example, to achieve a fast thermal response, the integral deviation between set-point temperature and actual system response can be defined as goal function. Optimization aims at reducing this value to zero. If the goal is to prevent overshooting, the minimum value of the difference between set point and response is a suitable objective, which should be minimized to zero as well. If both goals are to be achieved, a weighted combination of the above mentioned goal functions can be defined.

In the first application scenario, a step function acts as the set point. After optimization, the rise time shows an 18-fold improvement over the system's thermal time constant (Figure 9.18), thus indicating successful optimization and control performance. Proportional and integral gains were determined to be 0.3 and 750, respectively. The time constant of the cooling process, on the other hand, is inherent to the thermal microsystem and cannot be influenced with the current scheme. Only active cooling or operation at elevated temperatures would accelerate the cooling process.

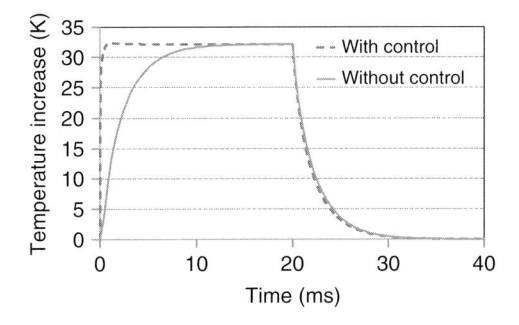

Figure 9.18 Controlled membrane temperature reaching set-point value with a rise time of 0.23 ms. The rise time of the uncontrolled system is 4.1 ms.

The drastic improvement of the rise time during the heating phase is achieved by excessive over heating during the initial phase. For the first 10 μs, the heating power equals 100 mW and drops to 2.5 mW at 1 ms, in contrast to a continuous supply of 2.5 mW without control scheme.

Secondly, a saw tooth signal is applied to the control loop. This shall lead to a likewise increase in membrane temperature. Optimization yields a different set of parameters to fulfill this task (proportional and integral gain equal 0.2 and 5000, respectively). Figure 9.19 shows how the membrane temperature of the controlled system tracks closely the linearly increasing set point. Small oscillations occur in the initial phase and decay after the first milliseconds. For comparison, we include results from an uncontrolled heating setup (as shown in Figure 9.15 with linearly increasing heating power) as well. The data are normalized to the respective maximum values. The response of the uncontrolled setup lacks behind the set point. As already mentioned above, the cooling process cannot be influenced by the present setup.

The control system, as shown in Figure 9.17 schematically, was implemented into system-level simulator by using operational amplifiers (Figure 9.20) to demonstrate that the system-level simulator is also capable of performing cosimulation of the reduced state-space model (same form as Eq. (9.2) but with much smaller dimension) of the hotplate, electronic, and electric components. The difference between actual membrane temperature and the set-point value is determined in OP1; no gain is added here. OP2 performs the time integration while simultaneously adding a gain factor. The heating resistor is driven by a unity gain buffer to decouple the electrical load of the heating resistor from the integrating operational amplifier. The dissipated electrical power is passed to the reduced model, which describes the dynamic behavior of the membrane. The temperature value at the location of the sense resistor is used to set the resistance value of the temperature sensing resistor. A constant current is fed through the sense resistor that yields in a temperature-dependent output voltage to be compared with the set-point value.

Note that the presented setup (reduced-order model of the MEMS device and the control circuitry) can be easily extended by coupling it with other physical domains and further analog and digital circuitry. In SIMPLORER, it is also possible to

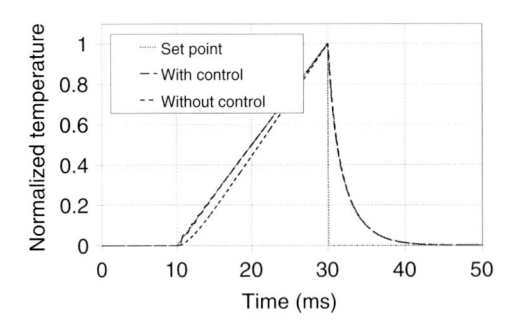

Figure 9.19 Temperature response to a linearly increasing heating power of an uncontrolled scheme and to a linearly increasing set-point value of a controlled system.

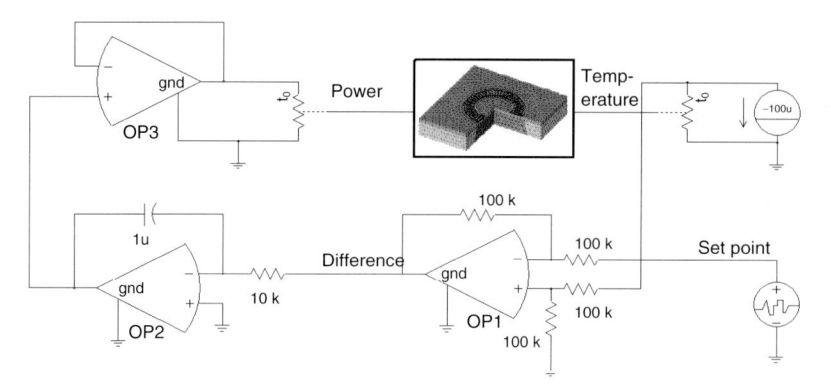

Figure 9.20 Reduced-order model implemented for cosimulation with a control circuit. Difference to the configuration of Figure 9.17 is that voltage difference, integrator, and unity gain buffer are realized using operational amplifiers. The reduced-order model is implemented as a state-space system (Eq. (9.6)).

incorporate VHDL descriptions of microcontrollers to give an all-encompassing description of embedded microsystems.

9.5
Application of Parametric Reduced-Order Model to the Extraction of Thin-Film Thermal Parameters

The design of new MEMS devices requires knowledge of properties of the used materials. This information is well known for most bulk materials. Monocrystalline silicon is the dominant material for MEMS fabrication and has been extensively studied so that its mechanical, thermal, and electrical characteristics are well known [10]. Devices that are purely made from silicon, for example, high-frequency resonators or oscillating mirrors, can be exactly modeled due to the well-known mechanical properties of the material. However, the fabrication of most MEMS devices involves the deposition of thin films, which are employed to fulfill specific functions, such as sensing, actuation, passivation, and so on.

Unfortunately, the material properties of thin films strongly depend on the fabrication parameters and subsequent process steps of the MEMS structure. To build an accurate MEMS model, the material properties of thin films have to be determined. In the case of electrothermal MEMS, the material thermal properties of thin-film materials are of special interest, as the transient characteristic of the device is determined by thermal conductivity and specific heat (parameters $\kappa(r)$ and $c_p(r)$ in Eq. (9.1)) of the employed materials. These two parameters determine how fast and to what extent the temperature changes in response to heating the device.

A conventional way to determine unknown thermal properties is to build and characterize dedicated test structures, which employ the thin-film material of interest as a functional component. A review of the state of the art in determining material thermal properties of thin films is given in [11]. Mentioned test structures preferably feature simple geometries [12, 13] and, hence, can be described by analytical models. If the applied heat and temperature distribution are measured, one can determine the material thermal properties of the employed thin films from analytical models. The main drawback of such an approach is that in these models the heat flow is mostly assumed to be one-dimensional, although heat conduction is a distributed phenomenon. Therefore, for higher accuracy, more precise numerical models are required, which, however, require considerable computational effort.

In the previous sections, we have seen that computational effort for simulating linear thermal models can be reduced by applying MOR. Once again, we consider the model to be linear, if the material properties can be assumed as temperature independent. It is further possible to use pMOR [14] for constructing reduced models that preserve material properties as parameters. Such parameterized small-size models can be used within a data fitting procedure to efficiently obtain those material parameters. We emphasize that the numerical model by itself might not cover all aspects of the original physical system, for example, it neglects inhomogeneity of the thin films and uncertainties of thermal measurements. However, the main goal at present is to speed up the simulation time of the numerical model to enable a time-efficient optimization. Hence, in the following, we assume an adequate accuracy of the numerical model.

9.5.1
Parametric Model Order Reduction

In its original form, MOR does not allow for preserving parameters within the system matrices, which arise in many applications in the form of, for example, boundary conditions, material parameters, or geometry parameters. For the silicon nitride membrane under convection boundary condition (Eq. (9.8)) and assuming T_{air} to be zero, Eq. (9.2) can be rewritten as

$$(C_0 + \rho c_p \times C_1)\dot{T} + (K_0 + \kappa \times K_1 + h \times K_2) T = F\frac{U^2(t)}{R(T)} \qquad (9.9)$$

$$y = E \times T$$

where the volumetric heat capacity $\rho \cdot c_p$, thermal conductivity κ, and the heat transfer coefficient h between the membrane and the ambient air are kept as parameters. It is important to emphasize that all system matrices depend linearly on these parameters and can, therefore, be factorized. MOR of Eq. (9.9) can also be performed by projection (similar to Eq. (9.6)), as follows:

$$(V^T C_0 V + \rho c_p \times V^T C_1 V)\dot{z} + (V^T K_0 V + \kappa \times V^T K_1 V + h \times V^T K_2 V) z \qquad (9.10)$$

$$= V^T F \frac{U^2(t)}{R(z)}$$

$$y_r = EV \times z$$

The task is now to find a parameter-independent projection matrix V, such that y_r is a good approximation of y in Eq. (9.9) for any parameter value. In multivariate moment-matching approach, V is constructed such that certain moments of the transfer functions of the full and reduced system are matched, as explained below.

9.5 Application of Parametric Reduced-Order Model to the Extraction

The transfer function of Eq. (9.9) can then be written as

$$H(s, p_i) = E(s(C_0 + \rho c_p C_1) + K_0 + \kappa K_1 + h K_2)^{-1} F \quad (9.11)$$

where p_i stands for parameters in general. From the numerical point of view, every parameter p_i in Eq. (9.11) is equivalent to the Laplace variable s. Hence, the idea investigated by several groups [14–26] is based on the generalization of moment matching, as follows: make multivariate Taylor expansions of Eq. (9.11) with respect to the Laplace variable s and parameters p_i. Thus, all p_i will be simultaneously preserved in symbolic form. In the early work [15], the authors investigated the problem with only two parameters and developed an algorithm to compute and match all moments of both transfer functions. That is, if the Taylor expansion of the full transfer function $H(p_1, p_2)$ and of the reduced transfer function $H_r(p_1, p_2)$ around, for example, (0,0) are written as

$$H(p_1, p_2) = H(0,0) + \frac{\partial H}{\partial p_1}(0,0) \cdot p_1 + \frac{\partial H}{\partial p_2}(0,0) \cdot p_2$$
$$+ \left(\frac{1}{2!} \frac{\partial^2 H}{\partial p_1^2}(0,0)\right) \cdot p_1^2 + \frac{\partial^2 H}{\partial p_1 \partial p_2}(0,0) \cdot p_1 \cdot p_2$$
$$+ \frac{\partial^2 H}{\partial p_2 \partial p_1}(0,0) \cdot p_1 \cdot p_2 + \left(\frac{1}{2!} \frac{\partial^2 H}{\partial p_2^2}(0,0)\right) \cdot p_2^2 +$$
$$+ \ldots$$

$$H_r(p_1, p_2) = H_r(0,0) + \frac{\partial H_r}{\partial p_1}(0,0) \cdot p_1 + \frac{\partial H_r}{\partial p_2}(0,0) \cdot p_2$$
$$+ \left(\frac{1}{2!} \frac{\partial^2 H_r}{\partial p_1^2}(0,0)\right) \cdot p_1^2 + \frac{\partial^2 H_r}{\partial p_1 \partial p_2}(0,0) \cdot p_1 \cdot p_2$$
$$+ \frac{\partial^2 H_r}{\partial p_2 \partial p_1}(0,0) \cdot p_1 \cdot p_2 + \left(\frac{1}{2!} \frac{\partial^2 H_r}{\partial p_2^2}(0,0)\right) \cdot p_2^2 +$$
$$+ \ldots \quad (9.12)$$

all encircled terms are equal. Generalization of the method in [15] to systems with an arbitrary number of parameters was proposed in [16] (referred to as *multiparameter moment matching*). However, the proposed computational procedure is potentially numerically instable due to the fact that the computation of moments (Taylor coefficients) is explicit and an orthonormalization step (for generating the orthonormal projection matrix V, as explained in Chapter 3), that is used in Krylov-subspace methods, is only performed after the moments are already computed. To improve the practical applicability of this method, several approaches with orthogonalization were proposed in [17–23]. Another method in [24] (referred to as *multidimensional moment matching*) can be seen as neglecting the mixed moments, that is, if the Taylor expansion of the full system's step response (in Laplace domain) $X(p_1, p_2)$ and of the reduced system's step response $X_r(p_1, p_2)$ around, for example, (0,0) are written as

$$X(p_1, p_2) = \boxed{X(0,0)} + \boxed{\frac{\partial X}{\partial p_1}(0,0)} \cdot p_1 + \boxed{\frac{\partial X}{\partial p_1}(0,0)} \cdot p_2$$
$$+ \boxed{\frac{1}{2!}\frac{\partial^2 X}{\partial p_1^2}(0,0)} \cdot p_1^2 + \frac{\partial^2 X}{\partial p_1 \partial p_2}(0,0) \cdot p_1 \cdot p_2$$
$$+ \frac{\partial^2 X}{\partial p_2 \partial p_1}(0,0) \cdot p_1 \cdot p_2 + \boxed{\frac{1}{2!}\frac{\partial^2 X}{\partial p_2^2}(0,0)} \cdot p_2^2 +$$
$$+ \ldots$$
$$X_r(p_1, p_2) = \boxed{X_r(0,0)} + \boxed{\frac{\partial X_r}{\partial p_1}(0,0)} \cdot p_1 + \boxed{\frac{\partial X_r}{\partial p_2}(0,0)} \cdot p_2$$
$$+ \boxed{\frac{1}{2!}\frac{\partial^2 X_r}{\partial p_1^2}(0,0)} \cdot p_1^2 + \frac{\partial^2 X_r}{\partial p_1 \partial p_2}(0,0) \cdot p_1 \cdot p_2$$
$$+ \frac{\partial^2 X_r}{\partial p_2 \partial p_1}(0,0) \cdot p_1 \cdot p_2 + \boxed{\frac{1}{2!}\frac{\partial^2 X_r}{\partial p_2^2}(0,0)} \cdot p_2^2 +$$
$$+ \ldots \tag{9.13}$$

then only the encircled terms are equal. This method does not suffer from numerical instabilities, because moments are computed recursively via disjoint Krylov subspaces, where one parameter is kept constant (at the chosen expansion point), while the Krylov subspace is generated for another (variable) parameter and *vice verse*. The orthogonal projection matrix V is then constructed by joining both subspaces. A generalization of this approach to an arbitrary number of parameters k is straight forward [25]. It is necessary to compute all derivatives of $H(p_1, p_2, \ldots, p_k)$ at, for example, $p_k = 0, k = 1, 2, \ldots, p$, respectively by assuming that other parameters are constant. Also in this case, the projection matrix V can be constructed by orthogonalization of a certain number of derivatives of H at each parameter via Krylov subspaces. Note that neglecting the mixed moments, while maintaining the accuracy, is only possible as long as a weak correlation exists between the parameters. This is the case for thermal problems, as demonstrated in [26] and [27]. Currently, no method exists to specify how many derivatives (moments) should be chosen for each parameter.

Figure 9.21 displays the results for transient simulations of the full-scale and parametric reduced-order model for the microhotplate model. The displayed temperature change is the arithmetic mean between the maximal and minimal node temperature of the sensing resistor. We have used the multidimensional model reduction approach from [26] and have reduced the time for a transient integration of the full-scale FE model by a factor of 100. The disjoint Krylov subspaces for the four parameters were created by standard Block–Arnoldi algorithm with deflation [1] and we have chosen to compute 30 moments around each parameter.

9.5.2
Parameter Extraction Methodology

Because material parameters and the heat transfer coefficient are preserved as parameters within the reduced model, they can be altered in each iteration of

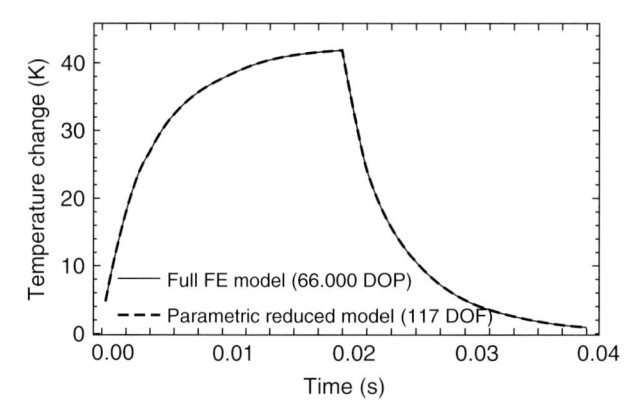

Figure 9.21 Comparison between the full FE models and the parametric-reduced models for the microhotplate. Full-scale model was computed with literature values for parameters κ, $\rho \cdot c_p$, and h from Ref. [28] and applied heating power was 2.49 mW.

the optimization process. By defining an objective function, which characterizes the quadratic difference between simulated and measured results, a data fitting procedure according to the algorithm in Figure 9.22 is performed. Note that, thanks to pMOR, no new FE model must be built and reduced in each iteration, as done in [27] and [29].

After 55 optimization cycles with the DOT® (Design Optimization Tools) software, the optimal values for all three parameters have been found (Table 9.1). Figure 9.23 shows the comparison between the measured temperature curve, the initial FE model, in which the thermal properties were set to literature values [28] from Table 9.1, and the optimized FE model. The temperature dependence of the material parameters c_p and κ has been neglected, as the measurement results were obtained near room temperatures. Note that it is not possible to independently determine the specific heat and mass density, as only their product is present in the system's heat capacity matrix.

In [11], the above procedure was successfully applied to extraction of thin-film thermal parameters of multilayer membranes, with up to 41 layers of silicon nitride (SiN_x), silicon oxide (SiO_x) and amorphous silicon (a-Si), with layer thicknesses of 200, 267, and 110 nm, respectively. The achieved parameter values agree well with the results presented in [32] for a film thickness of 500 nm and even continue the same decreasing slope to a thickness of 200 nm, as shown in Figure 9.24. The thermal conductivity of SiN_x thin films decreases for film thicknesses smaller than 1 μm. This is due to the changing layer composition during PECVD growth. The properties of the lower most layer portions differ from the upper regions, thus giving rise to a thickness-dependent thermal conductivity.

Note that the presented methodology can be equally applied to extraction of other material parameters than the thermal ones. The only present limitation is that the model must be linear in the sense that parameters of the finite element model in Eq. (9.2) must not be dependent on the state vector.

```
                    Start
                      |
         Build parametrized FE model
                      |
      Apply parametric model order reduction
                      |
              Integration of          <---+
              reduced model               |
                      |                   |
                  Evaluate                |
              objective function          |
                      |                   |
                           No        Change
                Convergence  ------> parameters
                      |
                    Yes
                      |
                     Stop
```

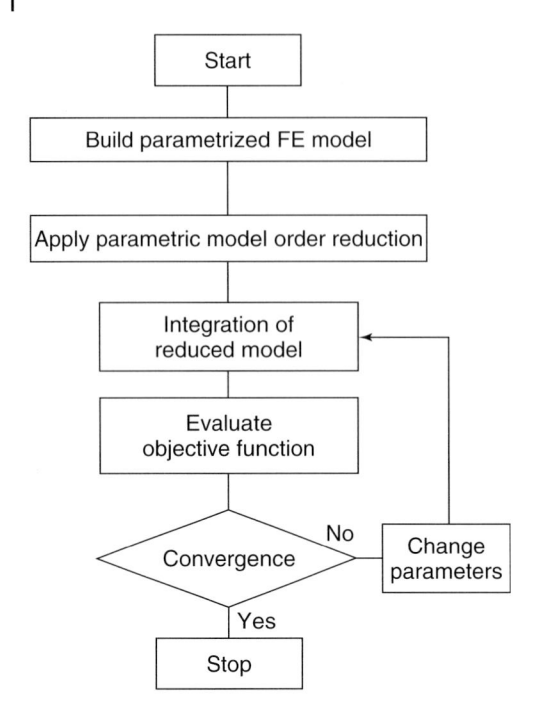

Figure 9.22 Algorithm for fast determination of material properties via parametric model order reduction. Parameterized FE model is created by extracting the system matrices from ANSYS®, via software tool MOR for ANSYS (Chapter 18). pMOR is done in mathematica [30], and optimization with DOT® – Design Optimization Tools [31].

Table 9.1 Material parameters.

		Initial guess from [28]	Optimization result SiNx (500 nm)
Heat conductivity	κ (W m^{-1} K^{-1})	2.5	4.2
Volumetric heat capacity	ρc_p (10^6 J m^{-3} K^{-1})	2.3	1.36
Heat transfer coefficient	h (W m^{-2} K^{-1})	10	11.4

9.6
Conclusion and Outlook

We have reviewed a methodology developed in [2] for applying moment-matching-based MOR to electrothermal MEMS models. Achieved results prove that this technology is mature to be used for linear numerical models. In the meantime, it is already available in the commercial software "MOR for ANSYS," as described in more details in Chapter 18. Still open issues are the mathematical proof of error bounds, as well as the coupling of several reduced-order models.

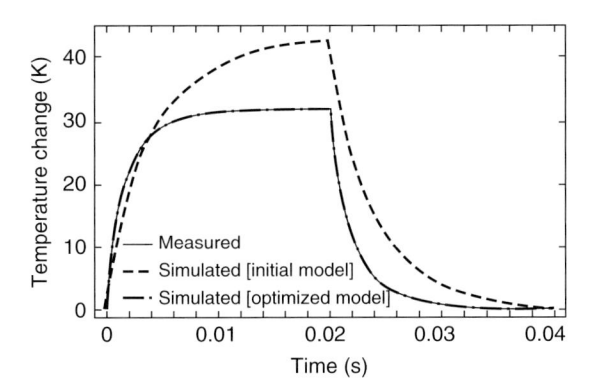

Figure 9.23 Measured and simulated transient curves (simulation was based on parametric-reduced model) of the sensor temperature for the microhotplate. The final model was gained after 55 optimization cycles.

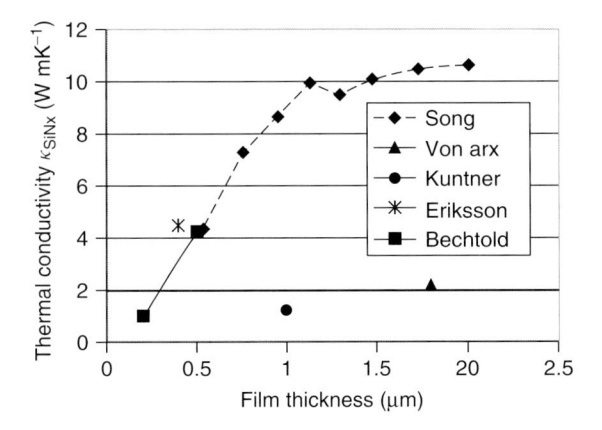

Figure 9.24 Comparison of the thermal conductivity values for PECVD-deposited silicon nitride gained with application of pMOR in Bechtold [11] and previous work (Song [32], von Arx [33], Kuntner [34], and Eriksson [35]).

We have further demonstrated the applicability of reduced-order models for an efficient system-level simulation of the device together with the control circuitry while preserving the accuracy of the device simulation. The reduced-order model can be used for parameterization of the controller for different application tasks.

We have also demonstrated a methodology for the efficient determination of material properties via pMOR and subsequent automatic parameter fitting. It is possible to extract thermal parameters of thin films from transient thermal characterization results, by using highly accurate three-dimensional numerical models of electrothermal MEMS test structures [11]. It is further possible to include thermal convection as a boundary condition to the model, while keeping the heat transfer coefficient between the hotplate and the surrounding air variable within the compact model, that is, pMOR allows to construct boundary-condition-independent compact thermal models.

Future work in this area should include the definition of the error bounds, reduction of thermal array structures, and the temperature dependence of material parameters. The latter leads to temperature-dependent system matrices C and K in Eq. (9.2) and requires application of MOR, that is, pMOR to a nonlinear system. Methods for nonlinear MOR are the topic of the following chapters.

References

1. Freund, R.W. (2000) Krylov-subspace methods for reduced order modeling in circuit simulation. *J. Comput. Appl. Math.*, **123**, 395–421.
2. Bechtold, T., Rudnyi, E.B., and Korvink, J.G. (2006) *Fast Simulation of Electro-Thermal MEMS: Efficient Dynamic Compact Models*, Springer, Berlin, ISBN-10: 3540346120.
3. Hohlfeld, D. and Zappe, H. (2007) Thermal and optical characterization of silicon-based tunable optical thin-film filters. *J. Microelectromech. Syst.*, **16**, 500–510.
4. Wöllenstein, J., Plaza, J.A., Cane, C., Min, Y., Böttner, H., and Tuller, H.L. (2003) A novel single chip thin film metal oxide array. *Sens. Actuators, B*, **93** (1–3), 350–355.
5. Graf, M., Juriscka, R., Barrettino, D., and Hierelemann, A. (2005) 3D nonlinear modeling of microhotplates in CMOS technology for use as metal-oxide based sensors. *J. Micromech. Microeng.*, **15**, 190–200.
6. Freund, R.W. (2011) in *Simulation and Verification of Electronic and Biological Systems* (eds P. Li, L.M. Silveira, and P. Feldmann), Springer, Dordrecht/Heidelberg/London/New York, pp. 43–70.
7. Bechtold, T., Hohlfeld, D., and Rudnyi, E.B. (2011) System-level model of electrothermal microsystem with temperature control circuit. Proceedings of the EuroSimE.
8. Ziegler, J.G. and Nichols, N.B. (1942) Optimum settings for automatic controllers. *Trans. ASME*, **64**, 759–768.
9. Chien, K.L., Hrones, J.A., and Reswick, J.B. (1952) On the automatic control of generalized passive systems. *Trans. Am. Soc. Mech. Eng.*, **74**, 175–185.
10. Hull, R. (ed.) (1999) *Properties of Crystalline Silicon*, Institution of Engineering and Technology.
11. Bechtold, T., Hohlfeld, D., and Rudnyi, E.B. (2010) Efficient extraction of thin film thermal parameters from numerical models via parametric model order reduction. *J. Micromech. Microeng.*, **20**, 045030.
12. Völklein, F. and Stärz, T. (1997) Thermal conductivity of thin films-experimental methods and theoretical interpretation. Proceedings of the Thermoelectrics, pp. 711–718.
13. Roncaglia, A., Mancarella, F., Sanmartin, M., Elmi, I., Cardinali, G.C., and Severi, M. (2006) Wafer-level measurement of thermal conductivity on thin films. Proceedings of the Sensors, pp. 1239–1242.
14. Feng, L.H. (2005) Parameter independent model order reduction. *Math. Comput. Simul.*, **68** (3), 221–234.
15. Weile, D.S., Michielssen, E., Grimme, E., and Gallivan, K. (1999) A method for generating rational interpolant reduced order models of two-parameter linear systems. *Appl. Math. Lett.*, **12**, 93–102.
16. Daniel, L., Siong, O.C., Chay, L.S., Lee, K.H., and White, J. (2004) A multiparameter moment-matching model-reduction approach for generating geometrically parameterized interconnect performance models. *IEEE Trans. Comput.-Aided Des. Integr. Circuits Syst.*, **23**, 678–693.
17. Feng, L.H., Rudnyi, E.B., and Korvink, J.G. (2005) Preserving the film coefficient as a parameter in the compact thermal model for fast electro-thermal simulation. *IEEE Trans. Comput.-Aided Des. Integr. Circuits Syst.*, **24** (12), 1838–1847.

18. Codecasa, L., D'Amore, D., and Maffezzoni, P. (2004) A novel approach for generating boundary condition independent compact dynamic thermal networks of packages. Proceedings of the THERMINIC, pp. 305–310.
19. Farle, O., Hill, V., Ingelström, P., and Dyczij-Edlinger, R. (2006) Multi-parameter polynomial model reduction of linear finite element equation systems. Proceedings of the MATHMOD.
20. Li, J.T., Bai, Z., Su, Y., and Zeng, X. (2007) Parameterized model order reduction via a two-directional Arnoldi process. Proceedings of the Computer-Aided Design, pp. 868–873.
21. Feng, L. and Benner, P. (2007) A robust algorithm for parametric model order reduction. *Proc. Appl. Math. Mech.*, **7** (1), 10215.01–10215.02.
22. Li, Y.-T., Bai, Z., Su, Y., and Zeng, X. (2008) Model order reduction of parameterized interconnect networks via a two-directional Arnoldi process. *IEEE Trans. Comput.-Aided Des. Integr. Circuits Syst.*, **27** (9), 1571–1582.
23. Li, Y.-T., Bai, Z., and Su, Y. (2009) A two-directional Arnoldi process and its application to parametric model order reduction. *J. Comput. Appl. Math.*, **226**, 10–21.
24. Gunupidi, P.K. and Nakhla, M. (2000) Multi-dimensional model reduction of VLSI interconnects. Proceedings of the Custom Integrated Circuits Conference, pp. 499–502.
25. Gunupidi, P., Khazaka, R., Nakhla, M., Smy, T., and Celo, D. (2003) Passive parameterized time-domain macromodels for high-speed transmission-line networks. *Microwave Theory Technol.*, **51** (12), 2347–2354.
26. Celo, D., Gunupidi, P.K., Khazaka, R., Walkey, D.J., Smy, T., and Nakhla, M.S. (2005) Fast simulation of steady-state temperature distributions in electronic components using multidimensional model reduction. *Compon. Packag. Technol.*, **28** (1), 70–79.
27. Bechtold, T., Hohlfeld, D., Rudnyi, E.B., Zappe, H., and Korvink, J.G. (2005) Inverse thermal problem via model order reduction: determining material properties of a microhotplate. Proceedings of the THERMINIC, pp. 146–150.
28. Shackelford, J.F. and Alexander, W. (2000) *CRC Materials Science and Engineering Handbook*, CRC Press.
29. Han, J.S., Rudnyi, E.B., and Korvink, J.G. (2005) Efficient optimization of transient dynamic problems in MEMS devices using model order reduction. *J. Micromech. Microeng.*, **15** (4), 822–832.
30. http://modelreduction.com/ModelReduction/Parametric.html.
31. Vanderplaats R&D, Inc. DOT® - Design Optimization Tools, User Manual Version 4.20, Copyright 2009, http://www.vrand.com/DOT.html.
32. Song, Q., Xia, S., Chen, S., and Cui, Z. (2004) A new structure for measuring the thermal conductivity of thin film. Proceedings of the Information Acquisition, pp. 77–79.
33. von Arx, M., Paul, O., and Baltes, H. (2000) Process-dependent thin-film thermal conductivities for thermal CMOS MEMS. *J. Microelectromech. Syst.*, **9** (1), 136–145.
34. Kuntner, J., Jachimowicz, A., Kohl, F., and Jakoby, B. (2006) Determining the thin-film thermal conductivity of low temp. PECVD Si_3N_4. Proceedings of the Eurosensors.
35. Eriksson, P., Andersson, J.Y., and Stemme, G. (1997) Thermal characterization of surface-micromachined silicon nitride membranes for thermal infrared detectors. *J. Microelectromech. Syst.*, **6** (1), 55–61.

10
Projection-Based Nonlinear Model Order Reduction
Amit Hochman, Dmitry M. Vasilyev, Michał J. Rewieński, and Jacob K. White

10.1
Introduction

Microelectromechanical systems (MEMS), as the name implies, involve energy conversion from one domain to another. As emphasized in the previous chapters, analyzing MEMS from first principles leads to complex multiphysics models expressed as coupled partial differential equations (PDEs). While such detailed models can capture most of the effects that might influence the dynamics of a device, combining these models to simulate an entire system is computationally impractical. The model order reduction (MOR) approach to this problem consists of deriving simple models from the detailed ones in a systematic way, by use of mathematical tools from fields such as control theory and signal processing (see Chapter 1.3 for more details).

The theory and methods of linear MOR have matured in recent decades (see Chapter 1.3 for references). Accordingly, MOR research has been turning to nonlinear systems, such as those encountered in coupled PDEs models of MEMS. This chapter focuses on reduction methods based on state projection. Projection methods are widely used for both linear and nonlinear MOR and provide a means of generalizing from the former to the latter. Important alternatives for nonlinear MOR are the Hankel-optimal [1], manifold construction [2], system-identification [3], and Volterra-like [4] methods. In this chapter, we focus on generalizing projection methods, explained in Chapter 1.3 for linear systems (see, e.g., Eq. (21)), to nonlinear systems.

This chapter is organized as follows: we begin by stating the nonlinear model reduction problem and the projection principle in Sections 10.2 and 10.3, respectively. We then describe three projection-based nonlinear reduction methods: the Taylor series method (Section 10.4), the trajectory piecewise-linear (TPWL) method (Section 10.5), and the discrete empirical interpolation method (DEIM) (Section 10.6).

System-level Modeling of MEMS, First Edition. Edited by T. Bechtold, G. Schrag, and L. Feng.
© 2013 Wiley-VCH Verlag GmbH & Co. KGaA. Published 2013 by Wiley-VCH Verlag GmbH & Co. KGaA.

10.2
Problem Specification

Consider the problem of reducing the system generated by spatial discretization of a coupled PDE model of a micromachined device, which typically has the form

$$\dot{x}(t) = f(x(t)) + Bu(t), \quad y(t) = Cx(t) \tag{10.1}$$

This system has n state variables, n_i inputs, and n_o outputs, represented in Eq. (10.1) by the vectors $x(t) \in \mathbb{R}^n$, $u(t) \in \mathbb{R}^{n_i}$, and $y(t) \in \mathbb{R}^{n_o}$, respectively. Also, in Eq. (10.1), we have $f : \mathbb{R}^n \mapsto \mathbb{R}^n$, a vector-valued function, $B \in \mathbb{R}^{n \times n_i}$ an input matrix, and $C \in \mathbb{R}^{n_o \times n}$ an output matrix. We derive our reduction method based on Eq. (10.1), as that is the form typically generated by finite-difference discretization of PDEs. We note that finite-element PDE discretization with dynamically adapted meshes [5] and networks of lumped-circuit elements are two examples of cases that may require the more general nonlinear descriptor form

$$\frac{dq(x(t))}{dt} = f(x(t)) + Bu(t) \tag{10.2}$$

but, for the methods herein, the extension from Eq. (10.1) to Eq. (10.2) is usually straightforward. A reduced model consistent in form with Eq. (10.1) is

$$\dot{x}_r(t) = f_r(x_r(t)) + B_r u(t), \quad y_r(t) = C_r x_r(t) \tag{10.3}$$

where $x_r \in \mathbb{R}^q$ is the reduced state vector, with q typically much smaller than n. For the reduced model, $f_r : \mathbb{R}^q \mapsto \mathbb{R}^q$, $B_r \in \mathbb{R}^{q \times n_i}$, $C_r \in \mathbb{R}^{n_o \times q}$, and $y_r(t) \in \mathbb{R}^{n_o}$ are the reduced function, reduced input matrix, reduced output matrix, and reduced model output, respectively.

The model reduction problem consists of finding a reduced model that is much cheaper to evaluate than the full model, yet accurately reproduces the full model's input/output behavior, a statement that can be made formal by defining cost and accuracy metrics (see chapter 1.3 for definitions of common error norms). For example, the model reduction problem can be cast as a minimax problem

$$\min_{C(f_r, B_r, C_r) < \alpha} \left(\max_{u \in \mathbb{U}} \|y - y_r\| \right) \tag{10.4}$$

where $\mathbb{U}$ is a, typically application specific, class of inputs and $C(f_r, B_r, C_r)$ denotes the cost of evaluating the given reduced model. In other words, the model reduction problem is to find a reduced model that costs less than α to evaluate but preserves input–output behavior over the range of inputs of interest. When $f(x)$ is linear and there are only a few inputs and outputs, a good surrogate for the evaluation cost is the order of the reduced system, q, and the term *MOR* is appropriate. For the nonlinear case, or even in the linear case with a large number of inputs and outputs, it may be better to use the term "model reduction," as reducing the order of the model alone may not significantly reduce evaluation cost [6]. This difficulty is explained in the following section.

10.3
Projection Principle and Evaluation Cost for Nonlinear Systems

The issue of evaluation cost can be best seen by considering reduction by projection [7]. As explained in Chapter 1.3, in projection methods, the n-dimensional state vector is approximated as a weighted combination of q length-n vectors, $\mathbf{U}_x \in \mathbb{R}^{n \times q}, q \ll n$, as

$$\mathbf{x} \approx \mathbf{U}_x \mathbf{x}_r \tag{10.5}$$

A dynamical system for the reduced state vector $\mathbf{x}_r$ can be obtained using the Petrov-Galerkin method (Chapter 1.3). When the approximation (Eq. (10.5)) is substituted into Eq. (10.1), an error, $\Delta(t) \equiv \mathbf{U}_x \dot{\mathbf{x}}_r - \mathbf{f}(\mathbf{x}(t)) - \mathbf{B}\mathbf{u}(t)$, is obtained. Forcing this error to be orthogonal to the columns of a testing matrix $\mathbf{V} \in \mathbb{R}^{n \times q}$, yields the reduced model

$$\dot{\mathbf{x}}_r(t) = \underbrace{\mathbf{V}^T \mathbf{f}(\mathbf{U}_x \mathbf{x}_r t)}_{\mathbf{f}_r(\mathbf{x}_r(t))} + \underbrace{\mathbf{V}^T \mathbf{B}\mathbf{u}(t)}_{\mathbf{B}_r \mathbf{u}(t)}, \quad \mathbf{y}_r(t) = \underbrace{\mathbf{C}\mathbf{U}_x \mathbf{x}_r(t)}_{\mathbf{C}_r \mathbf{x}_r(t)} \tag{10.6}$$

where we assume $\mathbf{V}$ is chosen so that $\mathbf{V}^T \mathbf{U}_x = \mathbf{I}$. If $\mathbf{f}(\mathbf{x})$ is linear, that is, $\mathbf{f}(\mathbf{x}) = \mathbf{A}\mathbf{x}$, with $\mathbf{A} \in \mathbb{R}^{n \times n}$, projection yields $\mathbf{f}_r(\mathbf{x}_r) = \mathbf{V}^T \mathbf{A} \mathbf{U}_x \mathbf{x}_r$. Since $\mathbf{V}^T \mathbf{A} \mathbf{U}_x$ is a $q \times q$ matrix and can be nearly diagonalized once during model construction (down to a 2×2 block diagonal matrix), model evaluation cost for the linear case is proportional to the order q. If $\mathbf{f}(\mathbf{x})$ is nonlinear and $\mathbf{f}_r(\mathbf{x}_r)$ is given by $\mathbf{V}^T \mathbf{f}(\mathbf{U}_x \mathbf{x}_r)$, evaluation of $\mathbf{f}_r(\mathbf{x}_r)$ entails computing the n-length vector $\mathbf{U}_x \mathbf{x}_r$, then evaluating the n elements of $\mathbf{f}(\mathbf{U}_x \mathbf{x}_r)$, and finally projecting down to a q-length vector $\mathbf{V}^T \mathbf{f}(\mathbf{U}_x \mathbf{x}_r)$, as shown in Figure 10.1. Hence, the evaluation cost in the nonlinear case is proportional to the full model order n, not the reduced order q.

Note that there are two issues associated with projection-based nonlinear model reduction: selecting the $2q$ vectors in $\mathbf{V}$ and $\mathbf{U}_x$ and finding a computationally inexpensive approximation to $\mathbf{f}_r(\mathbf{x}_r) = \mathbf{V}^T \mathbf{f}(\mathbf{U}_x \mathbf{x}_r)$. The strategies for selecting vectors in $\mathbf{V}$ and $\mathbf{U}_x$ fall into two broad categories: the first uses the singular value decomposition (SVD) to select vectors from time samples of the original model's state vector in response to carefully chosen inputs (the "snapshot" or proper orthogonal decomposition (POD) methods) [8, 9] and the second uses carefully chosen linearizations of the original model and then applies linear system methods such as the Krylov-subspace [10] or Grammian-based methods [1, 11, 12]. The strategies for finding approximations to $\mathbf{f}_r(\mathbf{x}_r)$ include using Taylor series [7, 13–15],

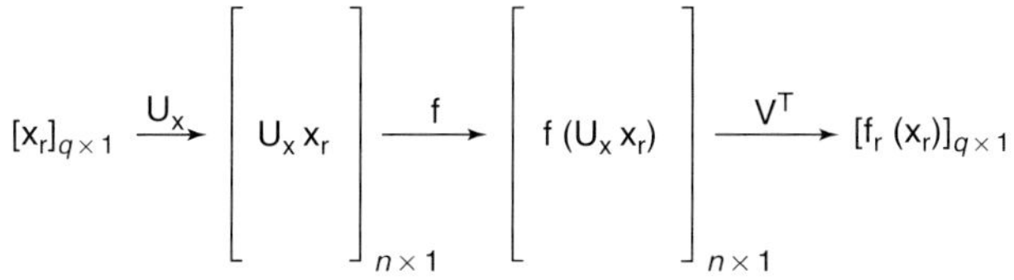

$$[\mathbf{x}_r]_{q \times 1} \xrightarrow{\mathbf{U}_x} \begin{bmatrix} \mathbf{U}_x \mathbf{x}_r \\ \\ \end{bmatrix}_{n \times 1} \xrightarrow{\mathbf{f}} \begin{bmatrix} \mathbf{f}(\mathbf{U}_x \mathbf{x}_r) \\ \\ \end{bmatrix}_{n \times 1} \xrightarrow{\mathbf{V}^T} [\mathbf{f}_r(\mathbf{x}_r)]_{q \times 1}$$

Figure 10.1 Sequence of operations for evaluating $\mathbf{f}_r(\mathbf{x}_r)$. Although $\mathbf{x}_r$ and $\mathbf{f}_r(\mathbf{x}_r)$ are both $q \times 1$, the operation count depends on n.

the TPWL and piecewise-polynomial methods [16, 17], and the DEIM [18, 19]. An important aspect to consider when deciding on these two issues is that the resulting model should preserve the stability of the original model and, in some cases, also its lack of stability for certain inputs. The stability of the methods described in this chapter is discussed in Sections 10.4.3, 10.5.2, and 10.6.2 for the Taylor series method, the TPWL method, and the DEIM, respectively.

10.4
Taylor Series Expansions

One of the early attempts at using nonlinear MOR for MEMS was based on a quadratic approximation of the nonlinear function [13, 14] and was found useful for weak nonlinearities, that is, ones for which the linear term is dominant and the nonlinearity is a small second-order effect. A natural way of extending this approach to stronger nonlinearities is to use higher-order polynomial approximations [13, 15, 20–22].

In the Taylor series expansion method (or equivalently, the Volterra method), the function $f(x)$ in Eq. (10.1) is approximated with a truncated multidimensional Taylor series about a nominal state, x_0

$$f(x) \approx A_0 + A_1 \Delta x + A_2(\Delta x \otimes \Delta x) \tag{10.7}$$

where $\Delta x = x - x_0$, the Kronecker product is denoted by $\otimes$, and

$$A_0 \equiv f(x_0) \in \mathbb{R}^n \tag{10.8}$$

$$A_1 \equiv \left.(\nabla^T \otimes f)\right|_{x=x_0} \in \mathbb{R}^{n \times n} \tag{10.9}$$

$$A_2 \equiv \tfrac{1}{2} \left.(\nabla^T \otimes \nabla^T \otimes f)\right|_{x=x_0} \in \mathbb{R}^{n \times n \times n} \tag{10.10}$$

and so on. Here, we have used the notation

$$\nabla \equiv \left[\frac{\partial}{\partial x_1}, \frac{\partial}{\partial x_2}, \ldots, \frac{\partial}{\partial x_n}\right]^T \tag{10.11}$$

although A_1 is just the Jacobian of f evaluated at x_0. Assuming $f(x)$ is differentiable to any order at $x = x_0$, and the states are sufficiently close to x_0, the truncated Taylor series can be an adequate approximation of $f(x)$.

The most commonly used, and lowest-order nonlinear, expansion is the quadratic expansion

$$\hat{f}(x) = A_0 + A_1 \Delta x + A_2(\Delta x \otimes \Delta x) \tag{10.12}$$

Inserting the quadratic expansion Eq. (10.12) in Eq. (10.6)

$$\dot{x}_r = V^T A_0 + V^T A_1 U_x \Delta x_r + V^T A_2(U_x \otimes U_x)(\Delta x_r \otimes \Delta x_r) + V^T Bu \tag{10.13}$$

which simplifies to

$$\dot{x}_r = A_{0r} + A_{1r}\Delta x_r + A_{2r}(\Delta x_r \otimes \Delta x_r) + B_r u \tag{10.14}$$

using the identity $(\mathbf{U}_x \Delta \mathbf{x}_r) \otimes (\mathbf{U}_x \Delta \mathbf{x}_r) = (\mathbf{U}_x \otimes \mathbf{U}_x)(\Delta \mathbf{x}_r \otimes \Delta \mathbf{x}_r)$. Here, $\Delta \mathbf{x}_r = \mathbf{x}_r - \mathbf{x}_{r0}$ in which it is assumed that $\mathbf{x}_0 \in \text{colsp}(\mathbf{U}_x)$ so that $\mathbf{x}_0 = \mathbf{U}_x \mathbf{x}_{r0}$. The coefficients of the constant, linear, and quadratic terms in Eq. (10.13) are, of course, independent of the reduced state and are calculated once during model construction. During evaluation, the cost of computing the quadratic term dominates, as $\mathbf{A}_{2r}$ is a, typically dense, $q^2 \times q^2$ matrix, so computing $\mathbf{A}_{2r}(\mathbf{x}_r \otimes \mathbf{x}_r)$ is proportional to q^4. Therefore, significant reduction in computation time and memory is only possible for very low-order reduced models, for example, $q < 10$. The situation is even worse for higher-order Taylor expansions, as the cost of evaluation increases exponentially with Taylor expansion order. We note that the Taylor series approximation of f can also be used to gain insight into how the nonlinearity should affect the choice of $\mathbf{U}_x$, as examined in [20].

10.4.1
Microfluidic Channel Example

The Taylor series expansion method is a natural choice for MOR of systems with a quadratic nonlinearity. An example of such a system was obtained in [23] by adapting a linear system described in [24]. The example consists of a U-shaped microfluidic channel of rectangular cross section, filled with a carrying, or buffer, fluid and a marker fluid (Figure 10.2). The buffer fluid is driven electrokinetically with a constant field, so the buffer flow is assumed to be steady. The total marker flux, $\vec{J}$ is modeled as a combination of electrokinetically driven buffer flow and isotropic diffusion, as in

$$\vec{J}(\vec{r}, t) = -\mu \left[\nabla_r \Phi(\vec{r}) \right] C(\vec{r}, t) - D \nabla_r C(\vec{r}, t) \tag{10.15}$$

where $C(\vec{r}, t)$ denotes the marker concentration at position $\vec{r}$ and time t, D the marker diffusion constant, and $-\nabla_r \Phi(\vec{r})$ the electrostatic field at position $\vec{r}$ due to the potential Φ satisfying $\nabla_r^2 \Phi = 0$ in the channel, $\Phi(\text{inlet}) = \Phi_0$, $\Phi(\text{outlet}) = 0$, and $d\Phi/dn = 0$ at the walls. For sufficiently high marker concentrations, Eq. (10.15)

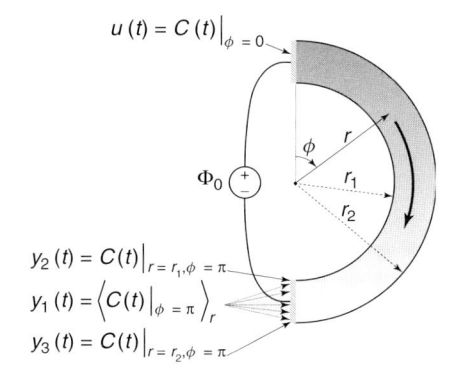

Figure 10.2 A microfluidic channel filled with buffer and marker fluids. The buffer fluid is driven electrokinetically.

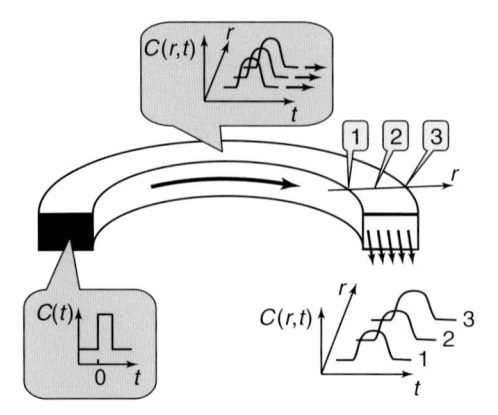

Figure 10.3 Propagation of a square pulse of marker concentration along a microfluidic channel. The pulse takes longer to reach points at the outer radius.

becomes nonlinear because the electroosmotic mobility and the diffusion constant become dependent on local concentration, $\mu \equiv \mu(C(\vec{r}, t))$ and $D \equiv D(C(\vec{r}, t))$. Enforcing conservation of marker concentration in Eq. (10.15) yields a convection–diffusion equation [25]

$$\frac{\partial C}{\partial t} = -\nabla_r \cdot \vec{J} = \nabla_r \Phi \cdot \left[C \nabla_r \mu(C) + \mu(C) \nabla_r C \right] + \nabla_r D(C) \cdot \nabla_r C + D(C) \nabla_r^2 C \tag{10.16}$$

As the walls are impenetrable, so normal flux is zero, $\hat{n} \cdot \vec{J} = 0$. The concentration at the inlet is determined by the given input, and $\partial C/\partial n = 0$ is assumed at the outlet.

A state-space system is derived from Eq. (10.16) by applying a second-order three-dimensional coordinate-mapped finite-difference spatial discretization to Eq. (10.16) on the half-ring domain in Figure 10.2. The states were chosen to be concentrations of the marker fluid at the spatial locations inside the channel. The time-varying concentration of the marker at the inlet of the channel is the input, and there are three outputs derived from the outlet concentration: the average, the inner, and the outer concentrations (Figure 10.3).

As a pulse of marker concentration propagates along the channel, diffusion smears the pulse and a "race track" effect is visible (Figure 10.3). That is, the marker concentration near the inner radius exits the channel first (point 1).

10.4.2
Model Reduction via Quadratic Taylor Expansion

Concentration-dependent mobility or diffusivity makes Eq. (10.16) nonlinear with respect to marker concentration. If they are affine functions of concentration, for example

$$\mu(C) = (28 + 5.6C) \times 10^9 \frac{m^2}{Vs}, \quad D(C) = (5.5 + 1.1C) \times 10^{-9} \; m^2 \; s^{-1} \tag{10.17}$$

then discretizing Eq. (10.16) leads to a quadratic dynamical system that is well suited to Taylor series-based reduction.

As is discussed in Section 10.4.3, quadratic models can have stability issues. Indeed, we obtained unstable models when $\mathbf{U}_x \neq \mathbf{V}$. However, models obtained with $\mathbf{U}_x = \mathbf{V}$, known as the *Galerkin method* (Chapter 1.3), were always stable in the examples we tested.

We generated a full model ($n = 2204$) and a reduced-order model ($q = 9$) for a channel with inner radius, $r_1 = 500\,\mu\text{m}$, outer radius, $r_2 = 800\,\mu\text{m}$, and height, $d = 300\,\mu\text{m}$. For the projection matrices, we used an Arnoldi method (see Chapter 1.3 for a detailed description), which yielded an orthonormal basis for the $q = 9$ Krylov subspace, that is

$$\text{colsp}(\mathbf{V}) = \mathcal{K}_9(\mathbf{J}^{-1}, \mathbf{J}^{-1}\mathbf{B})$$

where $\mathbf{J} \equiv \mathbf{A}_1$ denotes the Jacobian at $\mathbf{x} = 0$. The transient response to a square pulse input for the full model, the full model linearized, and the reduced quadratic model are plotted in Figure 10.4. As can be seen from the plots, the full model

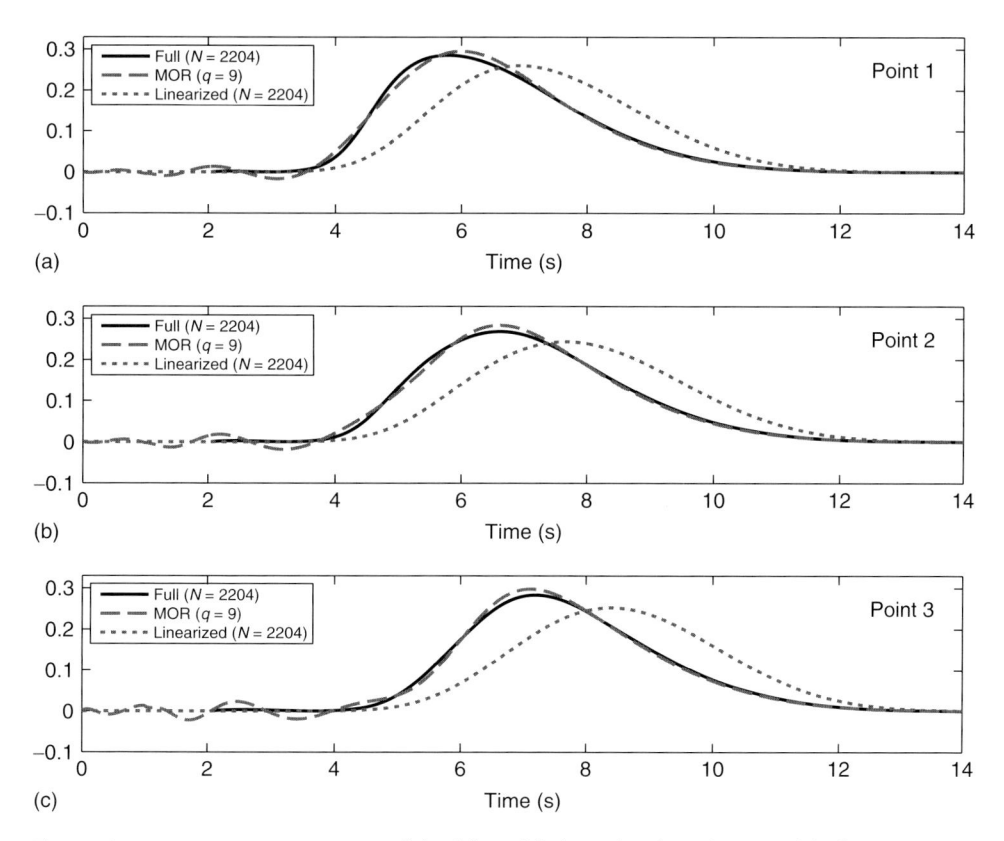

Figure 10.4 (a–c) Transient response of the full model, the reduced quadratic model of order $q = 9$, and a full linearized model. The input signal is a pulse of 1s duration.

is quite nonlinear (does not match the linearization) and the reduced quadratic model reproduces the full model behavior quite accurately. The oscillations visible in the reduced model are a result of the full model having a pure delay, which is hard to reproduce with a small reduced order.

10.4.3
Stability Issues

A significant problem with the Taylor series expansion method is that the reduced-order models obtained can be unstable. To see this, consider a function obeying the following *sector condition* [26]

$$\mathbf{x}^T \mathbf{f}(\mathbf{x}) \leq 0, \quad \forall \mathbf{x} \tag{10.18}$$

Then, the dynamical system $\dot{\mathbf{x}} = \mathbf{f}(\mathbf{x})$ is globally stable, as can be readily verified by use of the Lyapunov function $L = \mathbf{x}^T \mathbf{x}$. If $\mathbf{f}(\mathbf{x})$ is approximated by a polynomial, the sector condition may no longer hold. For a simple example, consider the scalar function $f(x) = \exp(-x) - 1$ that obeys $xf(x) \leq 0$. If it is replaced by its second-order Maclaurin series, $\hat{f}(x) = -x + x^2/2$, we find that $x\hat{f}(x) > 0$ for all $x > 1$. This situation is illustrated in Figure 10.5. Although global stability is not guaranteed in the Taylor series method, local stability can be. To see this, note that the Jacobian of the reduced model at $\mathbf{x} = 0$ is just the projected Jacobian, that is, $\mathbf{J}_r = \mathbf{V}^T \mathbf{J} \mathbf{U}_x$. Hence, if the full model is locally stable, $\mathbf{J}$ is a Hurwitz matrix and it is possible to derive $\mathbf{V}$ and $\mathbf{U}_x$ so that $\mathbf{V}^T \mathbf{J} \mathbf{U}_x$ is also Hurwitz. Methods for doing this are well known [27]: if the system matrix is negative definite, then taking $\mathbf{V} = \mathbf{U}_x$ preserves stability. If not, truncated-balanced realizations (TBRs) guarantee stability, although the computational cost of TBR is $O(n^3)$ and can be prohibitive for large problems, although methods for reducing this cost for sparse systems have been proposed [28]. In addition, an optimization-based alternative to TBR, whose cost is only $O(q^3)$, was proposed in [29].

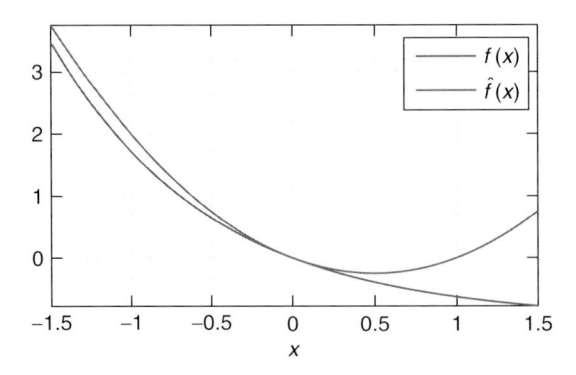

Figure 10.5 The function $f(x) = \exp(-x) - 1$ and its second-order Maclaurin series, $\hat{f}(x) = -x + x^2/2$. The sector condition no longer holds for $\hat{f}(x)$.

10.4.4
Taylor Series Expansions: Summary

The use of Taylor expansion methods has been limited to weakly nonlinear systems. The reason for this is that although the Jacobian and higher-order derivatives of the original system are sparse, no methods are known for preventing projection from destroying this sparsity. As a result, reduced model evaluation time and memory grow exponentially with the Taylor expansion order, making higher-order expansions impractical. Furthermore, global stability of the reduced-order model cannot be guaranteed, although local stability is usually easy to enforce. However, the need to analyze weakly nonlinear systems arises frequently in practice, often as the result of including second-order effects in a system designed to be linear.

10.5
Trajectory Piecewise-Linear Method

TPWL methods were proposed in [30] and further developed in [12, 16, 17, 29, 31–34] as an alternative to the Taylor expansion method. To obtain a TPWL model, a number of linearizations along a training trajectory are obtained. These are reduced by linear MOR and then the reduced-order models are interpolated for state vectors that fall between the linearizations. This scheme can be combined with any linear MOR method to yield a nonlinear function approximation that is cheap to evaluate.

To derive the TPWL method, assume that a set of "important" state vectors $\vec{x_i}$, $i = 1 \ldots l$, ones that are near common trajectories of the original system, have been obtained. The original nonlinear function, $\mathbf{f}(\mathbf{x})$, is then approximated as a weighted combination of the linearizations of $\mathbf{f}(\mathbf{x})$ at the $\vec{x_i}$. That is

$$\mathbf{f}(\mathbf{x}) \approx \hat{\mathbf{f}}(\mathbf{x}) = \sum_{i=1}^{l} w_i(\mathbf{x}) \left[\mathbf{f}(\vec{x_i}) + \mathbf{J}(\vec{x_i})(\mathbf{x} - \vec{x_i}) \right] \qquad (10.19)$$

where $\hat{\mathbf{f}}(\mathbf{x})$ denotes the approximation, $\mathbf{J}$ is the Jacobian of $\mathbf{f}(\mathbf{x})$ evaluated at the state vector $\vec{x_i}$, and the state-dependent weights, $w_i(\mathbf{x})$, satisfy

$$w_i(\mathbf{x}) \geq 0 \qquad (10.20)$$

$$\sum_{i=1}^{l} w_i(\mathbf{x}) = 1, \quad \forall \mathbf{x} \qquad (10.21)$$

$$\lim_{\mathbf{x} \to \vec{x_i}} w_i(\mathbf{x}) = 1 \qquad (10.22)$$

Note that the approximation in Eq. (10.19) is a convex combination of the linearizations of $\mathbf{f}(\mathbf{x})$. One example formula for the weights is

$$w_i(\mathbf{x}) = \frac{\exp\left(\frac{-\beta \|\mathbf{x} - \vec{\mathbf{x}_i}\|_2}{\min_k \|\mathbf{x} - \mathbf{x}_k\|_2}\right)}{\sum_{i=1}^{l} \exp\left(\frac{-\beta \|\mathbf{x} - \vec{\mathbf{x}_i}\|_2}{\min_k \|\mathbf{x} - \mathbf{x}_k\|_2}\right)} \quad (10.23)$$

where in the limit as $\beta \to \infty$, the weight of the nearest point is one and the rest are zero, and in the limit as $\beta \to 0$, all the weights are equal to $1/l$. Weighting schemes are discussed more generally in [16]. Given the weighting function Eq. (10.23), the reduced nonlinear function, $\mathbf{f}_r(\mathbf{x}_r)$, can be obtained by applying projection to Eq. (10.19)

$$\mathbf{f}_r(\mathbf{x}_r) \approx \sum_{i=1}^{l} w_i^r(\mathbf{x}_r) \left[\mathbf{V}^T \mathbf{f}(\vec{\mathbf{x}_i}) - \mathbf{V}^T \mathbf{J}(\vec{\mathbf{x}_i}) \vec{\mathbf{x}_i} + \mathbf{V}^T \mathbf{J}(\vec{\mathbf{x}_i}) \mathbf{U}_x \mathbf{x}_r \right] \quad (10.24)$$

with

$$w_i^r(\mathbf{x}_r) = \frac{\exp\left(\frac{-\beta \|\mathbf{x}_r - \mathbf{V}^T \vec{\mathbf{x}_i}\|_2^2}{\min_k \|\mathbf{x}_r - \mathbf{V}^T \mathbf{x}_k\|_2^2}\right)}{\sum_{j=1}^{l} \exp\left(\frac{-\beta \|\mathbf{x}_r - \mathbf{V}^T \vec{\mathbf{x}_i}\|_2^2}{\min_k \|\mathbf{x}_r - \mathbf{V}^T \mathbf{x}_k\|_2^2}\right)} \quad (10.25)$$

In Eq. (10.25), the weights are determined by the distance between the reduced state and the projected linearization points, which is equal to the projected distance between the approximate full-order state, $\mathbf{U}_x \mathbf{x}_r$, and the linearization points. That is, $\|\mathbf{x}_r - \mathbf{V}^T \vec{\mathbf{x}_i}\| = \|\mathbf{V}^T (\mathbf{U}_x \mathbf{x}_r - \vec{\mathbf{x}_i})\|$, which is obtained using $\mathbf{V}^T \mathbf{U}_x = \mathbf{I}$. A common method for generating the $\mathbf{V}$ and $\mathbf{U}_x$ matrices for TPWL is based on starting with a linear MOR method, such as Krylov subspace or TBR methods. The linear MOR is applied to the linearization at an equilibrium state or to some or all of the linearizations. A single pair of projection matrices $\mathbf{U}_x$ and $\mathbf{V}$ are obtained by using results from a single linearization or by concatenating the matrices of the various linearizations and using the SVD to discard vectors corresponding to small singular values. Alternatively, as in POD, $\mathbf{U}_x$ and $\mathbf{V}$ can be taken equal to one another and derived from the SVD of a snapshot matrix, $[\mathbf{x}(t_1), \mathbf{x}(t_2), \ldots, \mathbf{x}(t_{n_s})]$, where the number of snapshots $n_s > q$ [35]. The $\mathbf{x}(t_i)$ could be the $\vec{\mathbf{x}_i}, i = 1, 2, \ldots, l$ (the important state vectors used to collect linearizations) or some other set obtained as the system response to a different training input.

10.5.1
Nonlinear Transmission-line Model

A common MOR benchmark, which involves a dynamical system of the form Eq. (10.1), is a lumped element model of a nonlinear transmission line, shown in Figure 10.6. This example was considered without the inductors in [14, 36, 37] and with the inductors in [16]. The input is the current through the current source, and the output is the voltage across it. For this example, we used $n = 100$, and a training input $u(t) = (t/T)\mathrm{sgn}[\sin(2\pi t/T)]$, with $T = 10$ s and $t \in [0, 3T]$. Note that our training input is a square wave whose amplitude increases with time

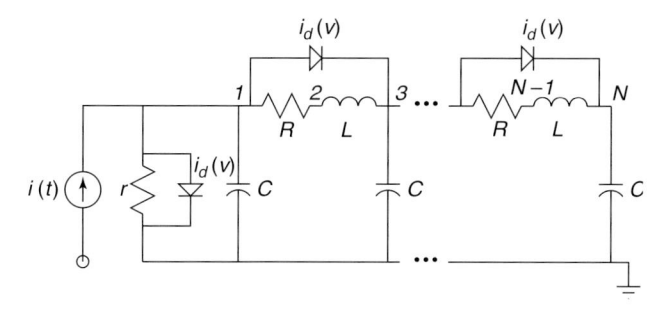

Figure 10.6 Nonlinear transmission line example. Lumped element values are $C = 1\text{F}, R = 1\,\Omega, L = 10\,\text{Hy}$. The I--V relation for all the diodes is $I = \exp(40\,V) - 1$.

(Figure 10.7, top-left). The snapshots, $\vec{x_i}$, were taken at 50 uniformly distributed time instants. In Figure 10.7, results for three testing inputs are shown. These are a square wave, $u_1(t) = I_0\text{sgn}[\sin(2\pi t/T)]$, a two-tone signal, $u_2(t) = I_0[2\sin(2\pi t/T) + \cos(4\pi t/T)]/3$, and a noiselike signal, $u_3(t)$, generated by fitting a spline through 60 uniformly distributed time points in $[0, 3T]$, where the time point values were randomly distributed between $[-I_0, I_0]$. We chose $I_0 = 1.5$ A that is high enough to excite a nonlinear response: for a harmonic input with a 1.5 A amplitude and period T, a third of the output power is in frequencies different from that of the input. As can be seen from Figure 10.7, accuracy of a few percent is achieved even with a very low order ($q = 4$) reduced model.

In Figure 10.7, the testing input amplitudes are well within the range of amplitudes of the training input. If the testing input amplitudes are increased, as in Figure 10.8, the accuracy suffers. For example, if the amplitude of the random signal is increased to 3.0 A, we obtain the results shown in Figure 10.8, in which loss of accuracy is evident.

10.5.2
Stability Issues

A TPWL reduced-order model of a stable nonlinear system may be unstable. First, linearized versions of a stable nonlinear system need not be stable. Second, even if all linearizations are stable, their reduction by projection can result in unstable reduced linearizations. As discussed in Section 10.4.3, methods for preserving stability of linear systems are well known and can be used to generate $\mathbf{U_x}$ and $\mathbf{V}$ matrices such that the reduced Jacobian at an equilibrium state is a Hurwitz matrix, and hence, local stability is obtained. Away from equilibrium, projection with the same $\mathbf{U_x}$ and $\mathbf{V}$ may result in reduced Jacobians that are not Hurwtiz.

An example of the foregoing problem appears in [12]. A micromachined switch, which is described in Section 10.7, was reduced using a two-step reduction procedure: first using a Krylov-subspace method and then using TBR, and it was found that when the order of the reduced-order model was odd, unstable reduced models were obtained. As explained in Chapter 4 of [23], when the order is odd,

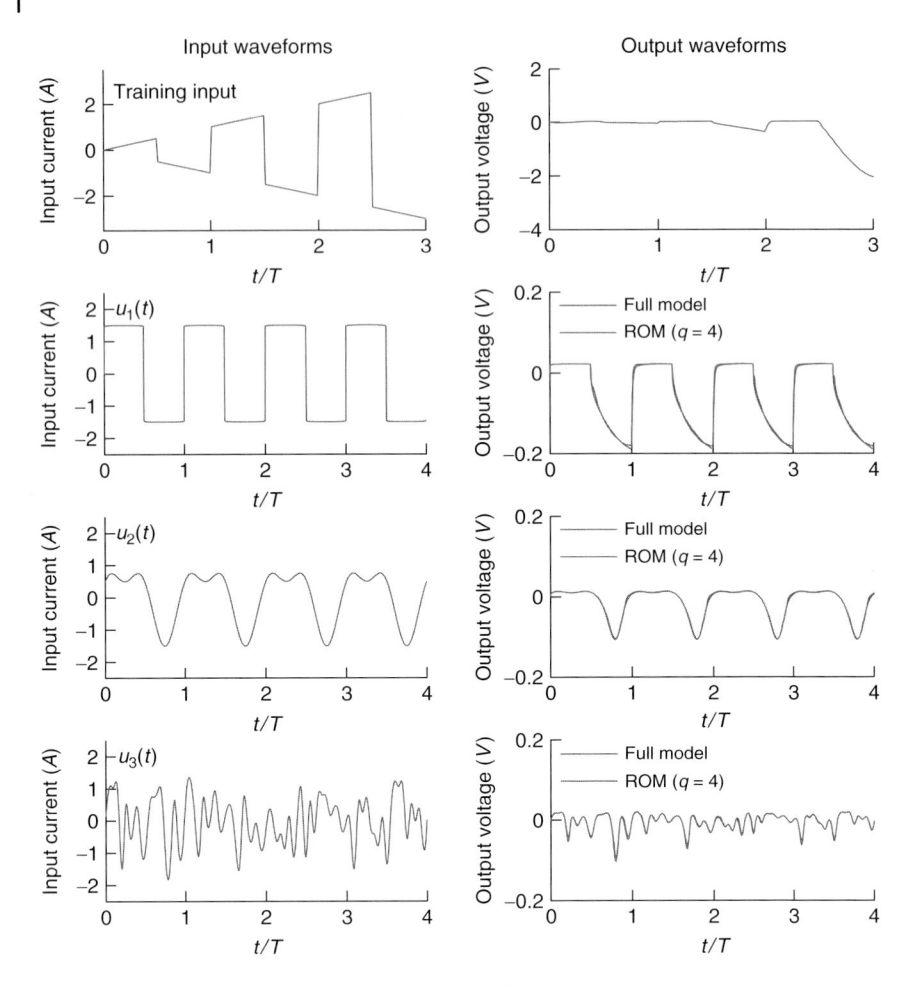

Figure 10.7 Training input and output (top row), three testing inputs, $u_1(t)$, $u_2(t)$,, and $u_3(t)$, and their respective outputs. Note that the output from the full model and $q = 4$ reduced model are nearly identical for all testing inputs.

the reduced equilibrium Jacobian has an eigenvalue with a small negative real part (Figure 10.9). Away from equilibrium, the Jacobian of the reduced system is a perturbed version of the equilibrium Jacobian and that small perturbation can push any eigenvalues close to the imaginary axis from the left-half to the right-half plane, and as a result, an unstable linearization is obtained. This problem can be largely avoided, when using TBR, by truncating the balanced realization at an order that does not separate two Hankel singular values that are close to each other [23]. When this rule is obeyed, the eigenvalues of the reduced system fall further away from the imaginary axis, and the reduced system is more robust to perturbations.

If all reduced linearizations are stable, the reduced-order model is usually stable for a large range of inputs, but it may not be globally stable. Global stability is

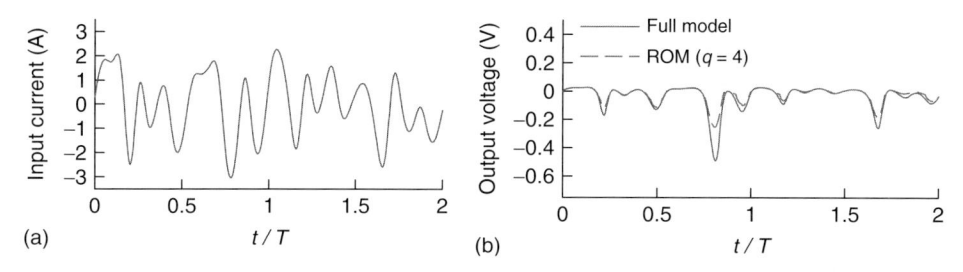

Figure 10.8 (a,b) Using the noiselike training input in Figure 10.7, whose peak amplitude is 1.8 A, and then testing using a signal with peak amplitude 3.0 A produces outputs from the full model and the reduced model that are visibly different.

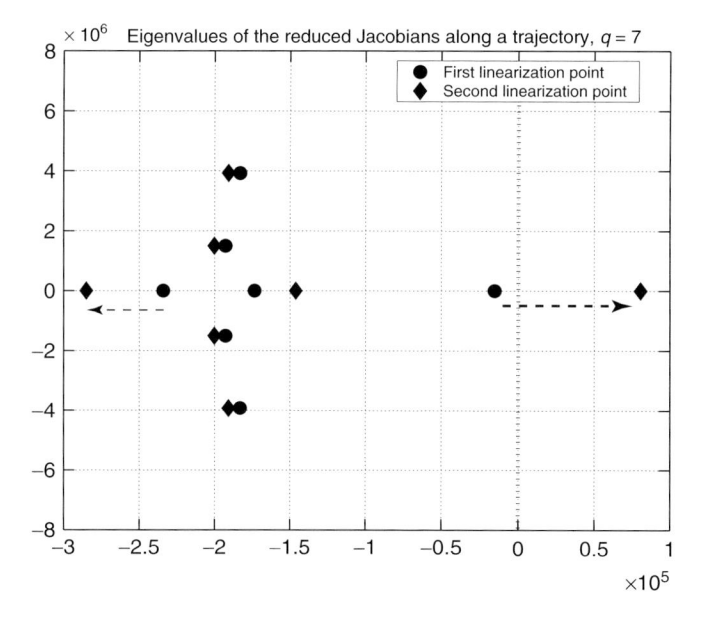

Figure 10.9 Eigenvalues of a $q = 7$ TBR model, at two linearization points.

guaranteed if $J(\vec{x_i})$ is negative definite for all i and $U_x = V$. Further techniques for generating stable TPWL models are developed in [29].

10.5.3
TPWL Summary

The main advantage of the TPWL method over the Taylor series expansion method is its applicability to strongly nonlinear systems, where the Taylor expansion would require too many terms to achieve reasonable accuracy. Issues to consider when implementing the TPWL are as follows. The expansion points $\vec{x_i}$ are usually samples of the system response to a given training input. This dependence on a training input can be advantageous if the range of inputs is known and is sufficiently limited. Then, the resulting reduced-order model is tailored to these inputs and can be more compact than reduced-order models intended for general inputs. The disadvantage of depending on a training input is that the response to dissimilar inputs can be poor. Also, it may be difficult to decide on the linearization points along the trajectory, and evaluating the Jacobians of the full system at each point can be expensive. A possible scheme for choosing linearization points, described in [38], is to simulate the reduced model and to add linearizations when the error becomes large, as determined from an *a posteriori* error bound, developed in [38]. If the number of linearizations becomes large, it is possible to use clustering algorithms to reduce it, and if necessary, to limit the interpolation to only a subset of the linearizations closest to the reduced state [39]. Though, the constant switching between linearizations may lead to artifacts such as artificial harmonic distortion.

10.6
Discrete Empirical Interpolation method

Another approach for approximating $\mathbf{f}(\mathbf{x})$ is the (DEIM, proposed in [19]). In the DEIM, partial evaluations of $\mathbf{f}(\mathbf{x})$ are used to generate a nonlinear reduced-order model that preserves the continuity properties of $\mathbf{f}(\mathbf{x})$ (important for harmonic analysis), yet whose evaluation cost typically scales with the *order of the reduced model*.

To briefly review the DEIM as presented in [19], consider evaluating $\mathbf{f}(\mathbf{x})$ in Eq. (10.1) at a set of n_s states, $[\mathbf{x}_1, \ldots, \mathbf{x}_{n_s}]$, selected to span the space in which x is likely to be found. Assembling the function evaluations in an $n \times n_s$ matrix, $[\mathbf{f}(\mathbf{x}_1), \mathbf{f}(\mathbf{x}_2), \ldots, \mathbf{f}(\mathbf{x}_{n_s})]$, and using the SVD to extract the $m \leq n_s$ dominant vectors, yields an $n \times m$ projection matrix for approximating $\mathbf{f}(\mathbf{x})$

$$\mathbf{U}_f = [\mathbf{u}_{F1}, \mathbf{u}_{F2}, \ldots, \mathbf{u}_{Fm}] \tag{10.26}$$

That is, $\hat{\mathbf{f}}(\mathbf{x}) \approx \mathbf{f}(\mathbf{x})$ is in the span of $\mathbf{U}_f$, or

$$\hat{\mathbf{f}}(\mathbf{x}) = \mathbf{U}_f \mathbf{a}(\mathbf{x}) \tag{10.27}$$

where $\mathbf{a}(\mathbf{x}) \in \mathbb{R}^m$.

If $\mathbf{U}_f$ is derived using the SVD, it will be orthonormal by construction, and the least-squares optimal values for $\hat{\mathbf{f}}(\mathbf{x})$ can be determined by projection, as in $\mathbf{a}(\mathbf{x}) = \mathbf{U}_f^T \mathbf{f}(\mathbf{x})$. Such an approximation is too expensive to evaluate because all n components of $\mathbf{f}(\mathbf{x})$ are required. Instead, in DEIM, the coefficients are determined from just m components of $\mathbf{f}(\mathbf{x})$, using a "gappy-least-squares" approximation to the least-squares solution [40]. If the selected m components are given by $\mathbf{P}_m^T \mathbf{f}(\mathbf{x})$, where the columns of $\mathbf{P}_m \in \mathbb{R}^{n \times m}$ are the corresponding m columns of the $n \times n$ identity matrix associated with the selected components of $\mathbf{f}(\mathbf{x})$, then the "gappy-least-squares" equation for $\mathbf{a}(\mathbf{x})$ is

$$\mathbf{P}_m^T \mathbf{f}(\mathbf{x}) = \mathbf{P}_m^T \mathbf{U}_f \mathbf{a}(\mathbf{x}) \tag{10.28}$$

Assuming, for the moment, that the matrix $\mathbf{P}_m^T \mathbf{U}_f$ is nonsingular, the approximation will be given by

$$\hat{\mathbf{f}}(\mathbf{x}) = \mathbf{U}_f (\mathbf{P}_m^T \mathbf{U}_f)^{-1} \mathbf{P}_m^T \mathbf{f}(\mathbf{x}) \tag{10.29}$$

In the DEIM, the m interpolation components are chosen by a simple and efficient greedy algorithm aimed at minimizing $\|(\mathbf{P}_m^T \mathbf{U}_f)^{-1}\|_2$. An upper bound on the norm achieved with this greedy algorithm was derived in [19]. Although it is $O(n^{m/2})$, our experience concurs with [19] that this bound is pessimistic and the interpolation is well conditioned in practice. It was also shown in [19] that the error of the approximation obeys

$$\|\mathbf{f} - \hat{\mathbf{f}}\|_2 \leq \|(\mathbf{P}_m^T \mathbf{U}_f)^{-1}\|_2 \|\mathbf{f} - \mathbf{U}_f \mathbf{U}_f^T \mathbf{f}\|_2 \tag{10.30}$$

which means that the error of the DEIM approximation is larger than the error of the optimal least-squares approximation by a factor that is at most $\|(\mathbf{P}_m^T \mathbf{U}_f)^{-1}\|_2$. The error of the optimal least-squares approximation can be decreased as much as desired by increasing m, thus enlarging the dimension of the subspace to which $\mathbf{f}(\mathbf{x})$ is restricted. Hence, the main assumption in the DEIM is that not only does $\mathbf{x}$ belong to a low-dimensional subspace but $\mathbf{f}(\mathbf{x})$ also belongs to a (different) subspace of comparable dimension.

Once the q-dimensional projection matrix, $\mathbf{U}_x$, and the q-dimensional testing matrix, $\mathbf{V}$, have been determined, the DEIM-based approximation to $\mathbf{f}_r(\mathbf{x}_r(t))$ is given by

$$\mathbf{f}_r(\mathbf{x}_r(t)) \approx \mathbf{V}^T \mathbf{U}_f (\mathbf{P}_m^T \mathbf{U}_f)^{-1} \mathbf{P}_m^T \mathbf{f}(\mathbf{U}_x \mathbf{x}_r t) \tag{10.31}$$

Although m and q need not be related, we will assume throughout that $m = q$ so as to have one parameter that determines the cost of simulating the reduced-order model. With this choice, it may be noted that $\mathbf{V}^T \mathbf{U}_f (\mathbf{P}_m^T \mathbf{U}_f)^{-1}$ collapses to a single $q \times q$ matrix. Also, the evaluation of $\mathbf{P}_m^T \mathbf{f}(\mathbf{U}_x \mathbf{x}_r t)$ does *not* require a full evaluation of $\mathbf{U}_x \mathbf{x}_r t$, only the values used to determine the q entries of $\mathbf{f}(\mathbf{x})$ are needed. Assuming the common case where the Jacobian of $\mathbf{f}(\mathbf{x})$ is sparse, only a small number of entries $\mathbf{f}(\mathbf{x})i$ need be computed. For example, if $\mathbf{f}(\mathbf{x})i$ depends on at most eight values of $\mathbf{x}$, then at most $8q$ entries of $\mathbf{U}_x \mathbf{x}_r t$ need to be computed. This situation is illustrated in Figure 10.10.

Lastly, the cost of generating the reduced-order models is quite moderate [19], and it is usually dominated by the need to calculate the snapshots from the full model. Computing a few dominant singular vectors can be done in $O(n_s^2 n)$ by use of Arnoldi methods, and the cost of finding the m interpolation coefficients is $O(m^4 + nm)$, which is usually negligible.

10.6.1
Thermal Analysis

The DEIM is particularly useful for nonlinear PDEs, for which it was originally proposed. The following example, which falls in this category, is a model for thermal analysis of an RF amplifier. The model has been adapted from [41],

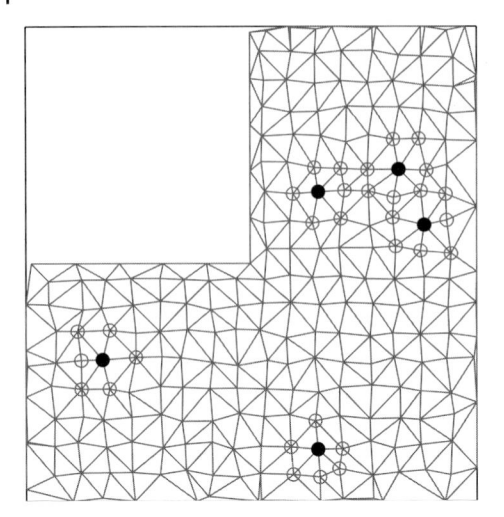

Figure 10.10 Components of **f**(x) selected as interpolation points (black full dots) and components of **x** needed for their evaluation (black hollow dots) in a finite-element discretization of an L-shaped region.

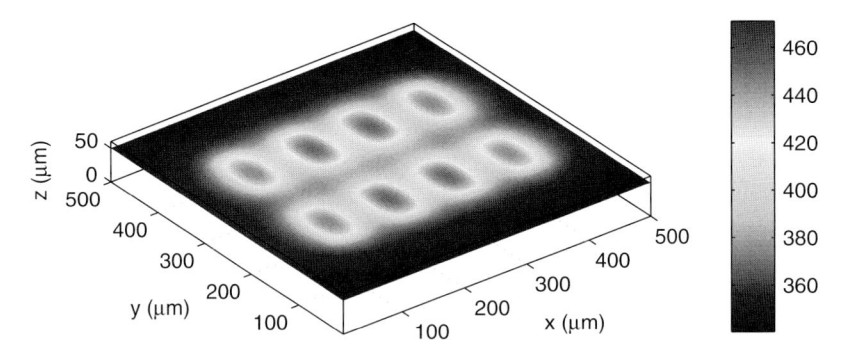

Figure 10.11 Temperature (K) on a plane just beneath the distributed heat sources representing the high-power transistors in an RF amplifier.

and it is shown in Figure 10.11. Compared with the model in [41], we have made a simplifying assumption, namely, that the bottom of the substrate is kept at ambient temperature; we also reduced the dimensions by a factor of two to allow a coarser discretization. The discretization is a finite-difference scheme and even after reducing the domain, the full model must be fairly large (we used $n = 21600$) to ensure convergence. The nonlinearity stems from the temperature dependence of the thermal conductivity of silicon, which varies by a factor of ≈ 2 for temperatures in the range of 30–230 °C. The input in this example is the power of each one of eight distributed heat sources representing high-power transistors; the output is the temperature at the center of one of the inner sources.

Like in the TPWL example, we tested this example with three test inputs: a square wave, $u_1(t) = 0.5P_0\{1 + \text{sgn}[\sin(2\pi t/T)]\}$, a two-tone signal, $u_2(t) =$

$P_0[2\sin(2\pi t/T) + \cos(4\pi t/T)]/6$, and a noiselike signal, $u_3(t)$, generated by fitting a spline through a set of random points in $[0, P_0]$, 20 points in each period of duration T. We chose $T = 2 \times 10^{-5}$ s, which is close to the estimated thermal time constant, and $P_0 = 7$ W so that the maximum temperatures approach 170 °C, which is as high as possible for safe operation. At these temperatures, the dependence of the thermal conductivity on temperature is significant, and neglecting it would result in an underestimation of the temperature by ≈50 °C. We used a single sinusoid as a training input, and set the training input amplitude to 1 W, much lower than the amplitudes of the testing inputs. The maximum temperature for this input was ≈45 °C. Also, the period of the training input was set to $2T$. The training and testing inputs are shown in Figure 10.12, along with their respective output, both of the full and reduced models. As can be observed, accuracy of a few percent is achieved even with a very small order ($q = 4$). However, like TPWL, the DEIM may become inaccurate if the training and testing inputs differ too much. This is shown in Figure 10.13, where the amplitude of the random signal has been doubled.

10.6.2
Stability Issues

Both the DEIM function approximation (Eq. (10.29)) and the subsequent projection (Eq. (10.31)) may lead, together or individually, to loss of stability. One can restore local stability about an equilibrium point when using DEIM with projection methods as follows. Let $\mathbf{f}(\mathbf{x}_0) = 0$, so that $\mathbf{x}_0$ is an equilibrium point. Then, $\mathbf{f}(\mathbf{x})$ can be decomposed as

$$\mathbf{f}(\mathbf{x}) = \mathbf{f}_p(\mathbf{x}) + \mathbf{J}(\mathbf{x} - \mathbf{x}_0) \tag{10.32}$$

where $\mathbf{J}$ is the Jacobian of $\mathbf{f}(\mathbf{x})$ at $\mathbf{x} = \mathbf{x}_0$ and $\mathbf{f}_p(\mathbf{x})$ is a nonlinear perturbation, the Jacobian of which is zero at $\mathbf{x} = \mathbf{x}_0$. Applying the DEIM approximation to $\mathbf{f}_p(\mathbf{x})$ but using only projection on $\mathbf{J}$ yields

$$\begin{aligned}\mathbf{f}_r(\mathbf{x}_r) &= \mathbf{V}^T \mathbf{J} \mathbf{U}_x \mathbf{x}_r + \mathbf{V}^T \mathbf{U}_{f_p}(\mathbf{P}_m^T \mathbf{U}_{f_p})^{-1} \mathbf{P}_m^T \mathbf{f}_p(\mathbf{U}_x \mathbf{x}_r) \\ &= \mathbf{V}^T \mathbf{J} \mathbf{U}_x \mathbf{x}_r + \mathbf{f}_{pr}(\mathbf{x}_r)\end{aligned} \tag{10.33}$$

It is easy to see that this DEIM approximation will preserve the equilibrium point because the Jacobians of $\mathbf{f}_p(\mathbf{x})$ and its reduced approximation $\mathbf{f}_{pr}(\mathbf{x}_r)$ are zero at $\mathbf{x} = \mathbf{x}_0$. Hence, if the linear term $\mathbf{V}^T \mathbf{J} \mathbf{U}_x$ is stable, the reduced model will be locally stable at $\mathbf{x} = \mathbf{x}_0$. For example, if $\mathbf{J}$ is negative definite, then using $\mathbf{V} = \mathbf{U}_x$ will ensure local stability. If $\mathbf{J}$ is indefinite, then TBR or the methods of [32] may be used. When using TBR, however, the issues discussed in Section 10.5.2 are relevant. Lastly, a multipoint stabilization scheme, which affords some control over the range of inputs for which a DEIM reduced-order model is stable, has been recently proposed in [42].

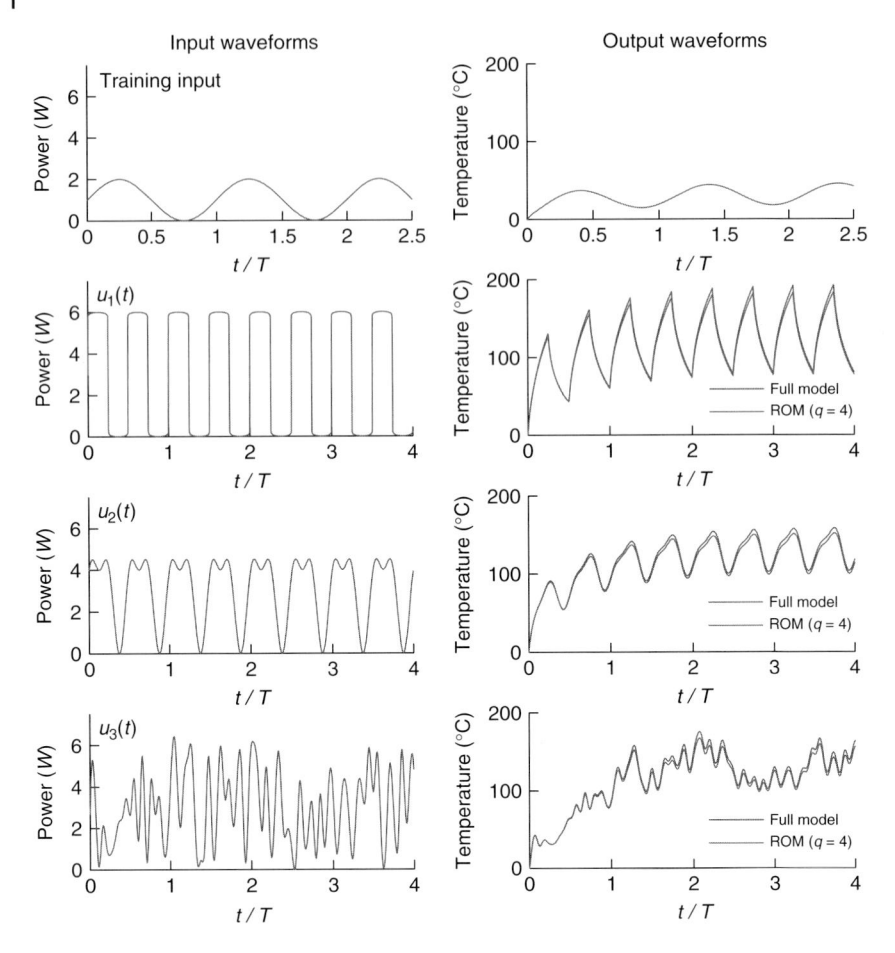

Figure 10.12 Training input and output (top row), three testing inputs, $u_1(t), u_2(t)$, and $u_3(t)$, and their respective outputs. Note that the output from the full model and $q = 4$ reduced model are nearly identical for all testing inputs.

10.6.3
DEIM Summary

The DEIM has a number of advantages and disadvantages compared with TPWL. The DEIM is somewhat simpler to implement and has fewer parameters that must be determined by the user. In particular, implementing an effective interpolation scheme in TPWL is not trivial, and these schemes also introduce harmonic distortion. In contrast, the DEIM function approximation preserves the continuity of the original function and its derivatives, so this does not occur. Like in TPWL, a training input is used to derive the reduced-order model, which may be inaccurate far away from the training input, and also a set of snapshots must be decided on. In some systems, evaluating $\mathbf{f}(\mathbf{x})$ is time consuming even when n is not large. This

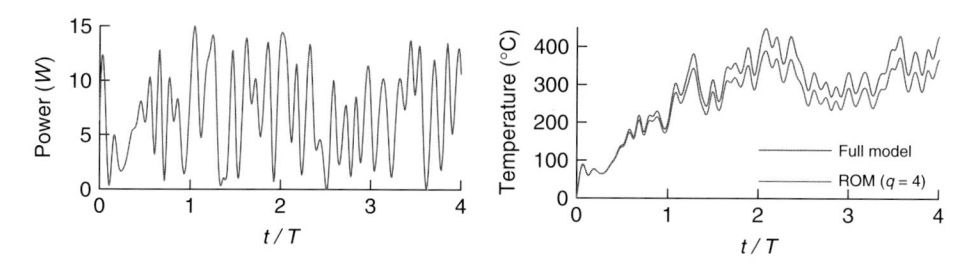

Figure 10.13 Using the noiselike training input in Figure 10.12, whose peak amplitude is 7 W, and then testing using a signal with peak amplitude 14 W produces outputs from the full model and the reduced model that are visibly different.

is the case, for example, in analog circuits, in which n corresponds to the number of nodes in the circuit and evaluating $\mathbf{f}(\mathbf{x})$ entails evaluating complicated transistor models. For these systems, DEIM would not yield a significant speed-up but TPWL would. Lastly, as in TPWL, the DEIM approximation may lead to unstable models, but some stabilization schemes have recently been proposed.

10.7
A Comparative Case Study of an MEMS Switch

In this section, we apply the three methods described to a multiphysics model of an MEMS switch. This model, shown in Figure 10.14, was proposed by Hung et al. [8] and subsequently investigated also in [8, 12, 16, 29]. In this example, a state-space model is obtained from the 1D Euler beam equation and the 2D Reynolds' squeeze film damping equation, together with the parallel plate capacitor approximation for the electric field. We have

$$\frac{\partial x_1}{\partial t} = \frac{x_2}{3x_1^2} \tag{10.34}$$

$$\frac{\partial x_2}{\partial t} = \frac{2x_2^2}{3x_1^3} - \frac{3\varepsilon_0 w}{2\rho} V^2(t) + \frac{3x_1^2}{\rho} \left[\int_0^w (x_3 - p_a) dy + S \frac{\partial^2 x_1}{\partial x^2} - EI \frac{\partial^4 x_1}{\partial x^4} \right]$$

$$\frac{\partial x_3}{\partial t} = -\frac{x_2 x_3}{3x_1^3} + \frac{1}{12\mu x_1} \nabla \cdot \left[\left(1 + \frac{6\lambda}{x_1} \right) x_1^3 x_3 \nabla x_3 \right]$$

where the state-space variables are the height of the beam above the substrate, x_1, the time derivative of x_1^3, $x_2 = \partial(x_1^3)/\partial t$, and the pressure distribution below the beam, x_3. The parameters are the Young modulus, $E = 149$ GPa, the moment of inertia $I = wh^3/12$, the beam width, length, and thickness, which are $w = 40$ μm, $l = 610$ μm, and $h = 2.2$ μm, respectively. The mean free path of air is $\lambda = 0.064$ μm, the stress coefficient is $S/(hw) = -3.7$ MPa, the density is $\rho/(hw) = 2330$ kgm^{-3}, the ambient pressure is $p_a = 1.013 \times 10^5$ Pa, the air viscosity is $\mu = 1.82 \times 10^{-5}$ kg (m s)$^{-1}$, and the permittivity of free space is $\epsilon_0 = 8.854 \times 10^{-12}$ F m^{-1}. The three continuous state variables are discretized by a finite-difference

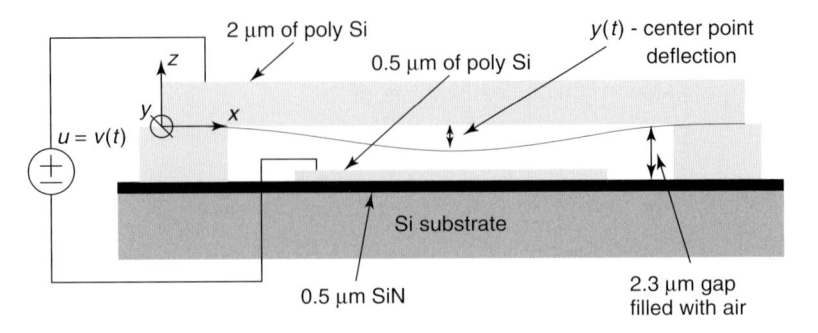

10 Projection-Based Nonlinear Model Order Reduction

Figure 10.14 MEMS switch, following Hung et al. [8].

Labels in figure: 2 μm of poly Si; 0.5 μm of poly Si; $y(t)$ - center point deflection; $u = v(t)$; Si substrate; 0.5 μm SiN; 2.3 μm gap filled with air.

approximation to obtain a dynamical system in the form of Eq. (10.1), where the input is the applied voltage squared, $u(t) = V^2(t)$, and the output is the height of the beam at its center. The number of degrees of freedom of this model is $n = 299$. For further details regarding this model, see [8].

10.7.1
Pull-In Effect

An important parameter of an MEMS switch is its pull-in voltage. As the input voltage is increased from zero, the beam is pulled toward the bottom electrode. The electrostatic force increases with increasing voltage not only because it is proportional to the voltage squared but also because it is inversely proportional to separation between the electrodes squared. Hence, beyond a certain critical separation, the electrostatic force increases rapidly; it overcomes the restoring force and the beam is pulled-in. The minimum input voltage for which this happens is called the *pull-in voltage*, and it is usually a DC voltage. However, one may also ask what is the AC pull-in voltage, that is, the amplitude of a harmonic input that will result in pull-in. This question is of significant practical interest because the larger motion amplitudes possible at nonzero frequencies may be used to improve measuring capabilities [43, 44].

A plot of the deflection of the beam center for harmonic inputs of increasing amplitude is shown in Figure 10.15. As can be observed in this figure, the dynamic pull-in voltage at $f = 300$ MHz is $V_\pi \approx 9.1$ V. An important feature that can be inferred from Figure 10.15 is that the full model used here is not globally stable. An input of finite amplitude that is greater than the pull-in voltage will result in an unbounded output. In reality, of course, the deflection of the beam is bounded, and this lack of global stability is a (common) artifact of the modeling scheme. Lastly, for voltages very close to the pull-in voltage, the slightest error when simulating the dynamical system may result in large output differences. Hence, we cannot expect to find a reduced-order model that will always yield moderate errors for this system. It makes sense, therefore, to examine the pull-in voltages predicted by reduced-order models obtained with various methods. To calculate the pull-in voltage, we use a rudimentary method: we begin with a voltage interval in which

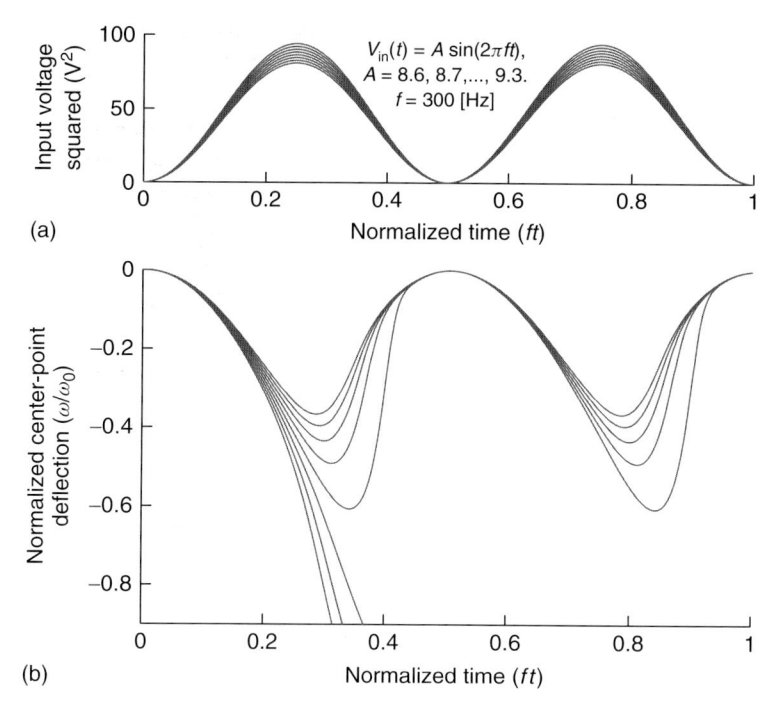

Figure 10.15 Dynamic pull-in effect. Harmonic inputs of increasing amplitude (a) and corresponding outputs (b).

we estimate the pull-in voltage to be and then simulate the system with a voltage at the center of the interval. We then update the interval to the appropriate half according to whether the beam pulled in or not. The results of this computation are shown in Figure 10.16 for reduced-order models obtained with the DEIM, TPWL, and quadratic approximations. The order of the reduced-order models is $q = 4$ in Figure 10.16a and $q = 8$ in Figure 10.16b. The number of snapshots in the DEIM and TPWL is also increased from $l = 50$ in Figure 10.16a to $l = 150$ in Figure 10.16b. The matrix $\mathbf{U}_x$ was obtained with the POD method, applied to a training input $u(t) = 5 \sin(2\pi ft)ft$, which was simulated for $t \in [0, 3/f]$. During this time interval, the system goes from small-amplitude oscillations to pull-in. The $\mathbf{V}$ matrix was obtained using the optimization approach described in [32]. This method yields a projection matrix $\mathbf{V}$ for which the reduced-order models are guaranteed to be locally stable. As can be observed in Figure 10.16a, the DEIM and TPWL method yield more accurate results than the quadratic approximation. When the number of snapshots and basis vectors is increased, both TPWL and DEIM converge to the results obtained with the full model, but the quadratic approximation is limited by its low order (Figure 10.16b). Increasing the quadratic model to a cubic one is impractical, as the length of $\mathbf{x}_r \otimes \mathbf{x}_r \otimes \mathbf{x}_r$ would be $8^3 = 512$ which is larger than the order of the full model.

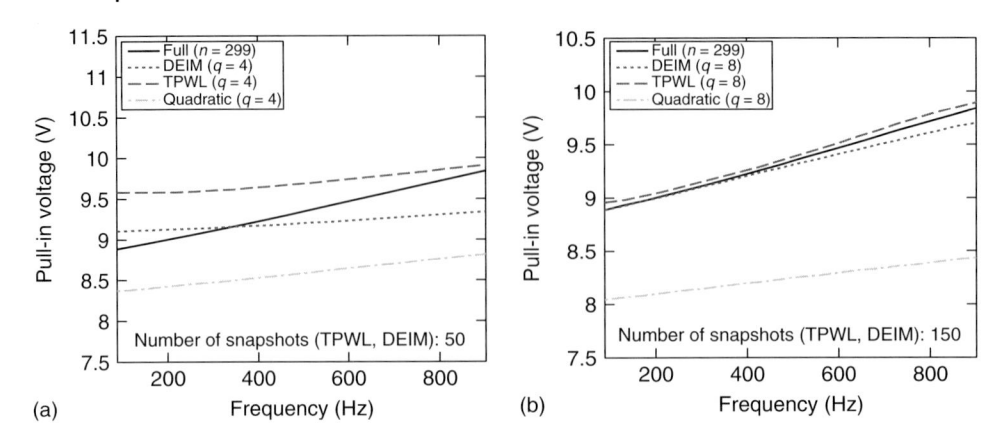

Figure 10.16 Pull-in versus input frequency, as predicted by DEIM, quadratic, and TPWL reduced-order models. In (a), fewer linearizations and basis vectors are used than in (b), in which convergence of TPWL and DEIM to the full model results is evident.

10.7.2
Generalizing from Training Inputs

An important aspect of the TPWL method and the DEIM is their dependence on training inputs. Some information about this dependence can be gleaned from Figure 10.16. The results shown in this figure were obtained with a narrowband training input, centered around $f = 300$ MHz, but for all frequencies in the range of $[f/3, 3f]$, the reduced-order models remain fairly accurate. Although narrowband, the training input used in this example had a large range of amplitudes. If instead we use as training input a relatively small-amplitude sinusoid, $u(t) = 7\sin(2\pi ft)$, we obtain the results shown in Figure 10.17, in which the DEIM appears to be significantly better than TPWL. This can be understood by noting that when the training input is of small amplitude, all linearizations are approximately the same and the TPWL approximation is not very different from a single linearization at the equilibrium position. In the DEIM, on the other hand, the selected components of the exact $\mathbf{f}(\mathbf{x})$ are evaluated and the approximation is as nonlinear as these components.

10.7.3
Harmonic Distortion

Another advantage of the DEIM over the TPWL method becomes apparent when the error in the harmonic steady-state response is considered. The DEIM model has as many continuous derivatives as the selected components of the original nonlinear function. Therefore, it can match higher-order harmonics resulting from the nonlinearity. In contrast, the TPWL models constantly switch between dominant linearizations, and this can introduce spurious harmonic distortion (Figure 10.18). In particular, the presence of the min operator in Eq. (10.25)

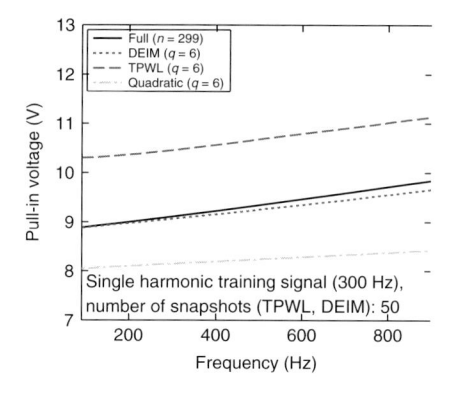

Figure 10.17 Pull-in voltage obtained with a low-amplitude training input. The DEIM approximation generalizes to large amplitudes much better than the TPWL approximation.

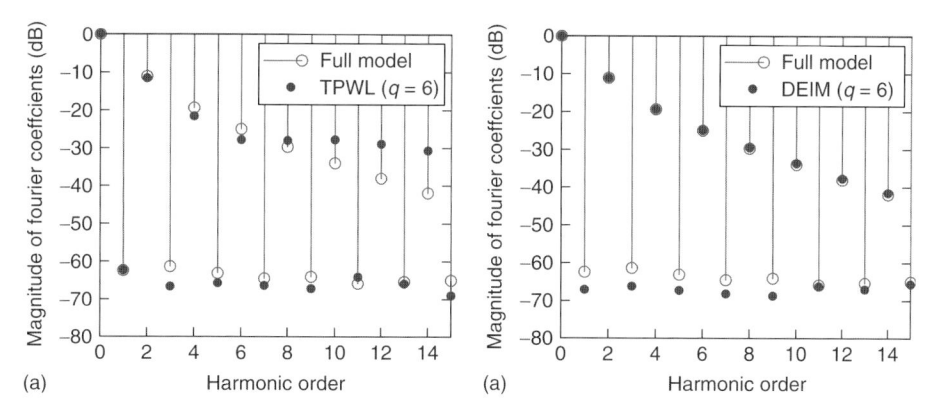

Figure 10.18 Comparison of harmonic distortion introduced by the (a) TPWL and (b) DEIM approximations, showing DEIM is better in this respect. The number of snapshots/linearizations used is 15. Odd harmonics are small because the input is squared.

implies that the approximation is not everywhere differentiable. Even if the min operator is avoided, as in [34], spurious harmonic distortion will occur if the transition between linearizations is not smooth enough. In Figure 10.18, only 15 linearizations were used. If this number is increased to 50, the harmonic distortion decreases (Figure 10.19).

10.7.4
A Note about CPU Times

The actual speed-ups that can be achieved with the various reduction methods depend heavily on the implementation of both the reduced and full models. Our implementations are nonoptimized, and in Matlab, in which many of the implementation details are hidden from the programmer. We do not cite CPU times, therefore, as they may be misleading.

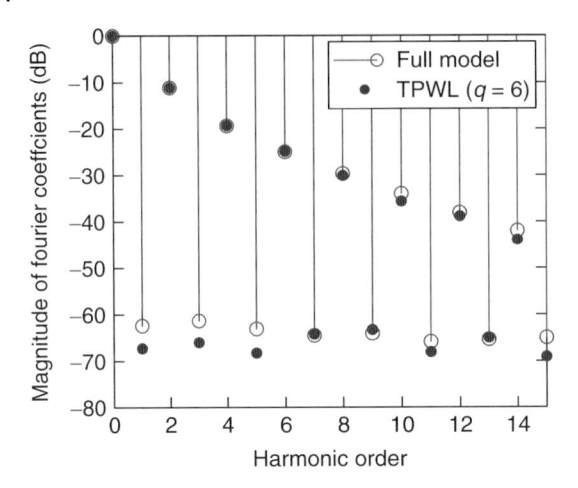

Figure 10.19 Reduced harmonic distortion introduced by the TPWL approximation when the number of snapshots/linearizations is increased to 50.

10.8
Summary and Outlook

In this chapter, we have reviewed and compared three methods for projection-based nonlinear model reduction: the Taylor expansion method, the TPWL method, and the DEIM. A key issue that is addressed differently by each of these methods is that in nonlinear systems, projection alone may not reduce the computational cost significantly. Therefore, each method involves approximating the nonlinear function $f(x)$ so as to render the reduced function $f_r(x_r)$ cheap to evaluate. Another important issue is preservation of stability, which can be lost when projection is applied or when $f(x)$ is approximated. Techniques for preserving stability are the subject of active research and references to the literature are given. A number of examples were used to show the advantages and limitations of the three methods. The Taylor series expansion method is useful for weakly nonlinear systems and does not require a training input as do the TPWL method and the DEIM. These methods are applicable to highly nonlinear systems, but their accuracy may suffer for testing inputs that are far from the training inputs. Of these two, DEIM appears to generalize better from the training inputs to dissimilar testing inputs, and it also avoids artifacts such as harmonic distortion that may occur in a TPWL method.

Acknowledgment

This work was supported by the Singapore-MIT alliance program in computational engineering, the Viterbi postdoctoral fellowship, and the Advanced Circuit Research Center at the Technion – Israel Institute of Technology.

References

1. Glover, K. (1984) *Int. J. Control*, **39** (6), 1115–1193.
2. Gu, C. and Roychowdhury, J. (2008) ManiMOR: Model reduction via projection onto nonlinear manifolds, with applications to analog circuits and biochemical systems. Proceedings IEEE International Conference on Computer Aided Design, pp. 85–92.
3. Bond, B.N., Mahmood, Z., Li, Y., Sredojevic, R., Megretski, A., Stojanovi, V., Avniel, Y., and Daniel, L. (2010) *IEEE Trans. Comput. Aided Des. Integr. Circ. Syst.*, **29** (8), 1149–1162.
4. Root, D.E., Verspecht, J., Sharrit, D., Wood, J., and Cognata, A. (2005) *IEEE Trans. Microw. Theory Tech.*, **53** (11), 3656–3664.
5. Gelinas, R.J., Doss, S.K., and Miller, K. (1981) *J. Comput. Phys.*, **40** (1), 202–249.
6. Silva, J., Villena, J., Flores, P., and Silveira, L. (2006) *Sci. Comput. Electr. Eng.*, **11**, 139–152.
7. Phillips, J.R. (2000) Projection frameworks for model reduction of weakly nonlinear systems. Proceedings of the 37th Design Automation Conference, pp. 184–189.
8. Hung, E.S., Yang, Y.J., and Senturia, S.D. (1997) Low-order models for fast dynamical simulation of MEMS microstructures. International Conference Solid State Sensors and Actuators.
9. Berkooz, G., Holmes, P. and Lumley, J.L. (1993) *Annu. Rev. Fluid Mech.*, **25** (1), 539–575.
10. Grimme, E.J. (1997) Krylov projection methods for model reduction. PhD thesis. University of Illinois at Urbana-Champaign.
11. Moore, B. (1981) *IEEE Trans. Autom. Control*, **26** (1), 17–32.
12. Vasilyev, D., Rewienski, M., and White, J. (2003) A TBR-based trajectory piecewise-linear algorithm for generating accurate low-order models for nonlinear analog circuits and MEMS. Proceedings of the Design Automation Conference, pp. 490–495.
13. Chen, J. and Kang, S.M. (2000) *An algorithm for automatic model-order reduction of nonlinear MEMS devices*. The 2000 International Symposium Circuits Systems.
14. Chen, Y. and White, J. (2000) A quadratic method for nonlinear model order reduction. Proceedings International Conference on Modeling and Simulation of Microsystems, pp. 477–480.
15. Phillips, J.R. (2000) Automated extraction of nonlinear circuit macromodels. Proceedings of the Custom Integrated Circuits Conference, p. 451.
16. Rewienski, M. and White, J. (2003) *IEEE Trans. Comput. Aided Des. Integr. Circ. Syst.*, **22**, 155–170.
17. Dong, N. and Roychowdhury, J. (2003) Piecewise polynomial nonlinear model reduction. Proceedings Design Automation Conference, pp. 484–489.
18. Barrault, M., Maday, Y., Nguyen, N.C., and Patera, A.T. (2004) *C. R. Math.*, **339** (9), 667–672.
19. Chaturantabut, Saifon. and Sorensen, Danny.C. (2010) *SIAM J. Sci. Comput.*, **32** (5), 2737–2764.
20. Phillips, J.R. (2003) *IEEE Trans. Comput. Aided Des. Integr. Circ. Syst.*, **22** (2), 171–187.
21. De Abreu-Garcia, J.A. and Mohammad, A.A. (1990) A transformation approach for model order reduction of nonlinear systems. Proceedings of the 16th Conference IEEE Industrial Electronics Society, vol. 1, pp. 380–383.
22. Innocent, M., Wambacq, P., Donnay, S., Tilmans, H.A.C., Sansen, W., and De Man, H. (2003) *IEEE Trans. Comput. Aided Des. Integr. Circ. Syst.*, **22** (2), 124–131.
23. Vasilyev, D.M. (2008) Theoretical and practical aspects of linear and nonlinear model order reduction techniques. PhD thesis. Massachusetts Institute of Technology.
24. Tang, Z., Hong, S., Djukic, D., Modi, V., West, A.C., Yardley, J., and Osgood, R.M. (2002) *J. Micromech. Microeng.*, **12** (6), 870–877.

25. Landau, L.D. and Lifshitz, E.M. (1977) *Fluid Mechanics*, vol. 6, Butterworth-Heinemann.
26. Khalil, H.K. and Grizzle, J.W. (2002) *Nonlinear Systems*, vol. 3, Prentice hall, New Jersey.
27. Bond, B.N. (2010) *Stability-preserving model order reduction for linear and nonlinear systems arising in analog circuit applications*. PhD thesis. Thesis, Massachusetts Institute of Technology.
28. Benner, P. (2009) *System-theoretic methods for model reduction of large-scale systems: simulation, control, and inverse problems*. Vienna International Conference on Mathematical Modelling, Vienna-Austria.
29. Bond, B.N. and Daniel, L. (2009) *IEEE Trans. Comput. Aided Des. Integr. Circ. Syst.*, **28** (2009), 1467–1480.
30. Rewieński, M. and White, J. (2001) A trajectory piecewise-linear approach to model order reduction and fast simulation of nonlinear circuits and micromachined devices. Proceedings International Conference on Computer Aided Design, pp. 252–257.
31. Rewienski, M. and White, J. (2002) Improving trajectory piecewise-linear approach to nonlinear model order reduction for micromachined devices using an aggregated projection basis. Proceedings International Conference Modeling and Simulation of Microsystems, pp. 128–131.
32. Bond, B.N. and Daniel, L. (2008) Guaranteed stable projection-based model reduction for indefinite and unstable linear systems. Proceedings 2008 International Conference on Computer Aided Design, pp. 728–735.
33. Dong, N. and Roychowdhury, J. (2008) *IEEE Trans. Comput. Aided Des. Integr. Circ. Syst.*, **27** (2), 249–264.
34. Tiwary, S.K. and Rutenbar, R.A. (2006) *Proceedings 2006 International Conference on Computer Aided Design*, ACM, pp. 876–883.
35. Bechtold, T., Striebel, M., Mohaghegh, K., and ter Maten, E.J.W. (2008) *PAMM*, **8** (1), 10057–10060.
36. Voss, T., Verhoeven, A., Bechtold, T., and Maten, J. (2006) Model order reduction for nonlinear differential algebraic equations in circuit simulation. Program in Industrial Mathematics, pp. 518–523.
37. Gu, C. (2009) QLMOR: a new projection-based approach for nonlinear model order reduction. Proceedings 2009 International Conference on Computer Aided Design, pp. 389–396.
38. MichałRewieński, M.S. (2003) *A trajectory piecewise-linear approach to model order reduction of nonlinear dynamical systems*. PhD thesis, Massachusetts Institute of Technology.
39. Tiwary, S.K. and Rutenbar, R.A. (2005) *Proceedings 42nd Design Automation Conference*, ACM, pp. 403–408.
40. Astrid, P., Weiland, S., Willcox, K., and Backx, T. (2008) *IEEE Trans. Autom. Control*, **53** (10), 2237–2251.
41. Mouthaan, K., Tinti, R., Arno, A., de Graaff, H.C., Tauritz, J.L., and Slotboom, J. (1997) *Solid-State Device Research Conference*, IEEE, pp. 184–187.
42. Hochman, A., Bond, B.N., and White, J.K. (2011) A stabilized discrete empirical interpolation method for model reduction of electrical, thermal, and microelectromechanical systems. Proceedings of the 48th Design Automation Conference.
43. Nayfeh, A.H., Younis, M.I., and Abdel-Rahman, E.M. (2007) *Nonlin. Dyn.*, **48** (1), 153–163.
44. Fargas-Marques, A., Casals-Terré, J., and Shkel, A.M. (2007) *J. Microelectromech. Syst.*, **16** (5), 1044–1053.

11
Linear and Nonlinear Model Order Reduction for MEMS Electrostatic Actuators

Jan Lienemann, Emanuele Bertarelli, Andreas Greiner, and Jan G. Korvink

11.1
Introduction

Microelectromechanical systems (MEMS) exhibit nonlinearities in a different manner than macroscopic devices owing to different actuation principles and scaling effects [1, 2]. With a length scaling factor s, areas scale with s^2 and volumes with s^3, so that the properties of the microsystem are determined by their surfaces and less by the "bulk."

One frequently used actuation principle is by electrostatic forces. On small scales, electrostatic forces show performances comparable to electromagnetic forces, while being characterized by low-power consumption and fast response time. Owing to the full CMOS (complementary metal oxide semiconductor) process compatibility [3, 4], electrostatic actuators are easier to fabricate along with the driving circuitry. The perspective of taking advantage of MEMS Integrated Circuit batch-processing technology to obtain affordable and reliable devices is of great interest [5, 6]. The direct integration with electronic components is a key feature for power supply and control of the device, as well as for the straightforward integration in a broad range of devices and systems from MEMS sensors and actuators to microfluidic devices, along the same fabrication process.

A common way to simulate these systems are 3D finite element method (FEM) models, but their complexity (thousands to millions of possibly nonlinear coupled ordinary differential equations) prevents their use in a circuit-design environment. A popular approach to handle this difficulty is the creation of an equivalent compact model.

For the model order reduction (MOR) of linear systems, a large number of results have been published, and robust, automatic MOR can be regarded as solved in principle (Refs [7–10] and Chapters 2-4) and is also accepted in industrial applications [11]. Latest research gives a priori error estimators even for large-scale systems [12], and it is possible to preserve parameters in the reduced system, so that design variations can be evaluated without having to perform MOR again (Ref. [13] and Chapter 9.4).

System-level Modeling of MEMS, First Edition. Edited by T. Bechtold, G. Schrag, and L. Feng.
© 2013 Wiley-VCH Verlag GmbH & Co. KGaA. Published 2013 by Wiley-VCH Verlag GmbH & Co. KGaA.

Yet, some engineering problems still pose a challenge to the common MOR methods. Among these are applications covering a wide frequency range, time-varying coefficients (an exception are parametric systems, where a few parameters can be distilled which describe the variation), devices that are coupled to the outside world via a large number of inputs and outputs, and simulation problems given by hyperbolic PDEs, as, for example, wave pulse propagation. Especially MEMS show strong nonlinearities due to large relative displacements. Apart from geometric and material nonlinearities, it is most often the coupling between the mechanical and the electrostatic domain, which causes, because of the deformation of the simulation boundary, nonlinear terms in an otherwise linear problem [14]. Nonlinear MOR still requires research to find a general, automatic procedure [15, 16].

In this chapter, we present results with different MOR approaches to tackle the $1/x^2$ nonlinearity that often appears in microscale electromechanical actuators: an approach based on a polynomial representation of nonlinearities, applied to parallel plate systems, and trajectory piecewise-linear (TPWL) MOR for platelike actuators.

First, we describe the physical effects that are the basis for electrostatic actuators and, for the parallel plate capacitor, show the consequences of these effects on the design of MEMS. Then, we review several Krylov-subspace-based MOR methods with regard to the example problems presented in the two subsequent sections. A presentation of results achieved with these methods concludes this chapter.

For a deeper understanding and an introduction to the terminology, we refer the interested reader to Chapter 3.7, where the Krylov subspace methods such as the Arnoldi method and the properties of the approximation are explained in detail. Also, we refer to Chapter 10, where the theory of nonlinear MOR methods is presented, including TPWL and higher-order MOR, which is the basis for the presented polynomial approach. Although our systems are not really parametric, they show mathematical similarities to those in Chapter 9.4.

The following discussion focuses on the specifics of the example problems, a scanning-probe data storage device and a micropump diaphragm. Both are actuated by electrostatic forces, that is, a voltage between different, electrically unconnected parts of the device leading to a mechanical force. In the case of the scanning probe, this force is used to write data to a polymer by structuring its surface (11.4). In the case of the micropump actuator (11.5), the force is displacing a diaphragm and thus generates a depression in the fluid chamber below, aiming to achieve fluid transport from a peristaltic motion.

11.2
The Variable Gap Parallel Plate Capacitor

The parallel plate capacitor with variable position can be seen as prototype, the *Form* of electrostatically actuated MEMS [14, 17, 18]. Already this very simple model shows the nonlinear effects that we address in this paper, as, for example, pull-in; therefore, we devote this section to an analysis of this model.

11.2 The Variable Gap Parallel Plate Capacitor

Given two parallel ideally conducting plates with gap d, overlapping area A, dielectric permittivity in the gap ε, application of a voltage V between the two plates results in a stored energy of

$$E_C = \frac{1}{2}\frac{\varepsilon A V^2}{d} \tag{11.1}$$

In MEMS, the parallel plate capacitor frequently appears in two different variants. In-plane actuation along a given axis is obtained by means of comb drives (Figure 11.1a), which results in an almost constant force independent of the displacement and does not suffer from pull-in effects along most of the actuation range, namely, it is always stable through its full range of motion.

However, here we focus our attention on the second alternative, a variable plate capacitor as depicted in Figure 11.1b, which is applied for out-of-plane motion or when one electrode can move sideways like in some rotating devices. Here, the distance of the gap d itself is varied, whereas the intersecting area A remains constant [19]. The lower (reference) plate is fixed, and the upper plate with mass m is held by a spring with spring constant k at a distance d. We neglect gravity and fringing fields and assume that the charges are evenly distributed. Spacers on the lower plate provide a lower limit d_{min} to the plates' distance d. When a voltage V is applied, the two capacitor plates are pulled toward each other with a force of

$$F_C = \frac{\partial E_C}{\partial d} = \frac{\partial \left(\frac{1}{2} C(d) V^2\right)}{\partial d} = \frac{1}{2}\frac{\partial \left(\frac{\varepsilon A}{d} V^2\right)}{\partial d} = -\frac{1}{2}\frac{\varepsilon A}{d^2} V^2 \tag{11.2}$$

The spring balances this force with

$$F_S = -k(d - d_0) \tag{11.3}$$

where d_0 is the equilibrium position of the spring. The gap d appears in a nonlinear way in this equation, which we call the $1/x^2$ nonlinearity in the scope of this chapter.

Finite element simulation packages have introduced special nonlinear elements to model this kind of actuation based on Eq. (11.2). For example, the commercial

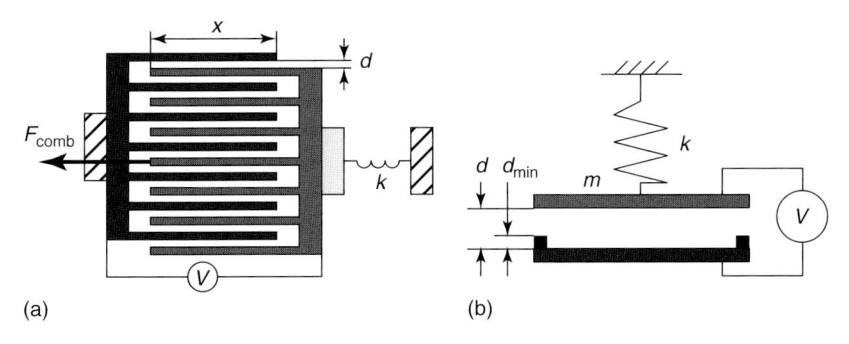

Figure 11.1 Electrostatic driving principles for MEMS. (a) A combdrive: the fingers are pulled toward the left. (b) A variable-gap capacitor: the upper, movable plate is pulled toward the lower.

FEM simulator ANSYS features a so-called two-node transducer element [19], where the distance between the nodes is used to calculate the capacity and the resulting force is nonlinear, as stated in Eq. (11.2). To increase the accuracy of the electrostatic pressure, larger areas are divided into smaller subareas, where each subarea is connected to is own transducer element. An extension to rotating plates was presented in Ref. [20].

In the following two subsections, we elaborate more on the effects of this nonlinearity on the static and dynamic behavior of the capacitor system. In contrast to linear systems, nonlinear systems show new characteristics that make it more difficult to find a comprehensive reduced model. The variable plate capacitor is a classical example of such a system, where force inversion changes the system trajectory completely and it is not even approximately linear.

11.2.1
Pull-In

We now compute the voltage which is necessary to achieve a certain displacement d and, the inverse question, what equilibrium displacement a certain voltage will result in. In the stationary case, the force balance $F_C + F_S = 0$ (Figures 11.2 and 11.3) yields

$$V = \sqrt{\frac{2kd^2(d_0 - d)}{\varepsilon A}} \tag{11.4}$$

There is an extremal value $V_{\text{pull-in}}$ at $2/3$ of the spring's equilibrium position

$$\frac{dV}{dd} = \frac{dk(2d_0 - 3d)}{\sqrt{2\varepsilon Akd^2(d_0 - d)}} = 0 \quad \text{for } d = 2d_0/3 \tag{11.5}$$

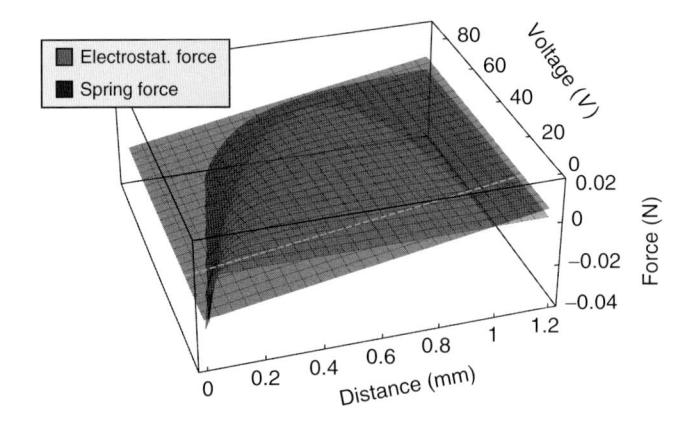

Figure 11.2 Plot of the forces on the upper plate versus voltage and distance. In this graph, the spring force is positive when it points toward the lower plate; the electrostatic force is positive when it points away from the lower plate; the two forces cancel each other out where the surfaces intersect.

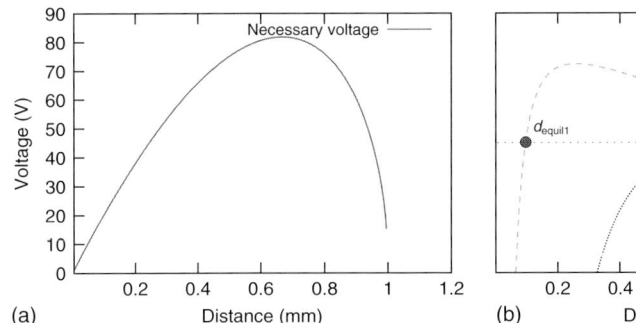

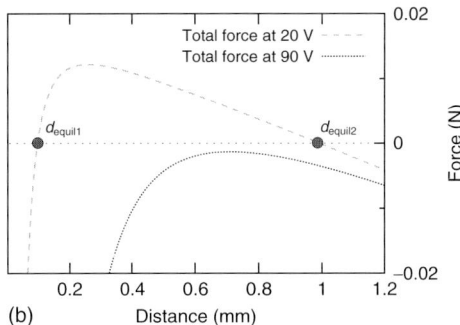

Figure 11.3 The equilibrium voltage at certain distances and forces at constant voltage (positive forces pointing away from the lower plate). As the force for 90 V is always below zero, there is no equilibrium position any more. The following parameters are used: $k = 20\text{N/m}$, $d_0 = 1\text{mm}$, $\varepsilon = 8.85\text{pF/m}$, and $A = 0.1\text{m}^2$. (a) Equilibrium voltage at different distances. (b) Total force (electrostatic plus spring force) at different distances.

and by virtue of Eq. (11.4)

$$V_{\text{pull-in}} = \sqrt{\frac{8}{27}\frac{kd_0^3}{\varepsilon A}} \tag{11.6}$$

When inverting Eq. (11.4) do determine the displacement at a given voltage, one observes that for a voltage above the *pull-in voltage*, $V > V_{\text{pull-in}}$, no real positive solution and thus no equilibrium exists. From a mechanical viewpoint, equilibrium in electrostatic actuators is achieved when the electrostatic forces are balanced by mechanical restoring forces. The stability of equilibrium states is determined by the electromechanical stiffness of the system. When the static pull-in voltage is reached, a critically stable condition is found, and the movable electrode spontaneously collapses onto the reference plate, as the spring elastic restoring force is unable to withstand any further increase of the electrostatic force. The operating range of such a voltage-controlled actuator is therefore limited to $d_0/3$ because of the requirement $V < V_{\text{pull-in}}$.

To illustrate the force equilibrium for the plate, we have plotted the force versus voltage and displacement as well as the force $-F_S$ in Figure 11.2. Equilibrium is the intersection of the two surfaces. This curve is also reproduced in Figure 11.3a. The maximum voltage where a stable position can be determined is clearly visible in the plot.

The dotted lines in Figure 11.3b show the sum $F_C + F_S$. At 20 V, two intersections with the zero line are present; the parts of the line above 0 represent total forces away from the reference plate, the parts of the line below represent total forces toward the reference plate. Two equilibrium points d_{equil1} and d_{equil2} exist, where only the point with larger distance (d_{equil2}) is stable, as a small displacement to positive d results in a negative retracting force. d_{equil1} is unstable, as a displacement in positive d results in a positive force and thus an even larger acceleration toward positive d.

For 90 V, on the other hand, there are no equilibrium points; the force always points toward the reference plate.

In the dynamic case, pull-in happens for smaller voltages than in the static case, as the kinetic energy can cause the system's displacement to overcome the force equilibrium point when the variable plate is moving toward the other. Let us assume that the system is initially in rest. If the voltage is large enough to drive the system's dynamic trajectory beyond the unstable equilibrium point d_{equil1} (and thus in the range of the negative effective tangential spring constant), the system is accelerated toward pull-in.

In *phase space* (position versus momentum/velocity, see Figure 11.4b), the nonlinear system shows deformed (because of the nonlinearity) ellipses in the stable range; pull-in happens when this trajectory touches the unstable range. When the voltage is above the pull-in voltage, no ellipses exist any more, no matter what the initial values of displacement and speed are.

The *dynamic* pull-in voltage is the voltage that, when suddenly applied to the system in rest, leads to a system displacement which is large enough so that the system moves below the unstable equilibrium point d_{equil1}. From an inspection of the conservation of energy, it can be concluded for a system with an initial position of d_0 and zero speed, and the pull-in voltage is [21]

$$V_{\text{pull-in,dyn}} = \frac{1}{2}\sqrt{\frac{kd_0^3}{\varepsilon_0 A}} \tag{11.7}$$

11.2.2
Regimes of the Trajectory

In Figure 11.4a, we can also recognize the different regimes of the possible trajectories.

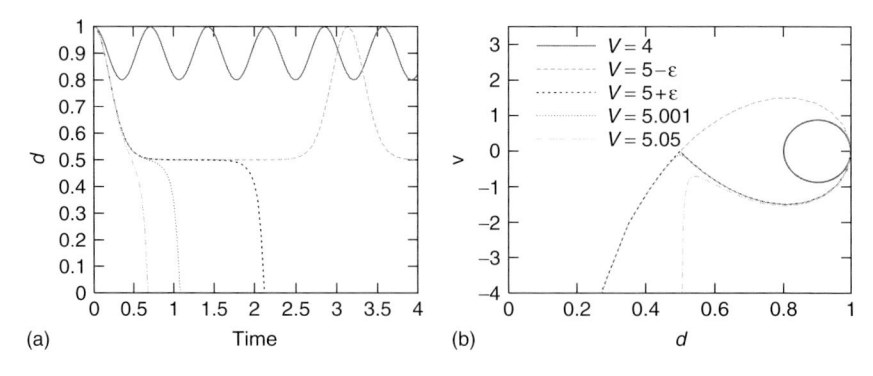

Figure 11.4 The behaviour of an undamped system near pull-in ($V_{\text{pull-in,dyn}} = 5\text{V}$, $k = 100\text{N/m}$, $d_0 = \varepsilon A = 1$). (a) Transient response. (b) Phase space diagram. ε in $5 - \varepsilon$ and $5 + \varepsilon$ represents a small deviation.

The solid line shows the small displacement, close to linear behavior. The step response shows a sinusoidal oscillation, and the shape in phase space is an ellipse. The first dashed line shows the trajectory for a voltage just below the pull-in voltage, showing a strong deformation close to the unstable equilibrium point d_{equil1}, where the total force drops down to almost zero. For a voltage just above pull-in, we see that finally the electrostatic actuation force gains over the spring's restoring force, and pull-in happens. The other two curves show pull-in for higher voltages, where the zero-force point is passed quite fast because of the momentum gained before.

An interesting fact is that a stable transient trajectory can reach a displacement larger than $d_0/3$. The reachable range is extended down to the lower, unstable equilibrium point d_{equil1}, but because of its instability, it is not possible to maintain a constant displacement at constant voltage there for real-world systems.

For MOR, this poses the problem that linear methods will have difficulties to switch between the different regimes, in particular, to capture the force inversion at the unstable equilibrium point. Also, as soon as the displacement brings the two plates close together, an infinite attractive force emerges.

Usually, designers try to avoid complete touchdown conditions of the movable plate, for example, with mechanical obstacles (stoppers) to impose a minimal gap or insulation layers. Reasons are the large currents that could melt the two plates together and other issues related to sticking. On the other hand, the contact changes the system in a very drastic, nonlinear way, a change that the reduced model must also be able to represent.

In conclusion, we can see that the $1/x^2$ nonlinearity gives rise to a number of effects that a designer and thus a reduced model must take into account. The MEMS must be designed to avoid sticking and electric breakthrough at pull-in, and contact must be modeled properly when the behavior at these conditions is to be simulated. Linear models fall short in predicting the behavior in these regimes. Therefore, we aim to trade off the complexity of the reduced model versus its accuracy to model the special conditions.

11.3
Model Order Reduction Methods

In this section, we explain in detail which MOR methods we use to treat the different nonlinear aspects of electrostatic actuation ($1/x^2$ force law, contact, and dependence of force on both voltage and displacement). After a short section on their representation in terms of nonlinear systems of ordinary differential equations, we discuss how projection methods can be applied to weakly nonlinear systems and how stronger, but localized nonlinearities, can be introduced via symbol isolation. Finally, TPWL methods are touched upon.

11.3.1
Representation of Nonlinearities

Let us first discuss how to represent weak nonlinearities. The general form for the equations that we are considering is as follows:

$$\dot{\mathbf{x}}(t) = \mathbf{f}(\mathbf{x}(t), \mathbf{u}(t), t) \tag{11.8}$$

where $\mathbf{x} : \mathbb{R} \mapsto \mathbb{R}^n$ is the vector of state variables, $\mathbf{u} : \mathbb{R} \mapsto \mathbb{R}^m$ is the vector of inputs, $t \in \mathbb{R}$ is time, and $\mathbf{f}$ gives the behavior of the system. The observables or outputs $\mathbf{y} : \mathbb{R} \mapsto \mathbb{R}^p$ are given by

$$\mathbf{y}(t) = \mathbf{g}(\mathbf{x}(t), \mathbf{u}(t), t) \tag{11.9}$$

In the following equation, we assume that the system is time independent, so that $\mathbf{f}$ and $\mathbf{g}$ do not show a dependency on t. Linearization at an operating point $(\mathbf{x}_0, \mathbf{u}_0)$ yields the following system, which exhibits the same small signal behavior at its vincinity

$$\begin{aligned}\dot{\mathbf{x}}(t) &= \mathbf{A}\,\mathbf{x}(t) + \mathbf{B}\,\mathbf{u}(t) \\ \mathbf{y}(t) &= \mathbf{C}\,\mathbf{x}(t) + \mathbf{D}\,\mathbf{u}(t)\end{aligned} \tag{11.10}$$

where the system matrix $\mathbf{A} \in \mathbb{R}^n \times \mathbb{R}^n$, the input matrix $\mathbf{B} \in \mathbb{R}^n \times \mathbb{R}^m$, the output matrix $\mathbf{C} \in \mathbb{R}^p \times \mathbb{R}^n$, and the throughput matrix $\mathbf{D} \in \mathbb{R}^p \times \mathbb{R}^m$.

On the basis of this linear approximation, weakly nonlinear systems or bilinear systems can often be very well approximated using a multivariate series expansion of the system state variables. Defining the set of p-tupels $I_m^p := \{(i_1, \ldots, i_p) : 0 \le i_1 \le m, \ldots, 0 \le i_p \le m\}$, we can write it in a very general form

$$\dot{x}_i(t) = \sum_{r_x=0}^{r_{x,\max}} \sum_{r_u=0}^{r_{u,\max}} \sum_{p \in I_n^{r_x}} \sum_{q \in I_m^{r_u}} A_{i\mathbf{p}\mathbf{q}} \prod_{s=1}^{r_x} x_{p_s} \prod_{t=1}^{r_u} u_{q_t} \tag{11.11}$$

for example, the corresponding term for $r_x = 3$, $r_u = 1$, $\mathbf{p} = (1, 1, 2)$, and $\mathbf{q} = (3)$ is

$$A_{i,1,1,2,3} x_1^2 x_2 u_3 \tag{11.12}$$

This form contains all the different powers of $\mathbf{x}$ and $\mathbf{u}$; in practical applications, $r_{x,\max}$ and $r_{u,\max}$ are often not larger than 2, as otherwise the matrix sizes become unfeasible ($A_{i\mathbf{p}\mathbf{q}} \in \mathbb{R}^{n^{r_x+1}} \times \mathbb{R}^{m^{r_u}}$).

One example is the parallel plate capacitor with moving plate, where the force between plates contains a $1/d^2$ term, which, although diverging at $d = 0$, can well be approximated by a polynomial at larger distances. However, nonsmooth or piecewise-defined force changes as, for example, contact, are difficult to model, as are trajectories further away from the series expansion point.

11.3.2
MOR Methods for Second-Order Linear Systems

Many mechanical and also electromagnetic problems are expressed by PDEs that are second order in time. In most cases, nonlinearities such as geometric nonlinearities, nonlinear material, and electrostatic coupling are only present in the zero-order part of the equations or in the first-order part. In the following, we consider the former case only, as damping effects will be approximated by Rayleigh damping [22] in the applications presented in this chapter. The system reads

$$\mathbf{M}\ddot{\mathbf{x}}(t) + \mathbf{E}\dot{\mathbf{x}}(t) + \mathbf{K}\mathbf{x}(t) = \mathbf{B}\mathbf{u}(t)$$
$$\mathbf{y}(t) = \mathbf{C}\mathbf{x}(t) + \mathbf{D}\mathbf{u}(t) \tag{11.13}$$

where $\mathbf{M} \in \mathbb{R}^n \times \mathbb{R}^n$ is the mass matrix, $\mathbf{E} = \alpha\mathbf{M} + \beta\mathbf{K} \in \mathbb{R}^n \times \mathbb{R}^n$ the (Rayleigh) damping matrix, $\mathbf{K} \in \mathbb{R}^n \times \mathbb{R}^n$ the stiffness matrix, and $\mathbf{x}, \dot{\mathbf{x}}, \ddot{\mathbf{x}}$ are the vector of unknowns and its first and second time derivatives, respectively.

In Chapter 3.7, Krylov-subspace-based algorithms for first-order linear systems were discussed. These methods can be generalized to reduced systems in the form of Eq. (11.13) while preserving the second-order structure [23]. For systems with Rayleigh damping (i.e., $\mathbf{E} = \alpha\mathbf{M} + \beta\mathbf{K}$), the damping matrix does not contribute new information to the computation of the subspace, so that the Krylov subspace can be computed for the undamped system; the resulting projection is also applied to the damping matrix [24–27]. The linearized undamped second-order system then reads

$$\mathbf{M}\mathbf{x}(t) + \mathbf{K}\mathbf{x}(t) = \mathbf{B}\mathbf{u}(t)$$
$$\mathbf{y}(t) = \mathbf{C}\mathbf{x}(t) + \mathbf{D}\mathbf{u}(t) \tag{11.14}$$

As in the first-order case, the projection introduces new reduced state variables $\mathbf{x}_r \in \mathbb{R}^{n_r}$, so that

$$\mathbf{x} = \mathbf{V}\mathbf{x}_r + \varepsilon \tag{11.15}$$

with an error ε, which should be minimized.

The transfer function $\mathbf{H}(s) = \mathscr{L}(\mathbf{y})/\mathscr{L}(\mathbf{u})$ after Laplace transformation $\mathscr{L}$ of Eq. (11.14) is

$$\mathbf{H}(s) = \mathbf{D} + \mathbf{C}[s^2\,\mathbf{M} + \mathbf{K}]^{-1}\mathbf{B} \tag{11.16}$$

We introduce a new variable $\tilde{s} := s^2$

$$\begin{aligned}\mathbf{H}(\tilde{s}) &= \mathbf{D} + \mathbf{C}(\tilde{s}\mathbf{M} + \mathbf{K})^{-1}\mathbf{B} \\ &= \mathbf{D} + \mathbf{C}((\tilde{s} - \tilde{s}_0)\mathbf{M} + \tilde{s}_0\,\mathbf{M} + \mathbf{K})^{-1}\mathbf{B} \\ &= \mathbf{D} + \mathbf{C}((\tilde{s} - \tilde{s}_0)[\underbrace{\tilde{s}_0\,\mathbf{M} + \mathbf{K}}_{\tilde{\mathbf{K}}}]^{-1}\mathbf{M} + \mathbf{I})^{-1}[\tilde{s}_0\,\mathbf{M} + \mathbf{K}]^{-1}\mathbf{B} \\ &= \mathbf{D} + \mathbf{C}((\tilde{s} - \tilde{s}_0)\tilde{\mathbf{K}}^{-1}\mathbf{M} + \mathbf{I})^{-1}\tilde{\mathbf{K}}^{-1}\mathbf{B} \\ &\approx \mathbf{D} + \mathbf{C}\sum_{i=0}^{n_r-1}(\tilde{s} - \tilde{s}_0)^i\,\mathbf{m}_i \quad \text{with moment vectors} \\ \mathbf{m}_i &= \left(-\tilde{\mathbf{K}}^{-1}\mathbf{M}\right)^i \tilde{\mathbf{K}}^{-1}\mathbf{B} \end{aligned} \tag{11.17}$$

where $\tilde{s}_0 = s_0^2$ and s_0 is the (possibly complex) expansion point. We now seek a projection $\mathbf{V}$ that provides a Padé approximation, that is, that yields the same first

n_r moments for the transfer function of the reduced system. As in the first-order methods, the moment vectors are used to span the Krylov subspace

$$\mathcal{K}_{n_r}\left(\tilde{\mathbf{K}}^{-1}\mathbf{M}, \tilde{\mathbf{K}}^{-1}\mathbf{B}\right) = \mathrm{span}\left(\tilde{\mathbf{K}}^{-1}\mathbf{B},\ (\tilde{\mathbf{K}}^{-1}\mathbf{M})\tilde{\mathbf{K}}^{-1}\mathbf{B},\ldots,(\tilde{\mathbf{K}}^{-1}\mathbf{M})^{n_r-1}\tilde{\mathbf{K}}^{-1}\mathbf{B}\right) \tag{11.18}$$

Orthonormalization of the Krylov vectors via the Arnoldi procedure results in a rectangular projection matrix $\mathbf{V} \in \mathbb{R}^n \times \mathbb{R}^{n_r}$, with which the damped system is reduced

$$\mathbf{V}^T \mathbf{M} \mathbf{V} \ddot{\mathbf{x}}_r(t) + \mathbf{V}^T \mathbf{E} \mathbf{V} \dot{\mathbf{x}}_r(t) + \mathbf{V}^T \mathbf{K} \mathbf{V} \mathbf{x}_r(t) = \mathbf{V}^T \mathbf{B} \mathbf{u}(t)$$
$$\mathbf{y}(t) = \mathbf{C} \mathbf{x}_r(t) + \mathbf{D} \mathbf{u}(t) \tag{11.19}$$

11.3.3
MOR Methods for Nonlinear Systems

Nonlinearity can be present in different manners, of which we discuss some important special cases in the following section. Especially the first example problem in Section 11.4 falls into more than one category.

Parametric systems (Chapter 9.4) are systems where one or more of the system matrices are a function of parameters **p**

$$\mathbf{M}(\mathbf{p}) \ddot{\mathbf{x}}(t) + \mathbf{E}(\mathbf{p}) \dot{\mathbf{x}}(t) + \mathbf{K}(\mathbf{p}) \mathbf{x}(t) = \mathbf{B}(\mathbf{p}) \mathbf{u}(t) \tag{11.20}$$

Time-dependent systems can sometimes be modeled this way.

The approaches to handle parameters in model reduction include interpolation techniques, multivariate moment-matching and series expansion [13, 28], but when a parametric dependency can be isolated from a system, symbolic isolation can be used during the assembly of the equations: The parametric or nonlinear part is isolated from the system, then the remaining system is reduced and the symbolic part is connected by input and output ports.

In addition, parametrized matrices can be linearized with respect to the parameter and then written as $M_{ij}(\mathbf{p}) = M_{ij}^{(0)} + \sum_k M_{ijk}^{(1)} p_k$ with a vector of parameters **p**.

Systems with nonlinear inputs have inputs to which, before they are further processed in the system, a nonlinear function is applied.

As this transformation can be separated from the rest of the equations, the system can be reduced by isolating the input transformation from the system and reducing the linear part with a linear reduction method.

Systems with few nonlinearities: We assume that these concern only a limited number of matrix entries. The solution is to isolate the nonlinearities by moving them into a vector $\mathbf{f}$ such that the system reads

$$\mathbf{M} \ddot{\mathbf{x}}(t) + \mathbf{E} \dot{\mathbf{x}}(t) + \mathbf{K} \mathbf{x}(t) = \mathbf{B} \mathbf{u}(t) + \mathbf{F} \mathbf{f}(\mathbf{x}, t)$$
$$\mathbf{y}(t) = \mathbf{C} \mathbf{x}(t) + \mathbf{D} \mathbf{u} + \mathbf{G} \mathbf{g}(\mathbf{x}, t) \tag{11.21}$$

Then, the components of $\mathbf{f}(\mathbf{x}, t)$ and $\mathbf{g}(\mathbf{x}, t)$ are defined as new inputs to the system; the arguments of $\mathbf{f}$ and $\mathbf{g}$ are recovered by the projection $\mathbf{x} = \mathbf{V}\mathbf{x}_r$; and then the system is reduced without considering **f** and **g**, which are reinserted later via the input vector

$$\mathbf{M}_r \ddot{\mathbf{x}}_r(t) + \mathbf{E}_r \dot{\mathbf{x}}_r(t) + \mathbf{K}_r \mathbf{x}_r(t) = \tilde{\mathbf{B}}_r \tilde{\mathbf{u}}(\mathbf{V}\mathbf{x}_r, t)$$
$$\mathbf{y}(t) = \mathbf{C}_r \mathbf{x}_r(t) + \hat{\mathbf{D}} \hat{\mathbf{u}}(\mathbf{V}\mathbf{x}_r, t) \tag{11.22}$$

with

$$\tilde{\mathbf{B}}_r = \mathbf{V}^T [\mathbf{B}\ \mathbf{F}] \quad \hat{\mathbf{D}} = [\mathbf{D}\ \mathbf{G}] \quad \tilde{\mathbf{u}} = \begin{pmatrix}\mathbf{u}\\ \mathbf{f}\end{pmatrix} \quad \hat{\mathbf{u}} = \begin{pmatrix}\mathbf{u}\\ \mathbf{g}\end{pmatrix} \tag{11.23}$$

It is crucial for a fast evaluation that the functions $\mathbf{f}(\mathbf{V}\mathbf{x}_r, t)$ and $\mathbf{g}(\mathbf{V}\mathbf{x}_r, t)$ do not use the full vectors $\mathbf{V}\mathbf{x}_r$ or $\mathbf{V}\dot{\mathbf{x}}_r$, so that $\mathbf{V}$ can be condensed to a few rows.

Systems with many nonlinearities have so many nonlinearities that it is not feasible to treat all of them as inputs, for example, because their number is in the same order of magnitude as the number of degrees of freedom of the model. Among the possibilities to proceed are the following: (i) use a trajectory based approach by taking system snapshots (proper orthogonal decomposition method [29, 30], balancing and optimization [31], system identification [32], and TPWL [16]); (ii) higher-order series expansion (weakly nonlinear polynomial approximation) or bilinearization [15, 33, 34]; and (iii) other methods [35–40].

11.3.3.1 Polynomial Projection

This approach works for systems that are represented by polynomials in the state vector [15, 33, 34, 41]. Let us for simplicity assume that only the stiffness matrix is nonlinear, then a second-order model reads

$$\sum_j M_{ij}^{(1)} \ddot{x}_j + \sum_j E_{ij}^{(1)} \dot{x}_j + \sum_j K_{ij}^{(1)} x_j + \sum_{j,k} K_{ijk}^{(2)} x_j x_k$$
$$+ \sum_{j,k,l} K_{ijkl}^{(3)} x_j x_k x_l + \cdots = \sum_j B_{ij}^{(1)} u_j$$
$$y_i = \sum_j C_{ij}^{(1)} x_j + \sum_j D_{ij}^{(1)} u_j \tag{11.24}$$

Now, the projection matrix $\mathbf{V}$ is found. For the matrices $\mathbf{M}^{(1)}$, $\mathbf{E}^{(1)}$, $\mathbf{K}^{(1)}$, $\mathbf{B}^{(1)}$, and $\mathbf{C}^{(1)}$ we proceed with the projection as in linear MOR methods. For the matrices $K_{ijk\ldots}^{(p)}$, we can write

$$\sum_{j,k,l,\ldots} V_{ji} K_{jkl\ldots}^{(p)} x_k x_l \cdots \approx \sum_{j,k,l,\ldots} V_{ji} K_{jkl\ldots}^{(p)} \left(\sum_m V_{km} x_{r,m}\right)\left(\sum_n V_{ln} x_{r,n}\right)\cdots$$
$$= \sum_{j,k,l,\ldots,m,n,\ldots} \left(K_{jkl\ldots}^{(p)} V_{ji} V_{km} V_{ln} \cdots\right) x_{r,m} x_{r,n} \cdots$$
$$= \sum_{m,n,\ldots} K_{r,imn\ldots}^{(p)} x_{r,m} x_{r,n} \cdots \tag{11.25}$$

The projected system then reads

$$\sum_j M^{(1)}_{r,ij}\ddot{x}_{r,j} + \sum_j E^{(1)}_{r,ij}\dot{x}_{r,j} + \sum_j K^{(1)}_{r,ij}x_{r,j} + \sum_{j,k} K^{(2)}_{r,ijk}x_{r,j}x_{r,k}$$
$$+ \sum_{j,k,l} K^{(3)}_{r,ijkl}x_{r,j}x_{r,k}x_{r,l} + \cdots = \sum_j B^{(1)}_{r,ij}u_j$$
$$y_i = \sum_j C^{(1)}_{r,ij}x_{r,j} + \sum_j D^{(1)}_{ij}u_j \qquad (11.26)$$

In the example given in this chapter, the main information of the system is contained in the matrix $\mathbf{K}^{(1)}$, so that the reduced solution almost lies in a subspace that is spanned by the moments vectors of the transfer function of the linear system in Eq. (11.14) [34, 42]. This means that we choose the columns of the projection matrix $\mathbf{V}$ for obtaining the reduced model of the nonlinear system to form a basis of the subspace spanned by the moments vectors of the transfer function of the linear system in Eq. (11.14).

One problem with this method is that matrices can become rather large ($\mathbf{K}^{(p)} \in (\mathbb{R}^{n_r})^{p+1}$) and are dense, which puts computational limits on the number n_r of reduced state variables. The ideas proposed here are to truncate the higher degree matrices or to delete small matrix elements which represent reduced states that couple only little and do not contribute much. Thus, it is possible to keep only all values above a certain threshold and thus increase the sparsivity of the matrix.

In 11.4.2, the application of these methods to the microelectromechanical system in 11.4 will be demonstrated.

11.3.3.2 Trajectory Piecewise-Linear (TPWL) Method

The *trajectory piecewise-linear method* (Chapter 10.5) represents the reduced nonlinear system by a weighted combination of several linear models, generated at different linearization points in the state space [16], which are chosen along typical (training) trajectories of the system for a particular *training input*. In the framework of the TPWL method, the Arnoldi procedure [43] can be applied to find a reduced subspace onto which the system is then projected.

In the following discussion, again the important assumption is made that as Rayleigh damping is used to describe the system, the damping matrix does not contribute to the subspace [43]. The projection matrix $\mathbf{V}$ is created by matching the moments of the initial state, that is, by considering only the first linearization performed around zero point, and the reduced system description is also enriched by adding the Krylov subspaces corresponding to other linearization points located along the system trajectory. It has been evidenced that this strategy generates models that are in general more accurate and with a lower order than the simple linearization around the initial point [16]. Details of the implementation and its application to the micropump example in Section 11.5 are given in Section 11.5.2.

11.4
Example 1: IBM Scanning-Probe Data Storage Device

The IBM MEMS-based parallel scanning-probe data storage device [22] is a novel data storage approach to allow for storage densities beyond the superparamagnetic limit of state-of-the-art hard disk recording [44]. These densities can be achieved by using local-probe techniques, inspired by atomic force microscopy, to write, read back, and erase data in very thin polymer films.

The thermomechanical scanning-probe-based data storage concept combines ultrahigh density and high data rates, which are achieved by parallel operation of large 2D arrays with thousands of micro/nanomechanical cantilevers (Figure 11.5). The cantilevers feature a tiny probe tip on their free side, a capacitive platform for an optimal electrostatic force and a thinned part near the support, which serves as a hinge with a spring constant of ≈ 0.05 N m^{-1}; lower-doped parts serve as resistors for heating the probe tip. An integrated micromagnetic x/y scanner [45, 46] moves a thin epoxy-based polymer below the tip (Figures 11.5 and 11.6).

The electrostatic actuation system deflects the cantilever until its heated tip touches the polymer medium: voltages below 20 V are sufficient to generate a force of 1 μN. There, it creates tiny indentations that represent the stored data bits. By moving the polymer with the scanner, a raster of bits emerges. Data densities of 641 Gbit in^{-2} with raw bit error rates better than 10^{-4} were successfully achieved [46].

The normal operation mode resembles a pull-in-like behavior. However, only a small part of the tip is in contact with the substrate. On the other hand, when the applied voltage is too high, the electrostatic actuation platform may also experience pull-in. The device and its circuitry design must avoid this situation.

For the design of the device and the codevelopment with the integrated analog front-end circuitry, an accurate compact model for dynamic simulations in a circuit

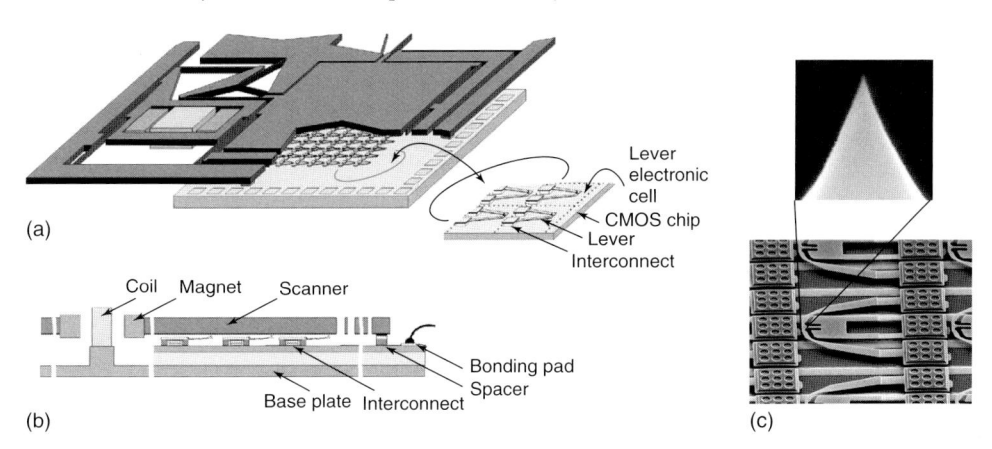

Figure 11.5 Setup of the cantilever array. (a) Setup of the complete system. (b) Cross section of the system. (c) Microscopy of the cantilevers and the read/write tip [47].

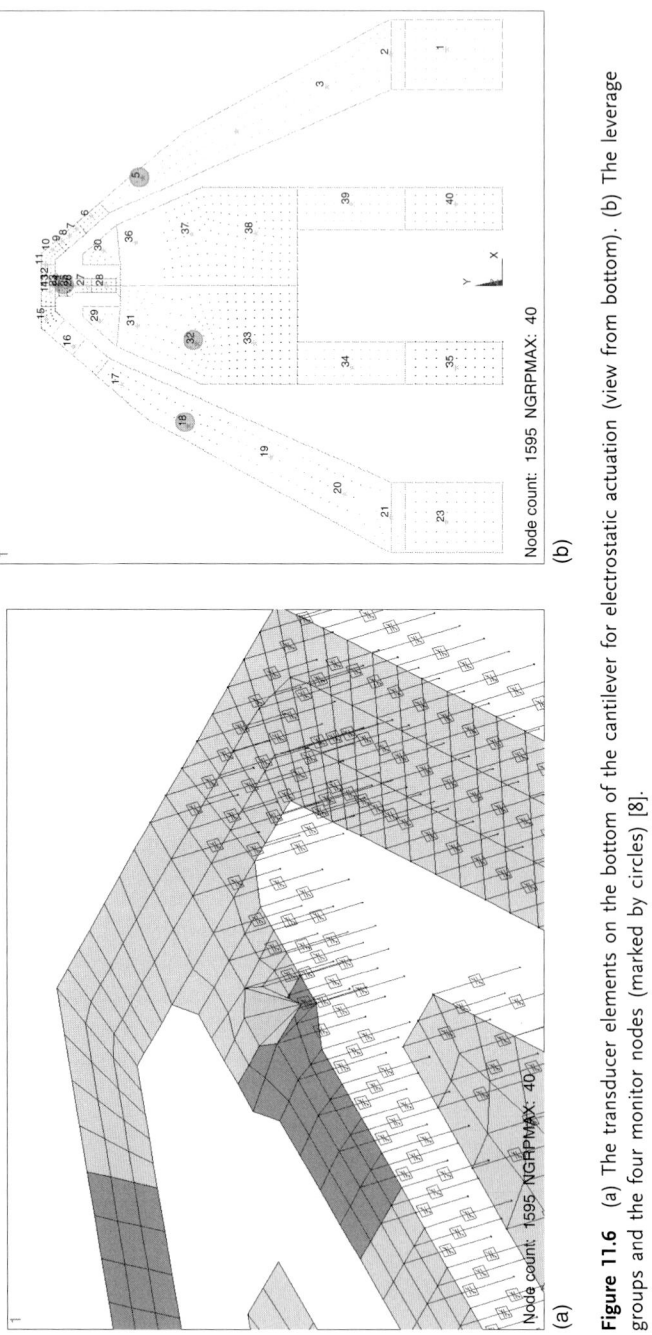

Figure 11.6 (a) The transducer elements on the bottom of the cantilever for electrostatic actuation (view from bottom). (b) The leverage groups and the four monitor nodes (marked by circles) [8].

simulator is required. The $1/x^2$-behavior of the electrostatic actuation force and the contact model of the tip introduce nonlinearities to the equations.

11.4.1
Cantilever Model

The scanning-probe data storage device model was created by IBM Research Zürich in Rüschlikon in the commercial finite element simulator ANSYS. The cantilever is modeled with shell elements. The tip is modeled with a single pyramidal volumetric element. The model has 9441 mechanical degrees of freedom. On the entire bottom surface, 1D transducer elements are used to couple the mechanical model to the electrostatic domain; every mechanical mesh node on the cantilever is assigned its own transducer (Figure 11.6).

The mechanical material properties are those of silicon and are assumed to be linear; geometric nonlinearities are also considered in the full model, but not in the reduced model, as their effect was found to be negligible.

The nodes are manually partitioned into *leverage groups*. Each leverage group is treated as separate voltage input V_j. The reason is that the underlying complex electroresistive model is calculated with a separate submodel and connected via input terminals (symbol isolation approach).

Finally, four monitor (i.e., output) nodes are defined as *system outputs*; one is placed at the tip, two on the legs, and one on the capacitive platform (Figure 11.6b).

11.4.2
Model Order Reduction with Polynomial Projection

The following steps were performed: (i) find a polynomial approximation based on the nonlinear FEM model; (ii) use polynomial projection to find the reduced matrices; and (iii) postprocess matrices to decrease complexity. The MOR was implemented in Mathematica (Wolfram Research) and is available for download within the IMTEK Mathematica Supplement [48].

11.4.2.1 Extraction of Nonlinear System
The linear mechanical part and the parameters for the nonlinear transducer model were extracted from the binary files of the ANSYS 10 model.

This process yields a system with the following form:

$$\mathbf{M}\ddot{\mathbf{x}} + \mathbf{E}\dot{\mathbf{x}} + \mathbf{K}\mathbf{x} = \mathbf{B}\mathbf{u} + \mathbf{f}(\mathbf{x}, \mathbf{u})$$
$$\mathbf{y} = \mathbf{C}\mathbf{x} \tag{11.27}$$

where $\mathbf{u} \in \mathbb{R}^m$ is the vector of inputs (squared voltages of the different leverage groups and force on the tip) and $\mathbf{f} : \mathbb{R}^n \times \mathbb{R}^m \mapsto \mathbb{R}^n$ represents the nonlinear transducer elements. We use Rayleigh damping so that $\mathbf{E} = \alpha \mathbf{M} + \beta \mathbf{K}$ with two constants $\alpha, \beta \in \mathbb{R}$, which are chosen as $\alpha = 0$, $\beta = 10^{-7}$s, resulting in a quality factor of $Q = 20$ at the cantilever's first resonance frequency of 86 kHz. The effect of squeeze film damping is small, as the cantilever operates in a regime away from touchdown. The tip element is connected to a separate contact model by providing an additional mechanical input and output and attaching the nonlinear 1 DOF model separately; its equation is, with $k_c = 100$ N m^{-1}

$$F_{\text{contact}} = -\max(0, k_c(-0.8\,\mu m - x_{\text{tip}})) \tag{11.28}$$

11.4.2.2 Polynomial Approximation

The nonlinear part $\mathbf{f}(\mathbf{x}, \mathbf{u})$ is now expanded into a Taylor series for the gap distance at operating point 1.25 μm. Moving all terms to the left-hand side of the equation, we can rewrite the first equation of the system as

$$\sum_j M_{ij}\ddot{x}_j + \sum_j E_{ij}\dot{x}_j + \sum_j K_{ij}x_j + \sum_{j,K} Q^{(1)}_{ijK} x_j u_K$$
$$+ \sum_{j,K,L} Q^{(2)}_{ijKL} x_j x_K u_L - \sum_j B_{ij} u_j = 0 \tag{11.29}$$

where $\mathbf{Q}^{(i)} \in (\mathbb{R}^n)^{i+1} \times \mathbb{R}^m$. The matrix coefficients were found by a series expansion of the capacities in terms of the gap sizes

$$C(x) = \frac{C_0}{x} \approx \frac{C_0}{x_0} - \frac{C_0}{x_0^2}(x - x_0) + \frac{C_0}{x_0^3}(x - x_0)^2 - \frac{C_0}{x_0^4}(x - x_0)^3 + \mathcal{O}((x - x_0)^4)$$

The force for node i with applied voltage V_j is then approximated (leaving out the rest term) by

$$F_i(x_i) = -\frac{1}{2}\frac{C_0}{x_i^2}V_j^2 \approx -\frac{C_{0,i}}{x_{0,i}^2}V_j^2 + 2\frac{C_{0,i}}{x_{0,i}^3}(x_i - x_{0,i})V_j^2 - 3\frac{C_{0,i}}{x_{0,i}^4}(x_i - x_{0,i})^2 V_j^2$$
$$= -6\underbrace{\frac{C_{0,i}}{x_{0,i}^2}V_j^2}_{B_{ij}} + 8\underbrace{\frac{C_{0,i}}{x_{0,i}^3}x_i V_j^2}_{Q^{(1)}_{jii}} - 3\underbrace{\frac{C_{0,i}}{x_{0,i}^4}x_i^2 V_j^2}_{Q^{(2)}_{jiii}} \tag{11.30}$$

As V_j always occurs squared, we use V_j^2 as input instead of V_j. As a result, we have two matrices $\mathbf{Q}^{(1)}$ and $\mathbf{Q}^{(2)}$ that we may include into our model as nonlinear terms. The force in off-plane direction of the contact model onto the tip is another input.

11.4.2.3 Model Order Reduction of the Polynomial System

With the classification given in Section 11.3.3, we can identify the following parts of the system:

- The contact point at the tip will not take part in the MOR but remain in a separate submodel that is connected to the rest of the electromechanical models via the tip's out-of-plane degree of freedom (symbol isolation).
- The voltages V_j are squared outside of the reduced model and then used as linear inputs.
- The remaining purely mechanical model can then be treated as weakly nonlinear model with a few inputs, where the number of inputs is reduced by the introduction of the leverage groups.

Using the Arnoldi procedure, a reduced model is found for all linear parts. The problem with this approach is that the $\mathbf{Q}_r^{(2)}$ matrix is of size $n_r^3 m$, where n_r is the size of the reduced system and m is the number of inputs, which in our case is the number of leverage groups. So if we assign n_v Arnoldi vectors to each input, the size of the final matrix is $n_v^3 m^4$. As, in the current model, there are 14 leverage groups and 1 tip input, we have a minimal size of $n_v^3 \cdot 15^4 = 50625 n_v^3$. This matrix is dense.

Numerical experiments show that because of the large number of inputs at different parts of the model, one Arnoldi vector is sufficient for each input. As the force pattern at the tip input is different from that of the electrostatic force inputs, five vectors were taken for this input, so that $n_r = 19$.

Small entries in the dense matrix $\mathbf{Q}_r^{(2)}$ lower than a threshold were dropped to simplify the equations. Finally, a behavioral Spectre-Verilog A model was created for use in a mixed-signal circuit simulator (results are reported elsewhere [22]).

11.4.3
Results

In the following solution graphs, the curves are (from top to bottom) out-of-place displacement for monitor node on the side supports, monitor node on the capacitive platform, and tip.

Figure 11.7a,b shows the result of a transient simulation using a step input of $(7.45\,\text{V})^2$ for all voltage inputs for the full, nonreduced system (9441 DoFs, solid curves) and the polynomial system (also 9441 DoFs, dotted curves). The simulation time in Mathematica for 200 time steps was 5128 s for the original system and 5220 s for the polynomial system. This transformation results in a small error in the transient curve that we attribute to the error of the $1/x^2$ series expansion.

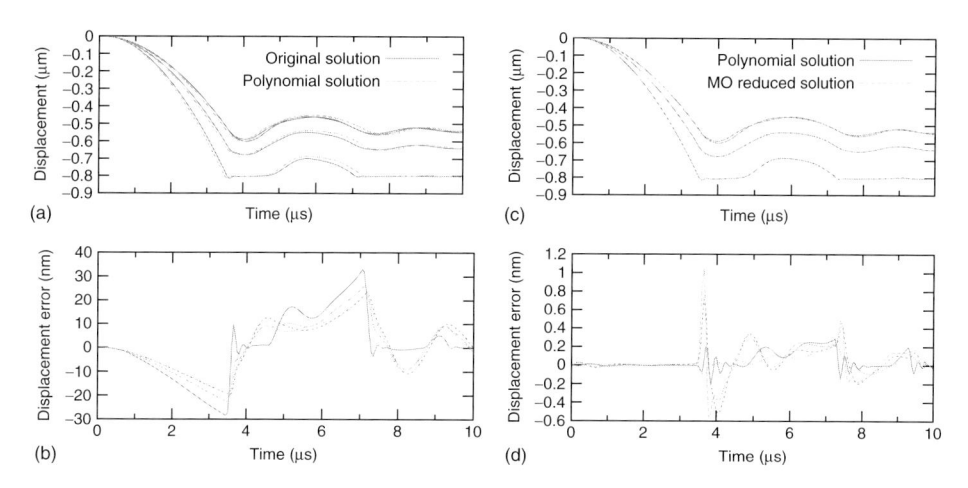

Figure 11.7 (a,b) Comparison of original nonlinear model and polynomial approximation. (c,d) Comparison of polynomial approximation and reduced model.

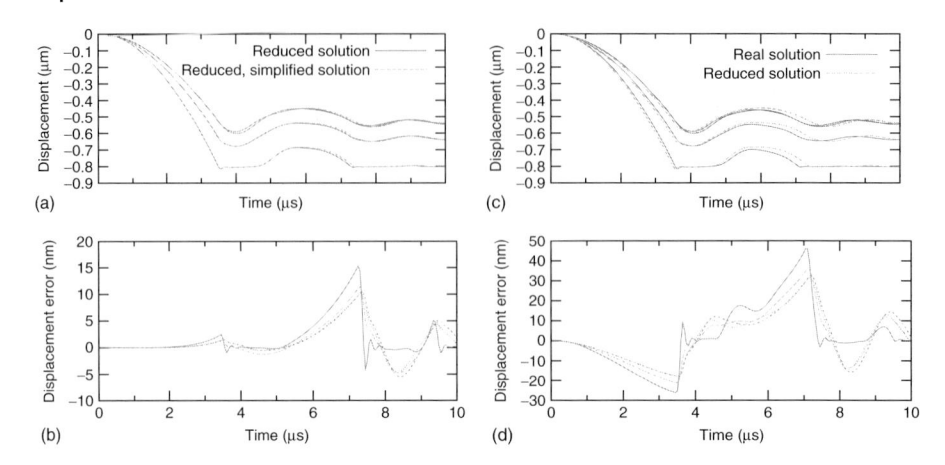

Figure 11.8 (a,b) Comparison of reduced model with small elements deleted ("simplified") and full reduced model. (c,d) Error of simplified reduced model compared to original system.

Figure 11.7c,d shows the transient response after model order reduction. It turns out that the error of this step is remarkably small. The simulation of the reduced system took 290 s, which is already a speed-up by a factor larger than 17.

Figure 11.8a,b shows the effect of removing all matrix elements in $\mathbf{Q}_r^{(2)}$ with absolute value smaller than 3×10^{-9} (Figure 11.9). The error of the reduced model is in the same order of magnitude as the error introduced by the polynomial approximation; the simulation time is reduced to 68.8 s, yielding a total decrease

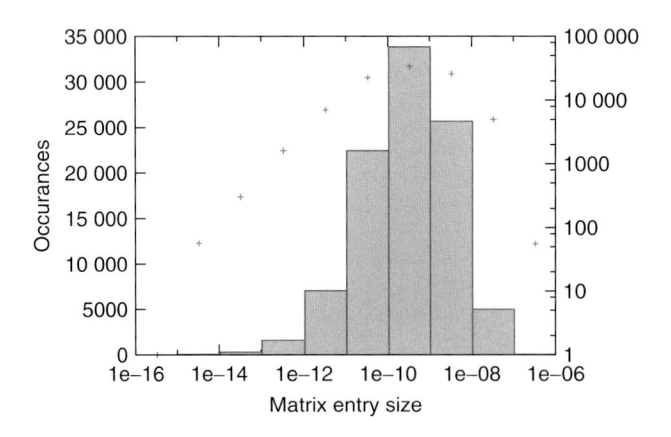

Figure 11.9 Distribution of matrix elements in $\mathbf{Q}_r^{(2)}$. Crosses: Logarithmic scale; bars: Linear scale.

in simulation time compared to the original system by a factor of about 75 with a relative error of about 6%.

Finally, Figure 11.8c,d shows a comparison between the final reduced model and the original system.

11.4.4
Discussion

The results are quite promising; however, there are certain limitations of this method. Apart from the approximation errors by using transducer elements (and therefore neglecting fringing fields and assuming always-parallel plate capacitors), it is mainly the polynomial approximation that is a source of errors. Methods to merge the transducer elements of one leverage group to a single transducer element and isolating this transducer element from the reduced system are under investigation.

11.5
Example 2: Electrostatic Micropump Diaphragm

Recently, micropumping is emerging as a critical research area for many applications [49]. This is undoubtedly motivated by the need to develop pumping mechanisms for biological fluid handling and drug delivery [50, 51]. Moreover, micropumps are being considered for applications in the thermal management of electronic components [52]. The first successful diaphragm micropump with electrostatic actuation was realized by Zengerle et al. [53].

Broadly speaking, the design procedure of this kind of device should be based on multiphysics simulations, capable of coupling electrostatic, structural, and fluid dynamic phenomena. Nevertheless, the actuator itself deserves particular attention, as it represents the core of the device. This is true in particular concerning electrostatic actuation, where the goal is to achieve an optimal balance of stroke volume – proportional to the maximum displacement, then to electrode gap [54] – and actuation force – inversely proportional to electrode gap. Accordingly, strong attention should be paid to pull-in phenomena also in the dynamic ambit, with the aim of controlling this feature and possibly of taking advantage of it. In fact, structure collapse may represent a limitation on the stable operation regime of a device such as a micropump [55], but in principle it could be mastered to maximize stroke volume.

MOR techniques can be conveniently applied to obtain a compact numerical model that describes the coupled electromechanical response to the diaphragm undergoing electrostatic actuation. In particular, the focus here is on the description of the dynamic response under step load application for both stable range and electrostatic collapse.

11.5.1
Plate Model Formulation

Let us consider the problem as represented in Figure 11.10. The upper flexible circular electrode (i.e., the diaphragm) with radius R and thickness t is made with an elastic, homogeneous, and isotropic material. It is clamped all along the boundary. The lower ground electrode has the same radius R. It is perfectly rigid and separated in the rest position from the flexible one by a uniform gap g_0, which is much smaller than the characteristic on-plane dimension R. Bodies are assumed to be perfect conductors, the space between them being filled with a homogeneous dielectric medium with permittivity ε.

Let us consider a nonuniform pressure p applied on the plate surface. If the thickness t of the diaphragm is such that it can be modeled as a thin plate, that is, $2R/t > 20$, and it undergoes small deformations, the displacement field $u(r)$ in quasi-static conditions can be obtained by the Germaine-Lagrange equation [56, 57]

$$D \Delta^2 u(r) = p(r) \tag{11.31}$$

plus suitable boundary conditions [58]. Here, $D = Eh^3/12(1-v^2)$ is the plate stiffness, where E is the Young's modulus and v the Poisson's ratio of the material, and $\Delta^2(\cdot)$ is the double Laplacian operator, in polar space coordinates. According to [59], the deflection of the circular electrode is assumed to be axisymmetric.

When a potential V is applied between the electrodes, assuming that fringing effects are negligible and that the electric field is uniform, the electrostatic pressure for the problem in Figure 11.10 can be written in the form [60]

$$p_e(r) = \frac{\varepsilon V^2}{2 g_0^2} \left(1 - \frac{u(r)}{g_0}\right)^{-2} \tag{11.32}$$

Dynamic behavior of the structure, accounting for inertial effects and damping, is described by the equation of motion [61, 62]

$$D \Delta^2 u(r) + \frac{2\pi f_0}{Q} \rho h \frac{\partial u(r,t)}{\partial t} + \rho h \frac{\partial^2 u(r,t)}{\partial t^2} = \frac{\varepsilon V^2}{2 g_0^2} \left(1 - \frac{u(r,t)}{g_0}\right)^{-2} \tag{11.33}$$

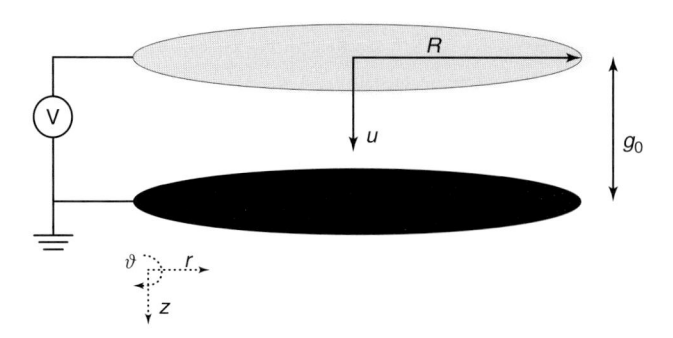

Figure 11.10 Schematic illustration of the microplate structure undergoing electrostatic actuation.

where ρ is the (constant) density of the material, Q the quality factor, and f_0 the natural frequency of the structure. In this case study, a Rayleigh damping is adopted [63], by considering mass damping only and by neglecting stiffness damping effects [56]. The relation between the quality factor Q and the fraction of critical damping xi_D is

$$Q = \frac{1}{2\,\xi_D} \tag{11.34}$$

and the general Rayleigh damping equation is

$$\xi_D = \frac{1}{2}\frac{\alpha_R}{\omega_0} + \frac{1}{2}\omega_0 \beta_R \tag{11.35}$$

where the angular natural frequency $\omega_0 = 2\pi f_0$. By assuming a stiffness damping coefficient $\beta_R = 0$, the mass damping coefficient α_R reads finally

$$\alpha_R = \frac{2\pi f_0}{Q} \tag{11.36}$$

A nondimensional approach is adopted herein, in order to obtain a more general and clear approach to the problem. A set of new variables is defined as

$$\hat{u} = \frac{u}{g_0}, \qquad \hat{r} = \frac{r}{R}, \qquad \hat{t} = t f_0 \tag{11.37}$$

The governing Eq. (11.33) can be written in the dimensionless form

$$\Delta^2 \hat{u}(\hat{r},\hat{t}) + \frac{2\pi m}{Q}\frac{\partial \hat{u}(\hat{r},\hat{t})}{\partial \hat{t}} + m\frac{\partial^2 \hat{u}(\hat{r},\hat{t})}{\partial \hat{t}^2} = \lambda\left(1 - \hat{u}(\hat{r},\hat{t})\right)^{-2} \tag{11.38}$$

where the mass coefficient m and the load parameter λ have been defined, respectively

$$m \equiv \frac{f_0^2 h \rho R^4}{D}, \qquad \lambda \equiv \frac{\varepsilon V^2 R^4}{2 D g_0^3} \tag{11.39}$$

Through (Eq. (11.38)) the behavior of the device is described by means of a minimum amount of parameters, namely, the quality factor Q, which explicitly represents the damping; the mass coefficient m, involving geometric and material parameters; and the load parameter λ, which also embeds electrostatic actuation parameters.

Through a suitable discretization, Eq. 11.38 is casted into the algebraic form

$$\mathbf{M}\ddot{\mathbf{x}} + \mathbf{E}\dot{\mathbf{x}} + \mathbf{K}\mathbf{x} = \lambda \mathbf{b} \tag{11.40}$$

where $\mathbf{b}$ is the (now nonlinear) input vector.

In particular, to discretize the system in space and to generate the stiffness matrix $\mathbf{K}$, a central finite differences scheme is adopted. For m elements, a number of $m+1$ equally spaced grid points are defined for the variable $\hat{r}$. Matrix $\mathbf{K} \in \mathbb{R}^{m+1} \times \mathbb{R}^{m+1}$, while vectors $\mathbf{x} \in \mathbb{R}^{m+1}$ and $\mathbf{b} \in \mathbb{R}^{m+1}$. Suitable boundary conditions, namely, axial symmetry and clamped rim, are then imposed. Finally, the diagonal matrices $\mathbf{M} \in \mathbb{R}^{m+1} \times \mathbb{R}^{m+1}$ and $\mathbf{E} \in \mathbb{R}^{m+1} \times \mathbb{R}^{m+1}$ are generated

$$\mathbf{M} = m\mathbf{I}, \qquad \mathbf{E} = \frac{2\pi m}{Q}\mathbf{I} \tag{11.41}$$

where mI is the identity matrix. The model is implemented in Mathematica 7.0 (Wolfram Research), with the IMTEK Mathematica Supplement [48].

11.5.2
Model Order Reduction by TPWL and Arnoldi Methods

By projection, the reduced form of the problem in the framework of TPWL method will be

$$\mathbf{V}^T\mathbf{M}\mathbf{V}\dot{\mathbf{x}}_r d + \mathbf{V}^T\mathbf{E}\mathbf{V}\dot{\mathbf{x}}_r + \mathbf{V}^T\mathbf{K}\mathbf{V}\mathbf{x}_r = \lambda \sum_{i=0}^{s-1} w_i \mathbf{V}^T \mathbf{b}_i \qquad (11.42a)$$

$$\mathbf{y} = \mathbf{C}\mathbf{V}\mathbf{x}_r \qquad (11.42b)$$

where $\mathbf{b}_i$ is the load vector for the ith linearization point and w_i is the corresponding state-dependent weight. The output matrix $\mathbf{C} \in \mathbb{R}^p \times \mathbb{R}^{m+1}$ selects the midpoint coordinate only ($p = 1$), that is, the model output $y = \hat{u}$ is a scalar. A compact form of Eqs. (11.41a) and (11.42b) for the problem at hand is then

$$\mathbf{M}_r \dot{\mathbf{x}}_r d + \mathbf{E}_r \dot{\mathbf{x}}_r + \mathbf{K}_r \mathbf{x}_r = \lambda \sum_{i=0}^{s-1} w_i \mathbf{b}_i^r \qquad (11.43a)$$

$$\hat{u} = \mathbf{C}_r \mathbf{x}_r \qquad (11.43b)$$

The novel idea introduced here is to perform the TPWL training for a characteristic loading condition for the system (the static pull-in voltage) and then to use the generated TPWL model to study generic load conditions. In practice, training is performed on the full nonlinear system undergoing a static pull-in load λ_{Spi}. Then, the nonlinear dynamic response of the structure is computed, for a generic load $\lambda = \eta \cdot \lambda_{Spi}$, by solving the equation

$$\mathbf{M}_r \dot{\mathbf{x}}_r d + \mathbf{E}_r \dot{\mathbf{x}}_r + \mathbf{K}_r \mathbf{x}_r = \eta \cdot \left(\lambda_{Spi} \sum_{i=0}^{s-1} w_i \mathbf{b}_i^r \right) = \eta \cdot \mathbf{b}_{Spi}^r(\mathbf{x}_r) \qquad (11.44)$$

where $\mathbf{b}_{Spi}^r(\mathbf{x}_r)$ represents the weighted combination of the linear models generated at different linearization points in the state space of the pull-in event used for training. This load vector is updated at every simulation step.

For an applied step load λ_{Spi}, the full system ($n = 500$ degrees of freedom) is solved first. A comparison is provided with finite element simulations reported in Ref. [56]: a good correspondence is found concerning the description of the pull-in event in nonlinear dynamics. The mass parameter m and the quality factor Q are here set accordingly with the FE model. It must be mentioned that the solution accuracy can be affected by the time stepping taken by the solver (here, $\Delta \hat{t} = 5 \times 10^{-4}$).

At this point, system training is performed on the pull-in event. The TPWL approximation is constructed with a number $n_p = 120$ of linearization points – equally spaced in time – and by generating a $\mathbf{V}_{n_p}$ projection matrix of order $n_r = 10$ at each step. The enriched projection matrix $\mathbf{V}$ is obtained by collecting the $\mathbf{V}_{n_p}$ matrices and by performing an SVD orthogonalization with drop tolerance $\epsilon_D = 10^{-3}$. The obtained projection matrix exhibits an order $n_r = 14$.

11.6
Results and Discussion

As clearly shown in Figure 11.11, the TPWL reduced model can effectively track the pull-in event. A small discrepancy is found concerning the pull-in time with respect to full order FDM solution as well as the FEM reference solution. This is essentially due to the number of linearization points adopted and on the solution time stepping. A study to optimize the s parameter and the time stepping can be envisaged. As expected, the linear reduced system fails to describe the nonlinear dynamics of the device.

For the case $\lambda = 0.2 \cdot \lambda_{Spi}$ (Figure 11.11b), both TPWL and linear reduced models can successfully represent device dynamics. This is obvious, as nonlinearities do not play a significant role in the experienced plate displacement. Instead, in Figure 11.12a, where $\lambda = 0.8 \cdot \lambda_{Spi}$ is imposed, the linear reduced model cannot reproduce the microplate response, while a good result is achieved by the TPWL approach. Concerning this example, the simulation speed-up is approximately in the order of 35–40% with respect to the full model. However, it is worth to point out that simulation speed is slowed down by the time required to build the load vector, as a high number of linearization points are chosen. An improved strategy of choosing the linearization points should be considered to decrease the number of linearized models and further reduce the simulation time.

In Figure 11.12b, a pull-in event triggered by the application of a load $\lambda = 2.0 \cdot \lambda_{Spi}$ is simulated. Similar to the previous examples, the TPWL model reproduces the

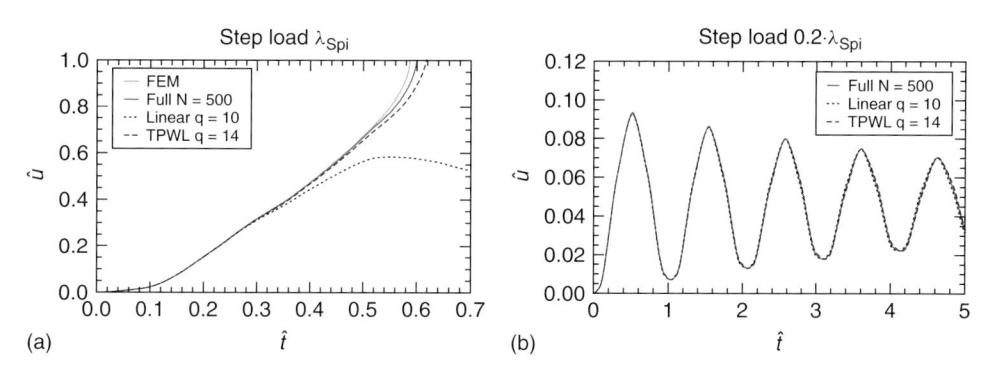

Figure 11.11 (a) Dynamic pull-in event for $\lambda = \lambda_{Spi}$ step load application. Comparison of full nonlinear model, reduced TPWL model (trained for λ_{Spi}), and reduced linear model. FE simulation results from Bertarelli et al. [56] are also reported. (b) Structure response for $\lambda = 0.2 \cdot \lambda_{Spi}$ step load application.

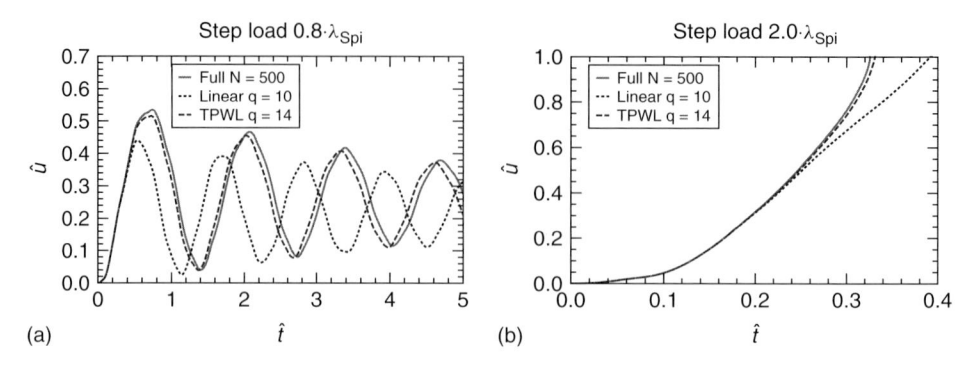

Figure 11.12 Structure response for $\lambda = 0.8 \cdot \lambda_{Spi}$ (a) and $\lambda = 2.0 \cdot \lambda_{Spi}$ (b) step load application. Comparison of full nonlinear model, reduced TPWL model (trained for λ_{Spi}), and reduced linear model (trained for λ_{Spi}).

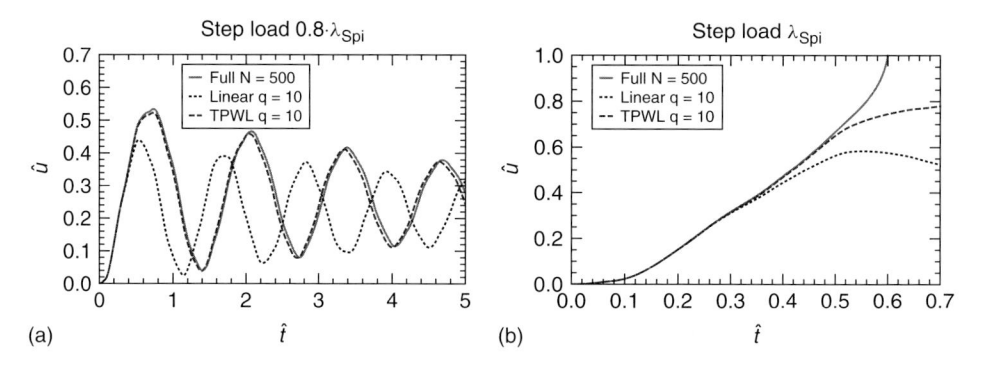

Figure 11.13 Reduced TPWL model trained for $\lambda = 0.8 \cdot \lambda_{Spi}$. Structure response for $\lambda = 0.8 \cdot \lambda_{Spi}$ (a) and $\lambda = \lambda_{Spi}$ (b) step load application.

dynamic pull-in event, while the linear reduced system result is poor. Remarkably, a training performed for a pull-in event has been used here to generate a piecewise-linear system that can apply to a variety of inputs. From a physical standpoint, the trajectories of the systems are found to belong to the region of the state space covered by the linearized model.

Finally, it is worth to show how a training performed below pull-in for $\lambda = 0.8 \cdot \lambda_{Spi}$ generates a TPWL model that provides a good representation of that precise loading condition (Figure 11.13a), while it fails to describe a pull-in event such that $\lambda = \lambda_{Spi}$ (Figure 11.13b).

11.7
Conclusions

Nonlinear MOR techniques have shown excellent performance in approximating a complex system with a much smaller and faster-to-solve system even in the case of electrostatic actuation, where the simulation domain being distorted causes the main effect of the device.

The reduced models show, due to the assumptions that make order reduction possible at all, several limits in the set of trajectories they can reproduce. For example, the transducer model for the cantilever does not allow to represent sideways motion or tilting of the capacitor plates; the polynomial expansion's validity breaks down close to pull-in; the polynomial approach is limited by the fast growing size of the resulting dense multidimensional matrices; the validity of the TPWL reduced model breaks down away from the training trajectory. However, due to sticking effects, many MEMS devices are designed to maintain a minimal gap at pull-in and to operate along a very limited path, so that these models despite their limitations enable designers to shorten design cycles or create model-based controllers, which can enrich their knowledge of the device's observed displacement with a fast-to-solve behavioral model of the device. Error estimates are more a concern for robust automatic MOR systems. Promising new results for control-theory-based methods could eventually provide this important building block.

Acknowledgments

We thank the DFG and the European Union for support over the past 10 years, in particular for the projects "MST-Compact" (KO-1883/6) and "Paramkompakt" (KO 1883/8) and within the Collaborate Research Center 499 (SFB 499) "Development, Production, and Quality Assurance of Primary-Shaped Micro Components from Metallic and Ceramic Materials." We also thank our industry- and university-based partners for the fruitful collaboration over the years. In particular, we wish to acknowledge the enjoyable cooperation with Dr Christoph Hagleitner of IBM Research Zürich in Rüschlikon. Finally, we thank the editors of this book for their very helpful review on this chapter.

References

1. Wautelet, M. (2001) *Eur. J. Phys.*, **22**, 601–611.
2. Wautelet, M. (2007) *Eur. J. Phys.*, **28**, 953–959.
3. Ko, W. (2007) *Sens. Actuators, A: Phys.*, **136**, 62–67.
4. Woias, P. (2005) *Sens. Actuat, B: Chem.*, **105**, 28–38.
5. Korvink, J.G. and Paul, O. (eds) (2006) *MEMS: A Practical Guide to Design, Analysis, and Applications*, William Andrew Publishing, Springer.
6. Wang, W. and Soper, S.A. (2007) *Bio-MEMS: Technologies and Applications*, CRC Press.
7. Rudnyi, E.B. and Korvink, J.G. (2002) *Sensors Update*, **11** (1), 3–33.
8. Lienemann, J., Korvink, J.G., Hagleitner, C., and Rothuizen, H. (2007) Nonlinear model order reduction of electrostatically actuated mems cantilever. Proceedings of the 3rd International Conference on Structural Engineering, Mechanics and Computation, SEMC, pp. 449–454.
9. Antoulas, A.C. (2005) Approximation of Large-Scale Dynamical Systems, No. 6

10. Lienemann, J., Rudnyi, E.B., and Korvink, J.G. (2006) *Linear Algebra Appl.*, **415** (2–3), 469–498.
11. Proceedings ANSYS Conference & 27th CADFEM Users' Meeting, Leipzig, Germany (2009).
12. Benner, P. and Saak, J. (2011) *Math. Comput. Modell. Dyn. Syst.*, **17** (2), 123–143.
13. Baur, U., Benner, P., Greiner, A., Korvink, J.G., Lienemann, J., and Moosmann, C. (2011) *Math. Comput. Modell. Dyn. Syst.*, **17** (4), 297–317.
14. Rochus, V. (2006) Finite element modelling of strong electro-mechanical coupling in MEMS. PhD thesis. Universitè de Liège.
15. Phillips, J.R. (2003) *IEEE Trans. Comput. Aided Design*, **22**, 171–187.
16. Rewienski, M. and White, J. (2003) *IEEE Trans. Comput. Aided Design*, **22**, 155–170.
17. Gad-el-Hak, M. (1999) *J. Fluid Eng.*, **121** (5), 5–33.
18. Sagi, H., Zhao, Y., and Wereley, S.T. (2004) *J. Vac. Sci. Technol.*, **22** (5), 1992–1999. doi: 10.1116/1.1776181.
19. Gyimesi, M. and Ostergaard, D. (1999) Electro-mechanical capacitor element for MEMS analysis in ANSYS. Proceedings MSM, Puerto Rico, pp. 270–273.
20. Avdeev, I. (2003) New formulation for finite element modeling electrostatically driven microelectromechanical systems. PhD thesis. University of Pittsburg, USA.
21. Rochus, V., Kerschen, G., and Golinval, J.C. (2005) Dynamic analysis of the nonlinear behavior of capacitive MEMS using the finite element formulation. Proceedings ASME IDETC.
22. Hagleitner, C., Bonaccio, T., Rothuizen, H., Wiesmann, D., Lienemann, J., Korvink, J.G., Cherubini, G., and Eleftheriou, E. (2006) Modeling, design, and verification for the analog front-end of a MEMS-based parallel scanning-probe storage device, in *Proceedings IEEE CICC*, IEEE, San Jose, CA.
23. Bai, Z.J. and Su, Y. (2005) *SIAM J. Matrix Anal. A*, **26** (3), 640–659. in Advances in Design and Control, SIAM.
24. Beattie, C.A. and Gugercin, S. (2005) Krylov-based model reduction of second-order systems with proportional damping. Proceedings 44th CDC/ECC, p. 2278–2283. Seville, Spain.
25. Rudnyi, E.B., Lienemann, J., Greiner, A., and Korvink, J.G. (2004) mor4ansys: generating compact models directly from ANSYS models. Proceedings Nanotech, vol. 2, pp. 279–282. Bosten, MA.
26. Su, T.J., Craig, J., and Roy, R. (1991) *J. Guidance*, **14** (2), 260–267.
27. Häggblad, B. and Eriksson, L. (1993) *Comput. Struct.*, **47**, 4/5.
28. Shi, G., Hu, B., and Shi, C.J.R. (2006) *IEEE Trans. Comput. Aided Design*, **25** (7), 1257–1272.
29. Moore, B.C. (1981) *IEEE Trans. Automat. Contr.*, **AC-26** (1), 17–32.
30. Willcox, K.E. and Peraire, J. (2002) *AIAA J.*, **40** (11), 2323.
31. Yousefi, A. and Lohmann, B. (2004) Balancing & optimization for order reduction of nonlinear systems. Proceedings American Control Conference, vol. 1, Boston, MA, pp. 108–112.
32. Kerschen, G., Golinval, J.C., and Worden, K. (2001) *J. Sound Vibrat.*, **244** (4), 597–613.
33. Phillips, J.R. (2000) Projection frameworks for model reduction of weakly nonlinear systems. Proceedings IEEE/ACM DAC, 2, pp. 184–189.
34. Chen, J., Kang, S.M.S., Zou, J., Liu, C., and Schutt-Ainé, J.E. (2004) *J. Microelectromech. Syst.*, **13** (3), 441–451.
35. Li, P. and Pileggi, L.T. (2003) NORM: compact model order reduction of weakly nonlinear systems. Proceedings IEEE/ACM DAC, San Diego, CA, pp. 472–477.
36. Troch, I., Müller, P.C., and Fasol, K.H. (1992) Modellreduktion für Simulation und Reglerentwurf. at – Automatisierungstechnik 40(2/3/4), 45–53/93–99/132–141.
37. Gunupudi, P.K. and Nakhla, M.S. (2001) *IEEE Trans. Adv. Pack.*, **24** (3), 317–325.
38. Liang, Y.C., Lin, W.Z., Lee, H.P., Lim, S.P., Lee, K.H., and Feng, D.P. (2001) *J. Micromech. Microeng.*, **11** (3), 226–233.

39. Tay, F.E.H., Ongkodjojo, A., and Liang, Y.C. (2001) *Microsyst. Technol.*, **7** (3), 120–136.
40. Gabbay, L.D., Mehner, J.E., and Senturia, S.D. (2000) *J. Microelectromech. Syst.*, **9** (2), 262–269.
41. Chen, Y. and White, J. (2000) A quadratic method for nonlinear model order reduction. Proceedings MSM, San Diego, CA, pp. 477–480.
42. Bai, Z.J. (2002) *Appl. Numer. Math.*, **43**, 9–44.
43. Antoulas, A.C. and Sorensen, D.C. (2001) Approximation of large-scale dynamical systems: An overview, Technical report, Rice University.
44. Kryder, M.H. (2005) Magnetic recording beyond the superparamagnetic limit. Proceedings INTERMAG, p. 575. doi: 10.1109/INTMAG.2000.872350.
45. Vettiger, P., Despont, M., Drechsler, U., Dürig, U., Häberle, W., Lutwyche, M.I., Rothuizen, H.E., Stutz, R., Widmer, R., and Binnig, G.K. (2000) *IBM J. Res. Dev.*, **44** (3), 323–340.
46. Pozidis, H., Häberle, W., Wiesmann, D., Drechsler, U., Despont, M., Albrecht, T.R., and Eleftheriou, E. (2004) *IEEE Trans. Magn.*, **40** (4), 2531–2536. doi: 10.1109/TMAG.2004.830470.
47. Hagleitner, C., Bonaccio, T., Rothuizen, H., Lienemann, J., Wiesmann, D., Cherubini, G., Korvink, J., and Eleftheriou, E. (2007) *IEEE J. Solid-State Circ.*, **42** (8), 1779–1789.
48. Rübenkönig, O. and Korvink, J.G. IMTEK Mathematica Supplement (IMS), http://portal.uni-freiburg.de/imteksimulation/downloads/ims.
49. Iverson, B. and Garimella, S. (2008) *Microfluid. Nanofluid.*, **5**, 145–174.
50. Zhang, C., Xing, D., and Li, Y. (2007) *Biotechnol. Adv.*, **25**, 483–514.
51. Nisar, A., Afzulpurkar, N., Mahaisavariya, B., and Tuantranont, A. (2008) *Sens. Actuators, B*, **5**, 145–174.
52. Garimella, S., Singhal, V., and Liu, D. (2006) On-chip thermal management with microchannel heat sinks and integrated micropumps. Proceedings of the IEEE, vol. 94, pp. 1534–1548.
53. Zengerle, R., Richter, A., and Sandmaier, H. (1992) A micro membrane pump with electrostatic actuation. Proceedings of the IEEE Micro Electro Mechanical Systems (MEMS) 92, Travemünde, Germany, pp. 19–24.
54. Bertarelli, E., Ardito, R., and Corigliano, A. (2011) Electrostatic diaphragm micropump electro-fluid-mechanical simulation. Proceedings Coupled Problems 2011, Rhodes, Greece.
55. Bertarelli, E., Ardito, R., Greiner, A., Korvink, J., and Corigliano, A. (2011) Design issues in electrostatic microplate actuators: device stability and post pull-in behaviour. Proceedings EuroSimE 2011. Lienz.
56. Bertarelli, E., Ardito, A., Corigliano, G., and Contro, R. (2011) *Int. J. Appl. Mech.*, **3**, 1–19.
57. Bertarelli, E., Corigliano, A., Greiner, A., and Korvink, J. (2011) *Microsyst. Technol.*, **17**, 165–173.
58. Timoshenko, S. and Woinowsky-Krieger, S. (1959) *Theory of Plates and Shells*, 2nd edn, McGraw-Hill International Editions.
59. Pelesko, J.A. and Chen, X.Y. (2003) *J. Electrostat.*, **57**, 1–12.
60. Batra, R.C., Spinello, D., and Porfiri, M. (2007) *Advances in Multiphysics Simulation and Experimental Testing of MEMS*, chap. Pull-in instability in electrostatically actuated MEMS due to Coulomb and Casimir forces, Imperial College Press.
61. Leissa, A.W. (1959) *Vibration of Plates*, National Aeronautics and Space Administration (NASA), Washington, DC.
62. Chakraverty, S. (2009) *Vibration of Plates*, CRC Press, Taylor and Francis Group, London.
63. De Silva, C. (2007) *Vibration Damping, Control, and Design*, CRC Press, Taylor and Francis Group, London.

12
Modal-Superposition-Based Nonlinear Model Order Reduction for MEMS Gyroscopes
Jan Mehner

12.1
Introduction

The goal of the transducer design is to find proper shape elements and structural dimensions of microelectromechanical system (MEMS), which fulfill all requirements at the given operational and environmental conditions. Generally, the design process starts with a conceptual design phase where several layouts have to be evaluated and optimized. A conceptual design is commonly based on lumped element models, which are derived from the fundamental physical background of involved engineering disciplines. Since lumped element models are directly represented by mathematical functions, a design optimization can be realized in the shortest time with reasonable accuracy. Results are the preferred layout with preliminary structural dimensions and related physical parameters.

Detailed investigations on the preferred layout follow in a subsequent component design phase. Component design is commonly based on finite element method or boundary element method (FEM/BEM), which capture the physical behavior with higher accuracy. There are two major advantages: component models based on FEM/BEM are not restricted to rigid body approximations (they capture flexible bending of masses and anchor regions inherently), and they allow for structural details where no acceptable analytical solution exists (electrostatic fringing fields at comb fingers, viscose damping of perforated plates). In practice, models of the preferred layout are greatly improved in accuracy and enhanced by new features. Beyond the basic transducer behavior, designers can evaluate vital performance parameters of sensors as resolution, linearity, bandwidth, cross-talk, and reliability issues before the manufacturing of prototypes. Unfortunately, FEM/BEM simulations are usually time consuming.

Model order reduction (MOR) procedures are primarily focused on reducing the computing time of the component models. As a matter of fact, thousands of Monte Carlo simulations have to be performed before the release of MEMS products, in order to assess the influence of tolerances, process variations, related materials property scattering, and changed environmental conditions on the sensor performance and the yield of manufacturing. Simulation runs often require

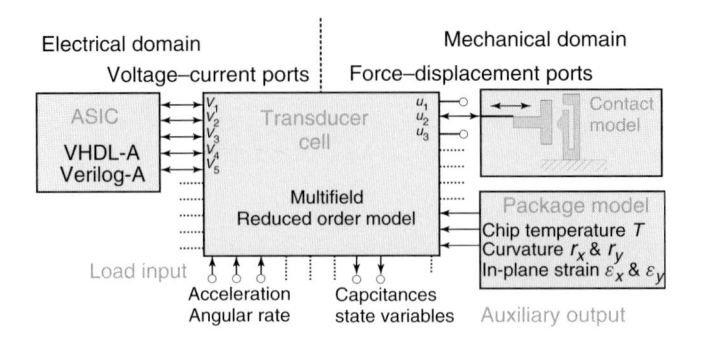

Figure 12.1 Schematic view of a reduced order model in a system simulation environment.

hundreds or thousands of time steps in a transient analysis to evaluate the settling time of sensors and their interactions to the electronic circuit or controller unit. Acceptable simulation costs are milliseconds per time step, which can hardly be achieved by ordinary FEM/BEM models. In addition to efficiency, MOR technologies allow designers to translate the governing equations into several design languages such as Matlab/Simulink, VHDL-A, or Verilog-A, which is needed to link the transducer cells to electronic or controller units.

The ultimate goal of MOR of MEMS is to provide an automated procedure to extract fast dynamic transducer models that can be directly inserted into a system design environment with the speed known from the lumped element models and the accuracy known from the FEM/BEM representations [1]. In general, reduced order transducer models should be able of being linked to electronic circuit models via voltage–current ports, to mechanical domain submodels via force–displacement ports (e.g., stopper with adhesion effects), and to packaging models to capture the influence of temperature, stress, and stress gradients on the performance. Typical input load data are acceleration vectors and angular rates for inertial sensors. Auxiliary output data are capacitances or state variables, which are helpful for debugging purposes (Figure 12.1).

12.2
Model Order Reduction via Modal Superposition

A common approach of MOR in computational mechanics is known as modal superposition (MSUP). The key idea of MSUP is to represent the mechanical deformation state at any load situation by a superposition of the eigenvectors or mode shapes, which correspond to the smallest eigenvalues of the undamped system. Instead of solving large FEM/BEM matrices with thousands of degrees of freedom (DOF) in dynamic simulations, MSUP represents the same model by a few ten modal DOF, which are referred to as *modal coordinates* or *weight factors of eigenvectors* with almost the same accuracy. MSUP technologies are state of the

12.3 MEMS Testcase: Vibratory Gyroscope

Advantages of MSUP technologies in modeling and simulation of MEMS are demonstrated on a vibratory gyroscope shown in Figure 12.2. The right and left sensor cells consist of two movable frames, which are driven in an antiphase vibration mode in x-direction (drive mode). The angular rate axis is out-of-plane in z-direction. Consequently, Coriolis forces act perpendicularly whereby motion in y-direction (sense direction) is detected capacitively on the inner frames. For the conceptual design where basic physical parameters of gyroscopes are specified, the mechanical behavior for the right cell can be defined by:

$$\begin{bmatrix} M_x & 0 \\ 0 & M_y \end{bmatrix} \begin{bmatrix} \ddot{x} \\ \ddot{y} \end{bmatrix} + \begin{bmatrix} D_x & 0 \\ 0 & D_y \end{bmatrix} \begin{bmatrix} \dot{x} \\ \dot{y} \end{bmatrix} + \begin{bmatrix} K_x & 0 \\ 0 & K_y \end{bmatrix} \begin{bmatrix} x \\ y \end{bmatrix}$$
$$= \begin{bmatrix} F_x \\ 0 \end{bmatrix} + \begin{bmatrix} +2M_y\Omega_z\dot{y} \\ -2M_y\Omega_z\dot{x} \end{bmatrix} \quad (12.1)$$

where M_x and M_y are effective masses in x- (inner and outer frame) and y-direction (only inner frame), D_x and D_y related damping and K_x and K_y stiffness coefficients, F_x electrostatic forces for actuation, and Ω_z the angular rate to be measured. The state vector components (x, y) represent mechanical displacements according to the specified coordinate system. Equation (12.1) represents the ideal case (perfect symmetric shape elements without any misalignment) and allows designers to evaluate basic performance characteristics of the gyroscope as the signal output or bandwidth.

In practice, imperfections as dimensional tolerances, electrostatic fringing fields, fluidic coupling, and mechanical stress cause disturbing interactions among motion DOF. Most important are interactions between drive and sense motion, which can be captured by nonzero off-diagonal terms in matrices of Eq. (12.1). Moreover, asymmetric etch profiles of springs and electrostatic levitation forces on comb

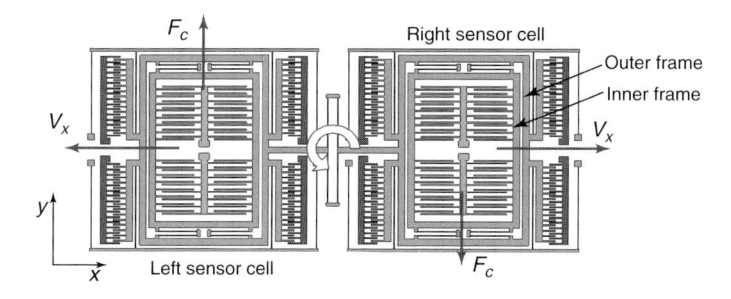

Figure 12.2 Simplified model of a vibratory tuning fork gyroscope for z-axis rate detection.

cells cause out-of-plane motion and tilt what likewise affects the performance of gyroscopes. For detailed investigations, all moving bodies should be modeled with 6 DOF (rigid body assumption) in order to correctly quantify cross-talk among structural components.

The true flexible nature of all structural components can only be captured by full 3D FEM/BEM simulations. Flexible bending of masses and anchor regions contribute to the deformation state and high-order warping modes of masses or springs might be in the operating range of sensors. Fortunately, the accuracy of FEM/BEM models can widely be maintained by MOR based on MSUP methods. In the simplest case, a two mode model can be established, which represents the drive and sense mode only. In contrast to rigid body representations, the drive and sense mode are not just acting in one direction. Mode shapes capture mechanical coupling of x- and y-motion as well as out-of-plane and tilt motion inherently if imperfections have been defined in the original FEM/BEM model. Imperfections are mapped to the eigenvectors and appear in the load terms on the right hand side of the governing equation system. Of course, further modes are usually added to the reduced order model to improve the accuracy, but their contribution is small since gyroscopes operate closely to the resonance (high Q-system). Even for nonresonant inertial sensor products (e.g., three-axes accelerometer), a small number of mode shapes is sufficient to capture the static and dynamic response correctly.

12.4
Flow Chart of the Nonlinear Model Order Reduction Procedure

Reduced order models are directly derived from a FEM/BEM representation with capabilities, potential limitations, and accuracy as specified in the user-defined FEM/BEM model.

The general reduced order modeling procedure consists of three major steps, that is, the generation pass, the use pass, and the expansion pass.

1) The *generation pass* is an automated procedure for MOR. In the first phase, essential physical data are extracted from FEM databases by APDL-scripts (ANSYS Parametric Design Language). In the second phase, all necessary equations for the reduced order model are translated in different languages such as Matlab/Simulink, VHDL-A, or Verilog-A by C^{++} scripts.

2) The *use pass* covers all subsequent simulations of the reduced order model, which are performed with system design tools. The sensor model can directly be linked to electronic components or controller units via voltage–current ports. External load situations as angular rates, acceleration loads, or thermal stress can be applied and superimposed in order to evaluate the performance of gyroscopes. In a system design environment, mechanical displacements can only be observed at specified points of interest referred to as *master nodes*.

3) The *expansion pass* makes a back-transformation of modal coordinates obtained in the use pass to a full order state vector of the finite element (FE) model. The

full order state vector is necessary to monitor the structural response of the entire microstructure in order to create 3D animation sequences or to perform FEM stress calculations at critical load situations for reliability investigations.

Reduced order models represent the physical behavior of a given FE database. Recently, MOR procedures have been extended to capture a set of global design parameters, which directly affect the FE model, referred to as *parametric model order reduction* (PMOR). In cases when the system matrices depend linearly on parameters, the idea of PMOR was explained in Chapter 9 and its references. In cases when system matrices do not depend linearly on parameters, several reduced order models with varying global design parameters are extracted in the generation pass. In the use pass, specific values for global design parameters must be set before the system simulations start. According to assigned parameters, a specific reduced order model is generated by interpolation of all system matrix coefficients derived in the generation pass. The computational efficiency is less affected, since the subsequent interpolation procedure is fast and the derived reduced order model claims the same effort as the nonparametric case. Global design parameters are typically dimensional tolerances (e.g., etch or thickness bias) or packaging interactions such as curvature radiuses and related anchor shift.

After loading the FE model into the ANSYS database, a prestressed modal analysis must be performed in order to compute a proper set of eigenvectors for the projection matrix. Special algorithms assist designers to select a subset of eigenvectors, which map typical load situations best [3]. The next step is to compute modal load vectors for nodal forces on master nodes and body loads as acceleration, angular rates, or other user-defined load cases. It follows a data sampling for physical properties that vary with structural displacements on the full order FE model. Essential properties are capacitances for the electrostatic domain, damping coefficients for the fluid domain, or strain energy values for mechanical stress-stiffening effects.

Capacitances for instance have to be determined at several linear combinations of eigenvectors in order to define multivariable capacitance–stroke functions where stroke must be understood as modal coordinates. Obtained capacitance–stroke functions are necessary to define the voltage–current relationship of conductors and electrostatic forces acting on modes. Finally, the reduced order model is exported into the system design languages of interest (Figure 12.3).

12.5
Theoretical Background of Modal Superposition Technologies

12.5.1
Linear Mechanical Systems

Using MSUP technology for analyzing linear mechanical systems has a long history. MSUP uses the m lowest eigenvectors as shape functions in order to represent the deformation state in a harmonic or transient analysis. The full order

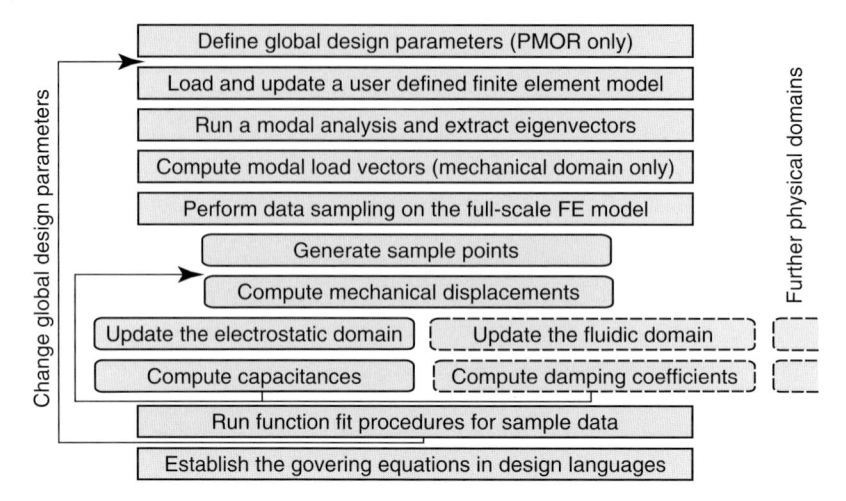

Figure 12.3 Flow chart of the reduced order model generation procedure.

FE matrix representation is transformed to a much smaller set of uncoupled equations, which represent the modal coordinates at given load situations [4]. In the time domain, the governing equations become

$$\begin{bmatrix} 1 & 0 & \cdots & 0 \\ 0 & 1 & \cdots & 0 \\ \vdots & \vdots & \ddots & \vdots \\ 0 & 0 & \cdots & 1 \end{bmatrix} \begin{bmatrix} \ddot{q}_1 \\ \ddot{q}_2 \\ \vdots \\ \ddot{q}_m \end{bmatrix} + \begin{bmatrix} 2\xi_1\omega_1 & 0 & \cdots & 0 \\ 0 & 2\xi_2\omega_2 & \cdots & 0 \\ \vdots & \vdots & \ddots & \vdots \\ 0 & 0 & \cdots & 2\xi_m\omega_m \end{bmatrix} \begin{bmatrix} \dot{q}_1 \\ \dot{q}_2 \\ \vdots \\ \dot{q}_m \end{bmatrix}$$

$$+ \begin{bmatrix} \omega_1^2 & 0 & \cdots & 0 \\ 0 & \omega_2^2 & \cdots & 0 \\ \vdots & \vdots & \ddots & \vdots \\ 0 & 0 & \cdots & \omega_m^2 \end{bmatrix} \begin{bmatrix} q_1 \\ q_2 \\ \vdots \\ q_m \end{bmatrix} = \begin{bmatrix} f_1 \\ f_2 \\ \vdots \\ f_m \end{bmatrix} \qquad (12.2)$$

$$f_k = \phi_k^T F \qquad (12.3)$$

where ω_k are the circular eigenfrequencies, ξ_k the modal damping ratios, f_k modal forces, and q_k the modal coordinates of mode k in the reduced order model. The modal force acting on mode k is computed from the scalar product of the eigenvector ϕ_k and the FE nodal force vector F, which depends on external loads (Eq. (12.3)).

$$u = \sum_{k=1}^{m} \phi_k q_k \qquad (12.4)$$

After solving the problem in a modal subspace, the obtained modal coordinates have to be transformed into true FE displacement vectors u by a summation (superposition) of modal coordinates q_k, which are multiplied with corresponding eigenvectors of the reduced order model (Eq. (12.4)).

All information needed to define the governing equations for MOR based on MSUP can be derived from the results of a FE modal analysis. Essential data are just the lowest eigenfrequencies and appropriate eigenvectors. In particular, nodal forces and structural displacements are just evaluated at the so-called master nodes in the use pass. This limits the number of eigenvector coefficients that actually have to be passed to the MOR databases essentially. Displacements at any other location of the FEM model are just needed for some special design tasks (e.g., stress analysis), which can be calculated in a more time-consuming expansion pass.

It should be noted that the damping matrix of Eq. (12.2) is not directly related to the mechanical domain. In MEMS, damping mainly results from energy dissipation in the surrounding fluid (e.g., air or nitrogen) and can be derived from squeeze and slide film theory [5]. Extraction of damping coefficients from FE fluid flow simulations and the transformation of FEM results into modal damping for MOR is reported in [6]. Inherently, the modal decomposition provides a numerical algorithm that allows the designers to assign damping ratios to each mode separately, which is of great importance for the design optimization. In contrast, the damping matrix of full order FE models is commonly based on the Rayleigh damping approach (alpha–beta damping) where damping can only be specified at two characteristic frequencies.

From the theoretical point of view, eigenvectors are orthogonal to mass and stiffness matrices only. Generally, an arbitrary damping matrix becomes off-diagonal terms after the transformation into a modal subspace. The same effect happens for some load situations that are related to modal velocity such as Coriolis forces at acting angular rates of gyroscopes. Coriolis forces transfer energy from the drive mode to the sense mode, and obviously modes become coupled in the modal representation. However, nonlinear interactions of electromechanical systems likewise create a coupling of modes. Since the total number of equations to be solved is small, the coupling terms affect the computational efficiency insignificantly.

12.5.2
Nonlinear Electromechanical Interactions and Body Loads of Capacitive Sensors

Electromechanical interactions for capacitive MEMS are inherently nonlinear, since electrostatic forces are parabolic functions of voltage and strongly depend on displacements. Linearized models cannot capture important effects as pull-in and hysteresis release of platelike capacitors or electrostatic softening, which is widely used for frequency tuning of vibration sensors, micromirrors, and vibratory gyroscopes [7].

Benefits of MOR based on MSUP become obvious, since the mechanical domain can be represented by a single equation for each involved mode and one equation for each conductor of the electrostatic domain. In total, the number of differential equations can be reduced to several tens, but coupling terms occur because of deflection- or velocity-dependent loads on the right hand side of Eq. (12.5). The mechanical domain is defined by

$$\ddot{q}_k + 2\xi_k\omega_k\dot{q}_k + \omega_k^2 q_k = \sum_r \frac{\partial C_{ij}(q)}{2\partial q_k}\left(V_i - V_j\right)^2 + \sum_s \phi_{k,s} F_s + \sum_t E^b(q,\dot{q}) S_t, \tag{12.5}$$

where the first term of the right hand side defines electrostatic forces acting on modes, the second term the nodal forces acting on master nodes, and the third term the body loads that act on the entire structure such as acceleration and angular rates. The voltage–current relationship at each conductor of the electrostatic domain is defined by

$$I_i = \sum_r \left(\left(\sum_{k=1}^m \frac{\partial C_{ij}(q)}{\partial q_k}\dot{q}_k\right)(V_i - V_j) + C_{ij}(q)\left(\frac{\partial V_i}{\partial t} - \frac{\partial V_j}{\partial t}\right)\right), \tag{12.6}$$

where C_{ij} are capacitance functions obtained from data sampling and function fit. V_i and V_j are appropriate conductor voltages, and I_i is the conductor current. Forces on master nodes F_s are simply multiplied by the eigenvector coefficients $\phi_{k,s}$ of mode k at the FE node s. The parameter E^b is referred to as *body load contribution vector* and defines how much a unit body load (e.g., acceleration, angular rate) stimulates individual modes. In the use pass, time-dependent body loads are multiplied by scale factors S_t, which map the current load situation of example files to the equations of the reduced order model. Body load contribution vectors or matrices might depend on modal displacements or modal velocities. The latter happens for angular rates of gyroscopes.

12.5.3
Parametric Reduced Order Models for Packaging Interactions

Parametric reduced order models are necessary for changed structural dimensions, for nonzero displacement constrains on anchors, and for changed location of fixed conductors of capacitances. Typically, parametric MOR have to be utilized for sensitivity analysis of structural tolerances and simulations of packaging interaction in order to evaluate the influence of substrate warpage on the sensor performance [8, 9].

Figure 12.4 shows a scaled deformation state of a vibratory gyroscope. Fixed comb conductors and anchors follow the bending line of the substrate surface. In

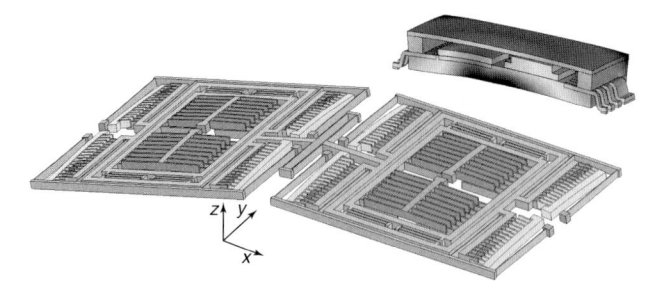

Figure 12.4 Enlarged deformation state of a simplified gyroscope due to packaging warp.

this manner, the outer parts of the fixed comb cells are going down. The movable masses behave differently in x- and y-directions. In x-direction, anchors are largely separated and masses tilt in opposite directions. In y-direction, anchors are close to the center. Hence, the masses rest in horizontal position.

Displacement constrains on anchors cause stress in suspension springs, which directly affects the eigenfrequencies and eigenvectors. Since eigenvectors are altered, three vital modifications of the algorithms become necessary. First, a prestressed modal analysis has to be performed in each outer loop of the MOR generation pass shown in Figure 12.3. Second, the changed position of fixed conductors must be taken into account for capacitance data sampling. Third, the static deformation state u_{eq} caused by an anchor shift is superimposed to the displacements obtained from MSUP. Consequently, the static deformation state has to be added to Eq. (12.4) in order to compute master node displacements and full order FEM results in the expansion pass.

12.6
Specific Algorithms of the Reduced Order Model Generation Pass

12.6.1
Extraction of Body Load Contribution Vectors for Modal Superposition

For MEMS design, typical body loads are accelerations, angular rates, and user-defined load cases as ambient pressure or thermal stress. Body loads appear on the right hand side of Eq. (12.5) whereby each load case is represented by a body load contribution vector (or matrix) and a scale factor that maps the current load situation in the use pass. In the following, body load contribution vectors are extracted at the initial position, which is appropriate for MEMS applications. Exemplarily, modal body loads for reduced order models can be expressed by

$$\begin{bmatrix} f_1^b \\ f_2^b \\ \vdots \\ f_m^b \end{bmatrix} = \begin{bmatrix} E_1^{ax} \\ E_2^{ax} \\ \vdots \\ E_m^{ax} \end{bmatrix} S_{ax} + \cdots + \begin{bmatrix} E_{1,1}^{cz} & E_{1,2}^{cz} & \cdots & E_{1,m}^{cz} \\ E_{2,1}^{cz} & E_{2,2}^{cz} & \cdots & E_{2,m}^{cz} \\ \vdots & \vdots & \ddots & \vdots \\ E_{m,1}^{cz} & E_{m,2}^{cz} & \cdots & E_{m,m}^{cz} \end{bmatrix} \begin{bmatrix} \dot{q}_1 \\ \dot{q}_2 \\ \vdots \\ \dot{q}_m \end{bmatrix} S_{cz}$$

$$+ \begin{bmatrix} E_1^u \\ E_2^u \\ \vdots \\ E_m^u \end{bmatrix} S_u + \cdots, \qquad (12.7)$$

where the first term on the right represents acceleration loads in x-direction and the second term includes Coriolis forces due to angular rates around the z-axis. The last term is an arbitrary user-defined load case.

Generally, modal forces f_k are the inner product of the eigenvector ϕ_k and a related FE force vector F (Eq. (12.3)). Both, eigenvectors and FEM force vectors can either be evaluated on FE nodes or directly by FEs. The latter one is preferred for inertial forces, since mass terms are primarily related to elements rather than to nodes.

For acceleration in x-direction, the element force vector is equal to the element mass vector multiplied by a unit acceleration load (e.g., 1 or 9.81 m s^{-2}). Consequently, the body load contribution factors are evaluated by summing up element data according to

$$E_k^{ax} = \sum_{i=1}^{elem} \phi_{k,i}^x M_i \, a_x \tag{12.8}$$

where $\phi_{k,i}^x$ is the x-component of eigenvector ϕ_k at the FE i, M_i is the element mass, and a_x is a reference acceleration load in x-direction.

Coriolis forces for z-axis angular rates induce FE force vectors in x- and y-direction as below

$$F_x^{cz} = +2M\dot{y}\Omega_z \tag{12.9}$$
$$F_y^{cz} = -2M\dot{x}\Omega_z \tag{12.10}$$

which are the element-wise product of the mass vector M, the element velocity vector, and the unit angular rate Ω_z (e.g., 1 rad s^{-1}). Element velocity vectors have to be expressed by modal velocities and appropriate x- and y-components of all m eigenvectors

$$\dot{x} = \sum_{j=1}^{m} \phi_j^x \dot{q}_j \tag{12.11}$$

$$\dot{y} = \sum_{j=1}^{m} \phi_j^y \dot{q}_j \tag{12.12}$$

Modal forces of z-axis angular rates become the sum of two scalar products where element force vectors in x-direction (Eq. (12.9)) and y-direction (Eq. (12.10)) are multiplied by the eigenvector x- and y-components, respectively. Finally, the body load contribution factors are defined by

$$E_{k,j}^{cz} = \sum_{i=1}^{elem} 2M_i \left(\phi_{k,i}^x \phi_{j,i}^y - \phi_{k,i}^y \phi_{j,i}^x \right), \tag{12.13}$$

which forms a skew-symmetric matrix E^{cz} ($E^{cz}{}_{k,j} = -E^{cz}{}_{j,k}$) with zero main diagonal terms ($E^{cz}{}_{k,k}$). E^{cz} is referred to as *Coriolis or gyroscopic matrix*.

Since Coriolis matrices are multiplied by the modal velocity vector, they are often added to the modal damping matrix on the left hand side of Eq. (12.2). Similar algorithms can be applied for arbitrary user-defined body loads (last term of Eq. (12.7)).

All data needed to extract body load contribution factors are directly derived from FEM databases. Even for large FE models of gyroscopes, the data extraction procedure takes just seconds on a PC, thanks to powerful vector operations provided by commercial FE software tools (e.g., *VGET, *VOPER in ANSYS/Multiphysics™).

The total number of parameters that have to be passed to the MOR database is the order of m or m^2 per load case, where m is the number of considered eigenvectors.

12.6.2
Extraction of Capacitances for Comb Cell Conductors and Platelike Capacitors

The relationship between capacitances and modal coordinates characterizes entirely electromechanical interactions of capacitive sensors and actuators. Multivariable capacitance functions with regard to modal coordinates are obtained from data sampling and function fit procedures. Therefore, capacitance data have to be determined at different linear combinations of eigenvectors, which are defined by modal coordinates.

Primarily, comb cell capacitors of different shape and platelike capacitors with and without perforation holes have to be utilized for MEMS design. While the quasi-homogeneous field between conductors can be captured by basic mathematical terms, a good approximation of fringing fields is rather complicated but vital for accurate behavioral models of MEMS.

Extraction of capacitances for MOR can be performed in the following ways:

1) **Analytical approach**: Overall capacitances of MEMS can be superimposed from several primitives, which are comb fingers or patches of platelike capacitors. Since primitives are simple, analytical capacitance–stroke functions can be derived for a rigid body motion based on the theory of electrostatic fields. In order to compute overall capacitances of MEMS, flexible bending of combs and platelike capacitors must be taken into account. Therefore, the local displacement state of each primitive at given modal coordinates are evaluated analytically and summed up to total capacitances. Unfortunately, analytical models of primitives are often not accurate enough.
2) **Numerical approach**: The capacitance data of the whole structure are directly determined by FEM/BEM simulations. In a first step, the mechanical domain elements are moved to the position of interest by displacement constrains. In a second step, mutual capacitances are extracted from a series of quasi-static field simulations. FEM/BEM simulations are highly accurate but time consuming.
3) **Hybrid approach**: The most promising way of capacitance data extraction is a combination of both, the numerical and analytical approach. Capacitance–stroke functions of primitives are extracted numerically by data sampling and function fit, and the overall capacitances are evaluated from the analytical results of cells. The hybrid approach is widely accepted for MEMS design.

Figure 12.5 illustrates the date sampling procedure of capacitances for comb cells. In x- and y-direction, analytical models agree well to expected results. Difficulties occur mainly for the capacitance relationship in z-direction. Simple analytical models that neglect fringing fields produce a bilinear capacitance–stroke

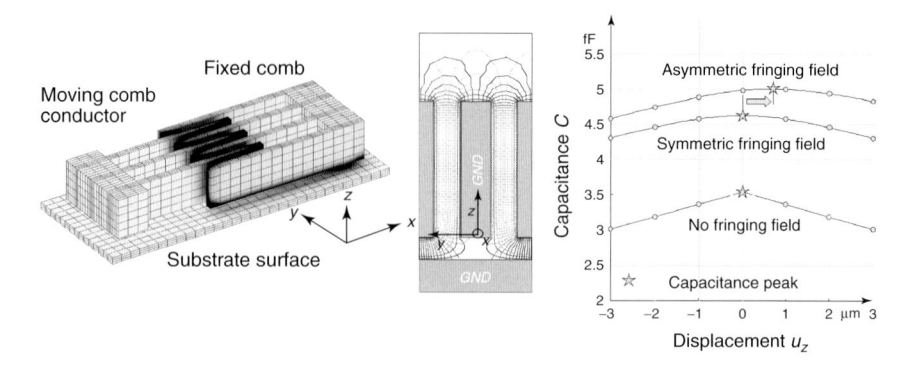

Figure 12.5 Capacitance–stroke function of a comb cell obtained from data sampling and fit.

function with discontinuous first derivatives. Hence, electrostatic forces jump from positive to negative values close to the initial position and transient simulations fail to converge. In contrast, numerical and hybrid models capture symmetric and asymmetric fringing fields at the top and bottom face of the moving comb fingers precisely. Asymmetric fringing fields shift the capacitance peak to the right and induce voltage-dependent z-axis forces, which lift comb cells out of the wafer plane (levitation effect). Quantification of levitation forces is essential to evaluate the resolution of vibrating gyroscopes and other sensor products.

12.6.3
Data Sampling and Function Fit Procedures for Multivariable Capacitances

Data sampling and function fit of multivariable capacitance equations are challenging because an excessive number of samples are required for a satisfying precision. For instance, a capacitance function of m modes that is represented by a Lagrangian polynomial of order n needs $(n+1)^m$ samples, which is unacceptable for practical applications.

Of course, there exist modern algorithms for data regression of large-scale systems in computational mathematics, but one should keep in mind that most electronic design tools neither support complicated mathematical functions nor vector or matrix operations.

Fortunately, there are several physical and engineering aspects that improve the efficiency of data sampling and function fit. Design evaluations of MEMS have shown that just a few modes become significant amplitudes during operation. Several other modes contribute to the deformation state but their modal coordinates are small. Consequently, the first group of modes is referred to as *dominant modes* and the second group as *relevant modes* [10]. Statistical evaluations of polynomial coefficients have shown that relevant modes behave fairly linear, and the interactions among relevant modes are negligible (zero coupling terms). In contrast, the capacitance–stroke relationship of dominant modes must be mapped by a higher polynomial order because of their nonlinear nature. Significant interactions appear among dominant modes and between dominant and relevant modes (nonzero coupling terms).

The above aspects must be considered by defining the number and location of sample points. The operating range of dominant modes is usually sampled with about five to nine points. Generally, relevant modes need exactly three samples in order to compute first and second derivatives of capacitances with regard to modal coordinates properly. For inertial sensor products such as vibrating gyroscopes, there are usually two dominant modes (drive and sense mode). Five to ten relevant modes contribute to highly accurate MOR results. The total number of samples S can be calculated by

$$S = (1 + 2M_R) N_D^{M_D} \tag{12.14}$$

where M_D is the number of dominant modes, M_R the number of relevant modes, and N_D the number of samples for dominant modes. Equation (12.14) results from the fact that all relevant modes must be varied by $\pm \Delta q_i$ at each dominant mode sample. Usually, 50–500 samples are sufficient for most applications, which can be extracted in a few minutes on a PC using the hybrid approach.

The operating principle of capacitive sensors is based on a motion-dependent change of capacitances, which is either realized by a varying electrode gap or a change in the overlapping electrode area. Consequently, the capacitance–stroke functions become nominator and denominator terms, which are commonly represented by rational functions [11]. In contrast to Lagrangian polynomials, rational polynomials (polynomials with rational coefficients) are able to map poles inherently. However, practical use of rational polynomials is complicated because the order of nominator and denominator terms is difficult to specify for multivariable functions. If the order of denominator terms is larger than necessary, unrealistic poles appear in the operating range of sensors (Figure 12.6a). This negative effect is enforced by scattering of capacitance sample data because of a changed FE mesh at samples. For this reason, more stable Lagrangian polynomials are utilized for data

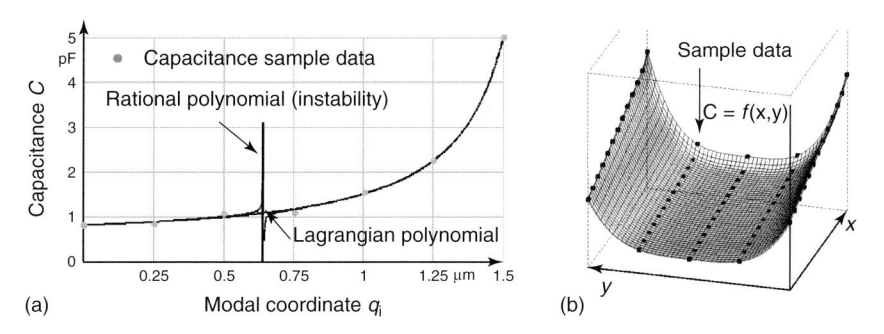

Figure 12.6 Fit functions for platelike capacitors ((a) with instability) and (b) comb capacitors.

regression based on least square fit algorithms. It should be noted that the order of polynomials can easily be changed in the use pass to adapt speed versus accuracy of reduced order models. For instance, high-order polynomials capture the nonlinear capacitance change at large displacements (large signal case), whereby low-order polynomials are sufficient for small displacements at the operating point known from closed-loop sensor applications (small signal case).

12.7
System Simulations of MEMS Based on Modal Superposition

12.7.1
Behavioral Analysis of Vibratory Gyroscopes in Matlab/Simulink

Reduced order models for Matlab/Simulink are based on signal flow graphs. The bidirectional voltage–current relationship known from Kirchhoffian networks must be separated into input- and output-only signals, whereby voltages are commonly considered as input and current as output. Electronic devices as resistors cannot be directly attached to voltage ports, rather feedback loops are utilized to model the voltage drop correctly. For this reason, signal flows are not appropriate for electronic design, whereas they are the preferred environment for controller design and system simulations.

Figure 12.7 shows an example of reduced order simulations of a gyroscope. The reduced order model at the center is primarily based on Eqs. (12.5) and (12.6), which links input data on the left to output data on the right. The governing equations of reduced order models can either be written in a computer language (Matlab, C^{++}, Fortran) based on "s-functions" or be represented by standard Simulink library elements. Both types of compact models have advantages and drawbacks depending on applications. Altogether there are two remarkable features

1) The Matlab/Simulink models can be generated automatically from MOR databases.
2) MOR models are efficient and accurate. Typical simulation costs are less than milliseconds per time step in a transient simulation on a PC.

The goal of system simulations is to investigate and optimize the transducer performance at different load situations. Voltage ports in Figure 12.7 are primarily utilized to assign drive and sense voltages. In combination with a controller unit, additional voltage signals are applied for frequency tuning, force feedback of closed-loop gyroscopes, and quadrature compensation.

The angular rate of gyroscopes has to be assigned at body load ports. Apart from quasi-static load situations, the designers can define chirp signals in order to evaluate the bandwidth or pulse functions to measure the settling time of sensors. Moreover, acceleration loads can be superimposed to angular rates in order to quantify drift effects known as *acceleration sensitivity*.

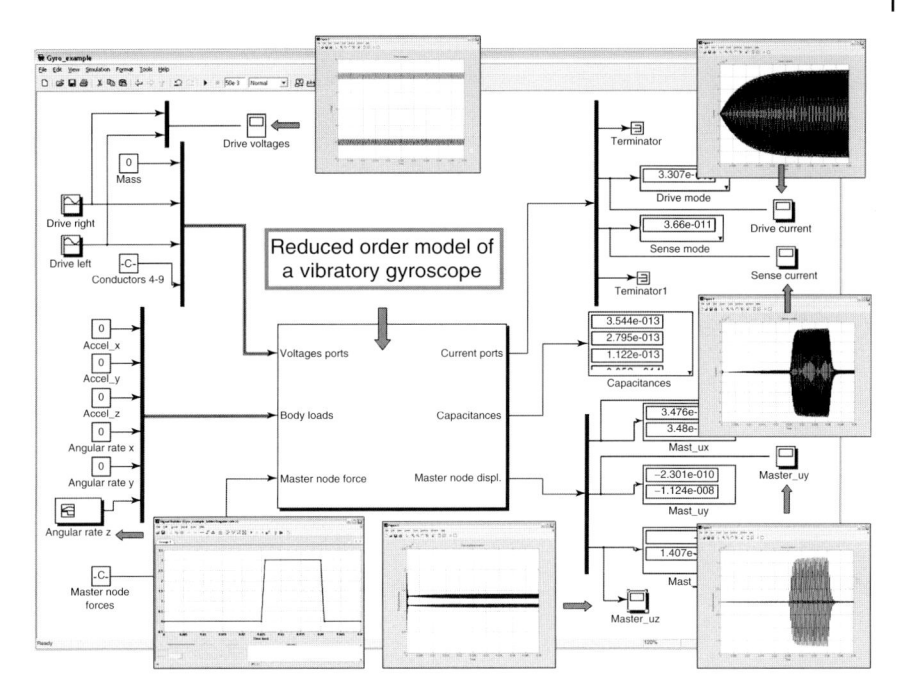

Figure 12.7 System simulations in Matlab/Simulink based on MOR of a gyroscope.

Primary output signals are displacements in all spatial directions on master nodes. Therefore, several master nodes are usually placed on the inner and outer frame in order to monitor drive and sense motion components. It should be noted that MOR models represent not only the functional behavior for the ideal case but also the influence of small disturbances due to imperfections that are mainly asymmetries in suspension springs (etch profile), asymmetries in electrostatic fringing fields, and asymmetric anchor warp due to packaging stress.

In practice, the resolution of gyroscope can be evaluated from the ratio of the sense motion signals with and without angular rates. At zero angular rates, simplified analytical models without imperfections do not show any motion in sense direction. In contrast, reduced order models exactly show mechanical displacements at zero rates that result from imperfections modeled in the original 3D FEM database, and which are of great importance.

Mechanical displacements can be evaluated by optical inspection, but controller units need electrical signals that quantify the drive and sense motion during the operation. Consequently, additional output quantities are capacitances and motion-induced current ports.

MOR simulations provide essential data to quantify drift and cross-talk, but the physical interpretation is often complicated. It requires experienced designers to identify design faults from transient simulation results. On the other hand, reduced order models offer powerful features to support design evaluations, since individual basis functions or specific effects (e.g., electrostatic softening) can be switched OFF or ON to compare results and to trace errors.

12.7.2
Simulations with Reduced Order Models Based on Kirchhoffian Networks

Beside signal flow graphs that are widely used for controller design, reduced order models can be likewise represented by generalized Kirchhoffian networks for electronic design. Typical design languages are VHDL-A or Verilog-A. In Kirchhoffian networks, interface ports act bidirectionally with across and through quantities. Across quantities are voltages and displacements. Through quantities are current and forces in the electrical and mechanical domain, respectively. In addition to bidirectional across–through interfaces, unidirectional signal ports are required to define body loads and to monitor output data as capacitances or modal coordinates. Figure 12.1 illustrates a typical reduced order model of a gyroscope based on a generalized Kirchhoffian network [12].

Equations (12.5) and (12.6) are primarily used to define the relationship between across and through quantities in both the electrical and mechanical domain. Capacitance polynomials are implemented by a set of algebraic equations, which are evaluated in each time step. Contrary to Matlab/Simulink, electronic design tools do support not only transient simulations but also harmonic response analysis (AC-sweep) and quasi-static simulations (DC-sweep).

The most important benefit of Kirchhoffian networks is that electronic devices can directly be attached to voltage ports. In the same way, mechanical domain elements can be linked to master node ports, which provide a force–displacement relationship. Master node ports are typically utilized for stopper modeling in order to limit the displacement range. Advanced contact models of stopper consider bouncing effects, friction, and adhesion forces, which can hardly be implemented in the MSUP representation. The highly nonlinear part of contact elements is modeled outside the MOR element and linked via interface ports. Master node ports can likewise be used to connect several reduced order models together. For example, the first MOR element represents the left sensor cell, and the other MOR element represents the right one of a tuning fork gyroscope model, which turned out to be an efficient way to evaluate asymmetries and dimensional tolerances.

12.7.3
Expansion of System Simulation Results to Full Order FEM Models

The expansion pass is vital to create 3D animation sequences or to visualize mechanical stress at critical load situations. Therefore, the modal coordinates calculated in the MOR use pass have to be stored in a disk. The obtained result file provides all information that is necessary to perform a full order FEM expansion pass. In the expansion pass, transferred modal coordinates are multiplied by the eigenvectors, which are afterwards superimposed to the displacement state at each time step according to Eq. (12.4). The expansion pass performs just simple vector operations in the postprocessor of FEM tools (e.g., *LCDEF, *LCOPER in ANSYS/Multiphysics™). For linear mechanics, there is no time-consuming FE analysis necessary.

Special features of the expansion pass implemented in ANSYS are

1) Support of interpolations among transient time step results to get a larger or a smaller number of frames for optimized animation sequences.
2) Support of FE submodeling technologies (local refinement of the FEM mesh) to investigate stress concentrations of notches at corners of the clamp.

Just as known from linear mechanics, eigenvectors form proper basis functions for special classes of mechanical nonlinearities such as the well-known stress-stiffing effect. Modal stiffness matrices can be approximated with reasonable accuracy by second derivatives of deflection-dependent strain energy functions with regards to modal coordinates. Unfortunately, the stress calculation fails if the eigenvectors are simply scaled and superimposed by modal coordinates.

Stress-stiffening effects mainly occur for bending beams with considerable axial forces or stress. Axial forces are caused unintentionally by packaging stress or by large displacements of double-sided clamped beams. Some applications generate axial forces on purpose for stiffness or frequency tuning during the operation. For the stress-stiffened case, a nonlinear static FE simulation is necessary to compute the stress situation from displacement constrains in the expansion pass. Likewise, displacement constrains are defined from superposition of eigenvectors, but data must only be applied to nodes on the neutral plane of beams and plates. Furthermore, displacement constrains are only applied in the normal (e.g., out-of-plane) direction. Nodal DOF in other directions (axial or in-plane) are free to move [10].

12.8
Conclusion and Outlook

The presented reduced order modeling methods and tools are particularly developed and tested for MEMS applications. The MOR procedures are applicable for coupled domain systems where nonlinear effects can be captured by displacement-dependent physical parameters. Those parameters are capacitances for electromechanical interactions, damping, and squeeze stiffness data for fluid–structural interactions in narrow gaps, or strain energy terms for mechanical stress-stiffening effects. Deflection-dependent parameters can be extracted by sampling and function fit procedures and results are transformed to multivariable equations depending on modal coordinates.

In contrast, load- or time-dependent material nonlinearities as plasticity or viscoelasticity in mechanics, saturation of magnetic materials, and non-Newtonian or turbulent flow in computational fluid dynamics (CFD) have not yet been investigated.

Target applications of reduced order models are mainly dynamic simulations of sensors and actuators in the frequency or time domain. The additional effort of MOR data extraction is rapidly compensated, since a large number of simulation steps can be processed in the shortest time. The generated MOR model can be utilized for a series of different load cases and types of analysis needed for MEMS design.

Reduced order models are exported to various design languages. Hence, MEMS manufacturers are able to provide sensor models to customers at different levels of abstraction. Confidential information about structural dimensions is not visible in a reduced order model, but it captures the physical behavior of the product.

Current research work in the field of reduced order modeling is mainly focused on efficient methodologies to consider global design parameters [13, 14]. As a matter of fact, a small number of design parameters that represent dimensional tolerances, material deviation, or changed packaging interactions can already be passed to the reduced order model. The databases of parametric reduced order models contain a large number of data. However, since the current global design parameters are set before the individual use pass simulations start, an adapted MOR element can be generated by interpolation procedures and exported to system design languages. Therefore, the MOR element requires the same computer resources and simulation time as the nonparametric counterpart.

References

1. Senturia, S.D. (1998) *Proceedings of the IEEE*, **86** (8), 1611–1626.
2. Mehner, J., Gabbay, L.D., and Senturia, S.D. (2000) *Journal of Microelectromechanical Systems*, **9**, 270–278.
3. ANSYS Documentation (2008) Theory Reference, Chapter 15.10, Reduced Order Modeling For Coupled Domains, Canonsburg, PA.
4. Varghese, M. (2002) Reduced-order modeling of MEMS using modal basis functions, PhD Dissertation, Massachusetts Institute of Technology, Cambridge, MA.
5. Veijola, T. (2007) Simple but accurate models for squeeze-film dampers. Proceedings of IEEE Sensors Conference, Atlanta, Georgia, October 28–31, pp. 83–86.
6. Mehner, J., Doetzel, W., Schauwecker, B., and Ostergaard, D. (2003) Reduced order modeling of fluid structural interactions in MEMS based on modal projection techniques. Proceedings of the International Conference on Solid State Sensors, Actuators and Microsystems, Transducers'03, Boston, MA, June 8–12, pp. 465–470.
7. Acar, C. and Shkel, A. (2009) *MEMS Vibratory Gyroscopes, Structural Approaches to Improve Robustness*, Series: MEMS Reference Shelf, Springer Science & Business Media.
8. Mehner, J., Schaporin, A., Kolchuzin, V., Doetzel, W., and Gessner, T. (2005) Parametric model extraction for MEMS based on variational finite element techniques. Proceedings of the International Conference on Solid State Sensors, Actuators and Microsystems, Transducers'05, Seoul, South Korea, June 5–9, 2005, pp. 776–780.
9. Mehner, J., Kolchuzhin, V., Schmadlak, I., Hauck, T., Li, G., Lin, D., and Miller, T.F. (2009) The influence of packaging technologies on the performance of inertial MEMS sensors. Proceedings of the International Conference on Solid State Sensors, Actuators and Microsystems, Transducers'09, Denver, Colorado, June 21–25, pp. 1885–1888.

10. Bennini, F., Mehner, J., and Doetzel, W. (2003) *International Journal of Computational Engineering Science*, **2** (2), 385–388.
11. Lancaster, P. and Salkauskas, K. (1990) *Curve and Surface Fitting*, Academic Press, London.
12. Schlegel, M., Bennini, F., Mehner, J., Herrmann, G., Mueller, D., and Doetzel, W. (2005) *IEEE Sensors Journal*, **5** (5), 1019–1027.
13. Kolchuzhin, V., Doetzel, W., and Mehner, J. (2008) *Sensor Letters*, **6** (1), 97–105.
14. Kolchuzhin, V., Mehner, J., Gessner, T., and Doetzel, W. (2007) Application of higher order derivatives, method to parametric simulation of MEMS. Proceedings of International Conference on Thermal, Mechanical and Multiphysics Simulation and Experiments in Micro/Nanoelectronics and Systems EuroSimE, London, April 16–18, 2007, pp. 588–593.

Part IV
Modeling of Entire Microsystems

System-level Modeling of MEMS, First Edition. Edited by T. Bechtold, G. Schrag, and L. Feng.
© 2013 Wiley-VCH Verlag GmbH & Co. KGaA. Published 2013 by Wiley-VCH Verlag GmbH & Co. KGaA.

13
Towards System-Level Simulation of Energy Harvesting Modules

Dennis Hohlfeld, Tamara Bechtold, Evgenii B. Rudnyi, Bert Op het Veld, and Rob van Schaijk

13.1
Introduction

In this chapter, we present a comprehensive modeling framework for system-level simulation of microsystems. We have chosen an application example involving microstructures and analog circuitry. The microstructure is capable of transforming mechanical to electrical energy through a piezoelectric material. As the mechanical energy is obtained from ambient vibrations, the term *energy harvester* has been coined for such microstructures, which collect otherwise unused forms of energy.

Energy harvesting is the process of converting unused ambient energy, for example, vehicle vibrations, tire acceleration, and motor vibrations in industrial applications, into usable electrical power. Harvesting ambient energy, for example, from mechanical vibrations or temperature gradients, is very attractive for wireless autonomous sensor networks. In addition, it is also very important for systems that do not allow battery replacement or are not able to have wired power coupling, for example, tire pressure sensors. Furthermore, devices have to be relatively small in order to be applicable in autonomous wireless transducer systems. As a consequence, harvesting devices are being investigated with a power density of $100\,\mu\text{W}\,\text{cm}^{-2}$ [1]. Energy harvesting can be performed, for example, through piezoelectric, electrostatic, and electromagnetic principles [2, 3] or by employing thermoelectric properties [4, 5]. In this chapter, a vibrational harvester concept is addressed together with the maximum power point (MPP) principle and an energy storage system (ESS).

Owing to the multiphysic nature a numerical model for the microstructure is essential in the design process. The inherent need for analog circuitry calls for coupled cosimulation of the numerical model together with the circuit model. We consider order reduction indispensable for efficient simulation. Therefore, we combine recently proposed order reduction techniques for the microstructure model with classical lumped element modeling for the circuit behavior. Previously, it has been shown [6] that model order reduction (MOR) is of key importance to

---

System-level Modeling of MEMS, First Edition. Edited by T. Bechtold, G. Schrag, and L. Feng.
© 2013 Wiley-VCH Verlag GmbH & Co. KGaA. Published 2013 by Wiley-VCH Verlag GmbH & Co. KGaA.

solution of complex microelectromechanical system (MEMS) models. It provides excellent accuracy at drastically reduced computational effort. For the first time, MOR has been established in application areas such as design, optimization, and parameter identification [7]. In this contribution, we used reduced models based on Krylov-subspace methods [8] within a simulation environment (ANSYS Simplorer®). The benefit of this approach is demonstrated in the example of a MEMS-based energy harvesting module. This module consists of a piezoelectric MEMS component for conversion of mechanical to electrical energy, a rectification circuit, and an energy storage mechanism.

As these energy harvesting devices shrink in dimensions, while still providing sufficient energy, they will be key enablers for wireless autonomous sensor network nodes. These sensor systems consist of numerous sensor nodes that are all capable of wirelessly communicating with each other, thus forming a network topology. Energy supply to these nodes is of crucial importance. For such a purpose, vibration harvesters are being investigated, which feature a footprint of $1\,cm^2$ and an average power harvesting level of $100\,\mu W$. Furthermore, the interaction between a piezoelectric vibration energy harvester, the power converter, and the ESS is investigated in this contribution. A system-level approach, including mechanical and electrical domains, is pursued for optimum energy transfer between both domains.

The focus of this chapter is on the micropower module consisting of the harvester and the principle of MPP operation. The basic principle of the vibrational energy harvester, experimental results in Section 13.3, and its modeling are given in the Section 13.4; followed by the MPP principle in Section 13.5.

13.1.1
Wireless Autonomous Sensor Nodes

Low power consumption is one of the key demands requested by a wireless autonomous sensor node (also called wireless autonomous transducer solution (WATS)). This demand has motivated the industry and research institutions to work on various advanced energy systems that can efficiently deliver power to demanding applications. Figure 13.1 shows a block diagram of the WATS system

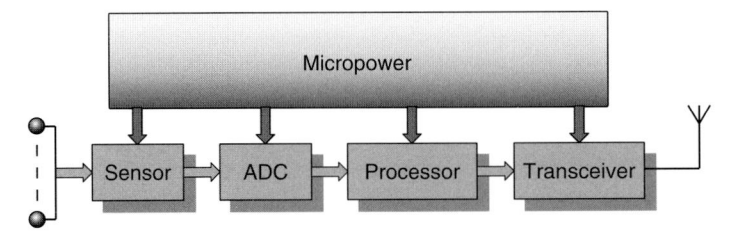

Figure 13.1 Block diagram of the wireless autonomous transducer system. The micropower module (shown in more detail in Figure 13.2) regulates and distributes the harvested power to the functional blocks: sensor, analog to digital converter, processor, and transceiver.

focused on the power distribution where the micropower module deals with rectification and level adaption of the voltage signal generated by the energy harvester.

13.1.2
Micropower Module

Figure 13.2 shows the micropower module already included in the block diagram in Figure 13.1 where the irregular energy of the vibrational harvester is stored in the ESS through an AC/DC converter after which several DC/DC converters are used to power the functional loads (analog to digital converters (ADC), processor, transceiver). Basic tasks for the power management circuit are as follows:

- Supply voltage control for all functional loads.
- Set the vibrational harvester at its MPP.
- Proper charging and protection of the ESS.
- If necessary, determining the State-of-Charge (SoC) of the ESS.

The MPP principle takes care that maximum possible power is extracted from the energy harvester. Power delivery depends, for example, on voltage level.

The functional loads consist of the sensor for amplification and filtering of the incoming signal(s), an ADC to convert the analog sensor signal into a digital one, the processor for processing the digital signal in order to extract relevant data, and finally the transceiver for taking care of the wireless communication.

In a WATS module, the ESS's basic task is to store the irregular energy obtained from the harvester and to cope with the high peak to average current ratio of the load current caused by duty cycling of the transceiver. The transceiver consumes a considerable amount of power during data transmission/reception, so duty cycling is applied to reduce the power to an acceptable level that leads to the large load

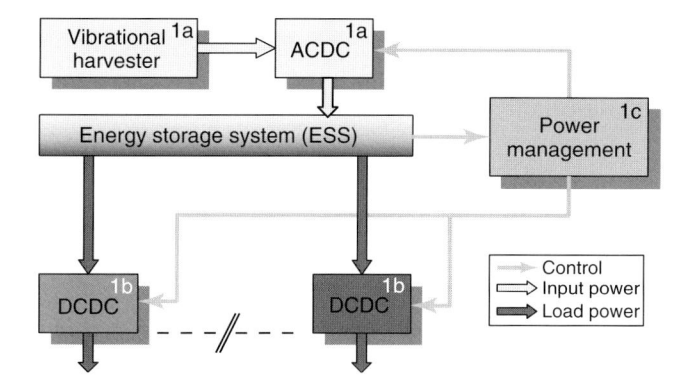

Figure 13.2 The micropower module as included in Figure 13.1. The vibrational harvester and AC/DC converter are part of the input power flow (1a) and the DC/DC converter(s) take(s) care for the output power flow (1b). All converters are controlled by the power management circuit (1c).

current crest factor. In some cases, it may be desirable to place a super-capacitor in parallel to the battery for providing load currents with a high crest factor.

13.1.3
Vibrational Harvesters

At present, energy harvesters are developed as valuable components of WATS [1]. Many industrial and automotive environments provide vibration frequencies ranging from a few tens of hertz up to several kilohertz. Normally, this energy remains unused or is even dissipated in mechanical dampers. Vibrational energy harvesters tap into this energy reservoir. The ambient vibration is used to excite a mechanical resonator consisting of a spring-suspended seismic mass. The amplitude of the mass under resonant excitation is used to drive an energy conversion mechanism. The most common ones are piezoelectric, electrostatic, and electromagnetic principles. The first principle received the most attention because of its ability to directly convert mechanical strain into electrical energy at usable voltage levels. The piezoelectric effect within a material is based on the generation of a dielectric polarization (electrical field) as a consequence to mechanical strain. The electrostatic conversion principle draws power from the work done against an electrical field present in a variable capacitor. This approach is most suitable for fabrication using silicon micromachining. Electromagnetic energy harvesters make use of an arrangement of strong permanent magnets and coils. A relative motion between magnet and coil leads to an electromotive force in the coil. When a load resistor is connected to the coil, one can extract power from the harvester. Despite being a proven energy conversion principle, this approach puts considerable challenges for miniaturization.

Most piezoelectric energy harvesters use a cantilever beam made of a structural material (e.g., steel or silicon) and a piezoelectric stack, which consists of piezoelectric material between two electrodes (Figure 13.3). While one end of this beam is clamped, a mass might be attached to the free end [9, 10]. The bending of the beam section during displacement of the mass causes mechanical strain in the piezopatch, which leads to an electrical polarization, respectively, a charge

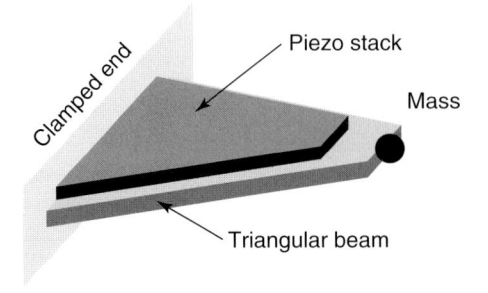

Figure 13.3 Schematic of a piezoelectric energy harvester using the out-of-plane dipole generation.

on the top and bottom electrodes. This charge is used to dissipate power in an attached resistive load. This approach uses the generation of an electrostatic field perpendicular to the applied strain. The beam section shown in Figure 13.3 shows a triangular shape in order to achieve a homogenous strain distribution along the beam length.

13.2
Design and Fabrication of the Piezoelectric Generator

The design of the piezoelectric device is depicted in Figure 13.4. The piezoelectric generator is located on top of the beam and consists of a piezoelectric layer sandwiched between two electrodes. The thickness of the bulk silicon is 675 µm, whereas the beam has a targeted thickness of 25 µm.

The devices are equipped with masses of different dimensions (3×3 to 7×7 mm^2). The designated resonance frequencies, ranging from 300 to 1000 Hz, are achieved by varying beam and mass dimensions.

The MEMS component for energy conversion as depicted in Figure 13.5 is based on a micromechanical resonator, which is composed of a cantilever with an extended mass section at its free end. Cantilever and mass section are fabricated by microstructuring of a silicon substrate. This resonator amplifies small vibrations of certain frequencies at the clamped edge of the cantilever to usable displacements of the mass section.

A parallel plate capacitor structure is fabricated on top of the cantilever by thin film deposition technologies. In this capacitor, a piezoelectric material, namely, aluminum nitride (AlN), is used as a dielectric between two metal electrodes. In resonant motion significant deformation of the cantilever occurs. The strain is also present in the capacitor and leads to an electric polarization of its dielectric material, which, for example, can be observed as a voltage across the capacitor electrodes or a surface charge. This voltage drives a current through a resistor connected to the capacitor, thus dissipating electrical power. In order to store the electrical

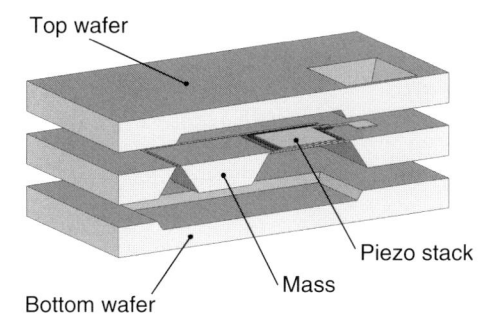

Figure 13.4 Schematic of the fabricated devices. A functional substrate carrying mass and cantilever is encapsulated between two substrates.

energy a storage capacitor is connected to the piezoelectric capacitor using a bridge rectifier. A comprehensive experimental characterization of this configuration is given in [11].

13.3
Experimental Results

Samples have been characterized on a vibration analysis system. The output power of the piezoelectric bender has been measured under various load conditions. For the tested samples, the thickness of the piezoelectric material and the silicon beam were 1000 nm and 45 µm, respectively. The dimensions of the AlN layer and the silicon beam were 1 and 5 mm, respectively. The mass attached to the tip was 35 mg, and the clamped capacitance of the piezoelectric capacitor was equal to 500 pF Section 13.2.

The average power dissipated in the resistive load was measured at the natural mechanical frequency for different acceleration levels of the input vibration. A maximum power output of almost 140 µW was achieved as shown in Figure 13.6. It should be noted that this output power compares well with the output levels of other micromachined vibrational harvesters.

13.4
Modeling and Simulation

In this section, we investigate the interaction between a piezoelectric vibration energy harvester, the power converter, and the ESS. A system-level approach,

Figure 13.5 Vacuum-packaged piezoelectric energy harvester. Maximum power delivery at resonance frequency is 140 µW (at 1.8 g excitation).

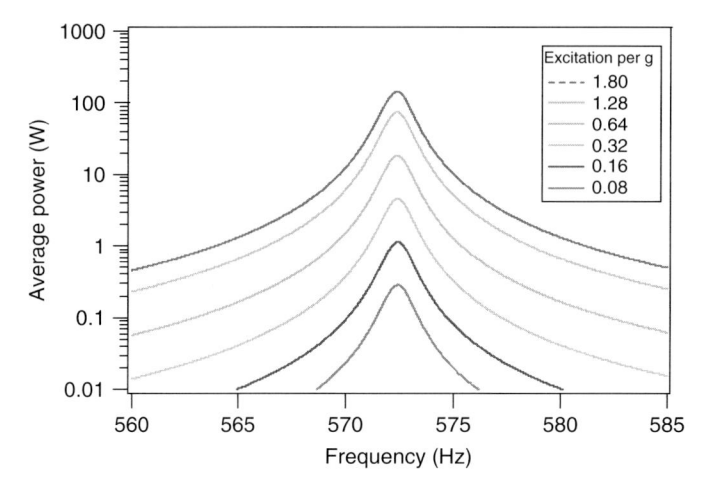

Figure 13.6 Average output power measured as a function of frequency for different acceleration levels [12]. The piezoelectric material is AlN.

including mechanical and electrical domains, is pursued in order to investigate the physical design aspects of the energy harvester on the electrical domain.

13.4.1
Lumped Element Modeling

If the behavior of the mechanical structure is reduced to a one-dimensional oscillator (giving only 1 DOF), we can build an equivalent circuit as a lumped element model (see also Chapter 2 of this book) depicted in Figure 13.7.

Although the above model is useful in evaluation of coupled-domain behavior, its mechanical parameters and transformer ratio cannot easily be derived from the device geometry and physics. Calculation of effective mass and effective stiffness is based on simplifying assumptions that do not hold under all devices dimensions;

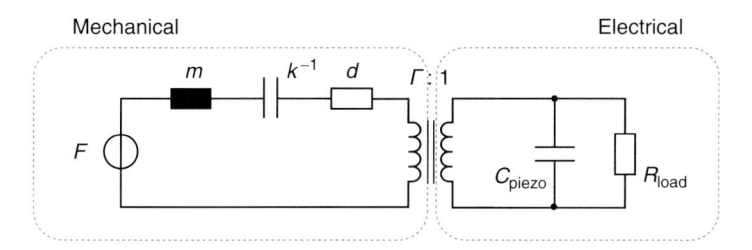

Figure 13.7 Equivalent circuit for a piezoelectric energy harvester. The mechanical resonator is represented through a series RLC circuit. Energy transfer between mechanical and electrical domain is modeled by a transformer with a given coupling ratio.

thus the model does not use physically based lumped elements. Therefore, we use numerical modeling based on the actual device geometry and physical material properties of the employed materials.

13.4.2
Finite Element Modeling

The device geometry has been implemented using ANSYS. The silicon and piezoelectric material's anisotropic properties have been considered. In contrast to the model shown in Figure 13.8, only one quarter was implemented using the device's symmetry.

The metal electrodes of the piezoelectric patch were modeled by constraining the potential of all nodes on the bottom side of the patch to 0 V. The potential of all top side nodes was coupled to a floating value. The finite element method (FEM) model was extended by implementation of a circuit element acting as a resistor. Its two electrical nodes were connected to the electrodes of the piezoelectric patch.

The model was analyzed in frequency domain only at its resonance frequency. A displacement amplitude of 1 µm was imposed at the device's anchor segment. A global damping factor was applied to the silicon material to reflect structural damping. The solution returns the power dissipated in the resistor, the oscillation amplitude of the free mass segment, and the resonance frequency. The load resistance value impacts on these solution parameters.

While changing the load resistance, one can observe the effect of load matching with the energy harvester in Figure 13.9. At a load value of 450 kΩ maximum power is dissipated in the resistor. Here, the resistance is closest to the harvester's internal impedance.

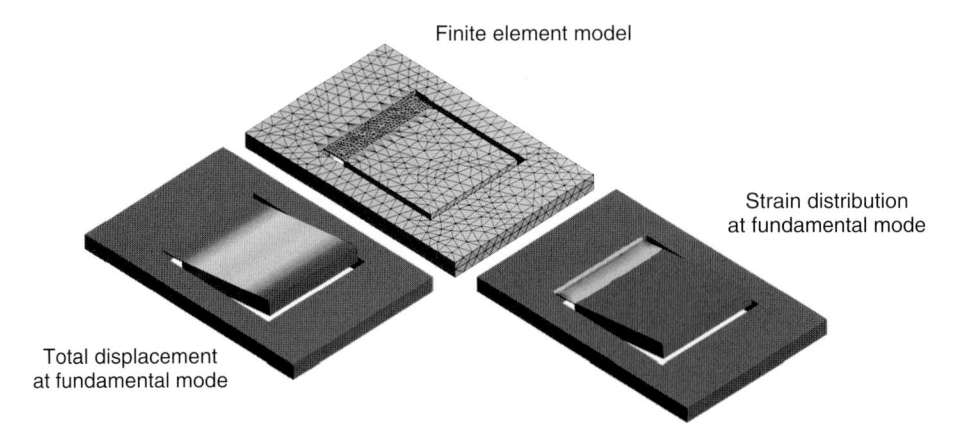

Figure 13.8 FE model with 2133 nodes and fundamental mode shape. The color scale indicates total displacement (left) and mechanical strain (right).

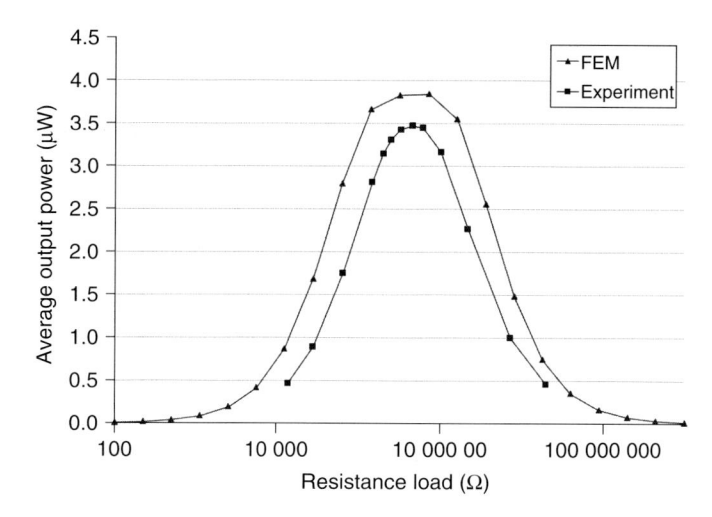

Figure 13.9 Load optimization for the piezoelectric energy harvester. A maximum load values is obtained at 450 kΩ.

The effect of load matching is also reflected in the mass oscillation amplitude as shown in Figure 13.10. Mass motion is not affected by energy extraction in the form of electrical power as long as load resistance is different from its optimum value. We consider small resistance values as "short circuit" and high resistance values as "open circuit" conditions. The varying oscillation amplitude can also be represented by changing resonator quality factors. Unaffected oscillation in

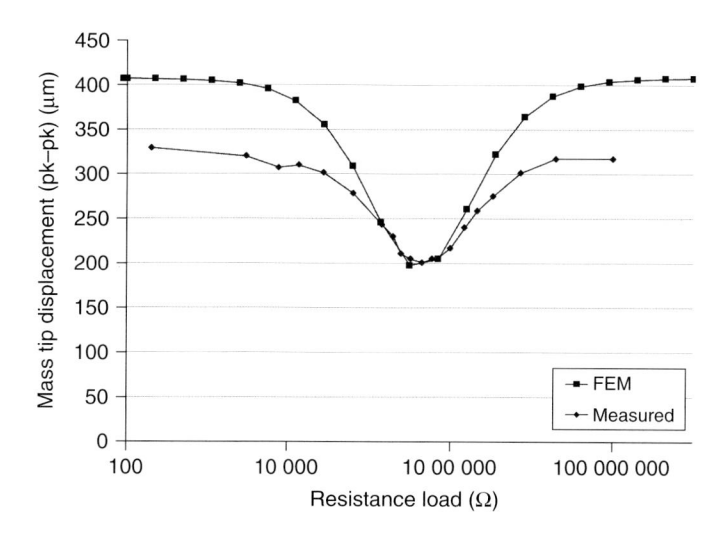

Figure 13.10 Oscillation amplitudes for various load resistance values. Optimum electrical energy happens at a load resistance of 450 kΩ. Here, maximum energy is extracted from the mechanical resonator. This results in reduced oscillation amplitude.

short or open circuit condition is solely determined by the mechanical quality factor. Attaching a resistor with correct resistance values to the harvesting capacitor introduced a further damping mechanism with its quality factor. Both effects superimpose and lead to the dip in oscillation amplitude, demonstrating that energy harvesting effectively extracts mechanical energy from the resonator and provides it to the electrical domain.

A short circuit condition effectively removes the capacitance of the piezoelectric patch. The resonance frequency is therefore a pure mechanical resonance. Under open circuit condition, the electrical capacitance of the patch is connected in series with the mechanical capacitance, leading to a new electromechanical capacitance. As a consequence, a slightly higher resonance frequency exists for this configuration. Transition between these states happens around the optimum resistance value as shown in Figure 13.11.

13.4.3
Model Order Reduction

In order to demonstrate the proposed method, we build a reduced model of the order 28 from a finite element (FE) model of the order 2133 by means of mathematical order reduction (see also Chapter 3 of this book). The FE model encompasses the silicon structure as well as the piezoelectric material on the cantilever. The electrode is modeled by coupling all nodes on the piezoelectric material's top and bottom side, respectively.

We consider now the system response as the mass segment is displaced by 750 μm from its rest position and is released to move freely subsequently. Owing

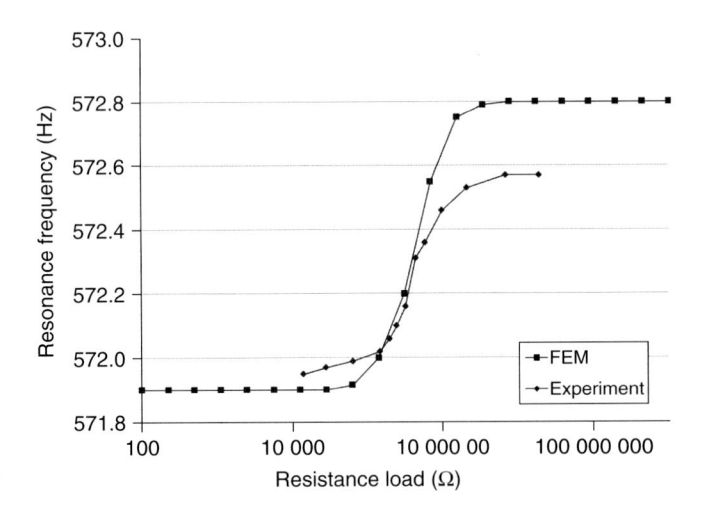

Figure 13.11 Resonance frequencies for various load resistance values. The electromechanical coupling through the piezoelectric element leads to a variation of resonance frequency with changing load resistance.

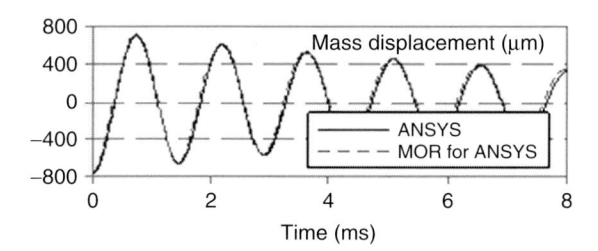

Figure 13.12 Comparison of transient simulation results of full FEM model and order-reduced model.

to the resonant properties and the small structural damping, this excitation results in a decaying oscillation (Figure 13.12). The solution for both models, the original FE models and the reduced order model, are in excellent agreement. The frequency of the oscillation matches the system's eigenfrequency.

13.4.4
Transient MEMS Circuit Cosimulation

In order to demonstrate the applicability of the proposed method to transient cosimulation of FE models and circuitry, we have chosen pulse excitation. Previous works [13, 14] consider harmonic excitation for calculation of expected power output. Realistic application scenarios require consideration of arbitrary excitation signals. We judge the proposed method as highly beneficial for design tasks, which involve coupled simulation of accurate numerical models and nonlinear circuitry.

The system depicted in Figure 13.13 includes the reduced model of the piezoelectric energy harvester. A pulse function acts as an excitation force signal, which is fed to the mechanical port. The electrical port of the reduced model provides two electrical pins giving access to the potentials of the capacitor's electrodes. These are connected to a half-wave rectifier circuit. The output of the rectifier is connected

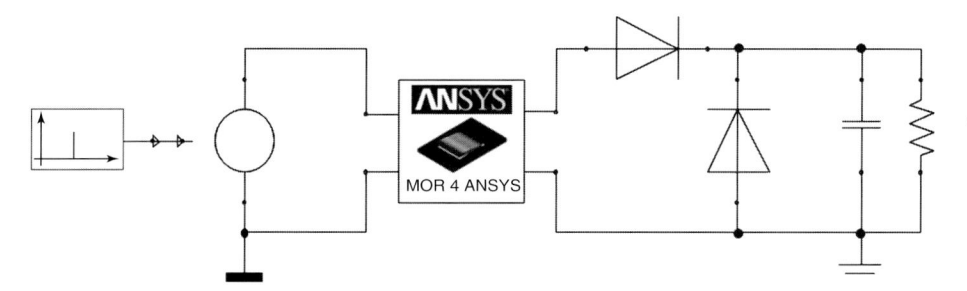

Figure 13.13 System-level model consisting of a pulse input function, the reduced model, and a half-wave rectifier. A parallel connection of a capacitor and a resistor represents a load case. The system-level model considers bidirectional coupling of the FE model and the circuit.

to a capacitor with a resistor in parallel. The capacitor acts as an energy storage element, while the resistor describes the power usage of a consumer (e.g., sensor, wireless transceiver, etc.).

In the beginning of the coupled simulation, the storage capacitor carries no charge. The pulse excitation (imposed as a force) at the mechanical port excites a decaying oscillation of the harvester's seismic mass and output voltage (Figure 13.14). This voltage drives some charge through the forward-biased diodes of the rectifier into the storage capacitor. This process is repeated by every period of the mechanical oscillation. As the collected charge increase in the storage capacitor, its voltage rises as well. A certain fraction of the stored charges continuously drains off the parallel resistor, thus reducing the capacitor voltage.

The harvester voltage surpasses the inevitable voltage drop of the two diodes a few times. Current supply from the harvester to the storage capacitor is limited to these periods (Figure 13.15). Only during these events, energy is transferred

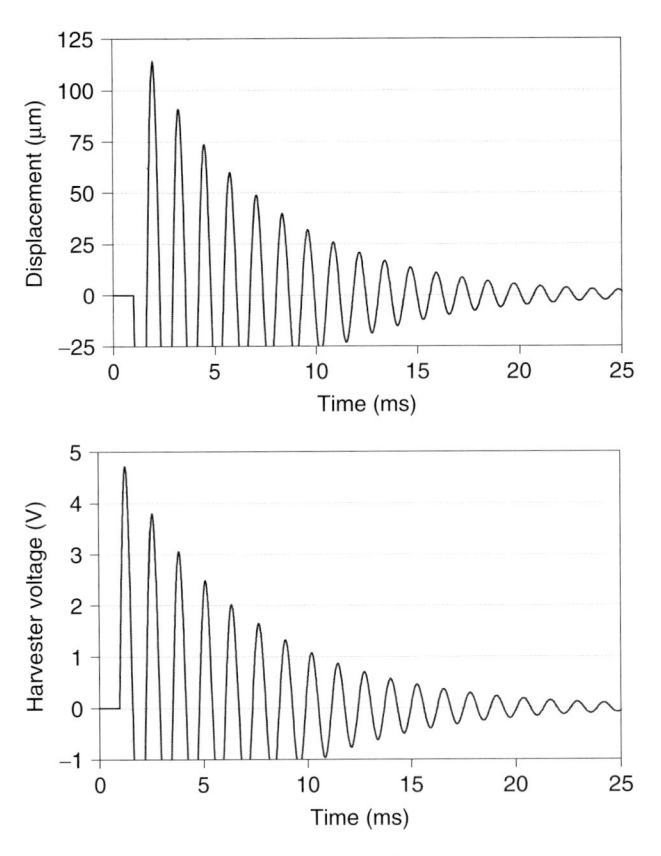

Figure 13.14 Simulation results: The pulse excitation at $t = 1$ ms triggers a decaying oscillation, which can be observed in the oscillation amplitude as well as in the voltage across the piezoelectric patch.

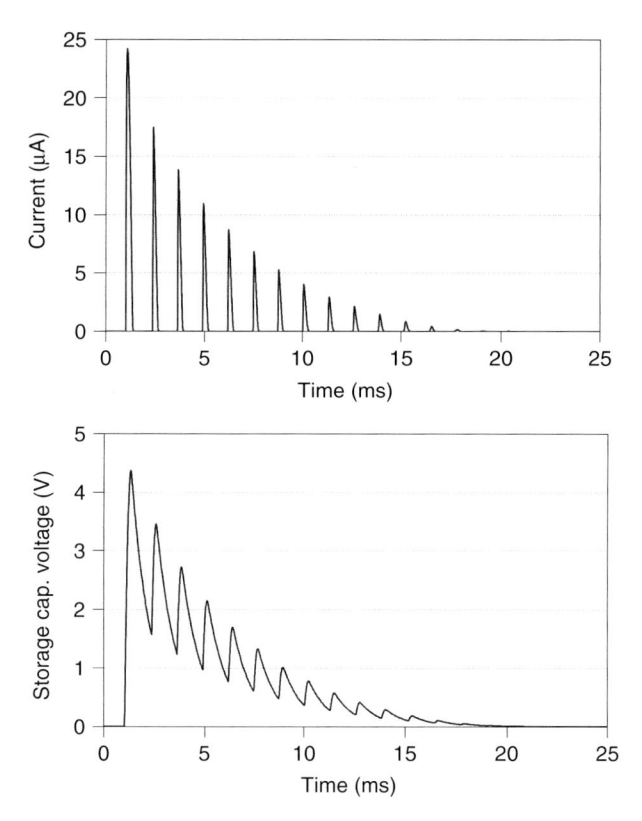

Figure 13.15 Simulation results: The pulse excitation triggers also a time-limited energy transfer to the storage capacitor. The discharge process is determined by the time constant of the parallel connection of the storage capacitor and the load resistor.

from the harvester to the storage capacitor. The subsequent discharge process is governed by the time constant of the parallel connection of the storage capacitor and the load resistor. Repeated pulse excitation would increase the output voltage up to the maximum harvester voltage.

13.4.5
Harmonic MEMS Circuit Cosimulation

In order to demonstrate bidirectional coupling of mechanical and electrical domain within the reduced order model, we complement the transient simulation from above with a harmonic analysis of a system-level model including a resistor as a linear circuit element (Figure 13.16).

The load resistance was used as a parameter and varied from 1 kΩ to 1 MΩ. For load resistance values below 1 kΩ, the output capacitance is nearly short circuited; effectively removing it from the equivalent lumped element model of Figure 13.7. This short circuit configuration results in resonance at the mechanical resonance

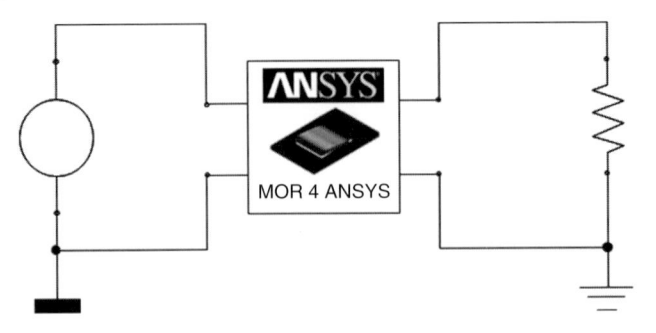

Figure 13.16 System-level model consisting of a harmonic force excitation, the reduced model, and a resistor as a linear circuit element. The system-level model considers bidirectional coupling of the FE model and the circuit.

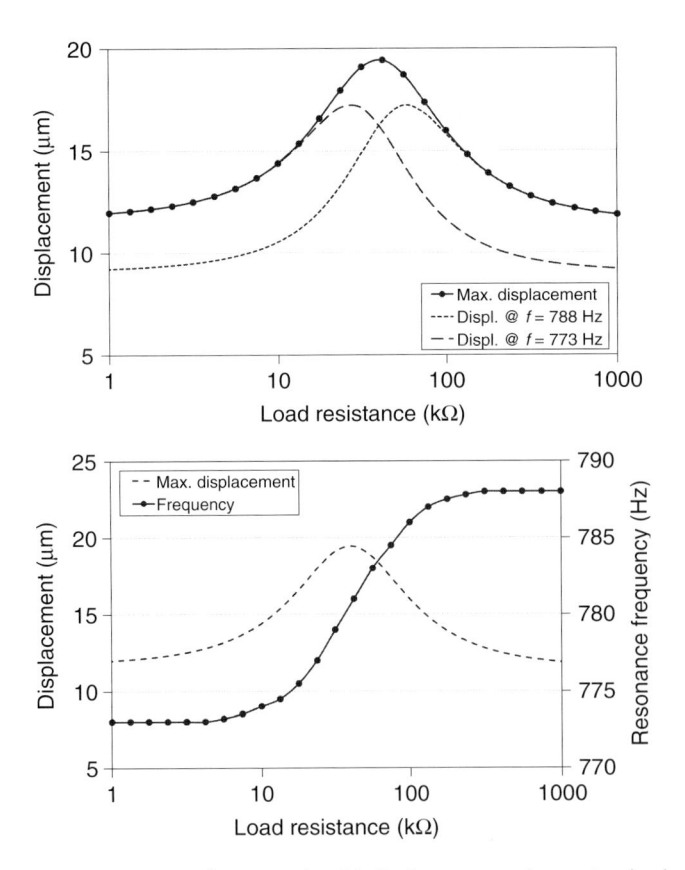

Figure 13.17 Simulation results: (a) displacement under varying load resistance; (b) shift in resonance frequency due to increase in load resistance.

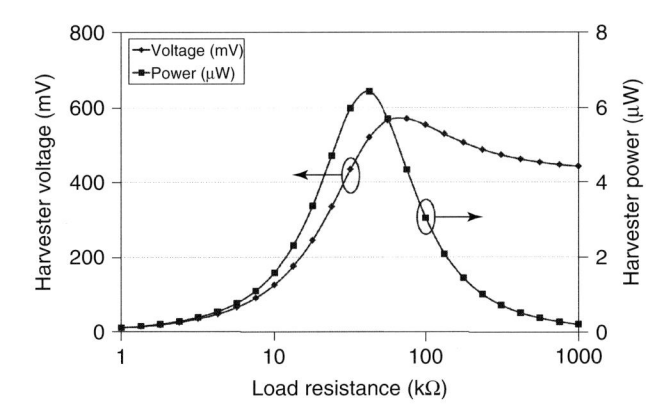

Figure 13.18 Simulation results: harvester voltage characteristic for various load resistance values. Voltages increases toward open circuit configuration. As harvester current decreases with increasing load, maximum power delivery occurs at optimum load resistance value.

frequency. At high load resistance value, termed as *open circuit condition*, the electrical capacitor is in series with the mechanical compliance of the equivalent circuit, thus leading to a reduced effective capacity and consequently to a higher resonance frequency. This situation is called *antiresonance*.

Harmonic analyses were performed over a frequency range covering series and antiresonance of the system, which in the present case are 773 and 788 Hz, respectively. Figure 13.17a shows the impact of the load resistance on the maximum displacement. In addition, displacement at resonance and antiresonance is given. Figure 13.17b demonstrates the transition from resonance to antiresonance situation as load resistance is increased.

Output voltage depends primarily on the strain in the piezoelectric material, which occurs because of the deformation of the beam or displacement of the mass segment. Figure 13.18 shows how the voltage amplitude depends on load resistance. Toward open circuit condition voltage reaches a maximum value. Not shown is the behavior of the harvester current. Similarly, it decreases from its maximum value at short circuit condition down to nearly zero in open circuit configuration. Combining both characteristics, power delivery from the harvester to the resistive load maximizes at a certain load resistance. In this case, this value is 50 kΩ.

13.5
Maximum Power Point for the Piezoelectric Harvester

Below we introduce two methods to obtain the maximum power from the piezoelectric harvester. In the first method (complex matching), a reactive and resistive load element are used, whereas in the second one (real matching), only a resistive load element is used.

13.5.1
Complex Matching

Figure 13.19 shows that the piezoelectric harvester and ESS are connected through an impedance matching network (Z_{MPP}) and AC/DC converter where the AC/DC converter consists of a rectifier and DC/DC converter part. The impedance of the matching network (Z_{MPP}) and input impedance of the AC/DC converter (R_0) form the load for the piezoelectric harvester and should be designed in such a way that the harvester operates at its MPP.

The linearized equivalent electrical circuit of the piezoelectric harvester is given in Figure 13.20, where the radial frequency of the external force on the frame is denoted by ω. Voltage V_S, inductance L_M, capacitance C_M, and resistance R_M represent the mechanical quantities: external force F_{EXT}, mass m, compliance of the beam k^{-1}, and viscous damping d. The intrinsic electrical capacitance of the piezoelectric material is denoted by C_E. The electrical equivalent and mechanical quantities are related by

$$V_S = \frac{F_{EXT}}{\Gamma}, \quad L_M = \frac{m}{\Gamma^2}, \quad C_M = \frac{\Gamma^2}{k}, \text{ and } \quad R_M = \frac{d}{\Gamma^2} \tag{13.1}$$

where Γ is a function of the piezoelectric coefficient and the piezoelectric coupling. It depends on the geometrical and mechanical aspects of the piezoelectric beam.

From circuit analysis, it is known that the maximum power delivery from source to load will occur when the source impedance (Z_S) is equal to the complex

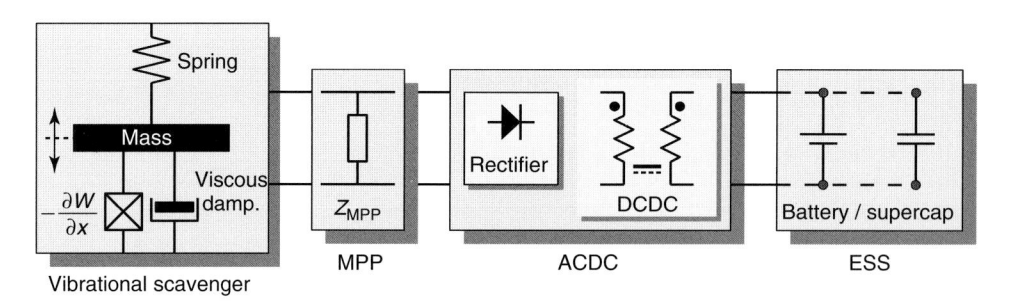

Figure 13.19 Interaction between a piezoelectric harvester, MPP, AC/DC converter, and ESS.

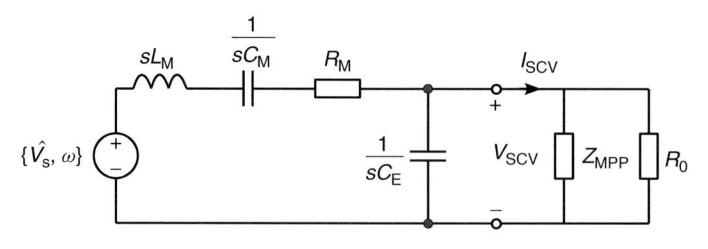

Figure 13.20 Equivalent electrical circuit diagram of the piezoelectric harvester, matching element, and load.

conjugate of the load impedance (Z_L), so $Z_S = \overline{Z_L}$. In the circuit from Figure 13.20, L_M, C_M, and R_M form the source impedance and C_E, Z_{MPP}, and R_0 the load impedance. In this particular case, the matching impedance (Z_{MPP}) will be purely reactive so either an inductor or capacitor can be used for matching. Regarding state-of-the-art techniques, integration of a qualitative good inductor in silicon is not feasible, therefore a capacitance is the most appropriate option for obtaining MPP what means $Z_{MPP} = 1/j\omega C_{MPP}$. The load is now represented by $R_0 \| (1/j\omega C_0)$ with $C_0 = C_E + C_{MPP}$. The source and load impedance in case of a capacitive matching element is given by Eqs. (13.2) and (13.3), respectively

$$Z_S = j\omega L_M + \frac{1}{j\omega C_M} + R_M \tag{13.2}$$

$$Z_L = \frac{R_0}{1 + j\omega R_0 C_0} \text{ with } C_0 = C_E + C_{MPP} \tag{13.3}$$

Substituting Z_S and Z_L in $Z_S = \overline{Z_L}$ and solving for R_0 and C_0 gives

$$R_0 = \frac{R_M^2 + \left(\omega L_M - \frac{1}{\omega C_M}\right)^2}{R_M} \tag{13.4}$$

$$C_0 = \frac{1}{\omega} \times \frac{\omega L_M - \frac{1}{\omega C_M}}{R_M^2 + \left(\omega L_M - \frac{1}{\omega C_M}\right)^2} \tag{13.5}$$

Matching of R_0 is always possible since it corresponds to a positive value. However, matching of C_0 depends on the following condition related to the mechanical quality factor (Q_M) and capacitor ratio k_C

$$Q_M^2 \geq \frac{1}{2 + k_C - 2\sqrt{1 + k_C}} \text{ with } k_C = \frac{C_M}{C_0} \text{ and } Q_M^2 = \frac{L_M}{R_M^2 C_M} \tag{13.6}$$

In practical applications, $k_C \ll 1$ that reduces the above-stated matching condition to

$$Q_M^2 \geq \frac{1}{4k_C^2} \tag{13.7}$$

Substituting Q_M and k_C from Eq. (13.6) in Eq. (13.7) gives

$$\frac{L_M}{R_M^2 C_M} \geq \frac{C_0^2}{4C_M^2} \Rightarrow R_M \leq 2\frac{\sqrt{L_M C_M}}{C_0} \tag{13.8}$$

Back substitution of all original mechanical and electrical harvester quantities (Eq. (13.1)) finally yields to

$$\frac{d}{\Gamma^2} \leq \frac{2}{\omega_M C_0} \text{ with } \omega_M = \sqrt{\frac{k}{m}} \text{ and } C_M \ll C_0 \tag{13.9}$$

where ω_M is the natural radial frequency of the mechanical part formed by the mass (m) and stiffness of the beam (k^{-1}). The harvester dimensions and material properties such as dielectric constant, piezoelectric constants, and overall mechanical coupling determine m, d, k, Γ, and C_E (part of C_0). This means that the geometry of the harvester and the material properties are crucial in order to reach the MPP in case of a capacitive matching element. The maximum power P_{MPP} one can obtain using complex matching is given by

$$P_{MPP} = \frac{\left(\frac{1}{2}V_S\right)^2}{R_M} = \frac{\left(\frac{1}{2\sqrt{(2)}}\hat{V}_S\right)^2}{R_M} = \frac{\hat{V}_S^2}{8R_M} = \frac{\hat{F}_{EXT}^2}{8d} \quad (13.10)$$

13.5.2
Real Matching

Maximum power (P_{MPP}) can only be obtained from the harvester if the impedance matching network (Z_{MPP}) is applied. In case this is not feasible, still the load resistance has to be set to a certain value in order to get a maximized, but suboptimal, power output. It is assumed that the system is excited at the natural radial frequency of the mechanical part ω_M formed by the mass (m) and stiffness of the beam (k^{-1}) (Figure 13.20), which means that the mechanical impedance is only represented by R_M as shown in Figure 13.21 (at ω_M, the inductive and capacitive elements compensate each other).

The power (P_0) in the load resistance (R_0) is now given by

$$P_0 = V_S^2 \frac{R_0}{R_M^2 + 2R_M R_0 + R_0^2 \left(1 + (\omega_M R_M C_E)^2\right)} \quad (13.11)$$

In order to get the maximum suboptimal power, the derivative from P_0 with respect to R_0 should be 0. Solving for R_0 leads to the expression for the load resistance in suboptimal condition $R_{0,sub}$

$$R_{0,sub} = \frac{R_M}{\sqrt{1 + (\omega_M R_M C_E)^2}} \quad (13.12)$$

Substituting R_0 by $R_{0,sub}$ in Eq. (13.11) gives the suboptimal power level $P_{0,sub}$ in case of omitting the matching element and operating at the natural radial frequency

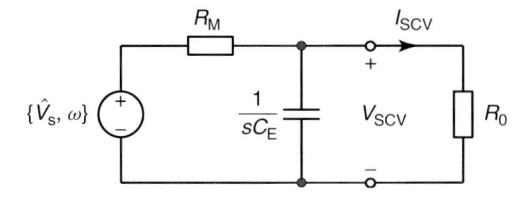

Figure 13.21 Electrical equivalent circuit diagram of the piezoelectric harvester and load, omitting the matching element.

of the mechanical part ω_M of the system

$$P_{0,sub} = \frac{V_S^2}{R_M} \frac{1}{2\left(1+\sqrt{1+(\omega_M R_M C_E)^2}\right)} = \frac{\hat{V}_S^2}{8R_M} \frac{2}{1+\sqrt{1+(\omega_M R_M C_E)^2}} \qquad (13.13)$$

13.5.3
Consequence of Matching

In the two previous sections, it is shown what power levels one achieves from a piezoelectric generator in case of matching with a capacitive element (P_{MPP}) and resistive matching ($P_{0,sub}$). In order to compare both, $P_{0,sub}$ will be expressed as a function of the maximum achievable power level P_{MPP}

$$\frac{P_{0,sub}}{P_{MPP}} = \frac{2}{1+\sqrt{1+(\omega_M R_M C_E)^2}} \qquad (13.14)$$

Substituting k_C and Q_M as defined in Eq. (13.6) with $C_0 = C_E$ gives

$$\frac{P_{0,sub}}{P_{MPP}} = \frac{2}{1+\sqrt{1+\frac{1}{(k_C Q_M)^2}}} \quad \text{with} \quad k_C = \frac{C_M}{C_E} \quad \text{and} \quad Q_M^2 = \frac{L_M}{R_M^2 C_M} \qquad (13.15)$$

Figure 13.22 shows the relation of $P_{0,sub}/P_{MPP}$ as a function of C_M/C_E with parameter Q_M. The mechanical design parameters and their derived electrical equivalents for the piezoelectric generator with lead zirconate titanate (PZT) and

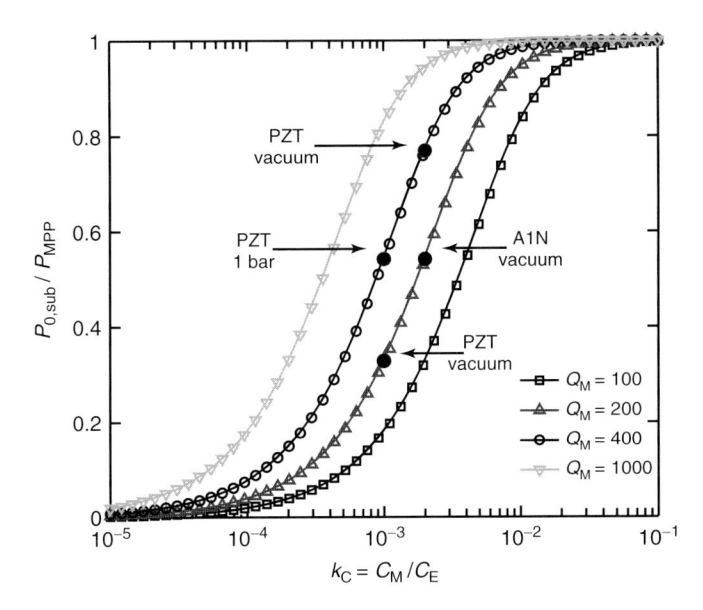

Figure 13.22 The power ratio $P_{MPP}/P_{0,sub}$ as function of the ratio C_M/C_E for different Q_M values.

Table 13.1 Mechanical design parameters and their derived electrical equivalents at 1 bar.

	m (kg)	k (N m^{-1})	D (Ns m^{-1})	Γ (N V^{-1})	C_E (F)	L_M (H)	C_M (F)	R_M (Ω)	Q_M (−)	k_C (−)
PZT	203×10^{-6}	2619	3.6×10^{-3}	1.62×10^{-3}	10^{-6}	77	10^{-9}	1.39×10^3	200	10^{-3}
AlN	203×10^{-6}	2619	3.6×10^{-3}	5.6×10^{-5}	6×10^{-10}	65×10^3	1.2×10^{-12}	1.16×10^6	200	2×10^{-3}

Table 13.2 Achievable power levels at 1g in case of matching and suboptimal matching.

		P_{MPP} (μW)	$P_{0,sub}$ (μW)
1 bar	PZT	136	45
1 bar	AlN	136	74
Vacuum	PZT	272	148
Vacuum	AlN	272	209

aluminum nitride (AlN) as piezoelectric materials are listed in Table 13.1. Ambient pressure of the device is 1 bar. The rather low k_C values for both PZT and AlN, when applied to the required condition of impedance matching with a capacitor (Eq. (13.6)), show that the design is not able to reach the maximum power level P_{MPP}. Figure 13.22 reveals that only 33 and 54% of P_{MPP} can be harvested in case of PZT and AlN, respectively ($Q_M = 200$). Under vacuum conditions, the damping factor d will decrease and hence Q_M can increase by a factor of 2. As a consequence, 54 and 77% of P_{MPP} can be reached where simultaneously the absolute P_{MPP} level is increased with a factor 2 because the power is inversely proportional to d (Eq. (13.10)). Table 13.2 gives the power levels for both matching methods at an acceleration level of 1g for the piezoelectric materials PZT and AlN at a pressure of 1 bar and in vacuum.

13.6
Conclusions and Outlook

Energy harvesters together with power management for wireless sensor nodes applications have been presented. After introducing a state-of-the-art numerical model of a piezoelectric energy harvester together with results obtained under varying load resistance, we proposed a new approach to system-level modeling of complete electromechanical systems including even nonlinear circuitry. In the first case, we have proven that multiphysic modeling is capable of correctly describing the effects of electromechanical coupling. The extension of numerical modeling via order reduction and coupling to electrical circuitry resulted in a highly efficient tool for transient cosimulation.

Harvester dimensions and material properties such as dielectric constant, piezoelectric constants, and overall mechanical coupling determine whether the MPP of a piezoelectric harvester can be obtained without the need of an inductive matching element. A general condition is derived where all basic lumped elements of the piezoelectric harvester are involved (mass, stiffness of the beam, viscous damping, intrinsic electrical capacitance of the piezoelectric material, and the electromechanical coupling).

Future work will cover advanced energy extraction schemes as active rectification and synchronized switching approaches as highly nonlinear circuit topologies. Integration of a higher order battery model will complete the system-level model. Only in the combination of order reduced multiphysic models together with state-of-the-art circuit simulation technology a consistent system design becomes possible.

References

1. Vullers, R.J.M., Leonov, V., Sterken, T., and Schmitz, A. (2006) *OnBoard Technology*, June 34–37.
2. Beeby, S.P., Tudor, M.J., and White, N.M. (2006) *Measurement Science and Technology Journal*, **17**, 175–195.
3. Beeby, S.P., Torah, R.N., Tudor, M.J., Glynne-Jones, P., Donnell, T.O., Saha, C.R., and Roy, S. (2007) *Journal of Micromechanics and Microengineering*, **17**, 1257–1265.
4. Leonov, V., Torfs, T., Fiorini, P., and Van Hoof, C. (2007) *IEEE Sensors Journal*, **7**, 650–657.
5. Torfs, T. et al. (2007) *Sensors and Transducers Journal*, **80**, 1230–1238.
6. Rudnyi, E.B. and Korvink, J.G. (2002) Review: Automatic Model Reduction for Transient Simulation of MEMS-based Devices, Sensors Update v. 11, pp. 3–33.
7. Bechtold, T., Hohlfeld, D., and Rudnyi, E.B. (2010) *Journal of Micromechanics and Microengineering*, **20**, 045030.
8. Rudnyi, Evgenii B., Model Order Reduction, http://modelreduction.com/, (accessed March 2011).
9. Glynne-Jones, P., Beeby, S.P., and White, N.M. (2001) *IEEE Proceedings-Science Measurement and Technology*, **148**, 68–72.
10. Goldschmidtböing, F., Müller, B., and Woias, P. (2007) Optimization of resonant mechanical harvesters in piezopolymer-composite technology. Proceedings of the PowerMEMS 2007, pp. 49–52.
11. Kamel, T.M., Elfrink, R., Renaud, M., Hohlfeld, D., Goedbloed, M., de Nooijer, C., Jambunathan, M., and van Schaijk, R. (2010) *Journal of Micromechanics and Microengineering*, 105023.
12. Elfrink, R., Goedbloed, M., Matova, S., Kamel, T.M., Hohlfeld, D., and van Schaijk, R. (2008) Vibration energy harvesting with AlN piezoelectric devices. Proceedings of the PowerMEMS 2008, pp. 249–252.
13. Zhu, M., Worthington, E., and Njuguna, J. (2009) *IEEE Transactions on Ultrasonics, Ferroelectrics and Frequency Control*, **56**, 1309–1317.
14. Schmitz, A., Sterken, T., Renaud, M., Fiorini, P., Puers, R., and Hoof, C.V. (2005) Piezoelectric scavengers in MEMS technology: fabrication and simulation. Proceedings of the PowerMEMS 2005, pp. 61–64.

14
Application of Reduced Order Models in Circuit-Level Design for RF MEMS Devices

Laura Del Tin, Evgenii B. Rudnyi, and Jan G. Korvink

The use of microelectromechanical system (MEMS) components for radio-frequency (RF) operations represents one of the currently most promising fields of application of MEMS devices. The resonant modes and bistability that characterize the dynamic behavior of mechanical structures can be exploited for the processing of harmonic signals in a wide range of frequency. Mechanical resonance offers an effective tool for frequency generation, filtering, and mixing operations. Bistable structures can be employed to realize switching components, which enrich the capabilities of this technology with regard to signal routing and reconfigurability. The mechanical properties of the devices are coupled with electrostatics, which emerges among the many actuation and sensing principles available at the microscale in terms of design flexibility and, above all, low power consumption. Performances of electromechanical signal processing in terms of high frequency selectivity and stability are promising.

The typical advantages of MEMS devices, intrinsic to their fabrication technology, are particularly beneficial in the RF application field. Small dimensions, monolithic integration with CMOS circuitry, wide choice of materials, batch fabrication, and low dispersion in design properties allow for more functionality per unit area at lower cost with respect to their current off-chip counterparts, such as surface and bulk acoustic wave devices. These factors make RF MEMS attractive building blocks for next-generation wireless transceiver front-end architectures.

One example of present implementation of a two-stage transceiver is shown in Figure 14.1a [1]. Off-chip devices performing frequency selection, generation, and switching operations are marked with gray blocks. The use of MEMS devices would allow not only the replacement of each of this block with an integrated component but also the concept of a completely new architecture by exploiting small area occupation and low power dissipation, with enhanced functionality and reconfigurability. A possible implementation is shown in Figure 14.1b, where gray blocks are used to identify RF MEMS components [2].

The multiple physical domains involved in RF MEMS and the complex electronic circuit in which these devices are integrated render their design untrivial. Many parameters have to be tuned in order to achieve the desired device performance. The development of accurate modeling and simulation tools for the characterization

System-level Modeling of MEMS, First Edition. Edited by T. Bechtold, G. Schrag, and L. Feng.
© 2013 Wiley-VCH Verlag GmbH & Co. KGaA. Published 2013 by Wiley-VCH Verlag GmbH & Co. KGaA.

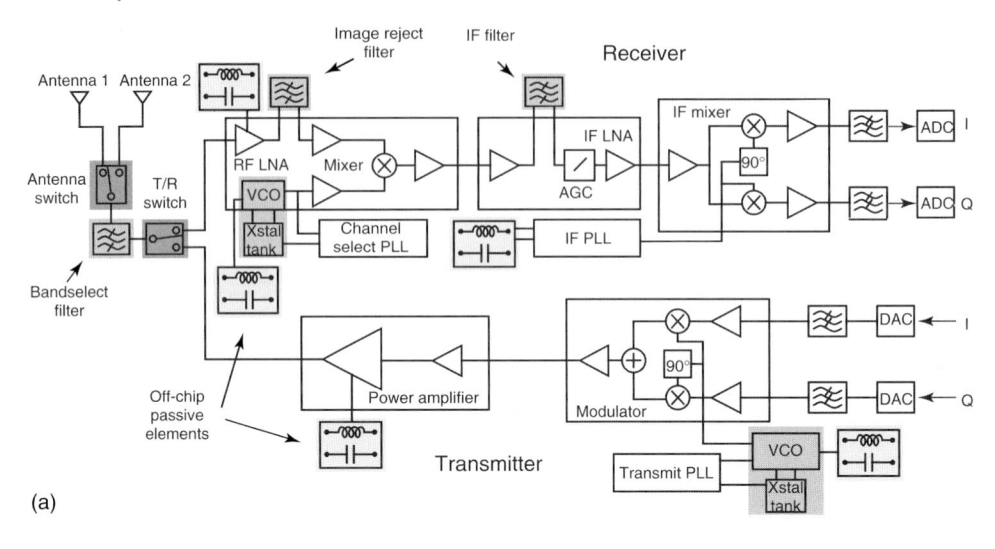

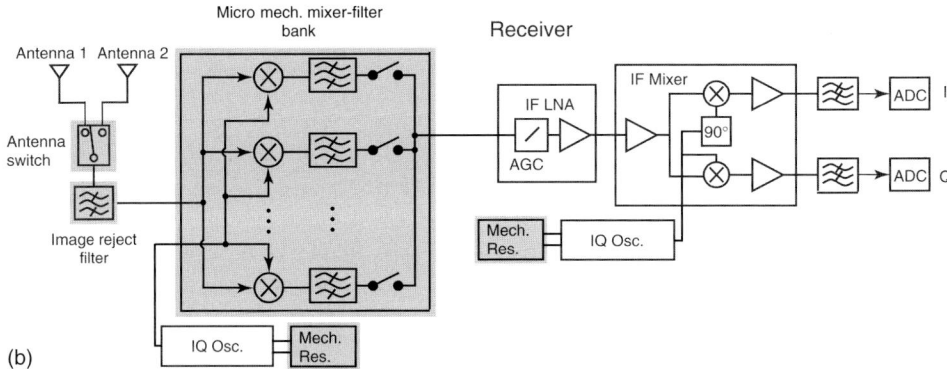

Figure 14.1 Block scheme of a transceiver architecture. A typical dual-conversion heterodyne transceiver front-end is shown in (a). Here, gray blocks represent components off-chips in current implementations (Source: From [1]). In (b) a new implementation, which makes use of RF MEMS devices is presented: gray blocks are realized using micromechanical devices (Source: From [2]).

and the efficient design optimization of these devices is therefore essential. A key point in the creation of device models is the accurate description of the energy domains involved and of their coupling. In addition, the integration of these models into a computer-aided design (CAD) flow is required in order to enable fast device simulation both at device level and at circuit level [3]. Cosimulation of MEMS with digital and analog circuitry is essential for assessing the real overall behavior of the device. This requires models that reproduce the behavior of MEMS devices at their electrical terminals and correctly interface with the state-of-the-art integrated circuit (IC) simulation framework, that is, system level models.

When mathematical modeling is used, the complexity in MEMS behavior translates into large systems of ordinary differential equations (ODEs). Analysis complexity adds to this, leading to long computational times. Switches require for their characterization transient simulation devices with high quality factor; that is, sharp resonance peaks need very fine sweep harmonic simulations. Under these conditions, iterative design optimization and simulation of large and complex systems pose a strict limitation on the maximum model dimension: even with powerful computational resources, a low-order model becomes mandatory for the analysis of networks such as the one in Figure 14.1b, in which several microresonators, switches, and variable capacitors appear, together with RF CMOS circuitry.

In the following sections, we demonstrate the application of moment matching model order reduction (MOR) techniques, which have been described in Chapter 3, to the extraction of system level models for RF MEMS and cosimulation with electrical circuitry. We focus exclusively on RF MEMS active devices, that is, microresonators and switches. The specific properties of these two device classes are taken into account in the extraction and use of the respective reduced models. A short description of the mathematical modeling of RF MEMS is first presented, followed by the MOR procedure. Compact models extraction is then applied to circuit-level simulation. In particular, a microresonator, a microfilter, and a switching device are considered.

14.1
Model Equations for RF MEMS Devices

The spatial discretization of the partial differential equation (PDE) describing the mechanical behavior of a generic microstructure, using the finite element (FE) method, leads to a system of ODE of the type

$$\mathbf{M}\ddot{\mathbf{u}} + \mathbf{E}\dot{\mathbf{u}} + \mathbf{K}_m \mathbf{u} = \mathbf{F} \tag{14.1}$$

where $\mathbf{M}, \mathbf{E}, \mathbf{K}_m \in \mathbb{R}^{n \times n}$ are, respectively, the mass, damping, and mechanical stiffness matrices, $\mathbf{u} \in \mathbb{R}^n$ is the vector of displacement degrees of freedom DOF, and $\mathbf{F}$ is the vector of applied nodal forces. The dimension n of the system for complex 3D structures reaches easily the hundreds of thousands of DOF. The system mass matrix $\mathbf{M}$ is constant and proportional to the density of the material.

In the general case, the mechanical stiffness matrix $\mathbf{K}_m$ is a function of the deformation of the structure and the system in Eq. (14.1) is nonlinear. For the device under consideration, the most important type of mechanical nonlinearity is geometrical nonlinearity because of large strains/rotation or initial stress/loading of the structure. In general, large rotations do not affect the behavior of RF MEMS. Vibrating microstructures are usually subjected to very small deformations, so that the effects of large strains and rotations can be neglected without introducing a significant modeling error [4]. This assumption is often valid also for switches.

However, it cannot be generalized because, depending on the device geometry, the deformations occurring during its operation may be sufficient to give rise to large deformation effects. The device dimensions and maximal displacement should therefore be analyzed, case by case, before adopting a small displacement/strain approximation.

Fabrication-related issues lead often to initial mechanical stress in the microdevices constitutive layers. The state of stress of the structure results in an increase in its stiffness. This effect, known as *stress stiffening*, is important for thin structures in which the bending stiffness is small compared to the axial stiffness, as in typical MEMS devices, and couples in-plane and transverse displacements. Stress stiffening has a remarkable impact on parameters that characterize the behavior of resonators and switches, such as resonance frequencies and pull-in voltage. Therefore, it cannot be neglected. The effect of stress stiffening can be included in the mathematical model in Eq. (14.1), by computing with a nonlinear analysis the so-called stress stiffening matrix $\mathbf{K}_\sigma$, which should be added to $\mathbf{K}_m$ [5]. As the initial stress in the devices is usually much larger than the one encountered during their operation, the stress stiffening matrix computed for the initial stress state can be kept assumed constant in further analysis. We therefore work with a constant mechanical stiffness matrix $\mathbf{K}_m$.

The damping matrix should describe the effect of all energy dissipation phenomena important in RF micromechanical devices. While in MEMS switches squeeze-film damping is generally the main damping mechanism, in microresonators, multiple phenomena can contribute to energy dissipation, such as viscous damping, thermoelastic damping, acoustic radiation, supports losses, and so on, each of them prevailing over the others in a specific area of the design space. It is therefore difficult to derive a mathematical model of damping phenomena, at the same time accurate and valid for a wide range of design parameters. In the following, dissipation is described using the so-called *Rayleigh mode preserving damping*, widely adopted by civil and mechanical engineers. Rayleigh damping assumes the damping matrix to be a linear combination of mass and stiffness matrix

$$\mathbf{E} = \alpha \mathbf{M} + \beta \mathbf{K} \tag{14.2}$$

with α and β as constant coefficients [6]. Such an assumption results in a unique formulation of different damping sources and greatly simplifies the problem. Moreover, it can lead to a reasonably accurate damping description, if the damping coefficients are properly chosen. In this regard, different approaches can be followed. If experimental data of the dynamical behavior of the device are available, these parameters can be extracted by data fitting. Alternatively, a certain device dynamical parameter (e.g., the quality factor) can be estimated on the basis of analytical models, and this value can then be used to extract the damping parameters. These two methods are applicable to microresonators, in which both dynamical measurements and simplified analytical models of the damping behavior are available. For microswitches, damping parameters should be, instead, extracted on the basis of accurate simulations of the physical behavior of the device. It is worth noting that the adoption of Rayleigh damping allows avoiding a nonlinear damping matrix that would complicate the application of MOR. On the other end, if Rayleigh damping does not guarantee the required modeling accuracy of damping phenomena, the designer should consider the use, directly at circuit level, of the physics-based damping compact models available in the literature [7, 8]. One example of such models is presented in Chapter 2.3 of this book.

In RF MEMS, the applied nodal forces **F** in Eq. (14.1) are not mechanical loads, but the result of electrostatic interaction between conductors at different potential. Their rigorous calculation requires the solution of the Poisson equation in the dielectric media surrounding the conductors. This approach would, however, lead to sequential coupling between the electrical and mechanical energy domains: the electrical and mechanical problems are solved in a cycle using the solution of one problem as input for the other, till consistent results are achieved. The extraction of a reduced compact model requires strong coupling between energy domains, in which the electrical and mechanical problems are solved simultaneously. For this reason, the electrostatic force acting between pairs of overlapping conductors is computed as force between the plates of a capacitor. Typical geometry and behavior of RF MEMS allow simplifying assumptions in the calculation of such a force. In most cases, the surfaces of the conductors are parallel to each other and are not subjected to large deformation during device operation. A parallel plate capacitor approximation can therefore be generally adopted, with an overlapping area, which remains constant with device deformation and a capacitance variation that is a function of the relative displacement of the conductors along a single direction. Moreover, we assume uniform charge distribution on the conductor surfaces. Under these conditions, the electrostatic force between a couple of conductors k and j is perpendicular to both conductors, and its amplitude is given by

$$F_k^{el} = -F_j^{el} = \frac{1}{2}\frac{\partial C_{kj}}{\partial u_k}(V_k - V_j)^2 \tag{14.3}$$

where u_k indicates the instantaneous displacement of conductor k in the direction of the force, C_{kj} is the capacitance between the two conductors, and V_k and V_j are their respective potential. The value of C_{kj} is inversely proportional to the distance $u_k - u_j$ between the two conductors, and this results in the nonlinear coupling between the electrical and mechanical energy domains.

The calculation of the electrostatic force with Eq. (14.3) and its application to the mechanical model can be done in different ways. A single capacitor can be used to describe the interaction of each pair of conductors. Alternatively, the external surface of the conductor can be divided into patches, and a force can be computed between facing patches of each couple. Eventually, one pair of patches can be defined for each couple of facing nodes in the mechanical model. The forces can then be applied as lumped forces to one node of the mechanical model for each conductor or patch, or as a pressure load on all the nodes of the conductor's surface/patches. The more patches are considered, the more accurate is the description of the electrostatic forces achieved. The assumptions introduced to derive Eq. (14.3) are in fact locally satisfied with a higher degree of accuracy.

Further improvement of the electrostatic description can be obtained by including the effects of fringing capacitance and surface deformation with deflection. This can be done by extracting a capacitance–displacement curve, from a series of full electrostatic simulations.

Another type of force that contributes to the load vector in microswitches in Eq. (14.1) is the contact force experienced by the switch on actuation. This force has a strong nonlinear dependency on the device displacement DOF and is normally described by modeling the contact plane as a very hard spring. If d is the distance at a certain instant in time of the two device parts possibly subjected to contact and d_{gap} is the value of such distance at rest, then the contact force is

$$F^{cont} = K_n(d_{gap} - d), \quad \text{for } d < d_{gap} \quad (14.4)$$
$$= 0, \quad \text{for } d > d_{gap}$$

where K_n is the stiffness of the contact. A single or multiple contact force can be derived by monitoring the displacement of a single node or of multiple nodes per moving surface. As for electrostatic forces, they can then be applied as pressure loads or lumped nodal forces.

A full description of the electrical behavior of RF devices requires the knowledge of the current flowing in each of the electrical terminals of the device. The current flowing in each capacitor C_{kj} can be simply computed by taking the total time derivative of the stored charge

$$i_{kj} = \frac{d}{dt}\left[C_{kj}(V_k - V_j)\right]. \quad (14.5)$$

The total current through a device terminal will be given by the algebraic sum of all the currents in the capacitors that are connected to a specific terminal. It is worth noting that the value of the potentials at the device terminals is established by the external circuitry to which the device is connected. For modeling purposes, voltages can therefore be handled as fixed input quantities. This leads to a voltage-controlled implementation of the device circuit model, but does not further limit the modeling capabilities. The expression of the vector of the terminal currents, as derived from Eq. (14.5), has the following form:

$$\mathbf{i} = \mathbf{D}_u \dot{\mathbf{u}} + \mathbf{D}_v \dot{\mathbf{v}} \quad (14.6)$$

where $\mathbf{D}_u$ and $\mathbf{D}_v$ are constant matrices. If voltages are input quantities, the second term on the right hand side is known a priori. After solving the system in (Eq. 14.1), computation of the currents is straightforward. Nevertheless, it represents an essential step toward the extraction of a compact model suitable to be used in circuit simulation.

14.2
Extraction of the Reduced Order Model

In Chapter 3, the mathematical methods for the extraction of reduced order models from first order linear ODE systems have been presented. Attention has been focused on Krylov-subspace-based moment matching MOR, results of which are numerically very efficient in large-scale ODE systems (state-of-the-art MEMS models might contain several hundreds of thousands DOF). In the following section, we see how to adapt these techniques to the second order damped linear ODE system with nonlinear input function in Eq. (14.1). Among the Krylov-subspace-based methods, we will make use of the Arnoldi algorithm (or the block Arnoldi algorithm for multiple inputs [9]), which offers sufficient accuracy for the problems considered, simplicity of implementation and numerical stability. The procedures presented can, however, be applied also in combination with the Lanczos algorithm, by using the corresponding projection subspaces, as defined in Chapter 3.

14.2.1
Second Order ODE Systems

Second order ODE systems can be reduced by transforming them to first order systems and then using the methods described for this kind of problems [10]. However, in this way, the structure of the original system is lost. If the system in (Eq. 14.1) is undamped, that is, $\mathbf{E} = 0$, its transfer function around a certain expansion point can be formally written as the transfer function of a first order system, so that the reduction methods described in Chapter 3 can be applied to derive the projection subspaces for the second order system [11]. The reduced order system is then derived by orthogonal projection of each system matrix on such a subspace. The structure of the system and the matching properties of the chosen algorithm are both preserved.

If the system is damped, structure preserving MOR is more complex. One exception is when the damping matrix has the form in Eq. (14.2). In the case of Rayleigh damping, it can be demonstrated that the damping matrix does not take part in the reduction process and the reduced damping matrix can be simply computed from the reduced stiffness and mass matrices as

$$\mathbf{E}_r = \alpha \mathbf{M}_r + \beta \mathbf{K}_r \tag{14.7}$$

where the suffix r indicates reduced matrices and the coefficients α and β are the same as those appearing in Eq. (14.2) [12].

14.2.2
Handling Nonlinearities in the Input Function

As motivated in the previous section, we will assume linear system matrices, with the damping matrix given by Eq. (14.2). Then the ODE system to be reduced has the form

$$\mathbf{M}\ddot{\mathbf{u}} + \mathbf{K}_m \mathbf{u} = \mathbf{F}(\mathbf{u}, \mathbf{V}(t)) \tag{14.8}$$

where the damping matrix is neglected, as it does not take part in the reduction process. The load vector $\mathbf{F}(\mathbf{u}, \mathbf{V}(t))$ represents the electrostatic forces and, for switches, the contact forces, and it is a nonlinear function of the nodal displacements **u** and the applied voltages **V**. Voltages are input quantities and are not a real source of nonlinearities for the system. The dependency **F(u)** introduces nonlinearity in the system. Considering the different behavior of frequency-selective devices and switches, two different approaches will be applied for handling this form of nonlinearities.

Devices for frequency generation and selection are brought into vibration by applying a large bias voltage V_0 and superimposing a harmonic voltage δV of small amplitude between one excitation electrode and their movable part. In this operating condition, we can assume that the vector **u** undergoes small variations around a certain bias point $\mathbf{u}_0$

$$\mathbf{u} = \mathbf{u}_0 + \delta \mathbf{u}(t), \quad \delta \mathbf{u} \ll \mathbf{u}_0$$

and the load vector can then be linearized around the operation point $(\mathbf{u}_0, \mathbf{V}_0)$, as follows:

$$\mathbf{F} = \mathbf{F}_0 + \mathbf{K}^{uu}\delta\mathbf{u} + \mathbf{K}^{uv}\delta\mathbf{V} \tag{14.9}$$

where $\mathbf{K}^{uu}$ and $\mathbf{K}^{uv}$ are sparse matrices whose nonzero entries are defined as

$$K_{kj}^{uu} = \frac{\partial F_k}{\partial u_j}\Big|_{\mathbf{u}_0,\mathbf{V}_0}, \quad K_{kj}^{uv} = \frac{\partial F_k}{\partial V_j}\Big|_{\mathbf{u}_0,\mathbf{V}_0}$$

By inserting Eq. (14.9) in Eq. (14.8) and reorganizing the terms, we obtain the system

$$\mathbf{M}\delta\ddot{\mathbf{u}} + \mathbf{K}_{tot}\delta\mathbf{u} = \mathbf{K}^{uv}\delta\mathbf{V} \tag{14.10}$$

with $\mathbf{K}_{tot} = \mathbf{K}_m - \mathbf{K}^{uu}$. As the voltages are given, the product $\mathbf{K}^{uv}\delta\mathbf{V}$ represents a known input vector of the system. Such a system can be reduced with linear MOR techniques. It can be noticed that $\mathbf{K}^{uu}$, also known as *electrical stiffness*, lowers $\mathbf{K}_m$, thus reducing the stiffness of the structure. This is the basis of the electrostatic spring-softening effect observed in MEMS devices, which is therefore automatically captured by the system in Eq. (14.10).

Linearization cannot be used for the modeling of microresonators, when they are used for intrinsically nonlinear operations, such as signal mixing. It is also never applicable to microswitches, always operated in large signal conditions. An extension of this approach could be adopted, the trajectory piecewise-linear method, which creates a reduced order model starting from a weighted linear combination of models linearized around different operating points [13]. We choose, however, a more straightforward approach.

The extraction of a reduced order model of system in Eq. (14.8) for large signal analysis can be done by regarding nonlinearities as inputs. The nonlinear load vector in (14.8) can be rewritten as

$$\mathbf{F} = \mathbf{B}\mathbf{f}(\mathbf{u}, t) \tag{14.11}$$

where nonlinearities are moved in the new input function $\mathbf{f} \in \mathbb{R}^m$, with m number of nonlinear inputs, and $\mathbf{B} \in \mathbb{R}^{n \times m}$ being the scattering matrix that distributes the load on the appropriate DOF. The system matrices $\mathbf{K_m}$, $\mathbf{M}$, and $\mathbf{B}$ can be simply reduced by means of linear MOR. The argument of function $\mathbf{f}$ has to be recovered from the reduced state vector $\mathbf{u}_r$ by back projection on the subspace $\mathbf{V}$, that is, $\mathbf{u} = \mathbf{V}\mathbf{u}_r$. The reduced order system will have the form

$$\mathbf{M}_r\ddot{\mathbf{u}}_r + \mathbf{K}_r\mathbf{u}_r = \mathbf{B}_r\mathbf{f}(\mathbf{V}\mathbf{u}_r, t) \tag{14.12}$$

The solution of the system in (Eq. (14.12)) is computationally advantageous with respect to the solution of Eq. (14.8) only if the evaluation of the function $\mathbf{f}(\mathbf{V}\mathbf{u}_r, t)$ is fast [14]. This implies that the number of nonlinear equations of the system should be small or possible to reduce to a small number. The requirement is also important for an efficient extraction of the model, since the time needed for the extraction of the model depends linearly on the number of inputs of the system. This poses some restrictions on the modeling of electrostatic forces. In order to have a reduced number of nonlinear inputs, the distributed electrostatic forces need to be replaced by a limited number of concentrated loads. Recalling the procedure introduced in the previous section for the calculation of such forces, if we partition the surface of the conductors in large-area capacitors, we will end up with a limited number of inputs. The description of the electrostatic domain will be less accurate, but there will be a remarkable speedup in the time needed for the extraction of the reduced model and its subsequent simulation. The same consideration is also valid for contact forces: the trade-off accuracy/efficiency has therefore to be taken into account during the decision of the number m of nonlinear forces introduced in the model.

14.2.3
Extraction Procedure

Figure 14.2 schematically represents the steps necessary for the extraction of compact models for RF MEMS, based on the aforementioned theoretical background. The procedure starts with the creation of a FE model of the mechanical part of the device with any software that allows for the extraction of the FE matrices. We made use of ANSYS®. The extraction of the system matrices from ANSYS and their reduction were performed with the command line tool MOR for ANSYS [15]. This first step should include stiffness and mass matrices assembly and, whenever necessary, an initial nonlinear static analysis for computing the device bias point and/or the stress stiffening matrix $\mathbf{K}_\sigma$. Further steps differ depending on the type of device under consideration.

For devices operated in a small-signal regime, the matrices appearing in the expression of the linearized force (Eq. 14.9) have to be computed for the derived bias point and system in Eq. 14.10 has to be formed. Within ANSYS, special transducer elements can be used to describe the electrostatic forces according to Eq. 14.3 and assemble the matrices $\mathbf{K}^{uu}$ and $\mathbf{K}^{uv}$. Other FE packages may require additional external programming. Once an appropriate projection subspace is computed via the Arnoldi algorithm, the system in Eq. 14.10 can be projected on it. A reduced system is thus derived that, together with equation (14.6) for the

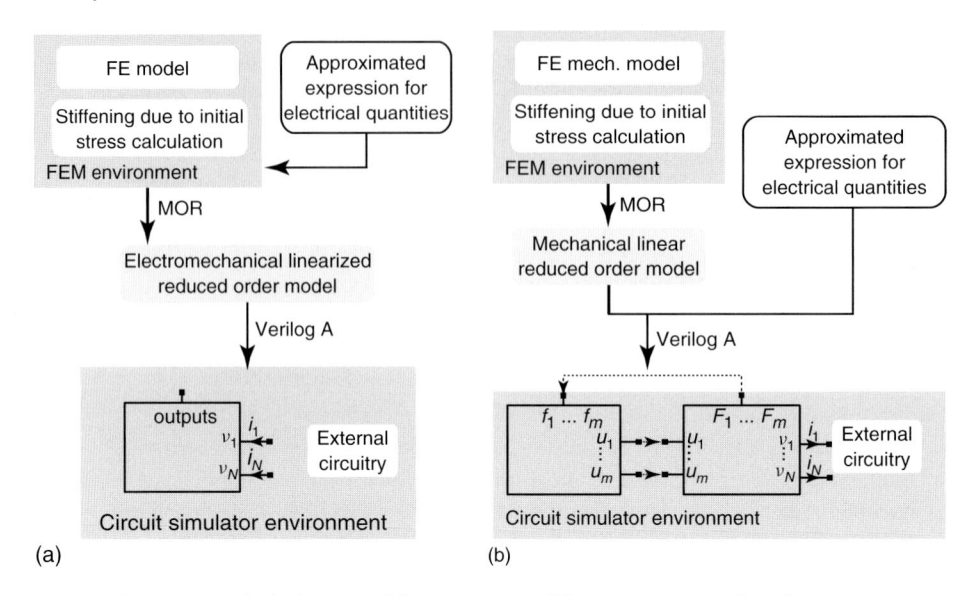

Figure 14.2 Block diagram of the compact models extraction procedure for microresonators (a) and microswitches (b).

currents, gives a full electrical description of the device at its terminal and can be translated into a hardware description language (HDL) and imported in a standard circuit simulator. We opted for the VerilogA language. The compact model is then ready to be used as a design component in any circuit layout.

Mechanical output terminals are added to the model to monitor the deformation of the device at the points of interest.

If the large signal transient or harmonic behavior is of interest, mechanical system matrices are extracted, reduced, and translated in HDL without taking the electrical domain into account. Electrical and mechanical models interface only via their input/output quantities. A correct coupling requires a proper definition of the scattering matrix **B** in Eq. 14.12 and eventually of the system outputs, which should include all the displacements necessary to compute the electrostatic forces[1]. A model for these forces is then implemented with the desired degree of approximation directly at circuit level, with the description of the currents at the device terminal. By connecting mechanical and electrical models, the nonlinear compact electromechanical model is obtained. Output displacements can also be used to monitor contact conditions. If contact occurs, contact forces can be computed according to Eq. 14.4 and added to the electrostatic forces.

The handling of the involved energy domains separately and with different methodologies renders the extraction procedures easy to implement and flexible in terms of modeling capabilities, accuracy, and complexity. The MOR step can be

---

1) The Arnoldi reduction algorithm is independent from the choice of the outputs. It is therefore in principle possible to retrieve the full vector of DOF

performed in an automatic manner, once two parameters are fixed. The first is the order of the reduced order model. Different approaches to fix this value according to the desired level of accuracy are presented in [16]. The choice of the expansion point, or eventually of multiple expansion points, for the transfer function of the original system (Chapter 3) is the second parameter and strongly depends on the dynamic range and the kind of analysis for which the model should be used.

Reduced models of microresonators are normally used for the harmonic analysis of the device around a frequency of interest, approximately known in forehand (typically one of the resonance frequencies of the microstructure). In this case, fixing the expansion point close to this frequency value generally results in a good description of the original system in a wide frequency band around it. However, it is difficult to exactly quantify the width of this band. If multiple frequency ranges are of interest, eventually disjoined as for devices performing mixing operations, one expansion point in each frequency range should be used.

The characterization of microswitches normally involves static pull-in analysis or transient response analysis to a square wave signal applied between their movable membrane and actuation electrode. For these cases, it is sufficient to fix the expansion point to zero. The resonance frequency of the switches is in fact generally low (in the range of 5–50 kHz) and the frequency of the applied signal is chosen even lower to limit oscillation of the device after pull out. Whenever high frequency signals are applied, such as in an intermodulation analysis, the expansion point should instead be chosen with the same criteria introduced for microresonators.

14.3
Application Examples

In this section, the application of MOR is demonstrated for the two typologies of RF MEMS devices: a microresonator and a micromechanical switch. The obtained results are compared with FE simulations of the full models.

14.3.1
Vibrating Devices

As an example of resonating device, we will consider a microresonator to be used for frequency selection and generation operations. A compact model has been derived for a silicon carbide square resonator, whose design and material parameters have been taken from literature [17]. The resonator is made of a square suspended plate anchored at its corners through four short beams. One electrode is laterally facing each side of the plate, so that in-plane vibration of the plate can be excited and sensed. The plate is biased through the beams. The gap between the electrodes and the plate sides is 0.1 µm in order to enhance electrostatic coupling. Despite its simple structure, the device can be used in various circuit configurations that have a strong influence on its response and performances. Under these circumstances, reduced order modeling is a useful tool for fast circuit design optimization.

By applying proper signals on the device terminals, either the Lamé or the extensional mode of the plate can be excited. The Lamé mode shape is shown in Figure 14.3a: the square plate extends along one direction and contracts along the perpendicular one, while conserving the total volume. The extensional mode is characterized by alternate contraction and extension of the plate in all directions, thus conserving the plate shape, as in Figure 14.3b. For the chosen values of dimensions and material properties, the resonance frequencies of the Lamé and the extensional modes are 171 and 177 MHz, respectively.

The full three-dimensional mechanical device model has a total of 3000 DOF. A linearized reduced order model has been extracted with MOR assuming a 5 V bias voltage applied between electrodes and moving structure and choosing the expansion point for the reduction process at 175 MHz, that is, between the two resonance frequencies of interest. The numerical complexity of a single harmonic simulation of the full model is for this example still acceptable. The order of the reduced model required for a good modeling accuracy has then been deduced by comparison of the simulation results obtained with low-order models of different dimensions and with the full model. Results for the full model and a 15 DOF model, in the electrical configuration of Figure 14.4a with the sensing terminals (i_{out}) grounded, are shown in Figure 14.5. Figure 14.5a shows the variation in the magnitude of the device currents at the excitation and sensing terminal with frequency, while in Figure 14.5b the displacement amplitudes of two points located at the center of two adjacent sides of the plate are reported. It can be noted that the relative error computed with the 15 DOF model is lower than 1% in a 50 MHz range containing the resonance frequencies of interest, for both electrical and mechanical quantities. A harmonic simulation of the reduced model has been used to adjust

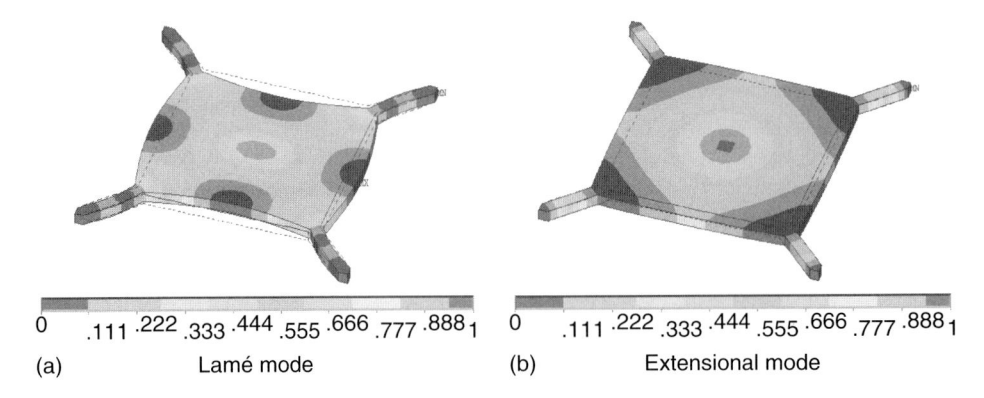

Figure 14.3 Mode shape amplitudes of a square resonator. The device total deformation is normalized to unity.

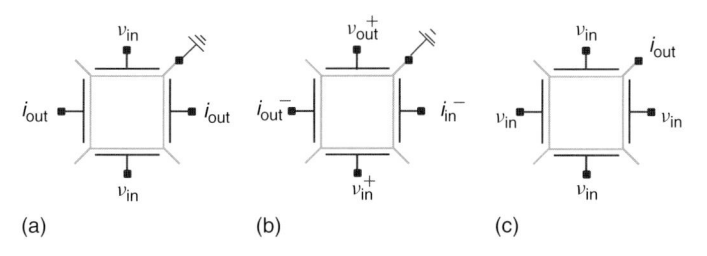

Figure 14.4 Two-port single-ended (a), two-port differential (b), and one-port single-ended electrode configuration used for the square resonator in its Lamé (a-b) and extensional (c) modes.

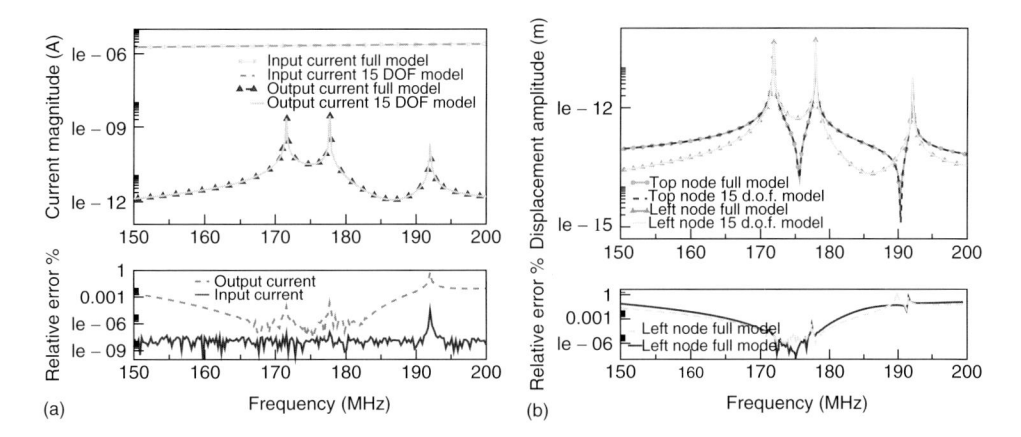

Figure 14.5 Harmonic simulation of the current magnitudes and the total displacement for the device in a single-ended configuration. Results achieved with a the reduced model are compared with one obtained with the full model.

the value of the damping coefficients so as to achieve an unloaded quality factor for the device equal to 9300.

The computational time required for all model extraction steps and the harmonic simulation with 200 frequency steps are reported in Table 14.1 both for the full and the reduced order model. The reduced model allows a considerable speedup of the simulation. Moreover, it can be extracted very fast after computation of the bias point and assembly of the matrices.

The 15 DOF model has been translated into a VerilogA behavioral model and has been imported into the Cadence Spectre simulator. The resulting circuit component has one input/output electrical terminal for each device excitation electrode, one for the resonator itself, plus two output-only mechanical terminals for monitoring the displacements at the middle points of two adjacent sides of the device. Moreover, two reference electrodes are present, one for electrical and other for mechanical quantities. In order to avoid floating terminals in the circuit, the mechanical outputs are always connected to the mechanical ground via a very high resistance.

Table 14.1 Computational time comparison for the square resonator.

Computation	Time (s)
Harmonic prestressed analysis (full model)	212
Bias point computation & matrices calculation	3.3
Reduced order model extraction	3.6
Harmonic analysis (reduced model)	1

Circuit-level simulation has been used to study the influence of the series resistance of the suspended resonator structure on the device performance in the resonant modes in Figure 14.3 and for the different electrode configurations. The Lamé mode is excited using either the single-ended or the differential two-port electrode configurations shown in Figure 14.4a,b respectively, in which each pair of neighboring electrodes is driven with a 180° phase offset. The schematic circuits in Figure 14.6 can be used for this purpose. Figure 14.6a shows the single-ended implementation, with each pair of facing electrodes connected to a port with an internal impedance of 50 Ω. Figure 14.6b represents the fully differential implementation, which makes use of two ideal balanced–unbalanced (balun) transformers whose balanced outputs are connected to the adjacent electrodes of the resonator. In both cases, the resonator terminal is grounded through the resistance R_{res}, representing the resistance of the vibrating plate and beams. Although offering a very high acoustic velocity, silicon carbide has a rather low conductivity, so that this has been estimated to be around 200 kΩ [17]. The model has been simulated for values of R_{res} ranging from 0 to 100 kΩ, and a fine frequency discretization has been used in order to capture the resonance peak. The device behavior has been observed through its mechanical outputs (left node and top node) and by computing the electrical scattering parameters.

The graph in Figure 14.7 shows the harmonic behavior of the device transmission parameter S_{21} in dB, for the single-ended configuration. It can be seen that, for low values of R_{res}, this shows a clear resonance peak at 171.3 MHz, while for increasing values of resistance the peak is gradually more difficult to detect and eventually disappears. This can be explained by looking separately at the displacement and motional components of the output current, plotted in Figure 14.8. At low resistances, the motional current is the predominant contribution to the output current. However, as soon as the resistance R_{res} increases to values higher than 100 Ω, the capacitive feedthrough between excitation and sensing ports gets more and more important and the displacement current prevails. With 200 KΩ resistance, the device would actually be unusable for frequency selection operation.

By using the fully differential electrode configuration, the behavior of the transmission parameter S_{21} is insensitive to the value of R_{res}, as shown in Figure 14.7b. With this setup, the resonator is virtually grounded and no current flows through its suspensions. The displacement current at the unbalanced input

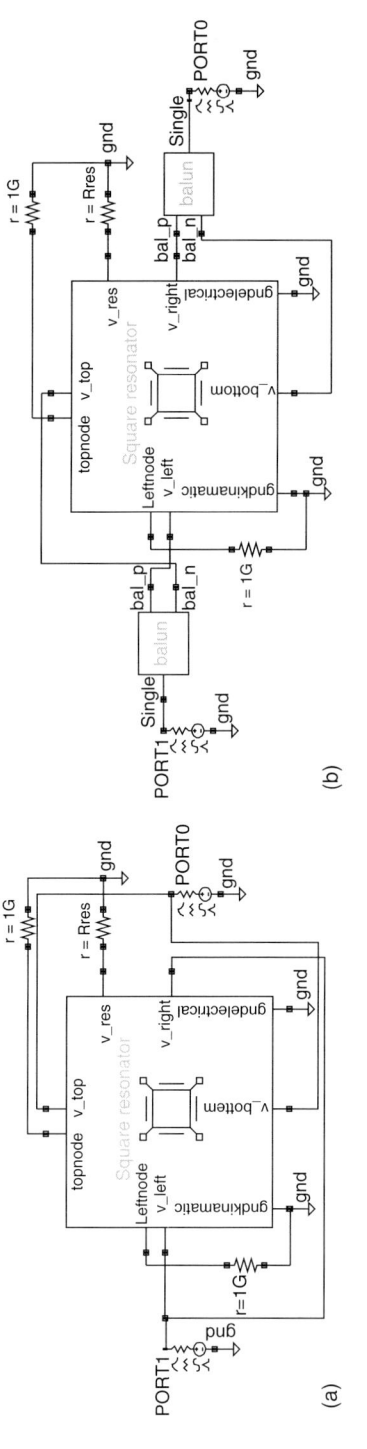

Figure 14.6 Schematic circuits used for the simulations of the square resonator in a two-port single-ended (a) and differential (b) electrode configuration.

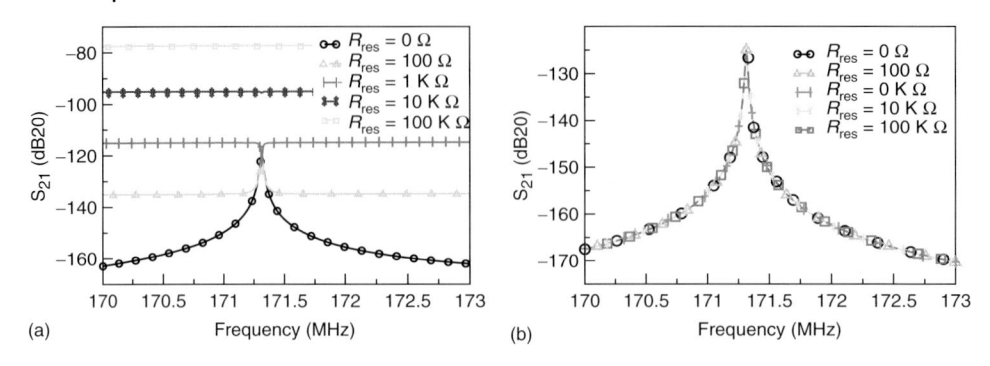

Figure 14.7 Circuit-level simulation of the square resonator in its Lamé mode with a single-ended (a) and differential (b) electrode configuration.

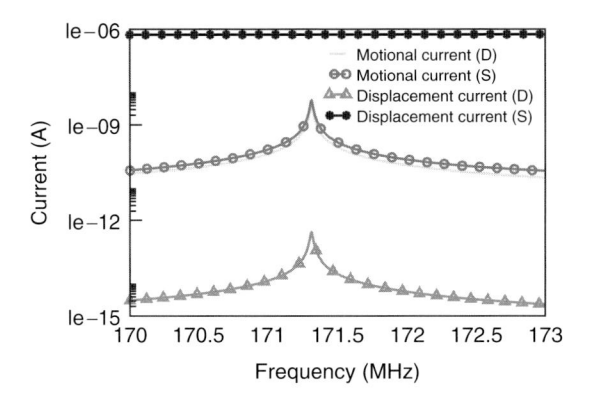

Figure 14.8 Motional and displacement components of the current at the terminal v_top for the single-ended (S) and the differential (D) configurations.

terminals of the balun is small and is completely eliminated at its output terminal, where, owing to common mode cancelation, only the motional current of the resonator is present. Hence, the device works properly, and its Q factor is not altered by the high resistivity of its material.

The obtained simulation results are in agreement with the experimental results presented in [17]. The very low values of the transmission parameter are due to the unmatched loading of the device with the standard 50 Ω impedance at the input and output ports and to the lack of an amplifying stage, which is always present in measurements setups to detect the tiny motional currents.

The performance of the device in its higher frequency extensional mode can also be analyzed, by making use of the electrical configuration in Figure 14.4c, in which four voltage signals with equal magnitude are applied in phase to the electrodes, and the output signal is read from the plate terminal. Simulation results for this setup are reported in Figure 14.9. Independently from the value of the series resistance R_{res}, a high capacitive feedthrough, typical of one-port measurement setups, is

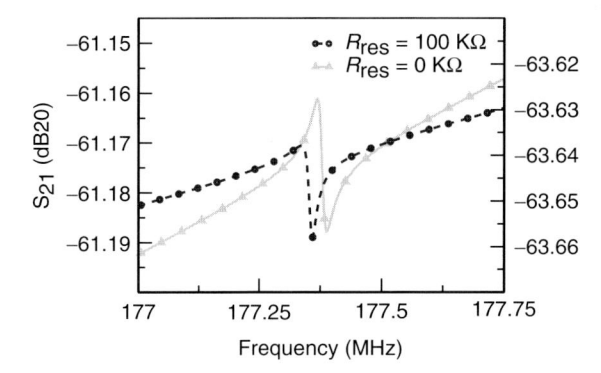

Figure 14.9 Simulated transmission parameter of the square resonator in its extensional mode.

present between excitation and sensing terminals, so that resonance becomes very difficult to detect. The large contribution of the displacement current to the output current justifies the much higher values of the transmission parameter with respect to the fully differential electrode configuration. With the increase in the value of R_{res}, the device Q factor drastically decreases. The use of the device under these conditions leads therefore to poor performance. Altogether, we can then conclude that the best use of the device is in its Lamé mode with a fully differential electrode configuration.

14.3.2
Microswitch

A microswitch model has been created for a capacitive switch, whose geometrical shape was inspired by the device presented in [18]. This has a rectangular membrane suspended 1.4 μm above an electrode, centered with respect to the membrane. The membrane is anchored through four beams, which form an angle of 30° degree with the longer edges of the membrane, as shown in Figure 14.10. We assumed the membrane to be made of a 3 μm thick aluminum layer. As the device has two symmetry axes, only one quarter of the structure has been modeled with and meshed with FE. Holes are present in the membrane to ensure its full release during fabrication. For such a geometry, fringing effects become important in the computation of electrostatic forces. For this reason, the external nodes of the membrane have been grouped into 15 batches, and the capacitance between each patch and the bottom electrode has been computed with an electrostatic analysis. In this way, also fringing effects due to the finite thickness of the membrane and the electrode are accounted for. Since during the actuation, the device deformation interests mainly the low-stiffness supporting beams, and hence the capacitances are computed only for the membrane in its undeformed state. On the basis of the electrostatic solution, lumped electrostatic forces are derived, which are applied at the center of each patch.

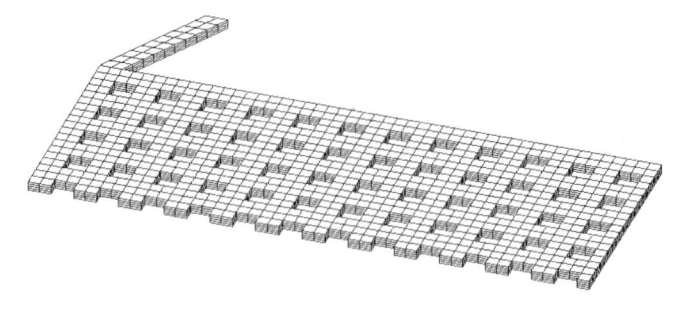

Figure 14.10 Finite element model of the switch. Only one quarter of the geometry has been considered.

A compact model of the mechanical domain has been extracted using a 15 DOF reduced order model with 15 input/output DOF for calculation and application of the electrostatic forces. Rayleigh damping coefficients have been parametrically included in the model. In order to capture the device pull-in characteristic, contact forces have been also included in the VerilogA model. This has then been complemented with electrostatic forces within the Cadence. The contact condition is checked for every output node of the reduced order model. If the absolute value of the nodal displacement is greater than the transduction gap of the device then both electrical and contact forces are applied to the node, otherwise only the electrical forces are considered. The computational time of each step necessary for model extraction are reported in Table 14.2. With respect to the resonator and filter models, owing to the larger dimension of the original model and the increased number of inputs, each step requires considerably more time. Nevertheless, the total extraction time is fully acceptable in the light of the great simulation speedup that the reduced model enables.

The static pull-in behavior of the device has been simulated at circuit level and compared with the results obtained with a sequentially coupled electromechanical simulation of the full model in ANSYS. The vertical displacements of two nodes of the membrane to which forces are applied have been observed. They correspond to the minimum and the maximum displacement of the area of the membrane

Table 14.2 Computational time for the extraction and simulation of a switch compact model.

Computation	Time (s)
Element matrix assembly	200
Reduced order model extraction	900
Electrostatic force computation (with FEM)	200
Electromechanical coupled simulation (full model)	10 100
Static & transient simulation reduced order model	<1

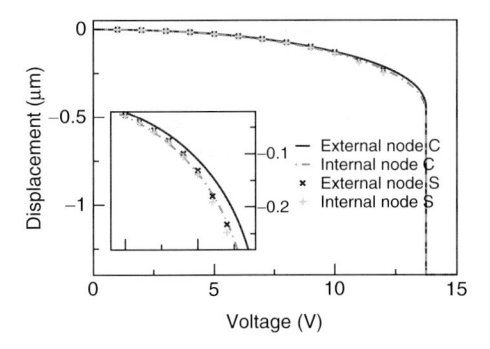

Figure 14.11 Static pull-in behavior of the switch. Full model sequential coupling results are compared with a 15 DOF reduced order model results.

above the electrode. Simulation results are plotted in Figure 14.11: continuous lines represent the circuit simulation results, while scatter points represent results obtained with the sequential solver. The latter does not describe the complete pull-in curve because convergence problems are encountered for vertical displacements of the membrane greater than 300 nm. For small displacements, the curves and the simulation points overlap almost exactly. At bigger displacement, a certain discrepancy between the simulation results arises, with the reduced order model underestimating the displacement of the device. The deviation from the simulated value is attributed to the fact that the horizontal force components arising as the membrane deflects are neglected. However, the resulting error on the device pull-in voltage is between 5 and 10%, which is considerably lower than the error introduced by uncertainty in the material properties and geometrical imperfections. Moreover, even taking into consideration the time needed to extract the compact model, simulation is much faster: while the computation of the electromechanical equilibrium for 12 voltage values requires approximately 10 100 s with the sequential solver, the reduced order model simulation in the Cadence uses less than a second.

The availability of a compact model is particularly useful for the characterization of the transient behavior of MEMS switches. A transient analysis of the full electromechanical model would in fact be prohibitive in terms of complexity and computational time. A parametric study of the dynamic pull-in voltage of the device has been performed at circuit level. The response of the device to a step voltage of increasing amplitude has been studied, for different values of the Rayleigh damping coefficients. In very low damping conditions, the device shows the behavior of Figure 14.12a. For values of the applied voltage lower than 12.75 V, the switch oscillates with a frequency that decreases with the applied voltage. This is due to the spring-softening effect. At 12.75 V, the switch reaches in its movement the instability region and collapses on the actuation electrode. This voltage value corresponds to the dynamic pull-in of the switch and is in good agreement with the expected analytical value. For a static pull-in voltage of 13.75 V, the predicted dynamic pull-in voltage is in fact equal to 12.78 V [19]. In addition, the expected

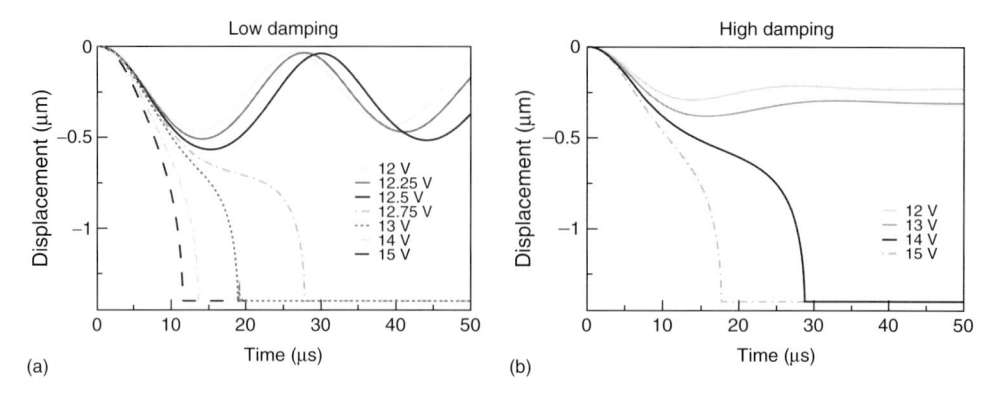

Figure 14.12 Dynamic pull-in behavior of the switch in low and high damping conditions for increasing value of the applied voltage.

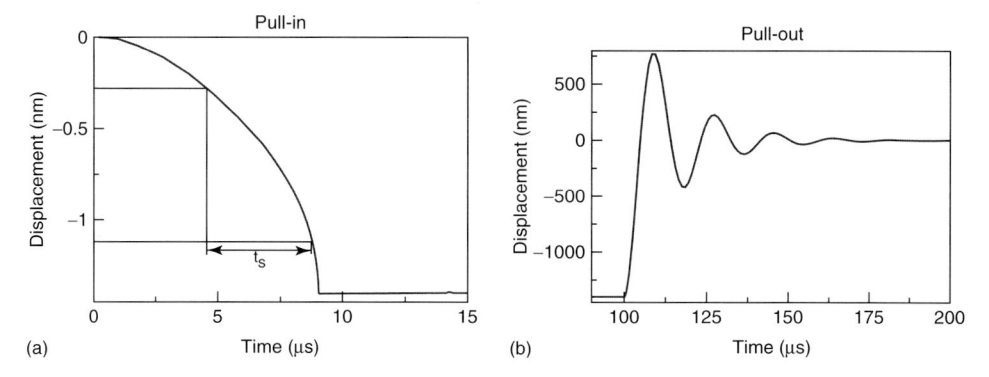

Figure 14.13 Dynamic pull-in (a) and release (b) behavior of the switch for an applied square wave with a 20 V peak value.

shortening of the pull-in time with the applied voltage is observed. If the damping parameters are incremented to values that represent the typical quality factor for a switch, static and dynamic pull-in voltage cannot be distinguished anymore and the pull-in characteristic of the switch has the behavior in Figure 14.12b. Transient analysis can also be used to assess the pull-in and pull-out time of the device. For a 5 kHz square wave actuation voltage with a peak value of 20 V, the expected pull-in and pull-out dynamic characteristics are shown in Figure 14.13a,b respectively. The pull-in time, measured as the time needed for the switch to goes from 0 to 80% of the top value can be easily measured and is approximately 7.5 μs.

14.4
Conclusion and Outlook

A procedure for the extraction of compact models from linearized FE models of active RF MEMS devices has been presented and demonstrated for vibrating devices and microswitches. The extraction process can be performed with minimal input information from the designer and is reasonably fast also for large original models. Other advantages are its capability of handling devices with arbitrary geometry, of capturing the effects of electrical and mechanical prestress, as well as its flexibility in modeling the interaction between the electrical and mechanical domains. These features render the approach applicable to most electromechanical devices.

Reduced models have been used for circuit simulation in simple networks, and their accuracy has been proved by comparison with the full models or, when possible, experimental data. The small complexity of the reduced model guarantees a substantial speedup of the simulation time, which results particularly beneficial for transient simulation. Fast and accurate circuital simulation is the premises for the use of these models for design optimization of more complex architectures.

Models extracted are not parametric in the device geometrical dimensions. Only parameters that enter linearly in the formulation of the FE matrices can be preserved in the compact model, as demonstrated in Chapter 3.1 for thermal MEMS models. Developments in the direction of parametric MOR would bring great benefit to the design of RF MEMS. Yet, the extraction of the reduced matrices is so fast that, when geometry optimization is necessary, is still possible to use MOR for a solution speedup [20].

References

1. Nguyen, C.T.C., Katehi, L.P., and Rebeiz, G.M. (1998) *Proceedings of IEEE*, 86 (86), 1756–1768.
2. Nguyen, C.T.C. (2001) Transceiver front-end architectures using vibrating micromechanical signal processors, Digest of Papers. Silicon Monolithic Integrated Circuits in RF Systems, pp. 23–32.
3. White, J., Senturia, S.D., and Aluru, N. (1997) *IEEE Computational Science & Engineering*, 4 (1), 30–43.
4. Kaajakari, V., Mattila, T., Kiihamäki, J., Oja, A., Kattelus, H., and Seppä, H. (2003) Nonlinearities in single-crystal silicon micromechanical resonators. Digest of Technical Papers, The International IEEE Conference on Solid-State Sensors Actuators and Microsystem (Transducers'03), pp. 1574–1577.
5. Zienkiewicz, C. (1977) *The Finite Element Method*, 3rd edn, McGraw-Hill.
6. de Silva, C.W. (2000) Vibration: fundamentals and practice, chap. Modal Analysis.
7. Veijola, T. (2004) Compact models for squeezed-film dampers with inertial effects. Proceedings of DTIP, pp. 365–369.
8. Schrag, G. and Wachutka, G. (2002) *Sensors and Actuators A*, **97-98**, 193–200.
9. Freund, R.W. (2000) *Journal of Computational and Applied Mathematics*, **123** (1-2), 395–421.
10. Antoulas, A.C. (2005) *Approximation of Large-Scale Dynamical Systems*, Society for Industrial and Applied Mathematic.
11. Su, T.J., Craig, J., and Roy, R. (1990) *Journal of Guidance*, **14** (2), 260-267.
12. Eid, R., Salimbahrami, B., Lohmann, B., Rudnyi, E.B., and Korvink, J.G. (2007) Parametric order reduction of proportionally damped second-order systems. *Sensors and Materials*, **19** (3), 149–164.
13. Rewienski, M. and White, J. (2003) A trajectory piecewise-linear approach to model order reduction and fast simulation of nonlinear circuits and micromachined devices. *IEEE Transactions on Computer-Aided Design of Integrated Circuits and Systems*, **22**, 155–170.

14. Lienemann, J. (2006) Complexity reduction techniques for advances MEMS actuators simulation. Ph.D. thesis. Albert-Ludwigs Universität Freiburg im Breisgau.
15. Rudnyi, E.B. and Korvink, J.G. (2006) *Lecture Notes in Computer Science*, **3732**, 349–356.
16. Bechtold, T., Rudnyi, E.B., and Korvink, J.G. (2005) *Journal of Micromechanics and Microengineering*, **15** (3), 430–440.
17. Bhave, S.A., Gao, D., Maboudian, R., and Howe, R.T. (2005) Fully-differential poly-SiC lamé mode resonator and checkerboard filter. Proceedings of 18th IEEE Micro Electro Mechanical Systems Conference (MEMS)'05, pp. 223–226.
18. Cusmai, G., Mazzini, M., Rossi, P., and Combi, C. (2005) *Sensors and Actuators A*, **A123-A124**, 515–521.
19. Rochus, V. (2007) Finite element modeling of strong electromechanical coupling in MEMS, Ph.D. thesis. University of Liege.
20. Han, J.S., Rudnyi, E.B., and Korvink, J.G. (2005) *Journal of Micromechanics and Microengineering*, **15** (4), 822–832.

15
SystemC AMS and Cosimulation Aspects

François Pêcheux, Marie-Minerve Louërat, and Karsten Einwich

15.1
Introduction

One of the great challenges of the next decade lies undoubtedly in the successful codesign of multidisciplinary systems exhibiting tight interactions between embedded hardware/software (HW/SW) systems and their analog physical environment. As time-to-market periods become shorter, the ability to model and simulate complex systems where digital HW/SW is functionally intertwined with analog and mixed signal (AMS, or Analog, Mixed-Signal in context with hardware description languages and system description languages) blocks (i.e., RF interfaces, power electronics, sensors, and actuators) becomes more and more essential. If such overall system and architectural level models are available as early as possible in the design cycle, the architecture exploration issues and design errors will be dramatically reduced.

This chapter presents a system-level design methodology and a use case for the modeling and simulation of complex heterogeneous systems based on SystemC and its AMS extensions, SystemC AMS. The chapter follows a bottom-up approach and is composed of two parts. First, the chapter starts with an introduction to the recently standardized SystemC AMS extensions and a short presentation of its currently supported modeling formalisms (MFs): timed dataflow (TDF), linear signal flow (LSF), and electrical linear networks (ELNs). A SystemC AMS description of the mechanical part of the MEMS accelerometer (reduced to a tractable spring mass damper problem) is given to illustrate the TDF and ELN models of computation (MoCs). Second, the described accelerometer is used as a building block in the SystemC AMS virtual prototype of a complex use case: a wireless sensor network (WSN) that aims at determining the two-dimensional coordinates of the epicenter of a seismic perturbation on a planar earth soil.

System-level Modeling of MEMS, First Edition. Edited by T. Bechtold, G. Schrag, and L. Feng.
© 2013 Wiley-VCH Verlag GmbH & Co. KGaA. Published 2013 by Wiley-VCH Verlag GmbH & Co. KGaA.

15.2
Heterogeneous Modeling with SystemC AMS

As technology enables the integration of MEMS with digital HW and SW to design an embedded application, the need to simulate the behavior of such complex AMS systems before going to silicon requires a smart methodology. On the one hand, MEMS simulations are time consuming when finite-element models are considered. On the other hand, the validation of the SW embedded in the heterogeneous system requires studying its interaction with MEMS. In order to decrease the computational cost involved in such use cases, on the MEMS side, mathematical model reduction method [1, 2] or lumped-element-based compact models are considered [3, 4] (cf. Chapters 2 and 3), while on the digital side, bit-true cycle-accurate models are [5] used.

The new SystemC AMS library of C++ classes is particularly well suited for modeling heterogeneous designs and more than the Moore systems [6–8]. This library is a compliant extension of SystemC that soundly manages several MFs and MoCs for AMS. AMS Parts of an heterogeneous system can currently be modeled in SystemC AMS using TDF, LSF, and ELN MoCs. The three MoCs have been designed to interact with each other or with other MoCs (i.e., discrete event of SystemC) by means of dedicated synchronization layers. Using this layered structure, interwoven parts of a complex heterogeneous system can be modeled and simulated within their optimal methodology and solution methods. From a system engineer's point of view, this approach provides a higher modeling efficiency and a faster simulation by several orders of magnitude.

15.2.1
SystemC AMS Timed Dataflow (TDF)

TDF can be used with great efficiency to model complex nonconservative (signal flow) behaviors at a functional level. The TDF model of computation is derived from the untimed synchronous dataflow MoCs [9]. In addition, the TDF MoC assigns specific time stamps to the sample values. The basic entities found in TDF are the TDF module, the TDF port, and the TDF signal.

The set of connected TDF modules forms a directed graph, called a *TDF cluster*. TDF modules are the vertices of the graph, and TDF signals correspond to edges. TDF signals are used to connect ports of different modules together. Each TDF module involved in the cluster contains a specific C++ member method, named **processing()**, that computes a mathematical function possibly involving port input values and the modules internal state. The overall functional behavior computed by the cluster is therefore defined as the mathematical composition of the functions of the involved TDF modules in the appropriate order. The *timestep* is defined as the time interval between two samples and can be set directly by the designer. It is also important to notice that integer *rates* and *delays* can be associated to ports, providing multirate capabilities and feedback loops. A TDF module may have several input and output ports, and a given TDF module function is calculated (or fired according to the Synchronous dataflow formalism) if and only if enough data samples are available at its input ports, depending on the involved port rates. In this case, the samples computed by a TDF module are written to the appropriate output ports. A time stamp is associated to each sample data, and a TDF module can produce more than one data sample when "fired."

Provided the associations performed on the ports or the modules of a TDF graph are compatible, the order and number of samples (sampling rate) in a TDF cluster is known for each calculation. Therefore, this order can be statically determined before the simulation starts and corresponds to a static schedule of the TDF cluster. The resulting static schedule, computed once during the simulator elaboration speeds up the simulation over traditional simulation kernels based on time-ordered event queues. Thus, and more formally, a TDF cluster can be defined as the set of connected TDF modules, which belong to the same static schedule.

15.2.2
Timed Dataflow Model of the Accelerometer

Figure 15.1 presents a simple example of an AMS accelerometer system, with a capacitance-based transducer. The mechanical part of the accelerometer (dashed lines) is a traditional mass spring damper system that can quite naturally be modeled in SystemC AMS TDF, using state space representation.

Free body dynamics states that the acceleration of the seismic mass is defined by the following equation:

$$\ddot{x} = -\frac{k}{m}x - \frac{b}{m}\dot{x} - \frac{1}{m}f(t) \tag{15.1}$$

where k is the spring constant, b is the damping (viscous) coefficient, m is the mass, $f(t)$ is the time-dependent force applied to the seismic mass, and x is the displacement of the seismic mass. The force applied to the seismic mass, $f(t)$, is simply modeled as a Gaussian pulse. This second-order differential equation can be rewritten as a couple of first-order differential equations that are functions of

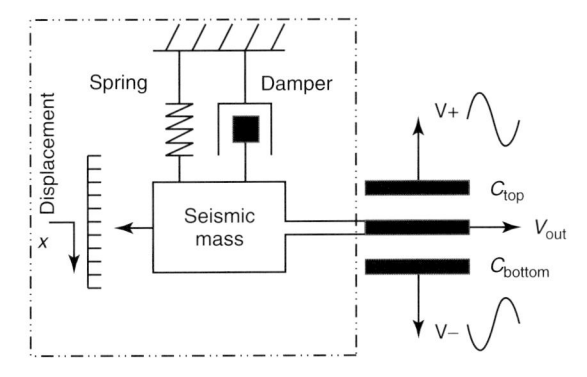

Figure 15.1 The MEMS accelerometer.

the seismic mass displacement and velocity

$$v = \dot{x} \tag{15.2}$$

and

$$\dot{v} = -\frac{k}{m}x - \frac{b}{m}v - \frac{1}{m}f(t) \tag{15.3}$$

In matrix form, these equations are rewritten as

$$\begin{pmatrix} \dot{x} \\ \dot{v} \end{pmatrix} = \begin{pmatrix} 0 & 1 \\ -\frac{k}{m} & -\frac{b}{m} \end{pmatrix} \begin{pmatrix} x \\ v \end{pmatrix} + \begin{pmatrix} 0 \\ -\frac{1}{m} \end{pmatrix} f(t) \tag{15.4}$$

where x and v are the variables of the state space representation of the system. If they are known at an initial time and if the values of the subsequent inputs are also known, the system can be completely determined at any time. Following are the four matrices of the state space representation of the system.

$$\mathbf{A} = \begin{pmatrix} 0 & 1 \\ -\frac{k}{m} & -\frac{b}{m} \end{pmatrix} \tag{15.5}$$

$$\mathbf{B} = \begin{pmatrix} 0 \\ -\frac{1}{m} \end{pmatrix} \tag{15.6}$$

$$\mathbf{C} = \begin{pmatrix} 1 & 0 \\ 0 & 1 \end{pmatrix} \tag{15.7}$$

$$\mathbf{D} = \begin{pmatrix} 0 \\ 0 \end{pmatrix}. \tag{15.8}$$

The TDF cluster that can be designed to model the accelerometer is presented in Figure 15.2.

The pulse is generated by a TDF module named **Gaussian pulse** and propagated to the downstream TDF module **accelerometer**. An optional module named **display**, connected to the output of the accelerometer acts as a sink that receives as input the x displacement values over time. The output port of **Gaussian pulse** is connected to the input port of **accelerometer** via a TDF signal that carries the sampled

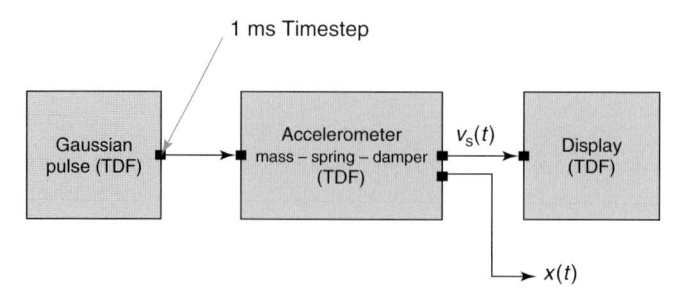

Figure 15.2 The MEMS accelerometer TDF cluster.

seismic pulse values. A 1 ms timestep association has been performed on the output port of **Gaussian pulse**; that is, this module generates a new seismic pulse sample every millisecond. This timestep is automatically propagated to the rest of the cluster modules during the simulator elaboration (i.e., each module is fired every millisecond in the following statically determined order: Gaussian pulse → accelerometer → display). The **processing()** function of this module is shown in Listing 1 in Appendix 15.A.

Each time the **Gaussian pulse** module is fired (every millisecond), the current value of simulated time is saved in the C++ double value t and used to compute the next value of the Gauss function. The computed Gauss value is written to the TDF output port **out**.

The complete code of the accelerometer TDF module, using Eqs. (15.4)–(15.8), is given in Listing 2 in Appendix 15.A and is briefly commented.

Line 1 defines the name of the TDF module, accelerometer. Lines 3–5 state that the accelerometer has one input port inp (representing the incoming seismic perturbation) and two output ports outx (representing the displacement x of the seismic mass) and outv (representing the velocity v). The **initialize()** function, beginning at line 7, is called once for each involved TDF module before the simulation starts. It allows to set member variables, values, and data structures for the TDF module. In this case, this function initializes the three equation coefficients k, b, and m, as well as the four state space matrices, with the appropriate values (lines 12–17). The four matrices are declared as C++ matrix objects in line 31. Accordingly, the initial state of the system is given by lines 18 and 19, and the corresponding C++ vector is declared in line 32. Thanks to a predefinite (yet powerful) operator for handling state space representation, the **processing()** function in line 21 remains quite readable. Each time the module is fired, the **processing()** function begins by reading the input force vector f value (line 23) and then use it as the last argument of the state space operator sca_ss function, named ss1() in the example (line 24). This function produces a vector y carrying the output values ($y(0)$ carrying the current x displacement, $y(1)$ carrying the current velocity v).

Provided the three TDF modules presented are instantiated and interconnected in a standard netlist main.cpp file, simulation plots such as the one presented in the upper part of Figure 15.3 can be obtained. As expected, the seismic mass reacts to the Gaussian pulse and is subject to a damped oscillation.

15.2.3
Electrical Linear Network Model of the MEMS Accelerometer

Theory on energetic equivalence between physical domains states that the mechanical accelerometer characterized by its mass, spring, and damper can be transposed into its electrical equivalent Resistance Inductance Capacitance (RLC) counterpart [3]. One can identify the mechanical displacement with the electrical charge of a capacitor and the velocity with the voltage of a parallel RLC circuit. Exact equivalences are detailed in Table 15.1.

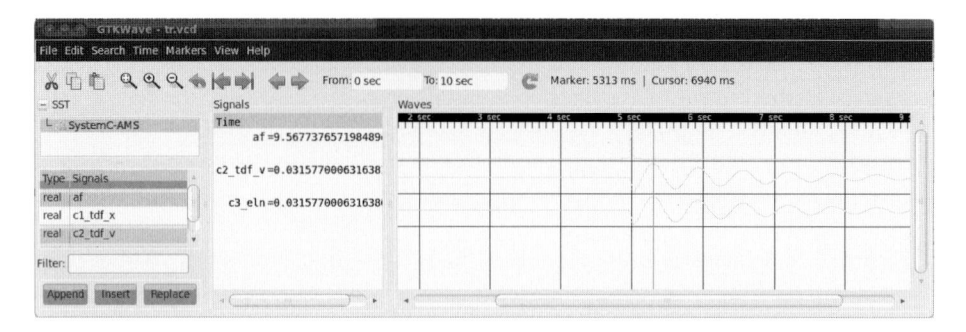

Figure 15.3 The accelerometer cluster including its RLC representation.

Table 15.1 Analogy between mechanical and electrical domains.

Mechanical	Electrical
Common value is velocity v	Common value is voltage v_s
Displacement, integral of velocity	Electrical charge
Mass m	Capacitor C
Spring constant $\frac{1}{k}$	Inductor L
Damping factor b	Resistor $\frac{1}{R}$
Force f	Current i_s

Figure 15.4 shows how the RLC representation of the accelerometer, modeled in SystemC AMS ELN, can directly be connected to the same TDF Gaussian pulse source.

The ELN representation is based on the use of four ELN primitives: a controlled current source generating a current i_s based on the value of the applied force $f(t)$, a resistor R, a capacitor C, and an inductor L. v_s is the corresponding voltage. All these primitives are connected in parallel according to the equation

$$i_s(t) + i_C + i_R + i_L = 0 \tag{15.9}$$

After integration, this equation can be rewritten as

$$\dot{v}_s(t) = -\frac{1}{LC}\int_0^t v_s(u)du - \frac{1}{RC}v_s(t) - \frac{1}{C}i_s(t) \tag{15.10}$$

The SystemC AMS model of this RLC representation is given in Listing 3 in Appendix 15.A.

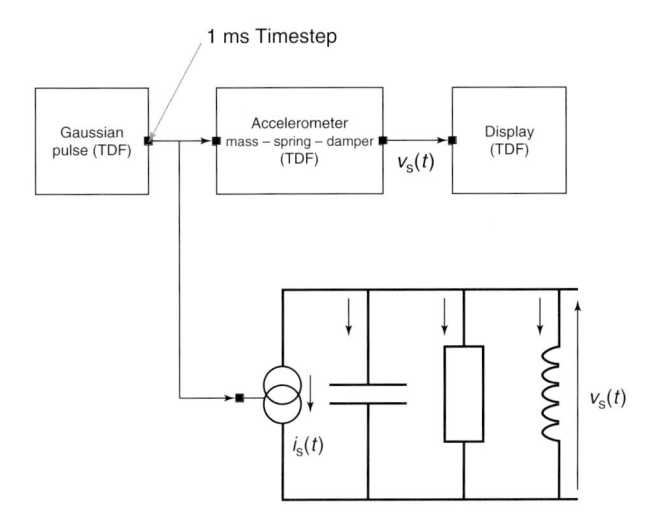

Figure 15.4 Plots showing the exact equivalence between TDF and ELN representations.

15.3
Case Study: Detection of Seismic Perturbations Using the Accelerometer

The described accelerometer has been integrated in a WSN use case that intends to show all the benefits of using SystemC AMS to model and simulate complex heterogeneous systems at a high level of abstraction. The presented system has been implemented in SystemC and SystemC AMS and is composed of approximately 100 C++ .cpp and .h files. Interesting results concerning some parts of the design have already been published [10]. As a point of interest, this section shows how a kinetic battery model and a seismic wave equation can be coded in SystemC AMS and illustrates some of the SystemC AMS structure used to smoothly interface the digital and AMS worlds.

The WSN in Figure 15.5 consists of four independent nodes N_0-N_3 located at well-known positions on a grid matrix representing the monitored ground surface. The nodes continuously monitor seismic activity with the previously described MEMS accelerometer. As shown in Figure 15.6, the whole system can be described as a set of interconnected modules belonging to five disciplines: physics (seismic perturbation generator and seismic sensor, modeled in SystemC AMS TDF), analog electronics (seismic ADC, modeled in SystemC AMS TDF and SystemC), digital (microprocessor, RAM, RF data serializer/deserializer, interrupts, modeled in SystemC), analog RF (2.4 GHz communications between nodes, modeled in SystemC AMS TDF), and chemistry (kinetic battery as power supply, modeled in SystemC AMS TDF). TDF clearly opens the gate to fast simulations and is necessary for systems of that complexity.

The seismic perturbation of Figure 15.5 is modeled as a Gaussian-like pulse with an initial amplitude on the grid, using the modeling principles detailed before (see

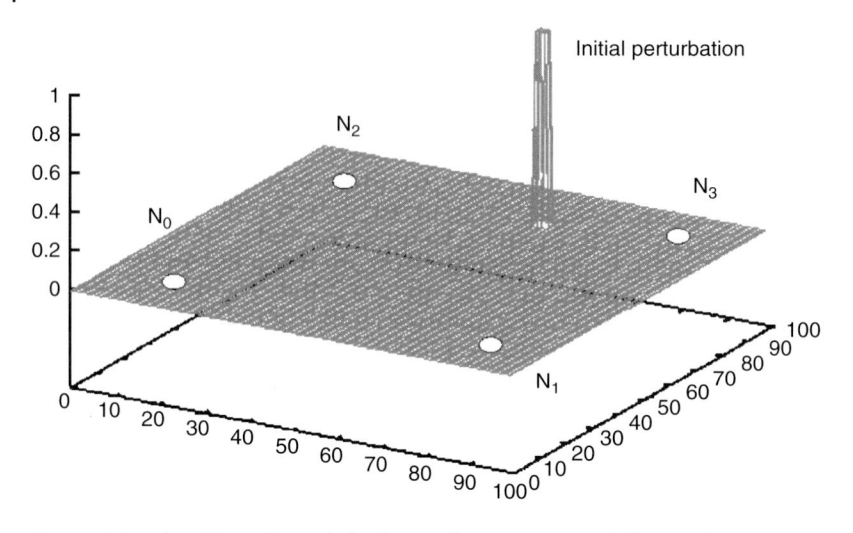

Figure 15.5 The 100 × 100 grid, the four nodes N_0 to N_3, and the initial pseudo-Gaussian perturbation.

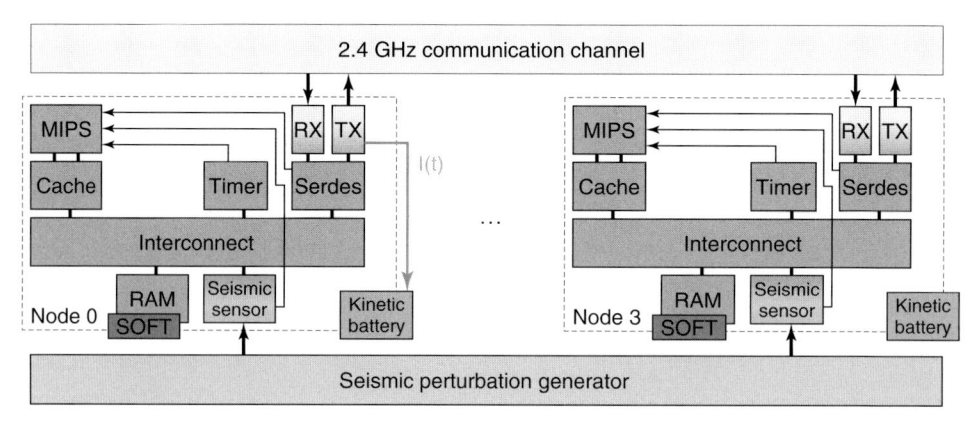

Figure 15.6 The complete WSN, consisting of four communicating nodes N_0 to N_3.

Listing 1 in Appendix 15.A). It is propagated both in time and space by the use of the wave equation with constant radial velocity. Once the acceleration sensor of a node N_i detects the presence of the seismic perturbation, N_i propagates its node identifier as well as its internal time stamp to other nodes, thanks to a quadrature phase shift keying (QPSK)-modulated 2.4 GHz radio-frequency (RF) transceiver and a noisy communication channel. Each node thus gathers information on all the consecutive time stamps of seismic perturbations. Thanks to a triangulation algorithm, and when it has accumulated enough information, each node is able to compute the epicenter using the embedded microprocessor to solve a system of nonlinear equations. Nodes that are totally equivalent from the HW and SW perspectives have an internal real-time clock initialized with the very same initial time stamp 0 (an unrealistic assumption to avoid node synchronization). The protocol retained to prevent collisions of RF messages is time division multiple access (TDMA).

15.3.1
Seismic Stimuli and Seismic Sensors in SystemC AMS TDF

The seismic perturbation is modeled as a pseudo-Gaussian pulse with an initial amplitude f that propagates to all points of the ground surface by means of the following wave equation:

$$\frac{d^2f}{dt^2} = c^2 \left(\frac{d^2f}{dx^2} + \frac{d^2f}{dy^2} \right) \tag{15.11}$$

The seismic stimulus generator is a SystemC AMS TDF module connected to the seismic sensor of each node with SystemC AMS ports. The environment is represented by a three-dimensional matrix (two dimensions for space and one for time). The amplitude of the wave at a given point is calculated using the discretized version of Eq. (15.11):

$$f_{x,y,t+1} = c^2 \cdot ((f_{x+1,y,t} + f_{x-1,y,t} + f_{x,y+1,t} + f_{x,y-1,t}) \\ (-4 \cdot f_{x,y,t})) - f_{x,y,t-1} + 2 \cdot f_{x,y,t} \tag{15.12}$$

This equation shows that it is possible to evaluate the amplitude of a point of the matrix at time t_{i+1} with the amplitude values of its neighbors at time t_i and the previous amplitude of the point at time t_{i-1}. Once the TDF module is activated by the SystemC AMS scheduler, the **processing()** function described in Listing 4 in Appendix 15.A is executed. First, all points of coordinates (x,y) on the grid are updated (lines 7–14). wave[x][y][1] and wave[x][y][0] represent the values at the previous iterations of the amplitude of point (x,y). After this update phase, the value corresponding to the position of the sensor is written to each node connected (lines 16–19). Then the current value becomes a previous one (lines 20–24).

The read amplitude incoming from the seismic perturbation generator is directly converted into its digital equivalent by means of a second-order sigma–delta 1 bit modulator with return-to-zero feedback [11] and a decimator using a third-order finite impulse response filter (Fir2) [12], which can be parameterized to generate a n-bit digital word [13].

15.3.2
Digital Controller in SystemC

The digital part of each WSN node has been designed with a dedicated library of digital models. All the models in the library are BCA compliant (Bit-Cycle Accurate), respect the SystemC description rules [5], and are interoperable, that is, they follow the standard Open Core Protocol (OCP) that allows any digital IP of the library to be interconnected with others through the use of a Network on Chip (NoC). To achieve this performance, all the models have been written as finite state machines (FSMs) (cf. Section 4.2.2). A given component may contain several Moore FSMs, and as such, can be fully represented as a set of two combinatorial functions (the transition function and the Moore output generation function) and a register containing the component current state. The simulation of a whole digital system merely consists, during elaboration of the simulator, in setting up a network of communicating synchronous finite state machines (CSFSMs). On the clock rising edge, all the transition functions of all the models in the digital design are called, and on the clock falling edge, all the Moore output generation functions are called. The simulation cycle of such a simulation is therefore quite canonical, and the discrete event algorithm performed by the SystemC simulation kernel has very low overhead because the event list is reduced to the rising and falling clock events. Processors are implemented as instruction set simulators (ISSs) that can simply be interconnected to the memory hierarchy (cache L1, cache L2, and main memory). Processor models include MIPS32, PowerPC405, ARM7, Sparc V8, and Nios. Furthermore, the library is freely available.

In the use case, the digital part mainly consists of a 32-bit MIPS32 processor and its associated instruction and data L1 caches, a RAM containing the embedded application code and data, a timer used to generate TDMA interrupts, the digital part of the seismic sensor, and the component responsible for the serialization/deserialzation (serdes) of the RF data. The MIPS32 may receive three interrupts, from the timer (to identify if the node is in its TDMA timeslot), serdes (when RF data has been received), and seismic sensor (when a given amplitude threshold has been reached, indicating the presence of the perturbation at the sensor location). The RAM model is smart enough to be able to initialize its code and data memory segments with the contents of a embedded Linux file (ELF) compliant binary file compiled with the GCC (GNU Compiler Collection) toolchain cross-compiled for MIPS processors.

15.3.3
2.4 GHz RF Transceiver

The RF transceiver is responsible for converting the digital bitstream emitted by the serdes into RF information (RF transmitter) and vice versa (RF receiver). The RF model uses a coherent QPSK transmission scheme, as explained in [14] and shown in Figures 15.7 and 15.8 with a $f_c = 2.4$ GHz carrier frequency and a $f_b = 2.4$ MHz data frequency. Additive white Gaussian noise (AWGN) allows taking into account channel noise in the modeling of the RF communication channel and is necessary for calculating the fundamental RF characteristic bit error rate (BER) with respect to signal-to-noise ratio (SNR). In this nearly ideal modeling, the power amplifier (PA) and the low-noise amplifier (LNA) models are very straightforward. Likely, the communication channel is considered as an ideal gain block with an AWGN.

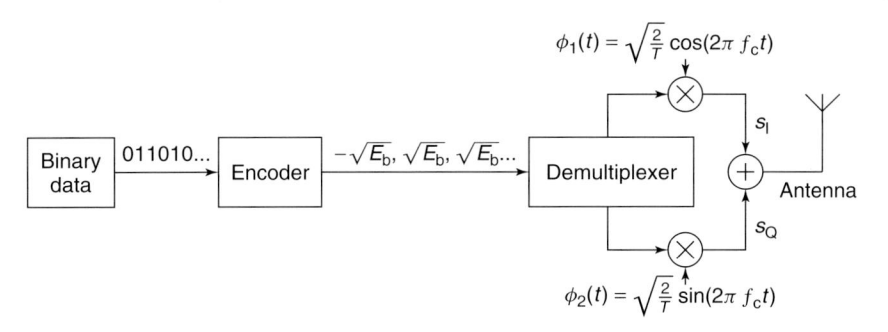

15.3 Case Study: Detection of Seismic Perturbations Using the Accelerometer

Figure 15.7 QPSK RF transmitter.

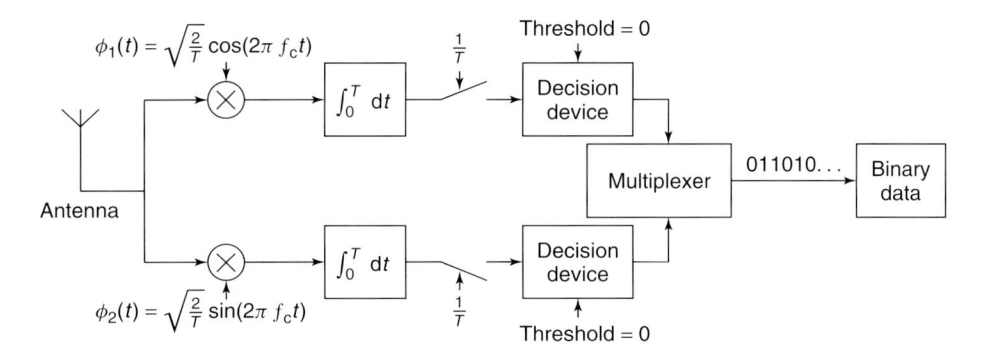

Figure 15.8 QPSK RF receiver.

To prevent this simulation time from becoming too prohibitive, and hence to validate and optimize parts of the WSN, a common technique [15] is to define an equivalence between the initial 2.4 GHz RF signal and its baseband representation to remove the carrier from the RF signal expression

$$x(t) = DC + I_1 \cos(\omega t) + I_2 \cos(2\omega t) + I_3 \cos(3\omega t)$$
$$+ Q_1 \sin(\omega t) + Q_2 \sin(2\omega t) + Q_3 \sin(3\omega t) \qquad (15.13)$$

In the baseband equivalent transmission scheme, the only data actually transmitted over the RF channel are the seven coefficients of Eq. (15.13) that represent signal harmonics and their associated second- and third-order distortions, at a rate that is 10 000 times smaller than that in a direct/naive simulation.

Because SystemC AMS uses its own representation of the simulation time, it is not recommended to directly interface a SystemC AMS TDF module with a SystemC module because of possible synchronization issues. This is why SystemC AMS provides dedicated converter ports (sca_tdf::sca_de::sca_out or sca_tdf::sca_de::sca_in) that ensure that the value written or read will be the right one at the right time. As an example, Listing 5 in Appendix 15.A shows the TDF module for the Power Amplifier pa, part of the RF transceiver. The RF transmission chain is modeled using baseband equivalent, which is represented in this module by the template parameter (lines 4 and 5). The input port *en* line (3) is connected to a control port from a SystemC module, which can (de)activate the RF transmitter. Lines 7–14 show how the processing function of the module uses this enable flag to send a valuated BB sample (corresponding to the case the node is actually emitting) or a null BB sample (when the RF transmitter is deactivated). All the required synchronization is done by the *sca_tdf::sca_de::sca_in* port, so the use of digital ports is transparent from the designers perspective.

15.3.4
Battery Modeling

For ecological and energetic reasons, the modeling of a WSN should take power estimation issues into account as early as possible. The described use case relies on the intuitive Kinetic Battery Model (KiBaM) from Manwell and McGowan [16] to address these issues. The model is called kinetic because it relies on the use of chemical kinetic differential equations. As shown in Figure 15.9, the battery charge is distributed over two wells: the available-charge well and the bound-charge well.

The available-charge well supplies electrons directly to the load, the WSN node, while the bound-charge well supplies electrons only to the available-charge well. The rate at which charge flows between the wells depends on the difference in heights of the two wells, and on a parameter k. The parameter c gives the fraction of the total charge in the battery that is part of the available-charge well. The change in the charge in both wells is given by the following system of differential equations:

$$\frac{dy_1}{dt} = -I(t) + k \cdot (h_2 - h_1),$$
$$\frac{dy_2}{dt} = -k \cdot (h_2 - h_1) \tag{15.14}$$

with initial conditions $y_1(0) = c \cdot C$ and $y_2(0) = (1-c) \cdot C$, where C is the total battery capacity. For h_1 and h_2, we have, $h_1 = y_1/c$ and $h_2 = y_2/(1-c)$. When

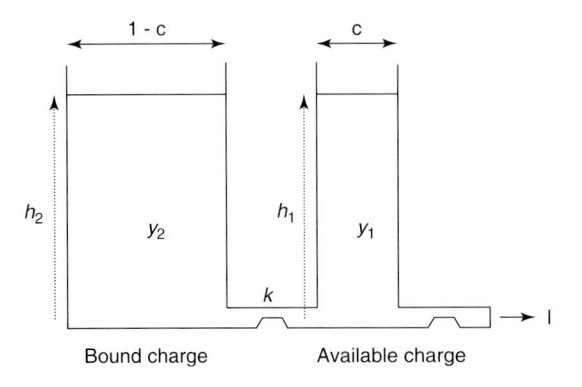

Figure 15.9 The battery model from Manwell and McGowan (Source: Taken from [17]) and built on the available-charge and bound-charge wells.

a load $I(t)$ is applied to the battery, the available charge reduces and the difference in heights between the two wells grows. When the load is removed, charge flows from the bounded-charge well to the available-charge well until h_1 and h_2 are equal again. So, during an idle period, more charge becomes available and the battery lasts longer than when the load is applied continuously. From a SystemC AMS perspective, a TDF module modeling the kinetic battery can be achieved by using the same state space modeling techniques presented earlier.

15.3.5
Embedded Software, Cross-Compiled GNU GCC Application for MIPS

The embedded application, responsible for initializing peripherals and managing interrupts, is cross-compiled with GCC for the MIPS32 processor. The application programmer can access the registers of the digital components (e.g., SERDES) directly with the dedicated set and get functions **periph_io_set()** and **periph_io_get()**. In this application, the four nodes communicate using a simple TDMA protocol implemented in SW and paced by an internal timer. One can examine the code of the main function given in Listing 6 in Appendix 15.A. Each node has a temporal slot associated, which can be used to broadcast a message to other nodes. A timer is set up to raise an interrupt that indicates that a new TDMA slot has begun (line 7). If this slot is the one associated to the node (line 10), a pending RF message can be sent (lines 11–21). If it is not the present TDMA timeslot, the node is set in a reception mode waiting for a message from the other nodes (lines 22–24). The message is a 32 bit word, 4 bits for the identification of the node sending it and the others for data (lines 14 and 15). The data in this case is the time at which the seismic sensor detects that the amplitude is over a given threshold. When an interrupt is generated by the seismic sensor of the node (line 11) or when a message from another node is received, the corresponding information is saved in a sensor table by a call to the **fill_table(...)** function. As soon as all the sensors have transmitted their information (line 9), the node computes the epicenter (lines 26–29).

15.3.6
Simulation Results

For a simulated microcontroller clock of 100 MHz, the simulation of these four nodes WSN lasts approximately 2 min on an AMD X2 PC running Linux at 2.5 GHz with 1 GB RAM. There are three TDF clusters, one for RF, one for the seismic perturbation, and one managing power issues. Owing to the MoCs used, it is not particularly difficult to obtain the functional mode of any component (AMS or digital) of the system at any time. Knowing this functional mode for each component and the corresponding electrical instantaneous current, it is possible to compute the global instantaneous current that can in turn be used as a subtracting input for the battery model. Figure 15.10 shows the evolution of the charge in each

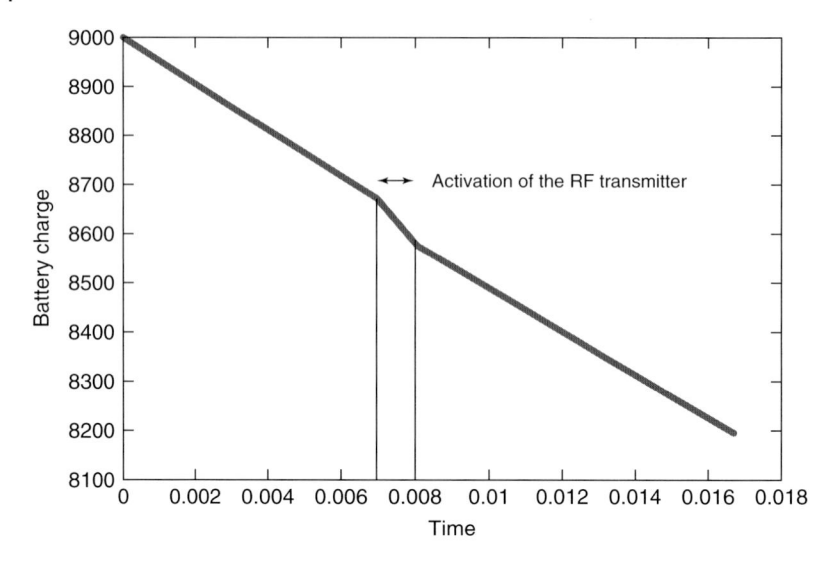

Figure 15.10 Battery charge evolution for one of the four nodes.

battery. As the current model is very simplified (only the functional mode of the RF TX component is considered), the recovery effect of the battery modeled by the two wells cannot be directly pointed out. The slopes of the plots correspond to the two cases, RF transmitter disabled (smooth slope) and enabled (steep slope).

In the RF part, an analysis has been performed to display BER according to SNR variation. A theoretical BER has been computed from AWGN characteristics and has been successfully compared to simulation results. This analysis has been extended in [13] to transceiver impairments.

15.4
Conclusion

The chapter shows that the system simulation of a complete WSN that encompasses several domains is actually possible with open source tools, with excellent accuracy compared to traditional commercial tools. Model interoperablity and performance obtained through the use of C++, SystemC and SystemC AMS, and simulation time (when using state-of-the-art RF modeling techniques) are extremely encouraging. One of the most interesting aspects of this work is the ability for the system designer to tune the system completely and to actually write the inner simulation loop according to his or her needs. The current version of this freely available HW-aware WSN simulator can easily be taken as a starting point to add real communication protocols, to handle synchronization issues between nodes, and to develop real power estimation methods.

Appendix

Listing 1 The processing function of the Gaussian pulse TDF module

```
void processing()
{
 double t=out.get_time().to_seconds();
 out.write(exp(-0.5*(((t-mu)/sig)*((t-mu)/sig))/(sig*sqrt(M_PI))));
}
```

Listing 2 The complete source code of the accelerometer TDF module

```
1   SCA_TDF_MODULE(accelerometer)
2   {
3       sca_tdf::sca_in<double> inp;
4       sca_tdf::sca_out<double> outx;
5       sca_tdf::sca_out<double> outv;
6       accelerometer(sc_core::sc_module_name nm){ }

7   void initialize()
8   {
9       double coefb = 10.0e-2 ; // viscous friction
10      double coefk = 10.0 ; // spring constant
11      double m = 0.1 ;      // seismic mass

12      a(0,0) = 0.0          ;        a(0,1) = 1.0 ;
13      a(1,0) = -coefk/m ; a(1,1) = -coefb/m ;

14      b(0,0) = 0.0          ;        b(1,0) = -1.0/m ;

15      c(0,0) = 1.0          ;        c(0,1) = 0.0 ;
16      c(1,0) = 0.0          ;        c(1,1) = 1.0 ;

17      d(0,0) = 0.0          ;        d(1,0) = 0.0 ;

18      state(0) = 0.0 ;
19      state(1) = 0.0 ;
20  }

21  void processing()
22  {
23      f(0) = inp.read() ;

24      sca_util::sca_vector<double> y = ss1(a,b,c,d,state ,f);
25      outx.write(y(0));
26      outv.write(y(1));
27  }

28  private:
29      sca_tdf::sca_ss ss1; // state-space equation
31      sca_util::sca_matrix<double> a, b, c, d; //matrices
32      sca_util::sca_vector<double> state; //state vector
33      sca_util::sca_vector<double> f; //input
34  };
```

Listing 3 The SystemC AMS netlist corresponding to the ELN representation of the accelerometer

```
1   int sc_main(int argc, char* argv[])
2   {
3     sca_tdf::sca_signal<double> disp_tdf_x,disp_tdf_v,f;
4
5     ...
6     sca_eln::sca_tdf_isource *iin_src;
7     sca_eln::sca_r *i_r;
8     sca_eln::sca_c *i_c;
9     sca_eln::sca_l *i_l;
10    sca_eln::sca_node_ref gnd;

11    sca_eln::sca_node disp_eln;

12    // m = C
13    // k = 1/L
14    // b = 1/R

15    double m = 0.1 ;
16    double coefk = 10.0 ;
17    double coefb = 10.0e-2 ;

18    iin_src = new sca_eln::sca_tdf_isource(``iin_src'',-1.0);
19    iin_src→n(disp_eln);
20    iin_src→p(gnd);
21    iin_src→inp(f);//incoming from TDF Gaussian pulse
22
23    i_r = new sca_eln::sca_r(``i_r'');
24    i_r→n(disp_eln);
25    i_r→p(gnd);
26    i_r→value=1.0/coefb;
27
28    i_c = new sca_eln::sca_c(``i_c'');
29    i_c→n(disp_eln);
30    i_c→p(gnd);
31    i_c→value=m;
32
33    i_l = new sca_eln::sca_l(``i_l'');
34    i_l→n(disp_eln);
35    i_l→p(gnd);
36    i_l→value = 1.0/coefk;
37
38    ...
39  }
```

Listing 4 The seismic stimulus generator in SystemC AMS

```
1   SCA_TDF_MODULE (wavegen)
2   {
3           sca_tdf::sca_out < double > out_sensor[4];
4   ...
5   void processing () {
```

```
6    int x,y,c;

7    for (x = 1;x<WAVE_SIZE-1;x++) {
8      for (y = 1;y<WAVE_SIZE-1;y++){
9        wave[x][y][2]=2.0*wave[x][y][1] -
10         wave[x][y][0] +
11         cd*cd*(wave[x+1][y][1] + wave[x-1][y][1] +
12                wave[x][y+1][1] + wave[x][y-1][1] -
13                4.0*wave[x][y][1]);
14   } }
15   ...
16   for (c=0 ; c < NB_SENSORS; c++)
17     out_sensor[c].write(
18       wave[pos_x_sensor[c]][pos_y_sensor[c]][2]
19                );

20   for (x=0;x<WAVE_SIZE;x++) {
21     for (y=0;y<WAVE_SIZE;y++) {
22       wave[x][y][0]=wave[x][y][1];
23       wave[x][y][1]=wave[x][y][2];
24   } }
25   }
26   ...
27   }:
```

Listing 5 The pa TDF module

```
1   SCA_TDF_MODULE (pa)
2   {
3     sca_tdf::sca_de::sca_in < bool > en;
4     sca_tdf::sca_in < BB > in;
5     sca_tdf::sca_out < BB > out;
6     ...
7     void processing () {
8       ...
9       BB input=in.read();
10      if (en.read()==true)
11        out.write (GAIN(input));
12      else
13        out.write (Nullbb);
14    }
15  ...};
```

Listing 6 The main() function of the embedded software

```
1   int main(int argc, char**argv)
2   {
3     uint32_t data;
4     init();
5     printf(``Sensor %d is monitoring...\n'', id);
```

```
 6   while (1) {
 7    if (tdma_trigger==1) {
 8     tdma_trigger=0;
 9     if(!flag_calcul){
10      if (tdma_slot==id) {
11       if(has_value) {
12        periph_io_set(
13           base(SERDES), SERDES_CTRL, TX_MODE_ONLY);
14        data=(id << 28)|
15           (sensor_cpt & 0x0fffffff );
16        printf(''sending data   %x\n$'',data);
17        periph_io_set(
18           base(SERDES), SERDES_DATA, data);
19        fill_table(id,sensor_cpt);
20        has_value = 0;
21       } }
22      else
23       periph_io_set(
24          base(PERIPH),PERIPH_CTRL,RX_MODE_ONLY);
25     }
26     else {
27      irq_disable();
28      compute_ epicentre();
29     } } }
30    return 0;
31   }
```

References

1. Bond, B. and Daniel, L. (2005) Parameterized model order reduction of nonlinear dynamical systems. IEEE ACM International Conference on Computer-Aided Design (ICCAD), San Jose, pp. 487–494.
2. Rudnyi, E. and Korvink, J. (2006) Model order reduction for large scale finite element engineering models. European Conference on Computational Fluid Dynamics (ECCOMAS CFD), Egmond aan Zee, Netherlands.
3. Galayko, D. and Kaiser, A. (2002) *Analog Integrated Circuits and Signal Processing*, **32** (1), 17–28.
4. Caluwaerts, K. and Galayko, D. (2008) SystemC AMS modeling of an electromechanical harvester of vibration energy. IEEE Forum on Specification, Verification and Design Languages (FDL), Stuttgart, pp. 99–104.
5. Accellera Systems Initiative: SystemC Language Reference Manual. IEEE Standard 1666-2005 (2006). *http://www.accellera.org/downloads/standards/systemc/*.
6. Accellera Systems Initiative AMS Working Group: Standard SystemC AMS Extensions Language Reference Manual. (2010), *http://www.accellera.org/downloads/standards/systemc/ams/*.
7. Vachoux, A., Grimm, C., and Einwich, K. (2004) Towards analog and mixed-signal SOC design with SystemC AMS. IEEE International Workshop on Electronic Design, Test and Applications (DELTA).
8. Vachoux, A., Grimm, C., and Einwich, K. (2003) Analog and mixed signal modelling with SystemC AMS. IEEE International Symposium on Circuits and Systems (ISCAS), Bangkok.
9. Buck, J., Ha, S., Lee, E.A., and Messerschmitt, D.G. (1992), Ptolemy:

A Framework for Simulating and Prototyping Heterogeneous Systems. Int. J. Comput Simula., **4**, PP. 155–182, (special issue on "Simulation Software Development" April, 1994).
10. Vasilevski, M., Pêcheux, F., Beilleau, N., Aboushady, H., and Einwich, K. (2008) Modeling an refining heterogeneous systems with SystemC AMS: application to WSN. The Design, Automation, and Test in Europe Conference (DATE), Munich.
11. Aboushady, H., Montaudon, F., Paillardet, F., and Louërat, M.M. (2002) A 5 mW, 100 kHz bandwidth, current-mode continuous-time sigma-delta modulator with 84 dB dynamic range. IEEE European Solid-State Circuits Conference (ESSCIRC), Florence.
12. Aboushady, H., Dumonteix, Y., Louërat, M., and Mehrez, H. (2001) Efficient polyphase decomposition of Comb decimation filters in sigma-delta analog-to-digital converters. IEEE Trandactions on Circuits and Systems-II (TCASII).
13. Vasilevski, M., Pêcheux, F., Aboushady, H., and de Lamarre, L. (2007) Modeling heterogeneous systems using SystemC AMS, Case Study: A Wireless Sensor Network Node. IEEE International Behavioral Modeling and Simulation Conference (BMAS), San Jose.
14. Haykin, S. (1994) *Communication Systems*, 3rd edn, Wiley.
15. Yee, D.G.W. (2001) A design methodology for highly-integrated low-power receivers for wireless communications. Ph.D. thesis. University of California, Berkeley.
16. Manwell, J. and McGowan, J. (1993) *Solar Energy*, **50**, 399–405.
17. Jongerden, M. and Haverkort, B. (2008), Battery modeling, http://doc.utwente.nl/64556/. (accessed 2008)

16
System Level Modeling of Electromechanical Sigma–Delta Modulators for Inertial MEMS Sensors

Michael Kraft

16.1
Introduction and Motivation

This chapter explores the system level modeling in Matlab/Simulink of microelectromechanical system (MEMS) sensors, which are incorporated in a closed loop force feedback control system. This approach has been widely and very successfully used for MEMS accelerometers and gyroscopes with a capacitive sensing element; and hence are used here exclusively as examples. MEMS accelerometers and gyroscopes are the so-called inertial sensors that have numerous applications. Their development was mainly driven by the automotive industry, and in modern vehicles, dozens of these sensors can be found for various safety and passenger comfort applications: MEMS accelerometers trigger the release of the airbags and tightening of the seatbelts in case of an accident; in electronic vehicle stability systems, they measure forces so that the control system can adjust the power distribution to the wheels, they can be used to measure noise and vibration for active damping control systems and to aid for short-term backup for GPS signal outages. Gyroscopes are also used for electronic stability control systems and for GPS backup, and for anti-roll-over protection systems. More recently, the applications of MEMS inertial sensors have broadened out, in particular, into the consumer market with examples of game controllers such as for Nintendo Wii, platform stabilization for camcorders and cameras, and in smart phones. These examples are extremely price conscientious and hence added complexity due to a force feedback control system is often prohibitive because of the cost increase. Also, for most of the applications mentioned so far, the highest performance is often not required. Hence this chapter deals with MEMS inertial sensors that are suitable for niche applications in which a higher price is justified if better performance can be achieved. Applications include, in the defence and aerospace industry, inertial navigation of missiles and airplanes, aviation instrumentations, and vehicle structural health monitoring; in robotics, in motion control and platform stabilization; in seismology and gas and oil exploration, as a replacement for geophones; and in high-speed train systems, for carriage stabilization. Other applications will emerge in the near to mid-term future once high-performance inertial MEMS sensors become more widely available (and

System-level Modeling of MEMS, First Edition. Edited by T. Bechtold, G. Schrag, and L. Feng.
© 2013 Wiley-VCH Verlag GmbH & Co. KGaA. Published 2013 by Wiley-VCH Verlag GmbH & Co. KGaA.

16 System Level Modeling of Electromechanical Sigma–Delta Modulators

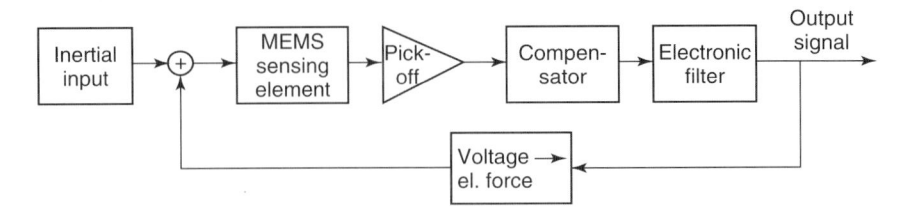

Figure 16.1 MEMS sensor incorporated in a closed loop force-feedback loop. The output signal is used to generate a feedback voltage setting up an electrostatic feedback force counteracting the inertial force to be measured.

as a result decrease in price) – in particular, vibration monitoring for structural health monitoring system of vehicles, machines, and buildings is expected to grow considerably. It is therefore justified to make efforts to further increase the performance of MEMS inertial sensors. Incorporating them into a force feedback loop, as shown schematically in Figure 16.1, is one way to realize a performance enhancement, especially in terms of linearity, bandwidth, and dynamic range. The most important argument for such a closed loop approach is that the performance parameter sensitivity of the overall sensor system on the micromachined sensing element is reduced, and hence issues due to microfabrication tolerances can be alleviated. Fabricating a MEMS sensing element made of silicon can often result in variations of an important design parameter such as the spring constant; variations of 10–20% are not uncommon, which, in turn, influence the resonant frequency and hence the bandwidth of the sensor. In a closed loop approach, the influence of fabrication tolerances on the sensor specification is considerably reduced.

Referring to Figure 16.1, the basic principle of a closed loop MEMS inertial sensor is to generate a feedback force that counteracts the inertial force (due to acceleration for a MEMS accelerometer, and Coriolis force in a gyroscope). The deflection of the proof mass is sensed capacitively by an interface circuit and then passed on to an electronic signal processing block that generates the output signal and the feedback voltage. The feedback voltage is applied to the sensing element to generate an electrostatic force that, ideally, cancels the input inertial force. In practice, there always will be a residual input force acting on the proof mass of the sensing element, which effectively becomes the error signal in control engineering terms. (Since this chapter focuses on simulation, it should be noted that the reader is assumed to be familiar with the fundamental operating principles of MEMS capacitive inertial sensors; if not, see, for example, [1, 2].)

There are two fundamental choices for the type or feedback control system: (i) analog force feedback and (ii) quasi digital based on sigma–delta modulators ($\Sigma\Delta$Ms). Here, we exclusively deal with the second choice, which has advantages in terms of stability as no electrostatic latch up is possible and resulting in a direct digital output in form of a pulse-modulated bitstream, which can directly interface to a DSP (digital signal processor) for further signal processing. The reader is referred to the literature [3, 4] for an in-depth discussion on the pros and cons of the two approaches for capacitive MEMS inertial sensors. A pragmatic reason, for the purpose of this book, to choose closed loop inertial sensors based on $\Sigma\Delta M$ is that their design strongly relies on system level simulations, as for their performance only approximate analytical solutions can be derived, and their loop stability can only inadequately be predicted by analytical means; the latter is particularly true for higher order $\Sigma\Delta M$. The most widely used simulation software for such system level simulations is Matlab and Simulink, albeit other software packages exist and can be used. Matlab/Simulink has the advantage that there is a wide range of literature about $\Sigma\Delta M$ simulation using this software, including some very useful toolboxes that can be downloaded from the Matlab file exchange server [5].

$\Sigma\Delta Ms$ were initially developed as analog to digital converters (ADCs) and have gained enormous popularity since their inception in the late 1970s and early 1980s. They have developed into a *de facto* standard for ADC requiring high resolution but relatively low bandwidth. A $\Sigma\Delta M$ ADC is a feedback system where one or several integrators (the number of integrators is usually referred to as the *order of the* $\Sigma\Delta M$) are used to filter the analog input signal, which is then passed to a sampled single or multibit ADC. The output is fed back to the input by passing it through a digital to analog converter (DAC) in the feedback path. This approach leads to quantization noise shaping where the quantization noise is reduced in the signal band and is more pronounced at frequencies outside the signal band, which can be removed by a digital low-pass filter (often referred to as *decimation filter*). This implies that the sampling frequency of the quantizer in the forward path is higher than the Nyquist frequency of the input signal that is being digitized. Therefore, $\Sigma\Delta M$ are also often called *oversampling ADC*. Many different architectures and topologies have been described in the literature, and several text books provide a good introduction to the topic, for example [6, 7].

A MEMS sensing element embedded in a $\Sigma\Delta M$ forms an electromechanical $\Sigma\Delta M$ (referred to as $EM\Sigma\Delta M$ in the following) and can be schematically represented as in Figure 16.2.

An $EM\Sigma\Delta M$ consists of the following building blocks: (i) the micromachined sensing element; (ii) the pickoff circuit that capacitively measures the displacement

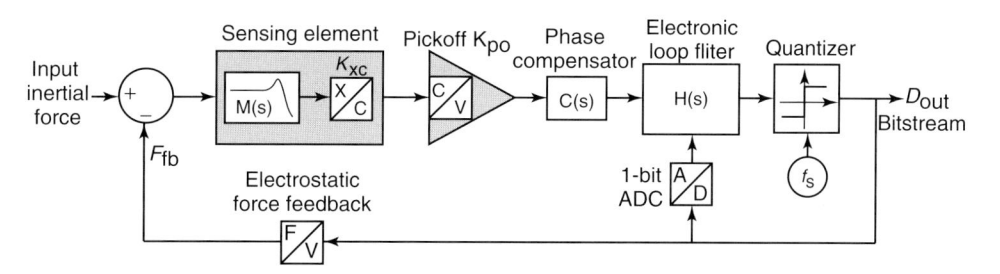

Figure 16.2 Generic block diagram of an electromechanical sigma–delta modulator. The micromachined capacitive sensing element (which can be for an accelerometer or a gyroscope) is cascaded with the pickoff interface circuit, a compensator, the electronic loop filter, and a 1 bit quantizer.

380 | *16 System Level Modeling of Electromechanical Sigma–Delta Modulators*

of the proof mass in response to an inertial force and converts it to a voltage; (iii) a phase compensator (which may not be required if the sensing element is overdamped); (iv) an electronic loop filter comprising several integrators and minor feedback or feedforward loops; (v) a clocked 1 bit quantizer; and (vi) a feedback block that converts the feedback voltage into an electrostatic force acting on the proof mass and counterbalancing the inertial force.

The task is now to develop a sufficiently accurate system level model of the sensing system shown in Figure 16.2 that allows parameter optimization, exploration of various architectures, and a prediction of performance and stability of the sensor.

16.2
Second Order Electromechanical $\Sigma\Delta M$ for a MEMS Accelerometer

16.2.1
Basic Model

Although the chapter title refers to higher order EM$\Sigma\Delta$M, we also discuss the system level modeling of a second order EM$\Sigma\Delta$M for a MEMS accelerometer in more detail, as many modeling and simulation issues can be demonstrated with this somewhat simpler control system. Such an approach was successfully applied to MEMS accelerometers by several researchers mainly in 1990s [3, 8–12]. In a second order EM$\Sigma\Delta$M, the sensing element provides the only dynamics of the control loop, thus H(s) in Figure 16.2 simplifies to unity, and the minor feedback loop "1 bit ADC" is not present. Here, we present the first, rather basic, Simulink model of such an EM$\Sigma\Delta$M in Figure 16.3.

Together with Table 16.1, which defines all system parameters, we describe the model shown in Figure 16.3 as following. The *input* source block provides a sinusoidal acceleration signal with an amplitude in G (1 G = 9.81 m s^{-2}) and is

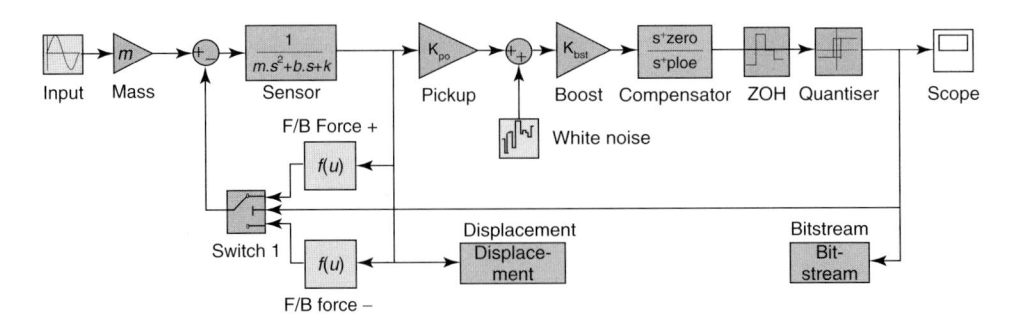

Figure 16.3 Simulink model of a MEMS accelerometer sensing element embedded in a second order sigma–delta modulator. The sensing element provides the only dynamic part of the loop, hence it is a second order electromechanical sigma–delta modulator.

16.2 Second Order Electromechanical ΣΔM for a MEMS Accelerometer

Table 16.1 Parameters of the accelerometer sensing element.

Parameter	Value	Comment
Mass, m (kg)	1e−6	Typical value for a bulk-micromachined accelerometer
Damping coefficient, b (N ms^{-1})	4e−4	Typical value
Spring constant, k (N m^{-1})	5	Low mechanical spring constant
Nominal sense capacitance, C_{nom} (pF)	5.5	High value, typical for a bulk-micromachined device
Nominal electrode gap, d_0 (μm)	5	Conservative value; easy to fabricate
Bandwidth (BW) (kHz)	1	Depends on the application
Max. acceleration, a_{max} (G)	±2 G	Depends on the application
Feedback voltage, V_{fb} (V)	10 V	Chosen to counterbalance 2G inertial force
Sampling frequency (kHz)	131 kHz	Low value, can be used on a PCB

then multiplied by the mass of the sensing element to convert the acceleration into an (inertial) force. The summation node calculates the difference between the (electrostatic) feedback force and the inertial force, which is applied to the transfer function of the sensing element. The sensing element is modeled as a simple second order transfer function $M(s) = (ms^2 + bs + k)^{-1}$, hence the output of the sensor block is the proof mass displacement in meters. The gain block *Pickup* with gain k_{po} converts this displacement into a voltage (which is obviously a very simplified model of a hardware implementation – more about this later). The *White Noise* block adds a noise contamination to the signal, which models the thermal noise introduced by the first amplifier of a capacitive pickup circuit. The *Boost* block represents a boost gain of value k_{bst}. Subsequently, a lead-lag filter (block *compensator*) is used to provide phase lead at lower frequencies, which is required to compensate the phase lag introduced by the sensing element. Then, the signal is sampled (block zero order hold (ZOH)) and subjected to a 1 bit quantizer (which is simply a comparator) digitizing the signal, and providing a ±1 output signal. The feedback consists of blocks that calculate the electrostatic feedback force; this is equivalent to the 1 bit DAC in an electronic ΣΔM, and hence can be regarded as a 1 bit digital to electrostatic force converter. The general equation for the electrostatic feedback force is

$$F_{el} = \operatorname{sgn}(u_1) \frac{1}{2} \frac{\varepsilon_o A_{fb} V_{fb}^2}{(d_0 + \operatorname{sgn}(u_1) \times u_2)^2} \tag{16.1}$$

In addition to the symbols defined in Table 16.1, sgn is the signum function, u_1 represents the value of the *Quantizer* block output, and u_2 the proof mass displacement. (u_1 and u_2 are used as parameter designators as these are standard for Simulink's function blocks.) It models the behavior of an EMΣΔM in the feedback path: depending on the state of the quantizer, an electrostatic feedback force is generated in either positive or negative direction along the sensitivity axis of the accelerometer. On average, it opposes the input inertial force, and hence reduces the proof mass deflection of the sensor compared to an open loop case.

As seen in Figure 16.3, for convenience, Eq. (16.1) is realized as three function blocks: $F/BForce+$, $F/BForce-$, and $Switch1$. $F/BForce+$ calculates the feedback force on the proof mass of the sensing element if the output of the comparator $u_1 = +1$ and $F/BForce-$ when $u_2 = -1$. The $Switch$ block is also controlled by u_1, switching through to its output either the top or bottom signal line for positive or negative values of u_1, respectively. Therefore, the electrostatic force acts on the proof mass in either positive or negative direction.

The sensing element assumed here (and the examples in the rest of the chapter, unless stated otherwise) is fabricated by bulk micromachining using an SOI (silicon on insulator) wafer, hence its axis of sensitivity is in the wafer plane. It has a relatively large (hence heavy) proof mass, a low spring constant, a high nominal capacitance and is underdamped (quality factor 5.6) with a resonance frequency of 356 Hz. The parameters of the sensing element are given in Table 16.1. The dynamic characteristic of the sensing element is shown in Figure 16.4a as a Bode plot, and the actual silicon structure is shown in Figure 16.4b. It is similar but not identical to the one described by Dong et al. [13]. The design and fabrication of the sensing element is not of great importance here; hence, details are not discussed, and the reader is referred to [14].

The choice of the feedback voltage is an important parameter, as it mainly determines the dynamic range of the sensor. This is a distinct advantage of closed loop control for a MEMS accelerometer: for an open loop sensor, the dynamic range is mainly determined by the spring constant and mass of the sensing element. For our sensing element, using the parameters from Table 16.1 and the steady state force balance equation $ma = kx$ together with a maximum value for the proof mass deflection of 10% of the nominal electrode gap,[1] it can be easily calculated that the open loop maximum acceleration is only 0.25G. Using closed loop control, the dynamic range can be extended; for example, if the maximum acceleration for the sensor is supposed to be 2G, the electrostatic force, F_{el}, needs to be equal to the inertial force, F_a:

$$F_{el} = F_a \Rightarrow \frac{1}{2}\frac{\varepsilon_0 a A E V_{fb}^2}{d_0^2} = ma_{max} \Rightarrow V_{fb} = \sqrt{\frac{2d_0^2 ma_{max}}{\varepsilon_0 A_{fb}}} \quad (16.2)$$

For the values assumed here, the required feedback voltage for a 2G dynamic range is 8.6 V. The feedback voltage for the EM$\Sigma\Delta$M needs to be chosen about 15% higher as a second order $\Sigma\Delta$M overloads when the input signal is higher than approximately 85% of feedback signal [15, 16], hence V_{fb} was chosen 10 V here.

There are some other important parameters that need to be chosen. The pickoff gain k_{po} was chosen as 330 000 and the boost gain $k_{bst} = 20$; these values are not

---

1) A maximum proof mass deflection of 10% is a somewhat arbitrarily chosen value. It depends on the linearity specification of the sensor as larger proof mass deflection increases the nonlinearity introduced by the relationship between the differential change in capacitance and deflection [15]. Ten percentage is a good compromise value.

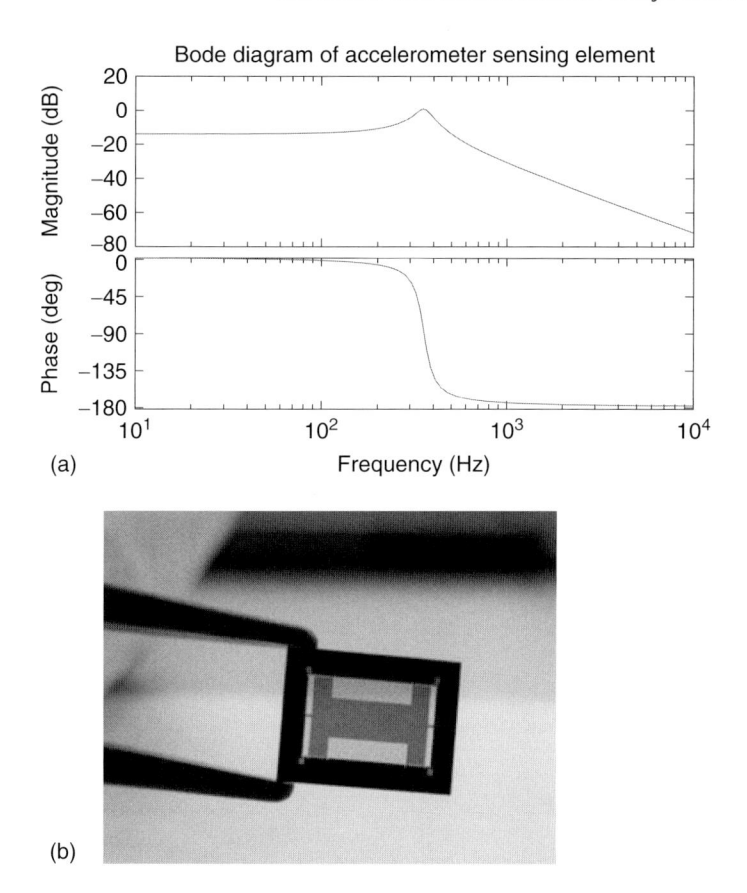

Figure 16.4 (a) Bode plot of the accelerometer sensing element used for the examples. It exhibits an under damped behavior. (b) The actual sensing element fabricated by bulk-micromachining in SOI technology.

very critical for the system level performance and were taken from the literature [17]. The compensator is a simple lead-lag filter that provides phase lead at frequencies below the sampling frequencies. In order to make it a realizable system, a pole at higher frequencies (above the sampling frequency) needs to be introduced. It was previously suggested to make the zero frequency a fifth of the sampling frequency, and the pole frequency five times [18]. However, it was found, by sweeping these two parameters, that the sensing system for the example considered here is optimized (in terms of maximizing the signal-to-noise ratio (SNR)) if a zero frequency of approximately a tenth of the sampling frequency and a pole frequency of four times is used. The variation in SNR is not very strong, although, at most about 4 dB. The methodology of sweeping one or several parameters and optimizing the performance is a powerful tool for the designer, and is one of the distinct merits of system level modeling and simulation.

Time domain simulations can now be carried out, which allow to assess the performance of the EM$\Sigma\Delta$M. The SNR of the output bitstream is a good performance criterion for the system. The SNR in the simulations presented here are calculated using functions available in the Delta Sigma ToolBox developed by Schreier, which can be freely downloaded at [19]. A positive SNR indicates working systems, a negative value instability. Instability in a $\Sigma\Delta$M is characterized by the output of the 1 bit quantizer not changing for a long time period, thus not by the bounded-input bounded-output criteria usually used for linear control systems. The noise in this example model comprises two components: the electronic noise originating from the pickoff noise modeled as a white noise source and the quantization noise that is introduced by the quantization process (mechanical or Brownian noise is neglected as the proof mass is relatively large). Figure 16.5 shows the input acceleration (first trace), the pulse-density-modulated output signal of the quantizer (second trace), the residual proof mass deflection (third trace), and the low-pass-filtered bitstream (fourth trace). This filtering process would be realized as a digital low-pass decimation filter, more details can be found, for example, in [20]. The input acceleration was assumed as a sinusoid with 2G amplitude and a frequency of 128 Hz. The sampling frequency of the modulator was set to 131 kHz,

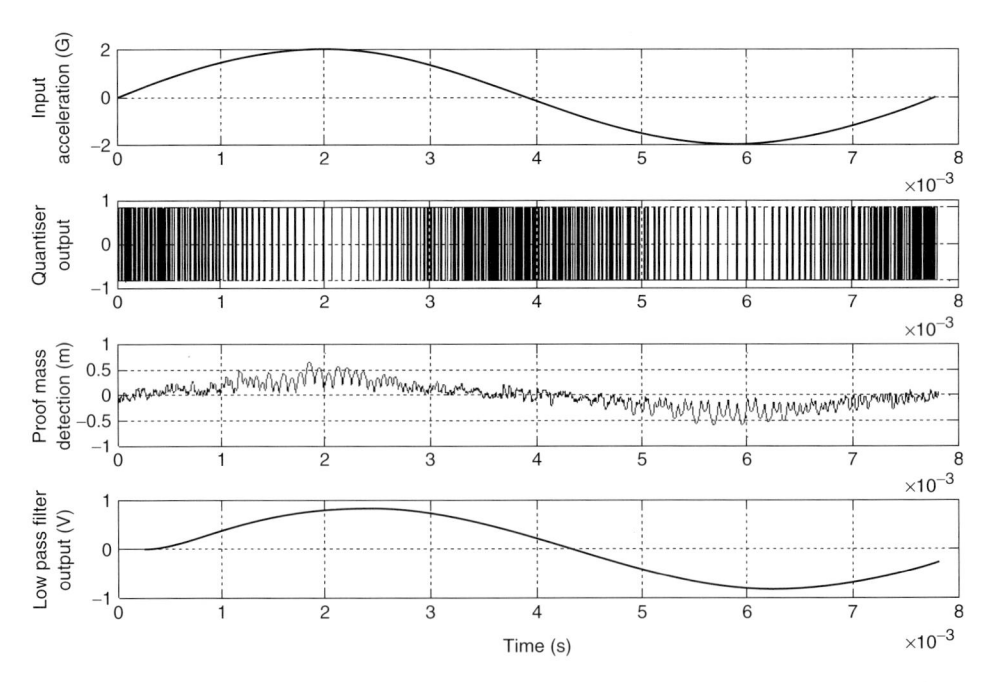

Figure 16.5 Time domain signals of a simulation run for the second order EM$\Sigma\Delta$M. first trace: sinusoidal input acceleration of magnitude of 2G and frequency 128 Hz; second trace: pulse-density-modulated output of the quantizer, the pulse density modulation is clearly visible; third trace: proof mass deflection, fourth trace: low-pass-filtered version of the bitstream shown in the second trace.

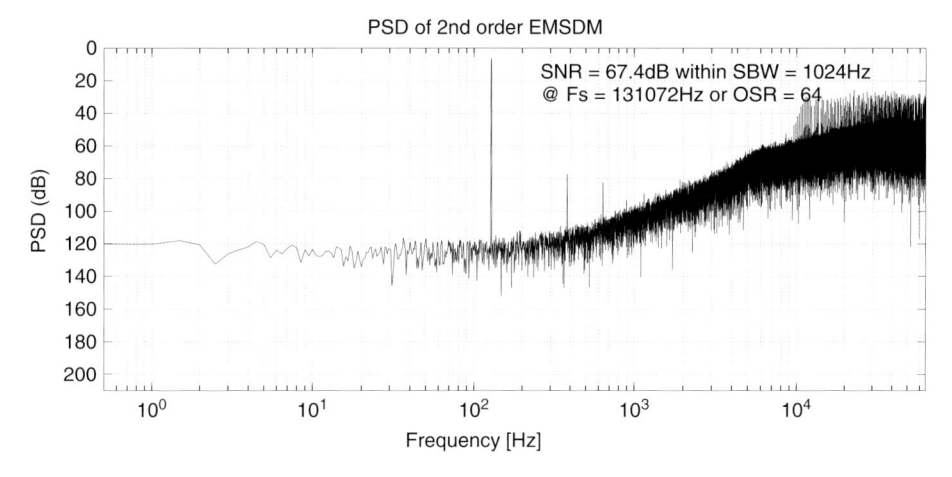

16.2 Second Order Electromechanical ΣΔM for a MEMS Accelerometer

Figure 16.6 Power spectral density (PSD) of the output bitstream. The signal peak is clearly visible. The other peaks are due to idle tones, which are typical for lower order ΣΔM.

which results in an oversampling ratio[2] of 64. The value of the sampling frequency was again chosen based on related studies presented in the literature [17]. It can be seen that the quantizer output is staying at its high (low) value for longer time periods when the input acceleration has a maximum (minimum). Any duration of a high or low period is always an integer multiple of the sampling time. The proof mass deflection is very small with a maximum of only about 60 nm. The open loop deflection would be 3.9 µm (calculated using Eq. (16.2)) for a 2G acceleration, which is impractical as the electrode gap is only 5 µm. This indicates another important advantage of the closed loop control approach: small proof mass deflection compared to the capacitive electrode gap ensures that nonlinear effects associated with a larger proof mass deflection are minimized and can be safely neglected. These include squeeze film damping [21], conversion from the proof mass deflection to a differential capacitance, which is, in turn, converted to a proportional voltage, and spring stiffening effects that could introduce a cubic term in the spring force equation [22]. The low-pass-filtered waveform is a good representative of the input acceleration, indicating correct coding. The exact quality of the coding is better evaluated in the frequency domain. This is usually done by calculating the power spectral density (PSD) of the bitstream, which is illustrated in Figure 16.6. The largest spike indicates the signal. The SNR is calculated using the Matlab functions available in the Schreier toolbox; for a bandwidth of 1024 Hz, the SNR $= 67.4$ dB. This value also can be theoretically predicted with the methodology and equations presented in [18], which calculates the in-band noise shaped by the quantization noise loop transfer

---

2) The exact value of the sampling frequency is 131 072 Hz $= 2^{17}$ Hz. The input frequency and bandwidth are also chosen as a power of two. Therefore, the oversampling ratio is also a power of two. Having all frequencies in the simulation file as powers of two ensures the correct use of the fast Fourier transform function in Matlab.

function (QNTF), and the usual assumption of linear $\Sigma\Delta M$ analysis that the 1 bit quantizer can be replaced by white quantization noise and a quantizer gain, k_Q [6]. The quantizer gain is very difficult to determine analytically, and hence it is best inferred from simulation. This is easily done by taking the average root-mean-square value of the positive quantizer output (which is unity) divided by the quantizer input. In the example, this gives a value of $k_Q = 3.56e-3$. The calculated SNR is 72.53 dB, which is in good agreement with the value determined by simulation.

The SNR determined by simulation is about 5 dB lower compared to the theoretically calculated value. This is due to the large spikes in the signal band at about three and five times of the input signal frequency. These are the so-called idle tones that are still relatively poorly understood in theory [7] but are correctly predicted by simulation. We will see that these idle tones are considerably reduced if higher order $\Sigma\Delta M$ are used.

Another interesting consideration is to determine the dominant noise source. As explained before, the model shown in Figure 16.3 contains two noise sources: the quantization noise from the 1 bit quantizer and the electronic noise from the first amplifier modeled as white noise source (this can relatively easily be changed to include flicker noise as well – there are building blocks available on the Matlab file-exchange). If the electronic noise is switched-off (easy to do in simulation, impossible in reality!), the signal to quantization noise ratio (SQNR) can be determined from simulation. In our example, omitting the electronic noise[3] results in a SQNR with a value that is virtually identical to the SNR. This indicates that the performance (at least in terms of minimum detectable signal) of the sensing system is determined by the quantization noise – in a well-designed EM$\Sigma\Delta M$, it should be the other way round. In fact, it can actually be shown that in a second order EM$\Sigma\Delta M$, the quantization noise always dominates [23], and experimental investigations have shown that such an interface degrades the noise performance by approximately 20 dB compared to an open loop sensor [24]. Again, using a higher order EM$\Sigma\Delta M$ solves this issue.

16.2.2
Advanced Model

As already stated, the model of the EM$\Sigma\Delta M$ shown in Figure 16.3 is only a very simplified model of an EM$\Sigma\Delta M$. The following two main simplifying assumptions are made:

- The sensing element is characterized by a second order lumped parameter model. In reality, any micromachined sensing element will have more than just the fundamental mode. For example, the sense fingers may have modal resonance frequencies that are below the sampling frequency of the $\Sigma\Delta M$, and

---

3) In the simulations, a white voltage noise rms value of 15 nV/$\sqrt{Hz}$ was assumed. For a circuit designer, this value may appear high, as a good noise low amplifier can have a noise rms value of about a factor of 10 or lower; however, it should be noted that the white noise source in the Simulink model is the input-referred noise, representing the entire electronic circuit – not only of the front-end amplifier.

this can have important consequences for the performance and stability of the loop [25].
- The conversion of the proof mass displacement to a voltage is a rather complicated circuit. In fact, it is one of the most important building blocks in a hardware realization of a EMΣΔM. It should be designed in such a way that the electronic noise floor is minimized, and the electronic noise primarily determines the SNR of the EMΣΔM; as a result, the quantization noise should be made sufficiently low so that the performance of the sensor is not degraded by it.

It is possible to derive more advanced system level models that incorporate these effects. However, the added complexity results in longer simulation times, which is particularly significant if optimization based on parameter variation studies are intended to be carried out. Good modeling and simulation practice involves developing a more sophisticated model that incorporates second order effects and then compare the results with the simplified model. If there is no significant difference in performance then the secondary effect modeled can be neglected – at least at system level; this may be different at transistor level when implementing the sensing system in hardware.

As a case study, a more advanced model of the sensing element itself shall be briefly considered here. The usual modeling and design methodology is to extract a reduced order model from finite element modeling (FEM) simulations and then incorporate it at system level. Such a model will incorporate modes other than only the fundamental one. In particular, any modes along the same axis as the fundamental mode are of significance. A reduced model extraction can be done particularly well in Coventor [26] that supports automated lumped parameter model extraction from FEM [27] with a direct interface to Matlab.

Here, we do not follow above methodology, as it would be beyond the scope of this chapter. Instead, we consider a case that still can be handled analytically by assuming that the comb fingers on the proof mass required for capacitive sensing introduce an additional resonance frequency at 12 kHz. This originates from the fact that a capacitive comb finger is not ideally rigid and has mode shapes as well. If only the first mode of the sense finger is considered, the lumped parameter model of the sensing element effectively consists of two masses coupled by mechanical springs and dampers. This is shown conceptually in the inset of Figure 16.7 together with the parameter values assumed in the example. The values are taken somewhat arbitrarily as they strongly depend on the design of the sensing element. It is assumed that 80% of the mass are in the main proof mass and 20% in the comb fingers. The transfer function of the sensing element becomes

$$M(s) = \frac{1}{m_1+m_2} \frac{m_1 m_2 s^2 + (b_1 m_2 + b_2 m_1 + b_2 m_2)s + m_1 k_2 + m_2(k_1+k_2)}{m_1 m_2 s^4 + (b_1 m_2 + b_2 m_1 + b_2 m_2)s^3 + (k_1 m_2 + b_1 b_2 + k_2 m_1 + k_2 m_2)s^2 + (k_1 b_2 + k_2 b_1)s + k_1 k_2}$$

(16.3)

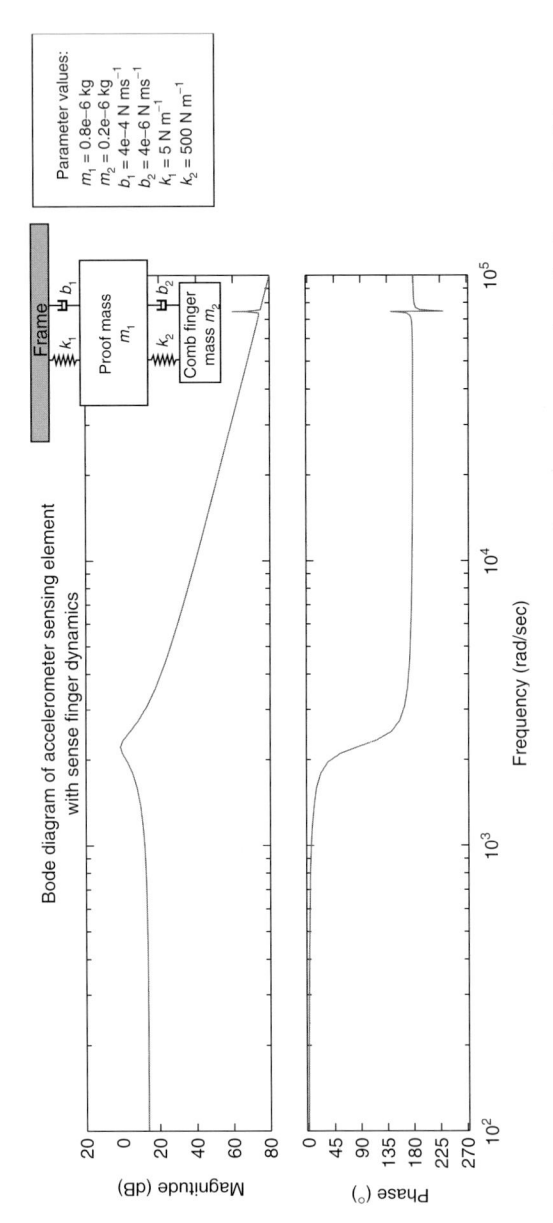

Figure 16.7 Bode plot of the sensing element if an additional resonance frequency originating from the first mode of the sense comb fingers is considered. The inset shows the improved lump parameter model of the sensing element consisting of two masses coupled by springs and dampers.

where s is the Laplace operator and all other symbols are defined in Figure 16.7.

It results in a Bode plot of the sensing element transfer function shown in Figure 16.7. The second resonance is clearly visible; otherwise the transfer function is identical to the one shown in Figure 16.4. The transfer function of Eq. (16.3) can be entered into the *Sensor* block of the Simulink model shown in Figure 16.3. For the values chosen in this example, the SNR of the second order EM$\Sigma\Delta$M does not significantly change, however, as we shall see, for a higher order EM$\Sigma\Delta$M, it can have a major impact. The same methodology can be employed to introduce further higher order modes; for example, the effect that the stationary sense fingers are not ideally rigid can be modeled in a similar way.

Next, let us consider an advanced model for the capacitive deflection interface measurement. The first improvement that can be made is to model correctly the conversion of a proof mass displacement to a change in capacitance. This is described by the following equation:

$$\Delta C = \frac{\varepsilon_0 A_S}{(d_0 - x)} - \frac{\varepsilon_0 A_S}{(d_0 - x)} = \frac{-2\varepsilon_0 A_S x}{(d_0^2 - x^2)} \qquad (16.4)$$

where A_S is the area of the sense capacitors. Equation (16.4) is easy to implement in Simulink using a *function* block. If $x \ll d_0$, the differential change in capacitance is proportional to the proof mass deflection, x, with a proportionality constant of $-2\varepsilon_0 A_S/d_0^2$, which provides a first justification as to why in the simple model a simple gain block was used to describe the pickoff circuit. In a stable EM$\Sigma\Delta$M, the displacement always should be much smaller than the nominal capacitor gap, but one may want to simulate the behavior of the loop if the accelerometer is momentarily exposed to an over-range condition in which the proof mass is deflected much further, and then study whether the control system can recover from such a shock condition. Therefore, implementing Eq. (16.4) in a Simulink may be useful; this should be done together with a *Limiter* block avoiding a physically impossible scenario that the proof mass is deflected more than the nominal capacitor gap. However, there is a severe problem: the differential change in capacitance ΔC is very small, as small as in the atto-Farad region. This can cause numerical problems and therefore very unexpected behavior in Simulink. Especially, as in our case, subsequent building blocks have to scale up the signal to values about unity.

The next building block in a real system is to convert the differential change in capacitance into a voltage. The details depend on the circuit implementation the designer chooses, but the principle is rather generic. It involves exciting the capacitive half-bridge formed by the interdigitated comb fingers of the sensing element with an AC signal acting as a carrier; the frequency of which should be chosen to be much higher (at least a factor of 10) than the sampling frequency of the $\Sigma\Delta$M. The change in capacitance is modulated (effectively multiplied) by the carrier signal. A charge amplifier is typically used to amplify the voltage generating at its output an amplitude-modulated signal, which then needs to be synchronously demodulated. From a system level point of view, this is achieved by multiplying the signal with the AC carrier resulting in a low-frequency component (which is

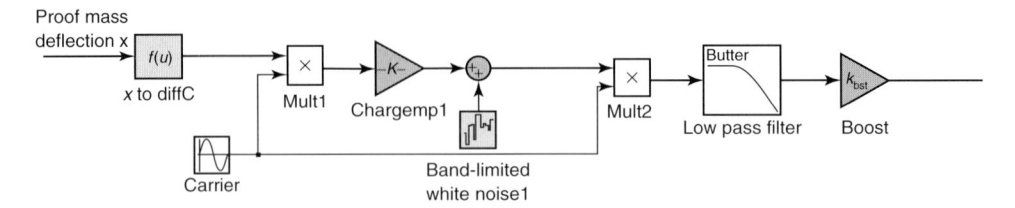

Figure 16.8 Simulink model representing a more realistic pickoff circuit, which, in principle, can replace the gain block k_{po} in the basic model of Figure 16.3. For more details on the operation, see main text. However, this model can produce unexpected results if the simulation parameters are not set very carefully, as the numerical dynamic range spans approximately 20 orders of magnitude; therefore, it is an example of bad modeling practice. Also, it is very slow as the AC carrier signal frequency has to be at least a factor of 10 higher than the sampling frequency of the EM$\Sigma\Delta$M.

the desired signal) and a component at twice the carrier frequency. The latter is subsequently removed by a low-pass filter. Again, building a model representing the aforementioned procedure at system level is relatively straightforward in Simulink, as shown in Figure 16.8, which in principle, can directly replace the gain block k_{po} of the basic model shown in Figure 16.3. The displacement of the proof mass is the input to the equation function block *x to diffC* containing Eq. (16.4); the differential change in capacitance is then multiplied by the carrier signal and amplified by gain block *ChargeAmp* (this has a value in the order of a million). This multiplication is numerically particularly problematic, since a value in the range of 10^{-18} is multiplied by a sinewave with an amplitude of typically 1–10 V. Even if the numerical issues could be solved, for example, by appropriate scaling (or by setting the simulation parameters of Simulink carefully enough), there is another significant disadvantage: the simulation time is increased considerably as it depends on the fastest frequency in the model. The carrier signal is a factor of 10 higher than the sampling frequency of the $\Sigma\Delta$M, therefore the simulation time is also increased by at least the same factor.

The conclusion from these considerations is that it should be very carefully evaluated which second order effects should be included in a system level model, and which not. The only building block of the capacitive position measurement interface that certainly should be included at system level is the low-pass filter, which removes the high-frequency component. There is no significant penalty in neither numerical complexity nor simulation time to add the filter; at the same time, inclusion of it may be of paramount importance for the loop stability because of the phase lag introduced. In the example of the second order EM$\Sigma\Delta$M, the low-pass filter has no effect on the loop stability and only insignificantly reduces the SNR (a reduction of 1.4 dB was observed.)

There are other second order effects that can be included, but the space restrictions for this chapter do not allow discussing them in more detail. For example, the feedback force may not be applied to the sensing element during a full clock cycle as one clock cycle may be divided into a sense and feedback phase to avoid electrical cross-coupling [28]. This can introduce another delay to the loop and influence stability.

16.3 Higher Order Electromechanical $\Sigma\Delta M$ for MEMS Accelerometer

16.3.1 Design Methodology for Higher Order EM$\Sigma\Delta$M

Second order EM$\Sigma\Delta$M suffer from the same drawbacks as their purely electronic equivalents such as tonal behavior, limit cycles, and relatively poor noise shaping. The micromachined sensing element additionally degrades the performance of the $\Sigma\Delta$M, as it replaces two (near-) ideal integrators in an electronic loop. However, the sensing element behaves far from an ideal integrator, which has a DC gain of infinity. In fact, the low frequency gain of the sensing element is quite low: it is the inverse of the mechanical spring constant which, in our case is $1/(5 \text{ N m}^{-1}) = 0.2 \text{ m N}^{-1}$ – obviously very far from infinity. As a result of these problems, several research groups have been developing higher order EM$\Sigma\Delta$M. This has the additional advantage that excellent quantization noise shaping can be achieved at lower sampling frequencies, which relaxes circuit implementation requirements and leads to a lower power consumption. The usual approach is to cascade the sensing element and the associated position measurement interface with one or several electronic integrators. The challenge of the design of such EM$\Sigma\Delta$M is to find a stable architecture, and a near-optimum solution. Especially, stability is challenging as the sensing element is a second order transfer function, which cannot be accessed internally. This makes it impossible to apply the typical approach for designing higher order electronic $\Sigma\Delta$M of using minor feedback (or feedforward) loops that access every node between each subsequent integrator. The design of high-order EM$\Sigma\Delta$M has nevertheless a lot in common with their electronic counterparts, and similar architectures have been explored.

A generic design methodology is based on an adaptation of the procedure described in [6], which needs to be modified for the specific requirements of an EM$\Sigma\Delta$M. The main steps are as follows. (i) Choose a higher order (order $N > 2$) purely electronic topology procedure. A topology should be chosen that does not contain feedback or feedforward paths directly from or to the input. Follow the "cookbook" procedure in [6]; this gives the coefficients of all gains by mapping the noise transfer function to an analog filter implementation function (often popular filter structures such as Butterworth or Chebychev are used); a stable higher order $\Sigma\Delta$M should result. (ii) Design a second order EM$\Sigma\Delta$M as described in Section 16.2.1 for the MEMS sensing element used for the design. (iii) Replace the first integrator of the electronic with the transfer function of the micromachined sensing element, the pickoff gain, and the compensator. This increases the order of the system by one as the sensing element is of second order, hence the EM$\Sigma\Delta$M will have an order of $N + 1$. Furthermore, replace the feedback 1 bit DAC of the electronic $\Sigma\Delta M$ with the building blocks for the conversion from a feedback voltage to an electrostatic force. (iv) Optimize and validate the design by using parameter sweeps and by including second order effects as described earlier. The methodology is illustrated in Figure 16.9; here a fourth order, discrete-time and feedforward topology of an electronic $\Sigma\Delta M$ is chosen as the starting point. The fifth order EM$\Sigma\Delta M$ is obtained by following the design methodology.

Since the original electronic $\Sigma\Delta M$ is discrete in time, so will be the high-order EM$\Sigma\Delta M$. It is also possible to follow the same methodology with a continuous time $\Sigma\Delta M$. Both approaches will result in similar performance at system level; however, in the hardware implementation, there will be important differences. As a general guideline, it can be stated that continuous time modulators are more suitable for printed circuit board (PCB) implementations, whereas discrete-time modulators lend themselves to be realized as switched capacitor circuits on an ASIC (application-specific integrated circuit). A detailed discussion of continuous versus discrete-time $\Sigma\Delta M$ is outside the scope of the chapter; the pros and cons primarily depend on the circuit implementation and noise performance (for a more in-depth discussion, see, for example, [24]). In [17], the methodology described above is applied to a wide range of architectures of discrete-time modulators up to an order of six.

16.3.2
Example Design

A fifth order, feedforward discrete-time architecture is chosen here as an example. For the simple reason that to the author's knowledge, such an architecture for an EM$\Sigma\Delta M$ accelerometer has not been presented in the literature to date. Starting from a fourth order $\Sigma\Delta M$, the design methodology is followed, which results in the Simulink model presented in Figure 16.10, and also shows all relevant simulation parameters. The same sensing element, sampling frequency, oversampling ratio (OSR), and input signal (2G, 32 Hz) as for the second order loop were chosen. The zero and pole frequencies and the boost gain value (k_{bst}) were determined by nested parameter sweep simulations. The feedback voltage was increased from 10 to 13 V as higher order $\Sigma\Delta M$ overload at lower input amplitudes; a fifth order loop at about 60% of the feedback signal amplitude. The spectrum of the output bitstream is shown in Figure 16.11. The noise floor has drastically dropped compared to a second order EM$\Sigma\Delta M$ and lies at approximately -160 dB. The SQNR[4] is 96.1 dB in a 1 kHz bandwidth, and the average proof mass deflection is about 2 nm. Also, there are virtually no signs of idle tones in the signal band, as were the case for the second order EM$\Sigma\Delta M$. The spectrum of the out-of-band noise has a slope of approximately 100 dB per decade – this clearly indicates fifth order noise shaping.

Once such a design has been established, it is of paramount importance to verify the design with extensive system level simulations. $\Sigma\Delta$Ms can become unstable

---

4) Quoting the SQNR here implies that the white noise source was switched off in the Simulink model.

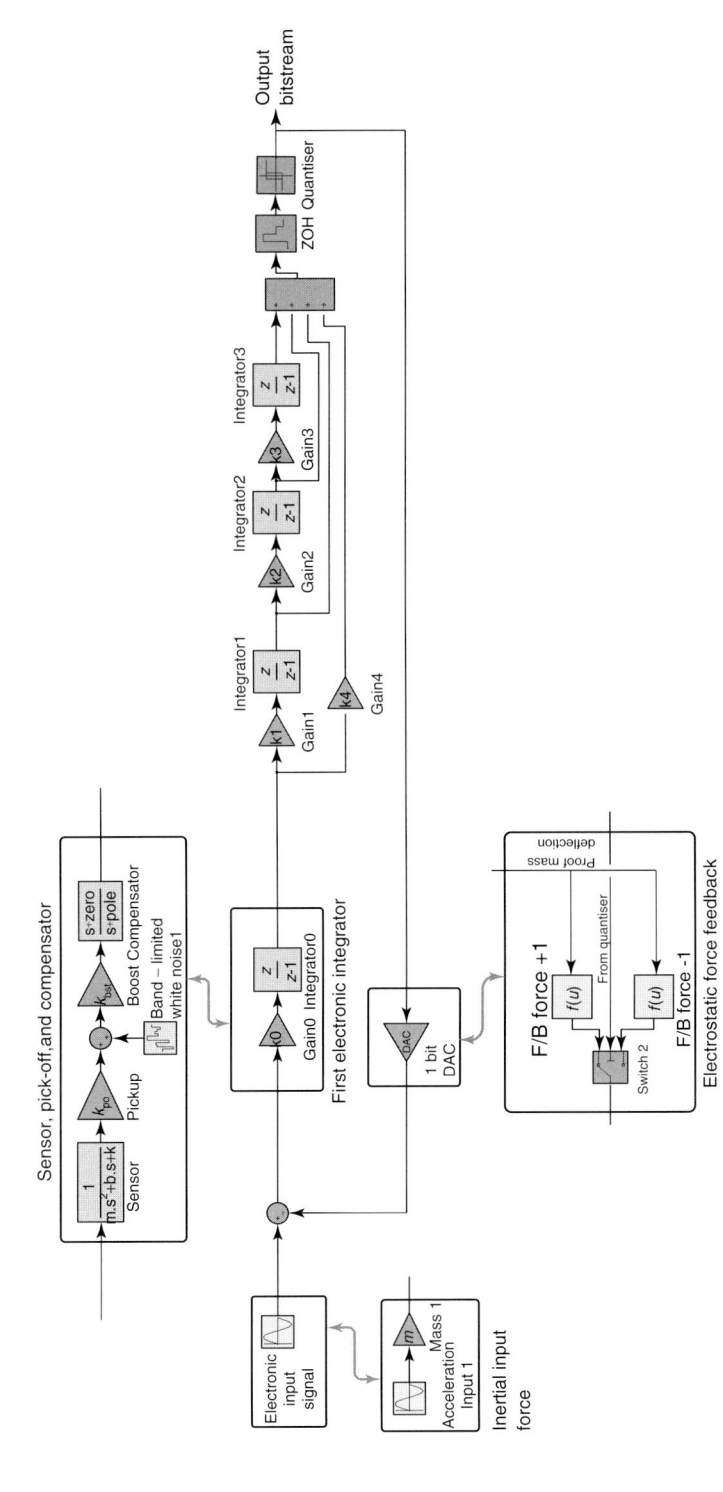

Figure 16.9 The design methodology for a high-order EM$\Sigma\Delta$M is illustrated for a fourth order, discrete-time and feedforward $\Sigma\Delta$M as starting point. The first integrator of the $\Sigma\Delta$M is replaced with the micromachined sensing element, the pickoff gain, and compensator. Likewise, the 1 bit DAC in the feedback path is replaced with the voltage-to-electrostatic force converter. It results in a fifth order EM$\Sigma\Delta$M.

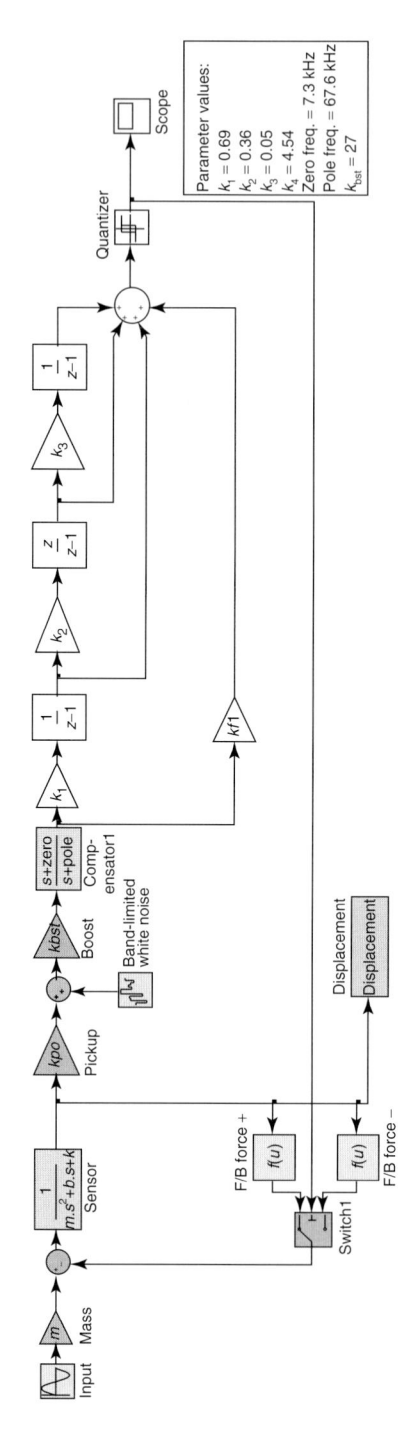

Figure 16.10 Simulink model of a fifth order EM$\Sigma\Delta$M with discrete-time feedforward architecture. The relevant parameter values are given in the inset box. All other parameters are the same as for the second order EM$\Sigma\Delta$M.

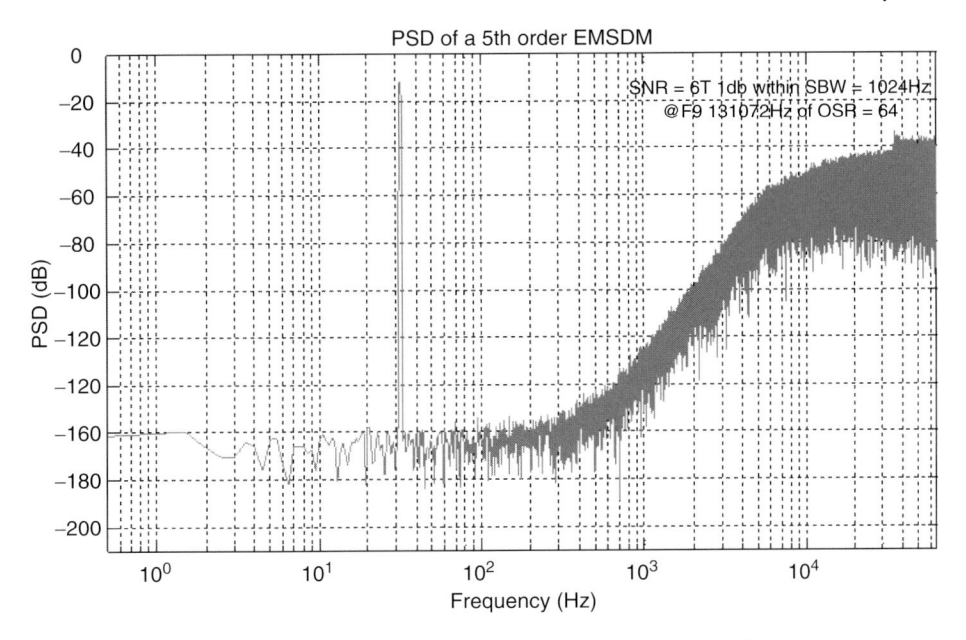

Figure 16.11 Power spectral density (PSD) of the output bitstream. The signal peak is clearly visible. The slope of the out-of-band quantization noise is 100 dB per decade indicating fifth order noise shaping.

only after a considerable time, so long simulations (>1 s in real time) should be performed. All nodes in the model should be monitored in such simulations to ensure that the signal levels are in a reasonable range, in particular, node values representing voltages have to lie within values that can be achieved with an electronic hardware implementation. Another important consideration is the robustness of an EM$\Sigma\Delta$M. Micromachined sensors often suffer from parameter uncertainties due to fabrication tolerances; for example, the spring constant can easily vary by $\pm10\text{--}20\%$ compared to the designed value. In [17], Monte Carlo simulations are performed that vary the proof mass by $\pm3\%$, the spring constant by $\pm20\%$, damping coefficient by $\pm10\%$, and all electronic gain constants by $\pm2\%$. The SNR is then plotted for each simulation. A good design has to remain stable for the worst case variation and should show a relatively small variation in SNR. We are not presenting such a parameter sensitivity analysis for the design discussed here owing to space restrictions and leave this to the interested reader.

16.3.3
Feedback Linearization

Another important consideration is the conversion of the feedback voltage to an electrostatic force. Ideally, the magnitude of the electrostatic force should be constant during one clock cycle. However, owing to the residual motion of the proof mass (which is small compared to the electrode gap but not zero) the force varies slightly (Eq. (16.1)). This introduces a nonlinearity in the feedback path affecting the performance of the EMΣΔM. In [29], it is suggested to apply a simple but effective feedback linearization scheme by introducing an additional parameter, k_{fbl}, that varies the constant feedback voltage based on the proof mass deflection to compensate for this effect. Equation (16.1) is modified to

$$F_{el} = \text{sgn}(u_1) \frac{1}{2} \frac{\varepsilon_0 A_{fb} \left(V_{fb} + \text{sgn}(u_1) \times k_{po} \times k_{fbl} \times u_2\right)^2}{\left(d_0^2 + \text{sgn}(u_1) \times u_2\right)^2} \tag{16.5}$$

where all symbols have the same meaning as in Eq. (16.1). The term $k_{po} \times k_{fbl} \times u_2$ does not exactly cancel out the variation in feedback force during a cycle but alleviates it to a great extent. The optimal value for parameter k_{fbl} can be found again by parametric sweeps. For our example, an optimal value of $k_{fbl} = 16.7$ was found. The PSD of the bitstream still looks similar to the one in Figure 16.11 (hence is not presented here) but with an almost 20 dB lower noise floor and a SQNR = 109.5 dB — which is an improvement of almost 14 dB! As described in [29], the hardware implementation scheme is also relatively straightforward requiring only an amplifier and summation circuit. It is therefore advisable to use such a linearization scheme when building a higher order EMΣΔM.

At such a low quantization noise level, the performance of the fifth order design of the EMΣΔM is now sensitive to whether or not the thermal noise is included in the simulation. If a white noise source with a square root spectral density of 15 nV/$\sqrt{\text{Hz}}$ is included in the simulation, the SNR drops approximately by 5 dB; this indicates that the loop now operates in a regime where the thermal noise dominates. It should be reminded here that this was achieved with a very modest sampling frequency of 131 kHz in a relatively wide signal band of 1 kHz.

16.3.4
Advanced Model

For further validation of the design, a higher order model of the sensing element extracted from FEM simulations should now be considered, as described in Section 16.2.2. Replacing the simple second order transfer function for the micromachined sensing element by the forth order transfer function of Eq. (16.3) including a second resonance peak from nonideally rigid sense fingers is straightforward in Simulink. Rerunning the simulation with no feedback linearization results in a SQNR = 95.5 dB, which is only marginally lower than that when using the simple second order transfer function for the sensing element. However, when using the same parameters as above including feedback linearization, the system becomes unstable, which is shown in Figure 16.12. The output bitstream first rapidly switches indicating a stable system, but from approximately 0.25 s onwards, it clearly becomes unstable. This example illustrates the often rather unpredictable stability properties of higher order EMΣΔM and the importance of including second order effects, which are often neglected. There are basically two

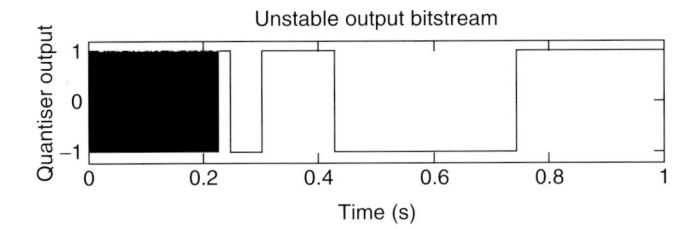

Figure 16.12 Output bitstream indicating unstable behavior after approximately 0.25 s for a system using the feedback linearization scheme and a sensing element transfer function including a higher order resonance.

approaches to fix this problem: First, the sensing element can be redesigned so that higher order modal frequencies occur only at higher frequencies (here the instability is caused by the fact that the second resonance frequency lies well below the sampling frequency of the $\Sigma\Delta M$, which, in general, should be avoided). Secondly, if redesigning the sensing element is not an option, the design procedure needs to be repeated by using a different filter mapping function of the noise transfer function. We leave the stability problem of the worked example deliberately open and invite the reader to work on a redesign.

16.4
Higher Order Electromechanical $\Sigma\Delta M$ for MEMS Gyroscopes

EM$\Sigma\Delta M$ can be used for MEMS gyroscopes as well, which is a more recent area of research and still very much ongoing. This section provides a brief review of developments of closed loop force feedback gyroscopes using $\Sigma\Delta M$. A MEMS gyroscope typically relies on a two degree of freedom system, which can be realized by suspending a silicon proof mass so that it is compliant to move along two orthogonal axes. The first degree of freedom is usually a driven mode; hence the proof mass is excited to oscillate along one axis (usually referred to as the *x-axis*) with constant amplitude and frequency. In presence of a rotation perpendicular to the drive and sense axes, the proof mass will start oscillating along the sense axes (*y-axis*) as a result of the Coriolis force. The amplitude of this oscillation is a measure of the rotation. There are many variations in this basic concept, for details the reader is, for example, referred to [30]. The sense mode of a MEMS gyroscope is suitable for inclusion in a force feedback control loop, as it can effectively be viewed similar to an accelerometer with the Coriolis force as input. One important difference is that this input force is at a known frequency, namely, the oscillation frequency of the driven mode (which typically coincides with the natural resonance frequency of the system). Furthermore, it is desirable to have the quality (Q) factor of both drive and sense modes as high as possible, since it is possible to effectively amplify the oscillation amplitude of the sense mode by Q, which is necessary as the Coriolis force is rather weak. (The magnitude of the Q-factor is primarily determined whether the MEMS gyroscope operates at ambient pressure or in a vacuum; for the former, Q-factors of typically 100 can be achieved, whereas for the latter, higher than 10 000 have been reported.) Therefore, the second important difference from an accelerometer is that the transfer function of the sensing element has a high resonance peak and that the input signal frequency lies at, or very close to, the resonance peak frequency.

The first successful application of incorporating the sense mode of a MEMS gyroscope in a $\Sigma\Delta M$ was reported in 2000 [31]. From a system level point of view, the system is very similar to the accelerometer EM$\Sigma\Delta M$ presented in Section 2. Only the dynamics of the sensing element provide the noise shaping resulting in a second order EM$\Sigma\Delta M$. In 2004, a forth order EM$\Sigma\Delta M$ interface for a MEMS gyroscope was presented by Petkov and Boser [32, 33]. The gyroscope sensing element was cascaded by two electronic integrators, and a discrete-time feedforward architecture was adopted, but an additional local feedback path was introduced to shape the electronic filter transfer function (and thus the quantization noise transfer function) so that a zero was introduced around the resonance frequency of the gyroscope sensing element and hence reduce quantization noise in the crucial frequency band for the gyroscope. The sampling frequency used for the experimental verification of their system was 850 kHz. A similar approach, together with a much more detailed design description, was presented by Raman et al. [34, 35].

The typical bandwidth of a MEMS gyroscope is 50–100 Hz, which means that the oversampling ratio OSR $= \text{fs}/(2 \times \text{BW})$ is in the order of 8500 (using the numbers from above), which is a very high value compared to the high-order EM$\Sigma\Delta M$ for the accelerometer. It was therefore suggested by Dong et al. [36] to use a band-pass $\Sigma\Delta M$ as the interface for MEMS gyroscopes. This approach cascades electronic resonators instead of integrators with the mechanical sensing element, and is schematically illustrated in Figure 16.13. To a large extent, using resonators is a more intuitive choice, since the sensing element is a mechanical resonator itself. The resonant frequency of the electronic resonators is made to coincide with the sensing element's resonant frequency. This leads to a dramatic reduction in the required sampling frequency, and hence a $\Sigma\Delta M$ with an OSR between 32 and 128 can achieve sufficient quantization noise attenuation in the signal band of the gyroscope. The design methodology relies on a transformation of a high-order low-pass (i.e., a design using integrators) EM$\Sigma\Delta M$ to a band-pass system; formally this can be achieved by: $z \rightarrow z^2$ [36]. The order of the system is increased as for each integrator with order one a resonator with order two is used; for example, a gyroscope sensing element cascaded with three electronic resonators results in an eighth order system. In [36], three different architectures are presented together with system level simulation results. Figure 16.14 shows, as an example, a system level model with a multifeedback and local resonators (MFLRs) architecture. It used a sampling frequency of only 63 kHz and achieved an SNR of 93 dB in a signal band of 256 Hz (thus the OSR was 128). The resulting spectrum of a system level simulation is shown in Figure 16.15. The deep notch in the spectrum is created by the resonant behavior of the sensing element in combination with the electronic resonators.

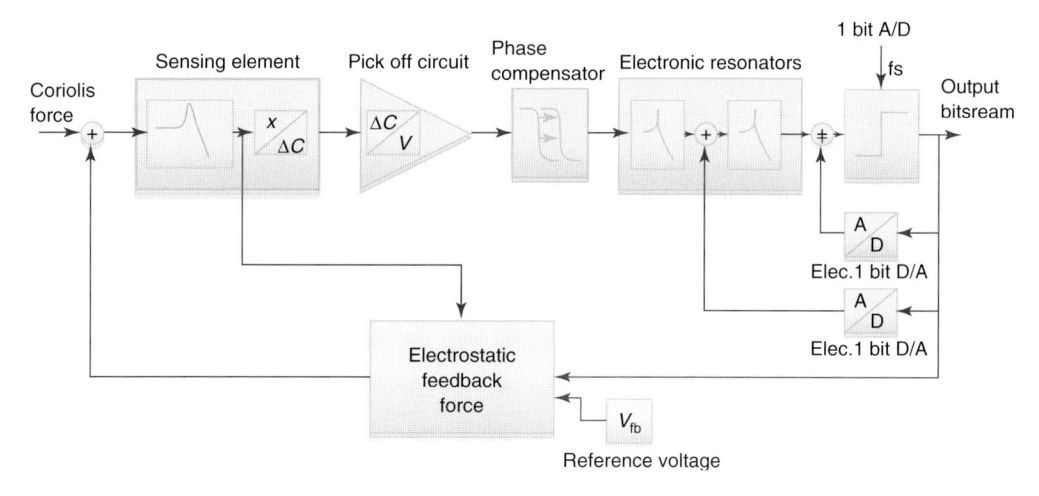

16.4 Higher Order Electromechanical ΣΔM for MEMS Gyroscopes

Figure 16.13 Schematic diagram for a band-pass EMΣΔM interface for a MEMS gyroscope. The sensing element is cascaded with electronic resonators, rather than integrators. This allows using a low sampling frequency, similar to the EMΣΔM for MEMS accelerometers.

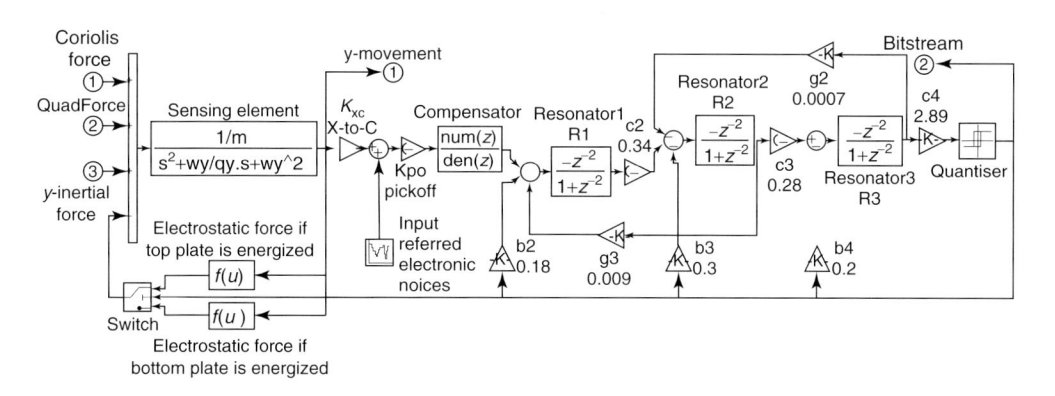

Figure 16.14 System level model of a band-pass EMΣΔM interface for a MEMS gyroscope. It employs a multifeedback and local resonators (MFLRs) architecture. The sampling frequency was only 63 kHz. (*Source*: Reproduced from [21].)

For the first time, the band-pass EMΣΔM approach was successfully implemented and applied to a gyroscope manufactured by SensoNor, described by Dong et al. in [37]. In this work, a sixth order architecture was described (sensing element cascaded by two electronic resonators) with the interface electronics realized with off-the-shelf components on a PCB. Since a PCB version was used, the EMΣΔM was a continuous time system; however, this has the disadvantage that it makes the electronic feedback paths to the resonators more complicated, requiring a half-return-to-zero DAC in parallel to a return-to-zero DAC; only then the same

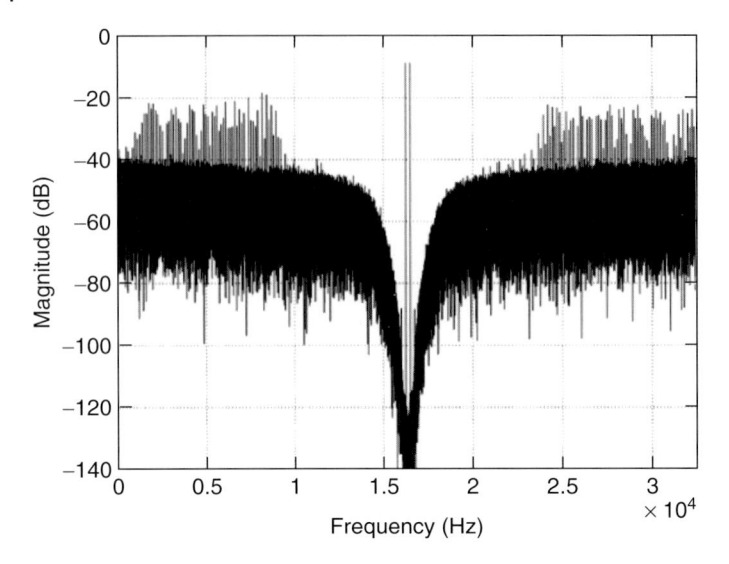

Figure 16.15 Output bitstream power spectral density of a simulation of the model presented in Figure 16.13. The deep frequency notch is produced by the sensing element and the electronic resonators. The input Coriolis signal manifests itself as a double-side band-frequency components. (*Source*: Reproduced from [36].)

frequency response as in a discrete-time band-pass $\Sigma\Delta M$ can be achieved [38]. The experimental results agreed well with simulation results and demonstrated an impressively low noise floor of -100 dB $/\sqrt{\text{Hz}}$, albeit no measurement data for a dynamic input was presented.

The area of designing EM$\Sigma\Delta$M for MEMS gyroscopes is still a very active research topic. The most recent results were published by Northemann *et al.* [39, 40], which also adapt the band-pass approach described earlier, but extends it to the drive mode of the gyroscope as well. Furthermore, they make extensive use of digital circuitry, therefore much of the interface electronics is implemented as a field programmable gate array (FPGA). Undoubtedly, more work will be forthcoming in this exciting research field in the next years.

16.5
Concluding Remarks

As this book is about system level simulation, deliberately this chapter barely discusses any analytical methods to determine the stability and performance of higher order EM$\Sigma\Delta$M. Very useful methods exist to predict stability (at least to some extent) and noise floor; mainly by using the standard linearization approach of replacing the quantizer by a white noise source and a gain. These methods have been extensively described in the literature on $\Sigma\Delta M$, and they can be applied directly to EM$\Sigma\Delta$M. For example, in [36], the analytical expression for noise and signal transfer functions are derived, and then root locus techniques are used to predict stability, which agree well with system level simulations. A root locus analysis allows determining a critical minimum value for the quantizer gain so that the poles lie within the unit circle. Nevertheless, such analysis relies on a linearized model of the $\Sigma\Delta M$, which inevitably makes simplifying assumption, and hence leaves always a considerable amount of uncertainty. This can only be addressed by system level simulations of which the most important techniques and approaches are described above. There are many issues which are not covered or briefly mentioned in this chapter, the reader is referred to the literature, which is relatively comprehensively quoted. The chapter, in particular, the section on the $EM\Sigma\Delta M$ accelerometer, tries to convey a "hands-on" approach, inviting the reader to try out the presented models and explore them further. Lastly, it should be noticed that all but the last of the referenced literature is mainly from academic research. However, while writing this chapter, it looks commercial interest in this field is growing. Colibrys is actively pursuing this approach for their high-performance accelerometers [41] and SensoNor for their gyroscopes [42].

References

1. Yazdi, N., Ayazi, F., and Najafi, K. (1998) *Proceedings of the IEEE*, **86** (8), 1640–1659.
2. Beeby, S., Ensell, G., Kraft, M., and White, N. (2004) *MEMS Physical Sensors*, Artech House, ISBN 1-58053-536-4.
3. Kraft, M., Lewis, C.P., and Hesketh, T.G. (1998) *IEEE Proceedings Circuits, Devices and Systems*, **145** (5), 325–331.
4. Aaltonen, L. (2010) Integrated interface electronics for capacitive MEMS inertial sensors. Doctoral Dissertation, Aalto University, School of Science and Technology, Finland.
5. http://www.mathworks.com/matlabcentral/fileexchange (accessed 4 February 2011).
6. Norsworth, S., Schreier, R., and Temes, C. (1997) *Delta-Sigma Data Converters: Theory, Design, and Simulation*, IEEE Press.
7. Reiss, J.D. (2008) *Journal of Audio Engineering Society*, **56** (1/2), 49–64.
8. Henrion, W. *et al.* (1990) Wide Dynamic Range Direct Digital Accelerometer Solid-State Sensor and Actuator Workshop, Hilton Head Island, SC, pp. 153–157.
9. Yun, W., Howe, R.T., and Gray, P. (1992) Surface micromachined, digitally force-balanced accelerometer with integrated CMOS detection circuitry. IEEE Solid-State Sensor and Actuator Workshop, Hilton Head Island, pp. 126–131.
10. Smith, T., Nys, O., Chevroulet, M., de Coulon, Y., and Degrauwe, M. (1994) Electro-mechanical sigma-delta converter for acceleration measurements. IEEE International Solid-State Circuits Conference, San Francisco, pp. 160–161.
11. Lu, C., Lemkin, M., and Boser, B.E. (1995) *IEEE Journal of Solid-State Circuits*, **30** (12), 1367–1373.
12. Lemkin, M. and Boser, B.E. (1999) *IEEE Journal of Solid-State Circuits*, **34** (4), 456–468.
13. Dong, Y., Kraft, M., and Gollasch, C.O. (2005) *Journal of Micromechanics Microengineering*, **15**, S22–S29.
14. Sari, I., Zeimpekis, I., and Kraft, M. (2010) A full wafer dicing free dry release process for MEMS devices, Proceedings of Eurosensors XXIV Conference, Linz, Austria, September 2010.

15. Rex, T.S.F. and Baird, T. (1993) *IEEE International Symposium on Circuits and Systems*, **2**, 1361–1364.
16. Thurston, A. and Hawksford, M. (1994) Dynamic overload recover mechanism for sigma delta modulators. 2nd International Conference on Advanced A-D and D-A, Conversion Techniques and their Applications, 1994, pp. 124–129.
17. Dong, Y., Kraft, M., and Redman-White, W. (2007) *IEEE Transactions on Instrumentation and Measurement*, **56** (5), 1666–1674.
18. Kraft, M. (2006) Digital feedback control for microsensors, *Smart MEMS and Sensor Systems*, Imperial College Press, ISBN 1-86094-493-0.
19. http://www.mathworks.com/matlabcentral/fileexchange/19 (accessed 4 January 2011)
20. Xie, P.Y., Whiteley, S.R., and Van Duzer, T. (1999) *IEEE Transactions on Applied Superconductivity*, **9** (2), 3632–3635.
21. Houlihan, R. and Kraft, M. (2005) *Journal of Micromechanics and Microengineering*, **15** (5), 803–902.
22. Lishchynska, M., O'Mahony, C., Slattery, O., and Behan, R. (2006) *Journal of Micromechanics and Microengineering*, **16**, S61–S67.
23. Jiang, X. (2003) Capacitive position-sensing interface for micromachined inertial sensors. PhD dissertation. Department of Electrical Engineering and Computer Science, University of California, Berkeley, CA.
24. Külah, H., Chae, J., Yazdi, N., and Najafi, K. (2006) *IEEE Journal of Solid-State Circuits*, **41** (2), 352–361.
25. Seeger, J.I., Jiang, X., Kraft, M., and Boser, B.E. (2000) Sense finger dynamics in $\Sigma\Delta$ force feedback gyroscope. Technical Digest of Solid State Sensor and Actuator Workshop, Hilton Head Island, June 2000, pp. 296–299.
26. www.coventor.com (accessed 25 January 2011).
27. Breit, S.R., Welham, C.J., Rouvillois, S., Kraft, M., and McNie, M. (2008) Simulation environment for accurate noise and robustness analysis of MEMS under mixed-signal control. Proceedings of ASME International Mechanical Engineering Congress, Boston, November 2008.
28. Lemkin, M.A. (1997) Micro accelerometer design with digital feedback control. PhD dissertation. University of California, Berkeley, CA.
29. Dong, Y., Kraft, M., and Redman-White, W. (2006) *Journal of Micromechanics and Microengineering*, **16**, S54–S60.
30. Acar, C. and Shkel, A. (2009) *MEMS Vibratory Gyroscopes*, Springer, ISBN: 978-0-387-09535-6.
31. Jiang, X., Seeger, J., Kraft, M., and Boser, B. (2000) A monolithic surface micromachined Z-axis gyroscope with digital output. Proceedings Symposium VLSI Circuits, June 2000, pp. 16–19.
32. Petkov, V.P. and Boser, B.E. (2004) A fourth-order $\Sigma\Delta$ interface for micromachined inertial sensors. ISSCC IEEE International Solid-State Circuits Conference, pp 320–329.
33. Petkov, V.P. and Boser, B.E. (2005) *IEEE Journal of Solid-State Circuits*, **40** (8), 1602–1609.
34. Raman, J., Cretu, E., Rombouts, P., and Weyten, L. (2006) A digitally controlled MEMS gyroscope with unconstrained sigma-delta force-feedback architecture. IEEE MEMS'06, Istanbul Turkey, pp. 710–713.
35. Raman, J., Cretu, E., Rombouts, P., and Weyten, L. (2009) *IEEE Sensors Journal*, **9** (3), 297–305.
36. Dong, Y., Kraft, M., and Redman-White, W. (2007) *IEEE Sensors Journal*, **7** (1), 59–69.
37. Dong, Y., Kraft, M., Hedenstierna, N., and Redman-White, W. (2008) *Sensors and Actuator, A*, **145**, 299–305.
38. Maurino, R. and Mole, P.A. (2000) *IEEE Journal of Solid State Circuits*, **35**, 959–967.
39. Northemann, T., Maurer, M., Rombach, S., Buhmann, A., and Manoli, Y. (2010) *Sensors and Actuators A: Physical*, **162** (2), 388–393.
40. Northemann, T., Maurer, M., Buhmann, A. He, L., and Manoli, Y. (2009) Excess loop delay compensated electro-mechanical bandpass sigma-delta modulator for gyroscopes. Proceedings

of the Eurosensors XXIII Conference, vol. 1 (1), pp. 1183–1186.

41. Pastre, M., Kayal, M., Schmid, H., Huber, A., Zwahlen, P., Nguyen, A.M., and Dong, Y. (2009) A 300 Hz19b DR capacitive accelerometer based on a versatile front end in a 5th-order delta-sigma loop. ESSCIRC, Athens, Greece, September 14–18, 2009.

42. Lapadatu, D., Blixhavn, B., Holm, R., and Kvisteroy, T. (2010) A high-precision high-stability butterfly gyroscope with north seeking capability. Proceedings of the IEEE/ION Position Location and Navigation Symposium (PLANS), pp. 6–13.

Part V
Software Implementations

17
3D Parametric-Library-Based MEMS/IC Design

Gunar Lorenz and Gerold Schröpfer

17.1
About Schematic-Driven MEMS Modeling

Microelectromechanical system (MEMS) engineers and integrated circuit (IC) designers frequently express the need to cosimulate their MEMS and IC designs in a common simulation environment. Cosimulation is required to verify the IC design and to predict yield sensitivity to manufacturing variations. The most obvious path is to do the cosimulation in the environment used by the IC designers, which requires that the MEMS designers deliver a behavioral model of the MEMS device expressed in a hardware description language (HDL). At present, MEMS engineers have a very limited ability to deliver behavioral models in these formats. In practice, the employed methods are to handcraft a model, usually in the form of a lookup table, to generate a reduced-order model from finite element analysis (FEA), or to use an existing library of predefined MEMS component models. This chapter describes a MEMS system design methodology that enables MEMS and IC engineers to design and simulate in the same environment.

The development of MEMS component libraries for circuit simulators started in the 1990s [1]. A general approach was developed by different groups [2–5], which led to the first commercially available tool ARCHITECT®, from Coventor, Inc. [6, 7], later followed by the tools MEMS Pro® from SoftMEMS and SYNPLE® from Intellisense.

Over the past decade, ongoing development efforts have dramatically increased both the variety of MEMS library components and the sophistication of the underlying behavioral models, enabling system-level simulation for a wide range of MEMS device types [8–15]. In some cases, the library components are closely linked with the process technology, from which material properties and geometric parameters such as layer thickness are obtained. More recently, 3D visualization of the complex models and simulation results has been added [7, 15].

Early on, lumped-element libraries for MEMS closely resembled their electronic counterparts. Similar to IC designers, MEMS designers needed to work in a schematic-driven environment using symbols that represent individual parametric

System-level Modeling of MEMS, First Edition. Edited by T. Bechtold, G. Schrag, and L. Feng.
© 2013 Wiley-VCH Verlag GmbH & Co. KGaA. Published 2013 by Wiley-VCH Verlag GmbH & Co. KGaA.

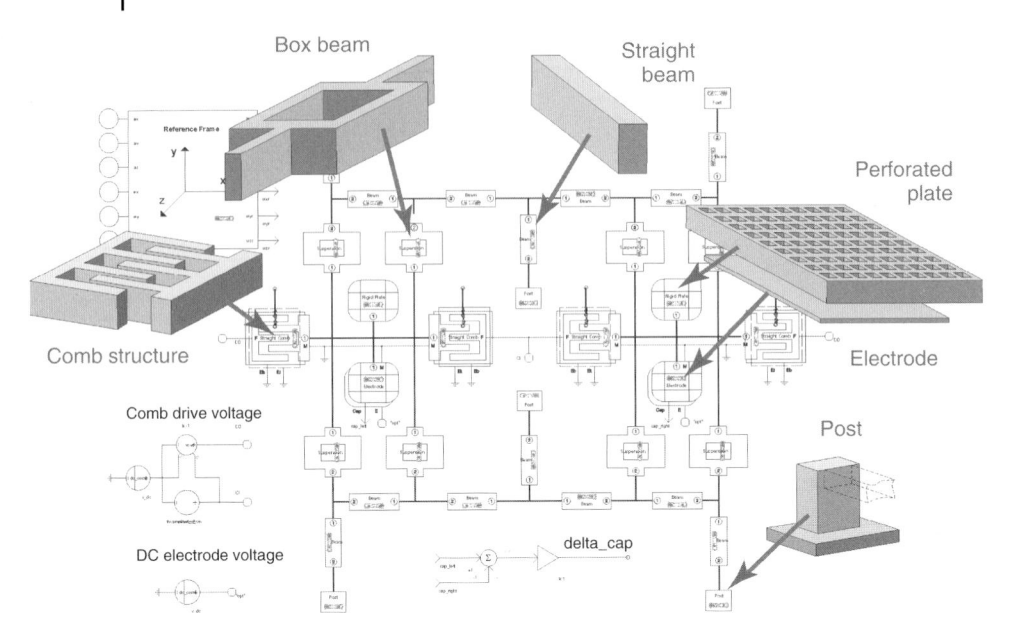

Figure 17.1 Schematic of a gyroscope assembled from MEMS components from Coventor's ARCHITECT behavior model library [7].

building blocks or components. These symbols were connected in the schematic to represent a three-dimensional MEMS device, as shown in Figure 17.1.

The schematic symbols and the mathematical models that they represent enable the exploration of the parametric design space in seconds or minutes with accuracy that rivals classic FEA [8, 9, 11, 14, 15]. Schematic-driven MEMS design environments, while extremely fast, parametric, and capable of incorporating nonlinear model properties, still face the following obstacles to widespread adoption:

- First, creating three-dimensional geometry using symbols and wires is often perceived as laborious and nonintuitive. MEMS designers, who are responsible for device model creation, traditionally prefer to use either 2D layout or 3D mechanical CAD tools for design entry. Therefore, using schematic-driven design entry instead requires a fundamental change in their preferred way of working.
- Second, IC designers and system architects, the principal clients for MEMS behavioral models, rely on either signal-flow simulators, such as MATLAB Simulink, or custom IC design and simulation environments from Cadence®, MentorGraphics®, or Synopsys®. None of the standard electronic design automation (EDA) environments provide a particularly attractive environment for MEMS design and, more importantly, there is no standard way to exchange behavioral models between signal-flow and circuit simulators.

Ideally, the MEMS designer should be able to create and modify the MEMS design in a 3D physical design environment that suits his needs, and then automatically generate required simulation models and layout views for either signal-flow or circuit simulators. The simulation models should be parametric and should accurately capture the complex behavior of the MEMS device while being sufficiently computationally efficient to allow simulation of the MEMS and IC together with reasonable CPU time.

17.2
A 3D Parametric Library for MEMS Design–MEMS+®

Coventor's answer to the given MEMS + IC design, simulation, and product development challenges is a novel design platform called $MEMS+^®$. Coventor's $MEMS+^®$ methodology allows the MEMS designer to work in a 3D environment that suits his needs and yet easily delivers parameterized behavioral models that are compatible with IC design and system simulation environments. The IC or system designer, meanwhile, will see no difference between including a MEMS device and any other analog or digital component. The parameters in the MEMS behavioral model may include manufacturing variables, such as material properties and dimensional variations, as well as geometric properties of the design. It has been proven that the sophistication and accuracy of these models will enable optimization of the system and the MEMS + IC design, both for performance and yield.

17.2.1
3D Design Entry for MEMS

The first step in the MEMS+ design methodology is to create a MEMS design by selecting MEMS building blocks from a parameterized 3D MEMS component library and assemble them into a MEMS device design. As part of this process, the MEMS designer can specify which parameters will be exposed in the IC or system design environment. Figure 17.2 shows the MEMS+ 3D graphical user interface (GUI) featuring the same gyroscope example shown in Figure 17.1.

One innovation of the MEMS+ methodology is that the MEMS designer can construct the behavioral model in a 3D view. Instead of creating an abstract schematic diagram, the user selects a component from the library, enters values for its parameters, and a corresponding 3D view is immediately presented on the canvas. This direct creation of a MEMS device in a 3-D view has proven to be much more natural to MEMS engineers, and therefore will save time in contrast to the schematic-based approach. Furthermore, providing a graphical design entry interface that is separated from the actual simulation environment allows for alternative designs creation methods, including assisted 3D geometry or 2D layout import and even free-hand drawing capabilities, which can be employed in parallel to a pure library-driven design approach.

It should be noted that the resulting 3D view differs from a traditional 3D CAD modeling tool in that there is an underlying behavioral model associated with each MEMS building block.

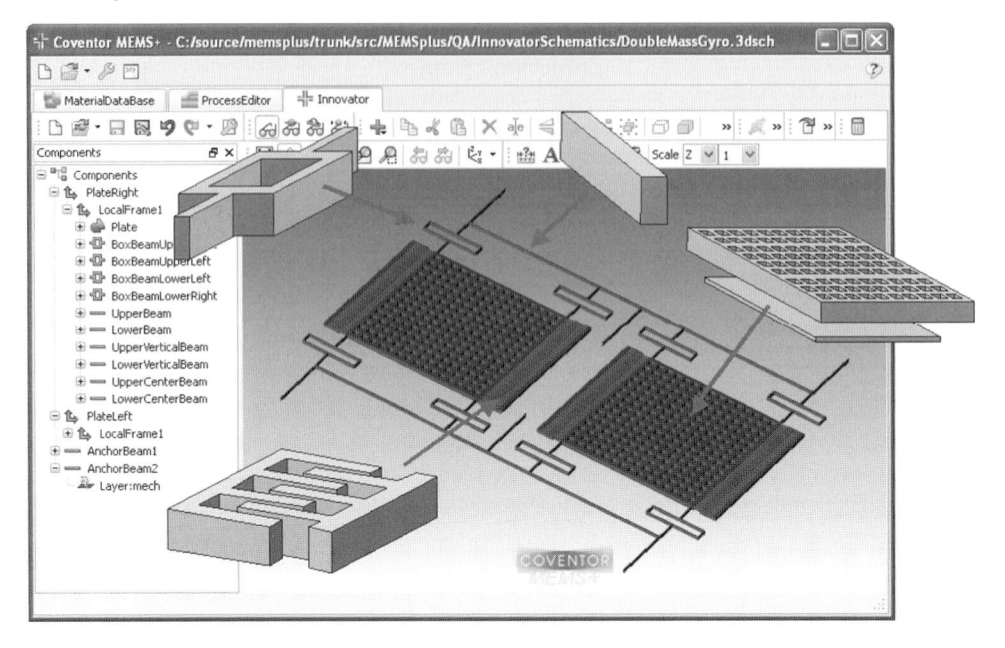

Figure 17.2 MEMS design environment that allows direct creation in 3-D.

17.2.2
MEMS Model Library

The MEMS+ component library is the product of many years of effort and can be thought of as the MEMS equivalent of the BSIM library in the IC design world [16]. The MEMS+ component library is strictly hierarchical and builds on top of three different mechanical model families: rigid plates, flexible beams and plates, and suspensions (Figure 17.3).

The rigid plate is used to create arbitrary shapes by assembling parametric geometrical primitives such as rectangular segments, triangular segments, straight and curved comb fingers, and so on. The final rigid plate is the result of a geometric Boolean operation on the individual segments. The behavioral model of the rigid plate is based on Newton's laws and the Euler equations.

The second group of mechanical elements, the flexible beams and plates, include basic shapes such as straight bars, circles, arcs, and quadrilaterals that can be used to create complex flexible structures, as seen in Figure 17.4.

The mechanical behavior of all flexible structures in MEMS+ is modeled by individual finite elements of variable order. The MEMS+ model library contains state-of-the-art finite beam, shell, and brick elements which will be chosen implicitly by the edge order parameters of the given library component. The rectangular plate component, for example, is represented by a shell if the user picks "one" for the vertical edge order parameter. A higher number yields a corresponding brick element.

17.2 A 3D Parametric Library for MEMS Design–MEMS+®

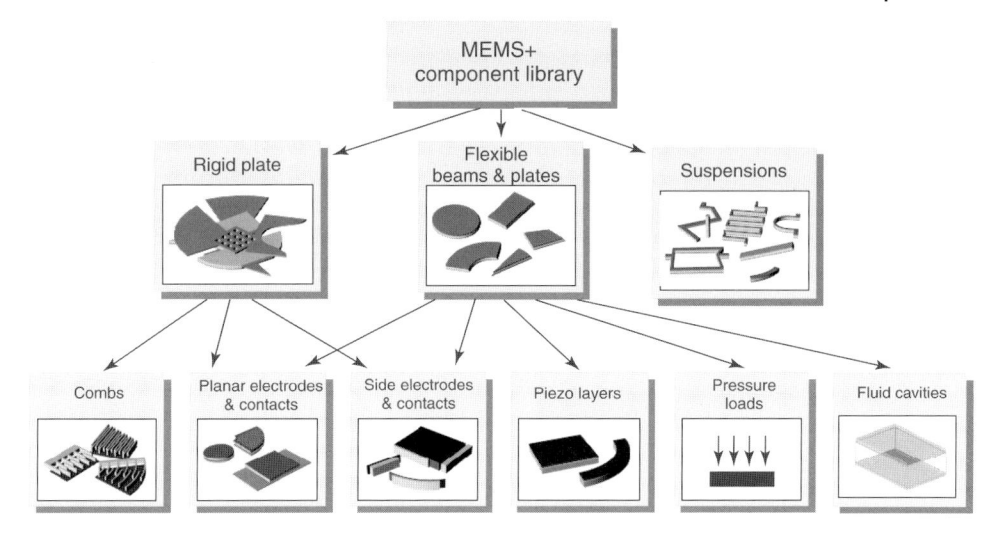

Figure 17.3 Library building blocks in MEMS+.

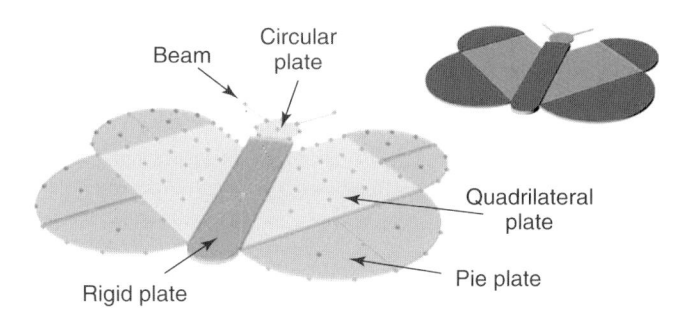

Figure 17.4 Butterfly created using standard mechanical elements from the MEMS+ library.

When it comes to mechanical modeling, the main difference of the MEMS+ methodology from standard FEM-codes such as ANSYS® or COMSOL® is not so much in the finite elements being used but rather in the creation of the model itself. While standard FEM tools rely on automeshers to fill an arbitrary geometry with low order finite elements, MEMS+ users assemble geometry by using parametric library components that are each associated to a corresponding specialized, height order finite element (Figure 17.4) or rigid plate.

The third group of mechanical elements, suspensions, includes parametric primitives for serpentines, beam paths, U-shapes, and many others. All suspension models are internally assembled from finite elements in conjunction with reduced-order techniques.

The finite elements used for mechanical modeling in MEMS+ include MEMS-specific and process-relevant effects such as support for high aspect ratios,

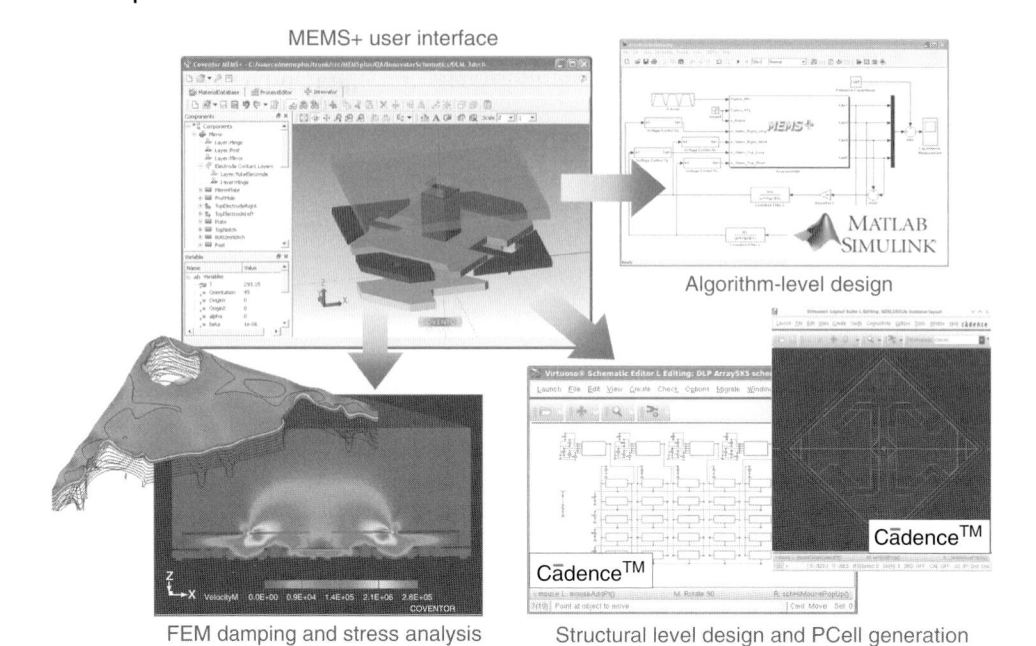

Figure 17.5 MEMS+ 3D design entry and interface options.

perforations, sidewall angles, prestress, and multiple material layers, as well as nonlinear behavior such as stress stiffening and buckling.

Most members of the mechanical model families can be "decorated" with sophisticated 3D electrostatic comb and electrode, as well as piezo electric, contact, or acoustic cavity models (Figure 17.3). Furthermore, the designer can augment the electrode and comb models with additional damping models (not shown in the diagram).

The corresponding behavioral models are based on various modeling techniques, including analytic formulae, numerical integration, conformal mapping, and finite elements. For details on general modeling theories, we refer the reader to earlier chapters. A detailed description of the underlying models in MEMS+ can be found in the comprehensive reference documentation that accompanies MEMS+ [17].

17.2.3
Integration with System Simulators

All MEMS designs created with the MEMS+ user interface can be shared between the different members of the MEMS design team. MEMS+ supports simple geometry export filters to FEA and layout tools, as well as sophisticated simulation interfaces to MATLAB and Simulink and IC design environments such as Cadence®, Virtuoso® (Figure 17.5).

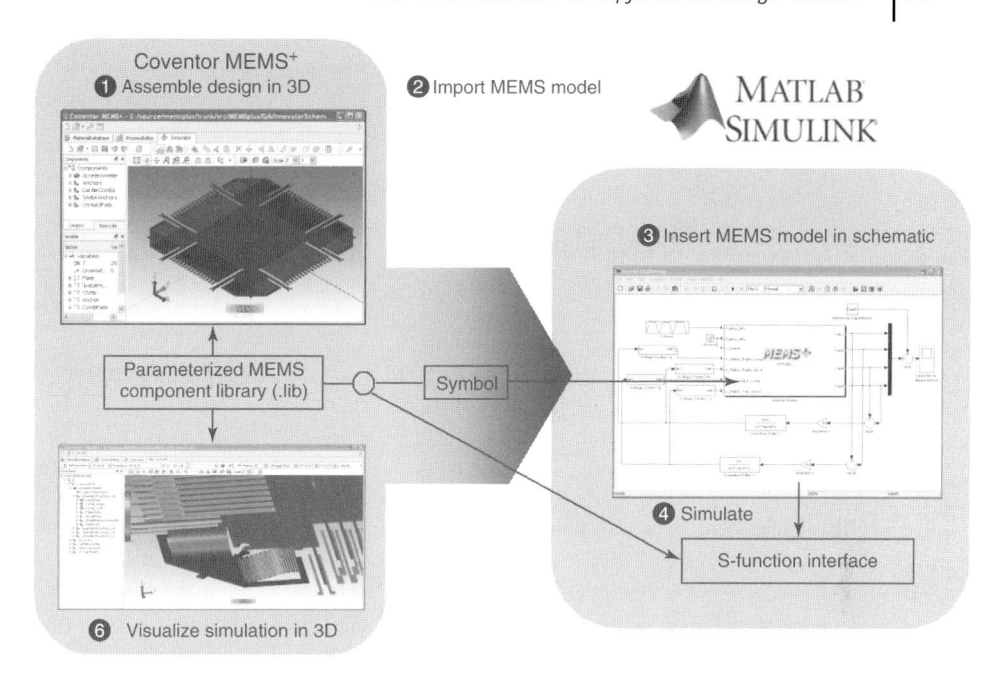

Figure 17.6 MEMS+ integration with MATLAB Simulink.

17.2.4
Integration with MATLAB and Simulink

MATLAB and Simulink from The MathWorks [18] are well known across all engineering disciplines as powerful tools for engineering innovation. These tools allow engineers to define system models specific to their domain and then simulate their behavior. With MEMS+, designers can import the parameterized model created in the MEMS+ design platform directly into MATLAB or Simulink. MEMS+ requires neither programming by the user of device physics (such as mechanical equations or capacitance extraction) nor FEA; it only requires the creation of the 3D design in MEMS+ using the tool's intuitive 3D graphical interface.

The MEMS+ MATLAB interface supports both the MATLAB scripting interface and device model importing into the Simulink schematic editor, as shown in Figure 17.6.

MEMS+ automatically generates a symbol that the user imports it into the Simulink model editor window. The number of symbol ports and parameters is automatically taken from the original MEMS+ design. The symbol representing the MEMS+ model can be inserted into larger systems and simulated with the native MATLAB solvers. During simulations, the simulator will connect, via Simulink's S-function interface, with the MEMS+ component library to evaluate the MEMS behavioral model at each time step. In addition to standard transient simulations, MEMS+ provides additional analyses such as DC, DC transfer, modal, and AC

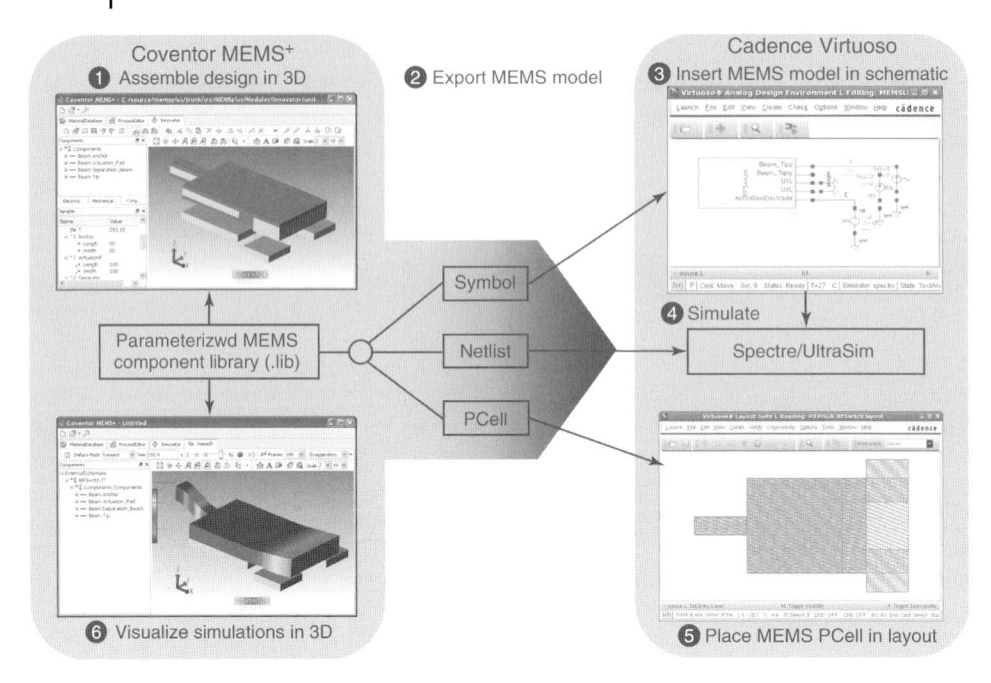

Figure 17.7 MEMS+ integration with standard EDA tools.

analysis. On completion of a simulation, the simulation results can be loaded into the MEMS+ user interface and visualized via 2D graphs and as fully contoured three-dimensional animations.

17.2.5
Integration with EDA Tools

IC engineers commonly use tools such as Cadence Virtuoso to design the analog/mixed-signal electronics that accompany a MEMS device [19]. In order to succeed, IC engineers require fast and accurate models of the MEMS device in the Cadence model library. MEMS+ facilitates the required model exchange by providing an easy way to import MEMS+ models to the Cadence model library (Figure 17.7). Every device created in MEMS+ can be imported into the IC design environment in the form of a netlist and a schematic symbol. Similar to the MEMS+ MATLAB Simulink interface, the number and names of the pins on the schematic symbol are controlled by the MEMS engineer and represent electrical connections to the MEMS device. The MEMS symbol can be placed into the IC schematic editor and surrounded by the complete IC design.

Simulations can be run in any of the Cadence circuit simulators that are compatible with Virtuoso, including Spectre, Spectre RF, Spectre APS, and UltraSim. The simulator will connect with the MEMS component library to evaluate the MEMS behavioral model at each simulation point, that is, time step or frequency.

It is important to highlight that all external solvers supported by MEMS+ (including MATLAB) use the same component library (Figure 17.3) during the actual simulation. All MEMS+ supported simulations can therefore expected to be of comparable accuracy.

On completion of a simulation, the designer can view the results in the MEMS+ 3D viewer, which can animate the motion of the MEMS device. At any time, but especially when the MEMS and IC designers are satisfied with the MEMS design, they can export a parameterized layout cell (PCell) that can generate a layout of the MEMS device.

17.3
Toward Manufacturable MEMS Designs

There are two categories of parameters that are relevant for MEMS design. The first category consists of the dimensional parameters of the MEMS building blocks (such as length, width, number of comb fingers, etc.) that are determined by the MEMS engineer. The second category consists of the material properties and geometric parameters that are determined by the selected manufacturing process.

17.3.1
Parameterization of Process and Material Properties

While MEMS and IC design share aspects related to manufacturing, they differ in the impact that manufacturing has on their design flows. In particular, the microfabrication processes for IC devices are standardized. IC components are fixed within a fabrication process, while MEMS components are not. For instance, a transistor (an IC component) is created out of specific layers implanted and deposited on the silicon substrate during the fabrication process, and these layers usually cannot be changed by the IC designer. But a mechanical beam component that is a part of a MEMS design might be placed on any of several "mechanical" layers, and that layer is a design choice. Also, the thickness of the mechanical layer might be changeable within certain limits. Conventional IC design tools do not offer the flexibility to change the location of a component within the various layers created during the fabrication process. Thus, the impact of the chosen fabrication process on an IC design is fixed from the beginning and does not change from one design iteration to the next. In comparison, the fabrication processes of MEMS devices are often not standardized.

In addition, it is sometimes necessary to tailor the fabrication process to a particular MEMS device in order to achieve the design goals for the device. Thus, the fabrication process is an important *free parameter* in a MEMS design that often needs to be refined as development progresses. The flexibility to change the description of the fabrication process is missing from conventional IC design environments. In addition, behavioral models of electrical IC components

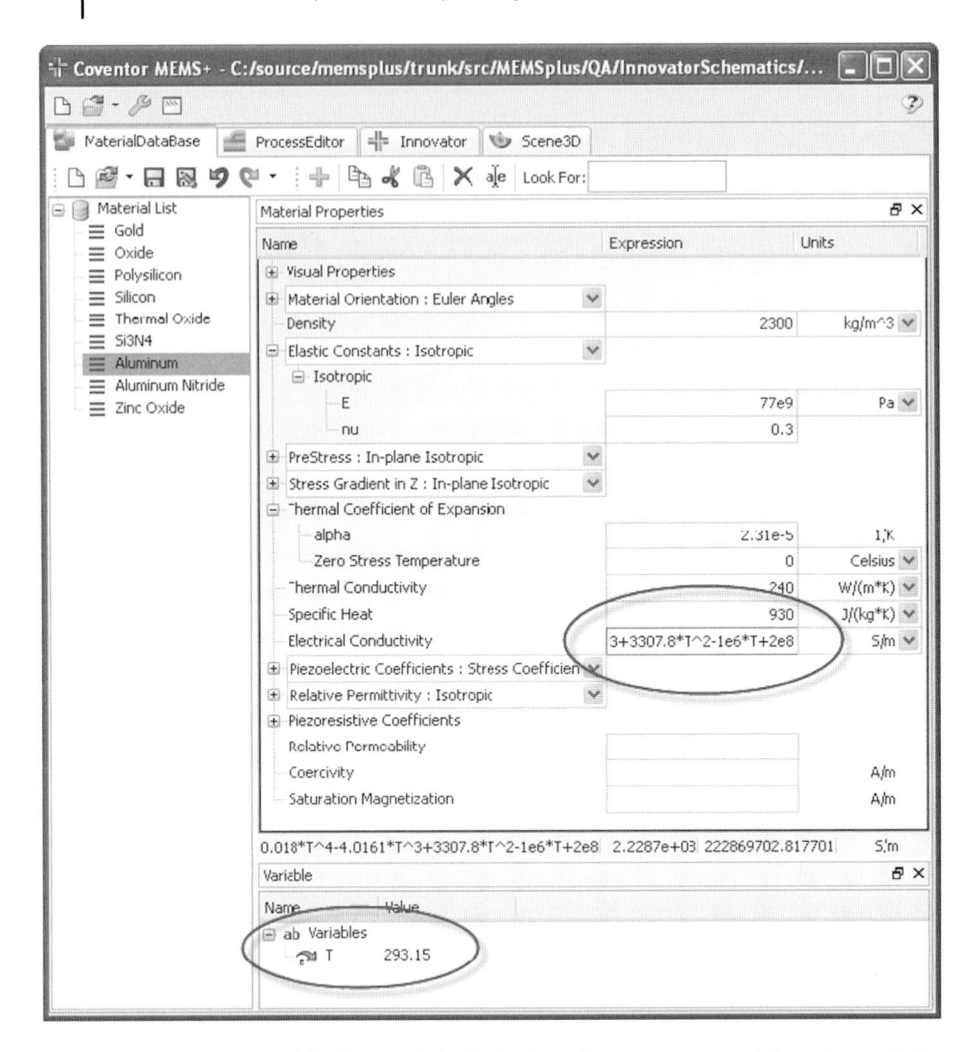

Figure 17.8 Material database with highlighted regular expressions and dependent variable.

cannot be parameterized in terms of the process parameters. In MEMS design, the parameters of the process description can be varied as a part of the design; thus, the models must be parameterized with respect to the process parameters.

Coventor's MEMS+ environment addresses the specific needs of MEMS designers by providing two built-in editors that are used to specify all relevant manufacturing-process-specific data. The material property editor shown in Figure 17.8 is used to create a material database that contains all relevant physical as well as visual properties used in the MEMS+ environment.

All *material properties* can be defined as absolute values, variables, or algebraic equations. A combination of variables and algebraic equations allows for

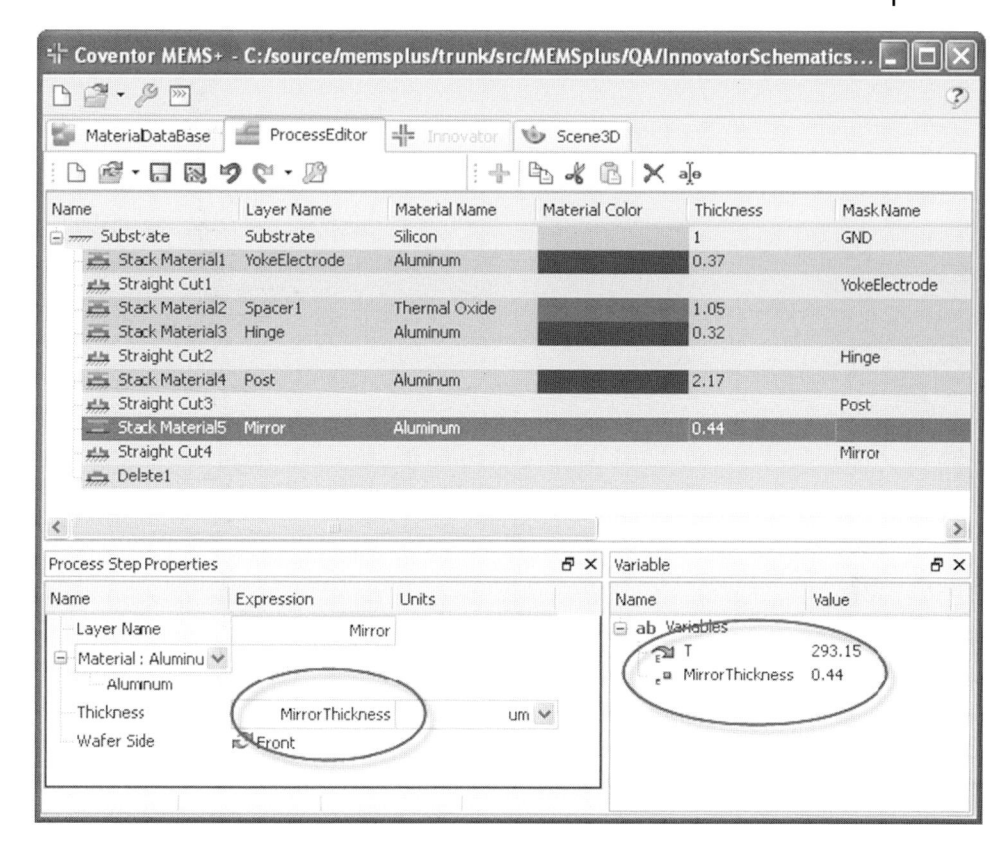

Figure 17.9 Process Editor for describing the sequence of steps in the MEMS fabrication process. Regular expressions and dependent variable are highlighted.

properties to be mutually dependent on other properties, environmental variables (e.g., temperature and humidity), or even entirely abstract variables such as the equipment settings of a given fabrication process. The MEMS engineer can choose which variables are to be exposed in the MEMS+ GUI and in the corresponding MEMS schematic symbol and layout. For instance, in Figure 17.8, the electrical conductivity of aluminum is given by an algebraic expression that depends on the temperature T. At the bottom of Figure 17.8, the icon beside the T variable indicates that the user has exposed it in the MEMS+ user interface and IC design environment.

The second built-in editor, the Process Editor, is used to define the sequence of MEMS fabrication steps, as shown in Figure 17.9. The underlying process data includes relevant information about the layer stack, such as the layer order, material type, thickness, and sidewall profile. The process data is dependent on a material database: each of the layers in the process data specifies a material type that must exist in the corresponding material database.

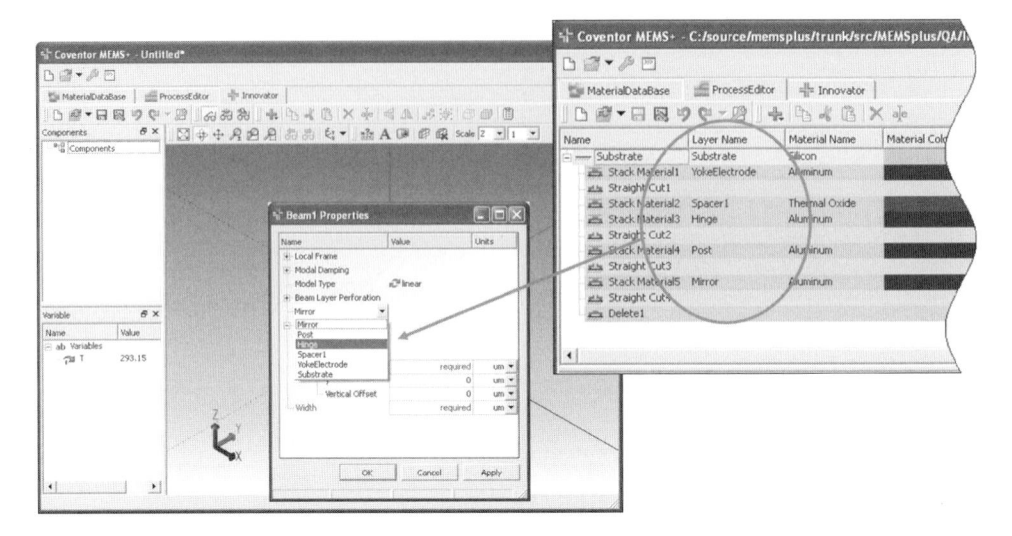

Figure 17.10 The parameter "Layer" associates a given component to one or more layers from the process editor.

The link between the MEMS+ library components (Figure 17.3) and the fabrication data is established by a common layer property that allows the user to assign a given component to one or more layer names (Figure 17.10).

The separation of the materials and process data from the MEMS design and simulation environment allows the model to have process-related parameters whose specifications are not fixed, but rather tied by reference to the manufacturing process data.

17.3.2
Process Design Kits

In practice, reusable process libraries or process design kits (PDKs) contain detailed material and process property information that has been previously correlated with simulation models of test structures and other test devices. This set of information should not only include nominal values for process geometries and materials, but manufacturing tolerances and process-related design rules. These constraints can be captured in the above-described editors and the resulting 3D models. At a minimum, a MEMS PDK should consist of a foundry and process-dedicated process-description file, a material database, and layout templates. Advanced versions of PDKs might also include a complete process-specific library of MEMS components or complete devices that represent the related manufacturing process. The availability of dedicated design kits is vital throughout the different stages of MEMS development.

With the growth of independent MEMS foundries and the corresponding emergence of fabless product companies [20], one can expect that the exchange of

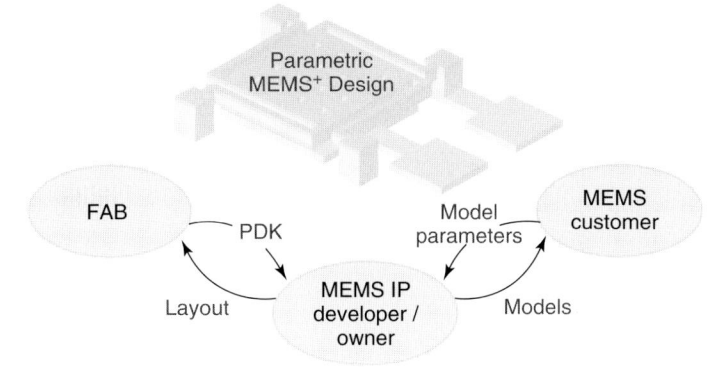

Figure 17.11 MEMS+ parametric design format provides a new standard to facilitate the communication between the partners of the MEMS ecosystem.

information between manufacturers and designers needs to be further facilitated, and MEMS-specific PDKs will play an increasing role in this exchange [21]. The different views of the MEMS component enabled by MEMS+ provide a customizable environment to transfer MEMS component intellectual property (IP) between various actors in MEMS development. Using the described methodology, the MEMS designer can provide a detailed accurate model to the manufacturing site, as well as to the system integrator, that is, the MEMS customer, as depicted in Figure 17.11.

17.4 Micromirror Array Design Example

The following chapter highlights how the MEMS+ methodology can be applied to a well-known real-world MEMS product, namely, a micromirror array. Texas Instruments' digital light processing (DLP) projection system is a well-known display technology that is built around a digital micromirror device (DMD), a chip that contains a two-dimensional array of MEMS micromirrors [22, 23]. Each mirror is fabricated on top of a static random access memory (SRAM) cell that determines its state. Thus, associated with the mirror array is an underlying array of SRAM cells that can be individually accessed by word and bit lines in the same way as the bits on an SRAM memory chip. The mirrors, being electromechanical devices, respond at a much slower timescale and with more complex behavior than the SRAM cells. For instance, on being flipped from one state to the other, the mirror may bounce before settling into the other state. The parasitic capacitance of the SRAM cells, as well as the interconnect parasitics (which vary with position in the array), can affect how the mirrors respond to control signals. Also, the vertical dimensions of the mirrors may vary with the position of the DMD die on the wafer, a manufacturing effect. The mirrors and control electronics together form a complex system, so there is a considerable potential for time and cost savings when simulating the complete DMD subsystem in advance of fabrication.

In this example, we demonstrate how MEMS+ and Cadence Virtuoso can be used to simulate a DMD mirror array together with its control electronics. The MEMS and IC design teams can use such simulations to identify potential issues, optimize the design, verify the functioning of the complete system, and verify robustness against manufacturing variations. It is worth noting that other MEMS-based display technologies involving arrays of MEMS exist for which this type of simulation would be useful. Thus, this example is but one of many ways that the use of MEMS+ with Cadence Virtuoso can provide useful design and verification functions.

The first step in design entry involves creating a behavioral model of a single DMD pixel, as seen in the left-hand side of Figure 17.12.

With the structured MEMS+ approach, the MEMS designer constructs the behavioral model of a single mirror in a 3D user interface (2). The MEMS designer selects building blocks from a library of parameterized MEMS components and assembles them to create a model of a single DMD mirror. The connection between the MEMS+ library components and the fabrication process is established by assigning the layer property of each component to one or more layer names specified in the process editor.

After completing the mirror design in MEMS+, it can be imported into the Cadence Virtuoso environment. MEMS+ automatically generates a MEMS symbol that can be placed in the Virtuoso schematic (4). This is more than a "black box" that only shows inputs and outputs – the MEMS designer can choose to expose parameters that may be of interest to the IC designer. In addition, MEMS+ automatically generates a netlist that contains references to the MEMS behavioral models. This netlist makes it possible for an IC designer working in the Virtuoso environment to simulate the MEMS device along with the analog circuitry around it without having any special MEMS or mechanical CAD knowledge.

The IC designer, meanwhile, creates a schematic of the SRAM memory cell underneath each mirror using the familiar components from readily available PDKs in the Cadence design environment (3). The CMOS SRAM cell can in turn be connected to the mirror to assemble the complete pixel cell (5). This pixel cell is then replicated to form an array and connected to the driving electronics, as seen in image (6) of Figure 17.13.

The complete mirror array can now be simulated with one of the Virtuoso simulators: Spectre, UltraSim, or Spectre APS. The simulator connects via the Cadence compiled-model interface (CMI) to the MEMS+ component library and evaluates the behavioral models of the MEMS devices during each iteration of the simulation.

Cadence Virtuoso's plotting capabilities can be used to analyze the coupling between the electromechanical device and the electronics, or even to probe mechanical motion and measure response time (7). On completion of the simulation, the user can, if desired, view a 3D animation based on the Cadence simulation results in the MEMS+ visualization plug in (8).

Furthermore, because the MEMS designer still needs the ability to verify the device on the physical implementation level, MEMS+ automatically generates 3D

17.4 Micromirror Array Design Example | 421

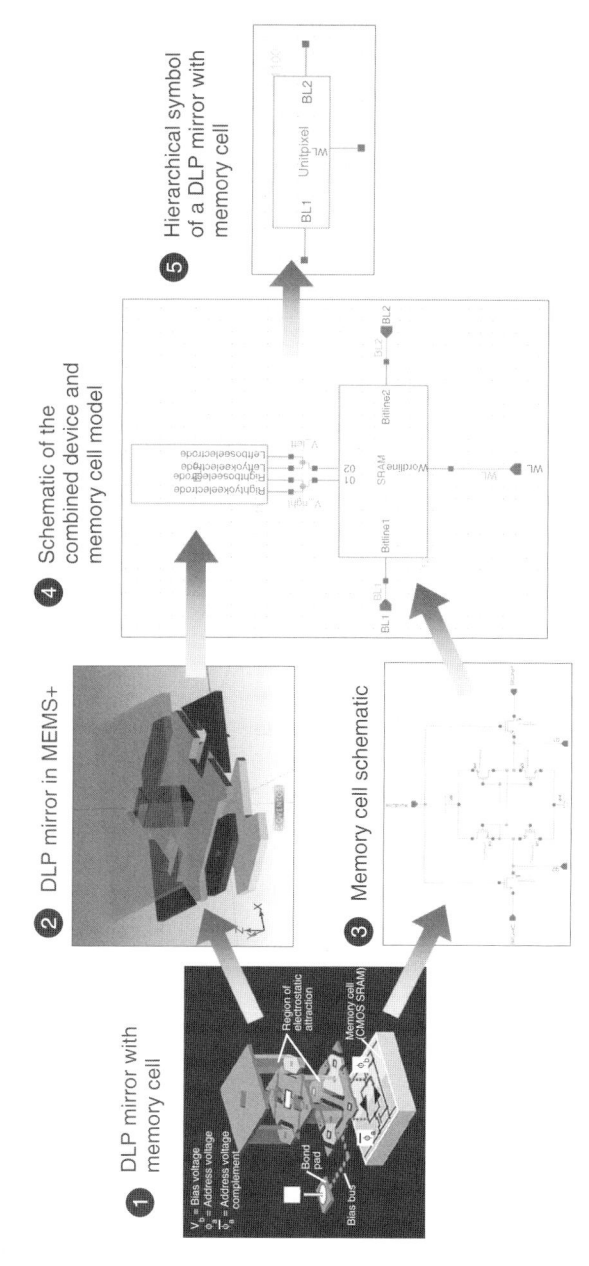

Figure 17.12 Model of a MEMS mirror created in MEMS+ combined with an SRAM cell in a Cadence Virtuoso schematic and then abstracted to a hierarchical symbol representing one pixel in the array.

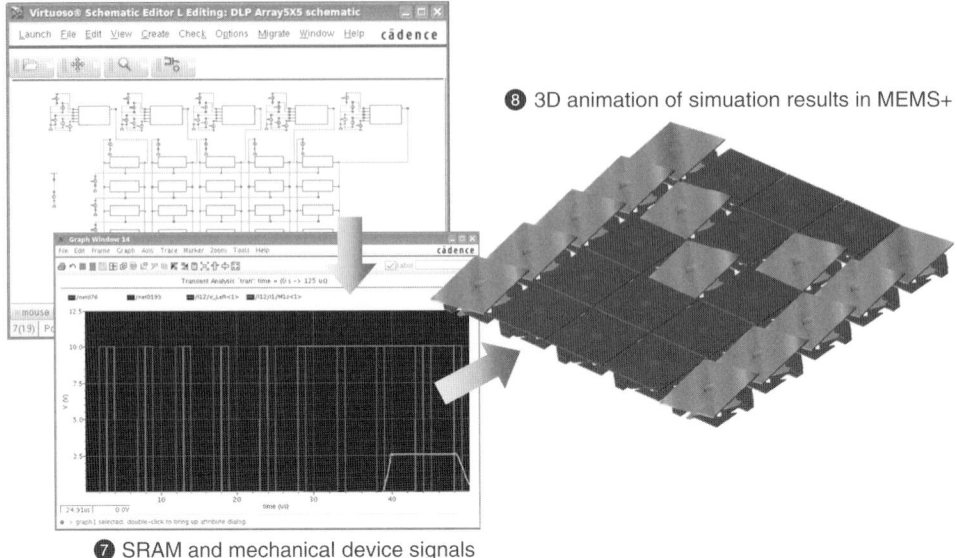

Figure 17.13 Complete schematic with 25 mirrors, CMOS SRAM and control circuitry (6), sample input and output signals (7), and 3D animation (8).

solid models that can serve as an input to FEA. MEMS+ can also automatically generate a fully parameterized PCell that has the same exposed parameters as the schematic symbol.

17.5
Conclusions

This chapter described a MEMS system design methodology that enables MEMS and IC engineers to design and simulate in the same environment. This methodology is implemented in a commercial design platform called *MEMS+* from Coventor, Inc. MEMS+ has substantial advantages over the traditional approach of transferring handcrafted or reduced-order behavioral models from the MEMS engineers to the IC engineers. First, the behavioral models of the MEMS device are sufficiently sophisticated to fully represent the MEMS behavior, capturing, for example, cross-coupling between the mechanical degrees of freedom. The accuracy of these behavioral models has been validated via other modeling techniques and measurements. Second, the 3D geometrical and behavioral models, as well as associated layout cells, are fully parameterized both with respect to manufacturing-dependent variations and geometric attributes of the design, which enable design and yield optimization studies in the EDA environment. Third, the automatic hand-off between the MEMS and IC design environments eliminates inevitable human errors that arise in any manual hand-off process.

While this chapter provides an example of applying the new methodology to a DLP MEMS mirror array with underlying electronics based on SRAM cells, the underlying MEMS component library has proven applicable to many types of MEMS. For instance, simulations of an accelerometer controlled by a sigma–delta modulator, as described in the previous chapter and in [8, 9].

References

1. Teegarden, D., Lorenz, G., and Neul, R. (1998) *IEEE Spectrum*, **35**, 66–75.
2. Romanowicz, B., Ansel, Y., Laudon, M., Amacker, Ch., Renaud, P., Vachoux, A., and Schröpfer, G. (1997) VHDL-1076.1 modeling examples for microsystem simulation, in *Analog and Mixed-Signal Hardware Description Languages* (ed.A. Vachoux), Kluwer Academic Publishers.
3. Zhou, N., Clark, J.V., and Pister, K.S.J. (1998) Nodal analysis for MEMS design using SUGAR v0.5. Proceedings of the International Conference on Modeling and Simulation of Microsystems, Semiconductors, Sensors and Actuators (MSM), Santa Clara, CA, pp. 308–313.
4. Vandemeer, J.E., Kranz, M.S., and Fedder, G.K. (1998) Hierarchical representation and simulation of micromachined inertial sensors. Proceedings of the International Conference on Modeling and Simulation of Microsystems, Semiconductors, Sensors and Actuators (MSM), Santa Clara, CA, pp. 540–545.
5. Lorenz, G. and Neul, R. (1998) Network-type modeling of micromachined sensor systems. Proceedings of the International Conference on Modeling and Simulation of Microsystems, Semiconductors, Sensors and Actuators, MSM98, Santa Clara, April 1998, pp. 233–238.
6. Lorenz, G. and Repke, J. *Sensors in Automotive Technology*, (2006) Jossey-Bass Publishers (A Wiley Company), vol. 4, pp. 58–72. ISBN: 978-3-527-60507-1.
7. Lorenz, G. and Kamon, M. (2007) A system-model-based design environment for 3D simulation and animation of micro-electro-mechanical systems (MEMS). Proceedings of APCOM'07 in conjunction with EPMESC XI, Kyoto, December 3–6, 2007.
8. Breit, S., Welham, C., Rouvillois, S., Kraft, M., and McNie, M. (2008) Simulation environment for accurate noise and robustness analysis of MEMS under mixed-signal control. Proceedings of the ASME International Mechanical Engineering Congress, Boston, MA, November 2008.
9. Welham, C., Rouvillois, S., King, M.D., Combes, D., and McNie, M. (2008) Modeling and simulation of multi-degree-of-freedom of micro-machined accelerometer with sigma-delta modulator. Proceedings of ESNUG Conference 2008, Munich, Germany, October 2008.
10. Ma, W., Chan, H.-Y., Wong, C.C., Chan, Y.C., Tsai, C.-J., and Lee, F.C.S. (2010), Design optimization of MEMS 2D scanning mirrors with high resonant frequencies. Proceedings of the 23rd International IEEE Conference on MEMS, Hong, January 24–28, 2010, pp. 823–826.
11. Casset, F., Welham, C., Durand, C., Ollier, E., Carpentier, J.-F., Ancey, P., and Aïd, M. (2008) In-plane RF MEMS resonator simulation. Proceedings of the MEMSWAVE 2008, Heraklion, Greece, 30 June–3 July 2008.
12. Judy, M. (2002) Computer-Aided Design (CAD) for Integrated, Microelectromechanical (MEMS) Devices. Final Technical Report AFRL-IF-RS-TR-2002-176, DARPA, approved for public release, August 2002.
13. Schröpfer, G., King, D., Kennedy, C., and McNie, M. (2005) Advanced process emulation and circuit simulation for co-design of MEMS and CMOS devices. Proceedings of the DTIP 2005, Montreux, Switzerland, June 1–3.
14. Matova, S., Hohlfeld, D., van Schaijk, R., Welham, C.J., and Rouvillois, S. (2009)

Experimental validation of aluminum nitride energy harvester model with power transfer circuit. Proceedings of the Eurosensors XXIII Conference, Lausanne, Switzerland, September 6–9, 2009, pp. 1443–1446.

15. Schröpfer, G., Lorenz, G., and Breit, S. (2010) *Journal of Micromechanics and Microengineering*, **20**, 064003.

16. Home page of BSIM (Berkeley Short-channel IGFET Model) Group, located in the Department of Electrical Engineering and Computer Sciences (EECS) at the University of California, Berkeley. *http://www-device.eecs.berkeley.edu/bsim/* (accessed 13 August 2012).

17. Company Homepage of Coventor Inc. *http://www.coventor.com* (accessed 13 August 2012).

18. Company Homepage of The Mathworks Inc. *http://www.mathworks.com* (accessed 13 August 2012).

19. Company Homepage of Cadence Inc. *http://www.cadence.com* (accessed 13 August 2012).

20. Eloy, J.C. (2007) *Sensors and Transducers Journal*, **86** (12), 1771–1777.

21. Schröpfer, G., Lorenz, G., Donnay, S., Rottenberg, X., Jansen, R., and Bienstman, J. (2011) SiGe MEMS process design kit for MEMS IC platform. Proceedings CDNLive! EMEA 2011 Conference, Munich, Germany, May 3–5 2011.

22. Hornbeck, L.J. (1998) Current status and future applications for DMD-based projection displays. Proceedings of the 5th Int Disp Workshops 1998, Japan, pp. 713–716.

23. Wilson, T., and Johnson, R. How DLP does work? *http://electronics.howstuffworks.com/dlp1.htm* (accessed 13 August 2012).

18
MOR for ANSYS
Evgenii B. Rudnyi

18.1
Introduction

In this chapter, software MOR for ANSYS to perform Krylov-based model reduction (Chapter 3) is described. The software uses the finite element models from ANSYS Mechanical as a starting point. It reads the system matrices of the original system and writes out the reduced matrices either in the Market Matrix format or in the form that could be directly imported in system simulator ANSYS Simplorer.

MOR for ANSYS [1] (the former name mor4ansys) emerged as a spin-off during the European project MicroPyros at Department of Microsystems Engineering of Freiburg University (IMTEK) [2]. The starting point was an engineering problem that required developing a compact thermal model for system level electrothermal simulation. A circuitry should deliver power dissipation into a compact thermal model and simultaneously receive back couple of temperatures at different positions of a microthruster. As starting point, however, a finite element model was employed to develop a reliable thermal model.

The above-mentioned problem demonstrates a common gap in simulation practice. On the one hand, there is an accurate finite element model that has been already developed; on the other hand, it is still necessary to invest time and efforts to develop a behavioral model for system level simulation. Hence, the goal was to find a general solution suitable for a wide class of simulation problems, not only for a particular task to be solved in the current project.

A thorough interdisciplinary review of available options [3] revealed that mathematicians have developed new methods [4] to approximate large-scale dynamical systems (see Figure 18.1 and also Chapter 3), but the finite element community was yet unaware of such a development. Rapid prototyping in Mathematica has shown us that the methods are working extremely well, and at the same time, this allowed us to choose the best practical way (bold line in Figure 18.1).

Eventually, it was decided to develop a scalable standalone software that implements the block Arnoldi algorithm for ANSYS finite element models. The original name was mor4ansys, but then it has been changed to MOR for ANSYS. The software helped us to continue research, as it has opened new opportunities for

System-level Modeling of MEMS, First Edition. Edited by T. Bechtold, G. Schrag, and L. Feng.
© 2013 Wiley-VCH Verlag GmbH & Co. KGaA. Published 2013 by Wiley-VCH Verlag GmbH & Co. KGaA.

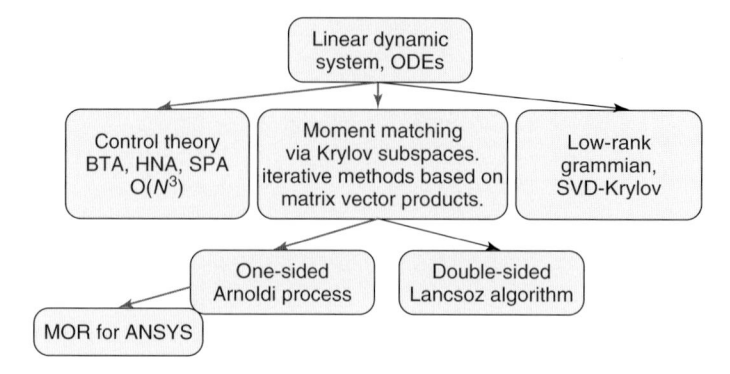

Figure 18.1 Approximation methods of large-scale dynamical systems [4].

collaboration with other groups in academia and industry and in turn brought back the knowledge of real-life needs. This information in turn led to the further development of the software.

The practice-oriented development needs solid theoretical foundations but at the same time requires finding a reasonable compromise between different tasks such as learning, programming, making research, and so on, as available resources are always limited. In this chapter, we review how it was done in the case of MOR for ANSYS. The chapter starts with the description of MOR for ANSYS related research in the next section, then considers the programming issues, and finally describes some open problems that come from practice.

18.2
Practice-Oriented Research during the Development of MOR for ANSYS

As already mentioned, the MOR for ANSYS related research was practice oriented. This means that the main goal was to solve a particular engineering problem that requires a compact model derived from a finite element model. What was necessary is to translate an engineering problem to the language of model reduction (see, for example, [4]), to employ the model reduction algorithm, and then to transfer reduced matrices into the simulation environment conventionally used for that kind of engineering task.

First, results have shown that the Arnoldi algorithm works pretty well for a variety of engineering problems. For example, we were first among the engineering groups to demonstrate that model reduction is a perfect tool to generate automatically a compact thermal model [5]. This may sound naive from a viewpoint of fundamental science, yet even now one can find papers where people derive compact thermal models basically manually. The same concerns applications with piezoelectrics and fluid–structure interaction at the acoustics approximation where MOR for ANSYS helped to prove for the first time that model reduction is a perfect tool for such applications.

18.2 Practice-Oriented Research during the Development of MOR for ANSYS

It is important to remember that model reduction as such is not new for finite elements. Methods such as modal superposition, Guyan reduction, or component mode synthesis (CMS) are known for ages among mechanical engineers. However, they are working for finite element models in structural mechanics only, and the attempts to use them (or their analogs) to the above-mentioned problems have been unsuccessful. Still it took noticeable efforts to convince engineers that mode superposition may not be the only model reduction method. We will come back to this problem in the Section 18.4.

The next simple thing was an observation that with the Arnoldi algorithm model reduction could be considered as a fast solver [6]. When something is changed in a finite element model (most often geometry) then a reduced model must be generated again. Yet, the process to generate a reduced model is much faster than to run full-scale transient or harmonic response simulation. This means that even in the optimization process where a reduced model is used once only, it make sense to use model reduction mere to speed the process up (Figure 18.2). Such a solution may look not elegant as compared with parametric model reduction, but

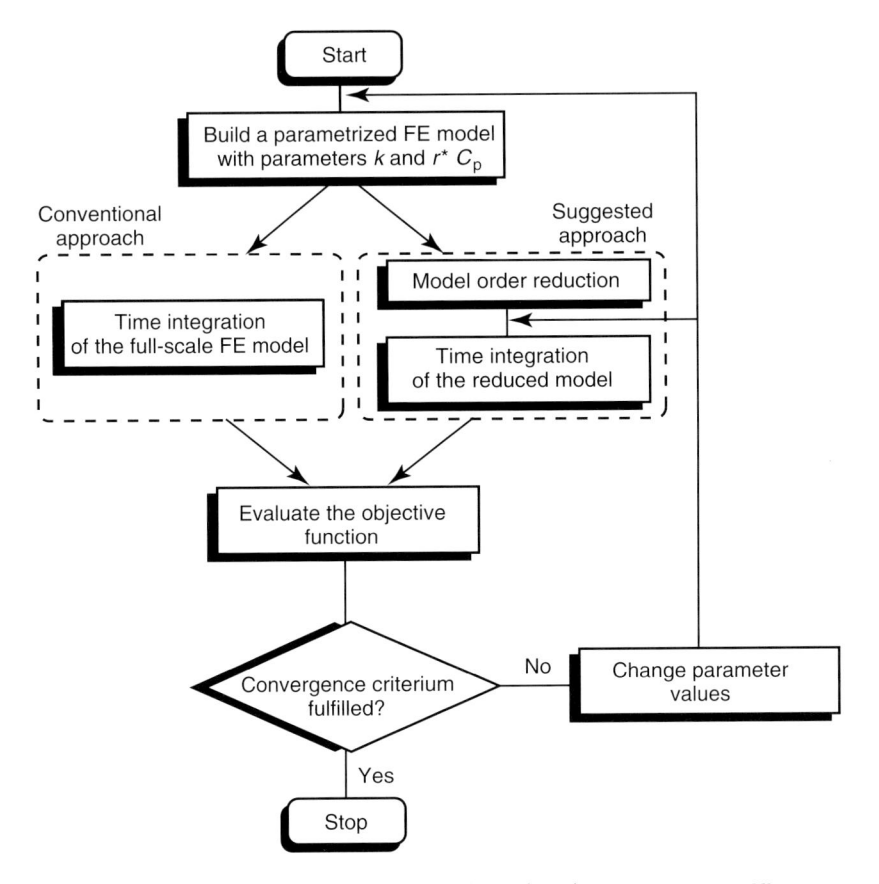

Figure 18.2 The use of model reduction as a fast solver during optimization [6].

on the other hand, it gives a simple but robust way to achieve engineering goals in many cases.

The above-mentioned expression that model reduction works well means that a relatively low-dimensional reduced model already achieves good accuracy. However, a choice of the dimension of the reduced model is an open question. At the beginning of the research, we have started using the magical number 30 for the dimension of a reduced model and it went well for many models. Yet in the general case, this is an open question, as moment matching algorithms do not have global error estimates. After several trials, we have found an error indicator [7] that brings some intelligence into the choice of the optimal dimension. It does not eliminate the need for certain know-how to be developed but allows us to simplify the choice considerably. The finding is of an empirical nature and I am unaware of any mathematical proof that supports it. Yet during our work, we have tried it for many different engineering systems, and the agreement between the exact local error and its estimates was always very good.

Model reduction has been developed by mathematicians for the first-order systems, and in practice, most systems are of the second order. In this case, it is always possible to transform dynamic system to the first-order system by increasing the dimension of the state vector twice (Figure 18.3). The disadvantage here is that a reduced system is obtained in the form of the first-order system and that computational requirements increase because of the increase in the dimension of the state vector. The use of the second order Krylov subspaces (for example, the algorithm Second Order ARnoldi (SOAR) [8] implemented in MOR for ANSYS) removes the disadvantages mentioned earlier. However, in many important applications, the proportional damping is employed.

$$E = \alpha M + \beta K \tag{18.1}$$

It happens that in such a case the damping matrix can simply be ignored during the process of constructing the projection basis. In this case, only the mass and stiffness matrices together with the input matrix are employed to generate the required Krylov subspace. The damping matrix is projected afterwards, and because of Eq. (18.1), it can actually be computed from reduced mass and stiffness matrices. This method first has been found by trial and error, inspired by the use of modal superposition in structural mechanics. Interestingly enough, recently,

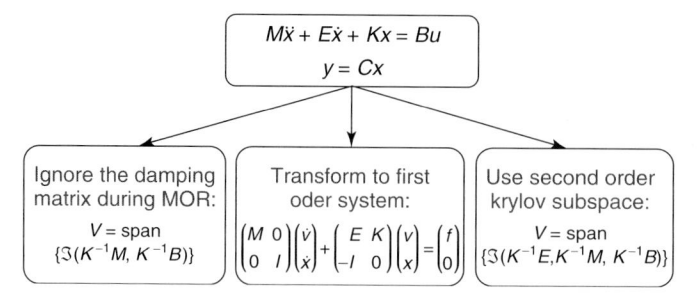

Figure 18.3 Model reduction for second order systems.

this result has been proved in the general case of Eq. (18.1). In [9], it has been shown that moment matching properties in the case of proportional damping for any values of α and β are always preserved.

In practical applications, the number of inputs can be significant, for example, 100 or even more. This brings an extra problem forward, as it happens that the number of degrees of freedom per input needed in many applications is roughly constant. In other words, the deflation that is possible in the block Arnoldi algorithm does not happen in real-life applications with multiple inputs. That is, the dimension of the reduced model in this case is a product of number of DOFs per input required by the number of inputs and could achieve 1000 or more. The system matrices for a reduced model are already dense, and hence the computational cost to simulate the reduced model increases by N^3 of its dimension.

The practical problem mentioned above has stimulated thoughts on how it could be possible to make the reduced system matrices not completely dense. Along this way, the superposition Arnoldi algorithm has been suggested, experimentally tested, and then mathematically analyzed [10]. Here, the reduced matrix has a block diagonal form. Let us consider an example of system with 100 inputs where the number of degrees of freedom per input needed in the reduced model is equal to 10. The dimension of the reduced model is then equal to $100 \times 10 = 1000$. In the case of the block Arnoldi algorithm, the system matrices are dense with dimensions of 1000×1000 (the number of nonzero values in a matrix is equal to 1 million). The superposition Arnoldi algorithm produces the block diagonal system matrices that also are of dimension 1000×1000. However, in this case, the matrices are block diagonal with 100 blocks each of dimension 10 (the number of nonzero entries in a matrix now is $100 \times 10 \times 10 = 10\,000$). The number of nonzero values in a matrix is 100 times smaller and this allows us to speed up system simulation with the reduced model.

It was mentioned earlier that in many cases, the practical goal is the optimization. An elegant solution would be to preserve some parameters as symbols during model reduction in the reduced model and then perform optimization over these parameters at the level of the reduced model. Research in this direction has been also done [11, 12] (see also Chapter 9).

18.3
Programming Issues

A mathematical algorithm should be implemented in some environment to be useful. To this end, rapid prototyping within an integrated environments such as Mathematica, MATLAB, and so on, is actually enough for research. Yet it brings some limits as well. If you decide further to distribute the developed code, this will imply that users must have the same environment and this may limit the number of potential customers. Another problem is that operations with high-dimensional matrices within such an integrated environment may not be optimal (Section 18.3.2).

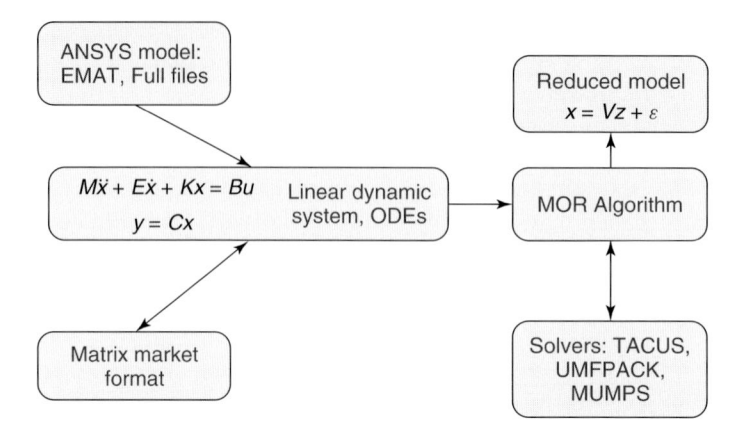

Figure 18.4 MOR for ANSYS block scheme.

Programming with compiled languages such as C++, C, or Fortran offers more flexibility, but on the other hand, it requires bigger investment of time. Generally speaking, the programming language for a similar programming task must allow easy use of the scientific libraries written in Fortran and C.

MOR for ANSYS is written in C++ and its block scheme is displayed in Figure 18.4. It is a relatively simple application that needs to read files and then write other files. As a result, it was decided to write as a standalone command line application when the user controls the applications through command line arguments. Such a decision has simplified the development considerably, and at the same time, this also makes it easy for the integration of the tool in a specific environment (for example, to integrate model reduction in ANSYS Workbench).

18.3.1
Obtaining System Matrices from ANSYS

Model reduction starts when one defines the original dynamic system through its system matrices. We have decided to interface ANSYS Multiphysics, and this happened to be a good choice, as this has opened a door to diverse engineering finite element models. MOR for ANSYS can also read the dynamic system to be reduced in the Matrix Market format.

ANSYS keeps element matrices in the EMAT file and assembled matrices in the FULL file. Both formats are documented [13], and one can access both binary files by means of the documented library supplied with ANSYS. When we have started the development with ANSYS 5.7, the FULL file did not have the load vector as well as other supplementary information necessary to interpret it, and we have been working with the EMAT file. ANSYS has a special command, called a *partial solve, PSOLVE*, with which one can evaluate element matrices for a given state vector without going through the real solution stage. This allows us to generate an EMAT file efficiently for a given model. However, it was necessary to overcome the following problems:

- The EMAT file does not contain the information about either Dirichlet boundary conditions or equation constraints. They should be extracted separately.
- The EMAT file has a contribution to the load vector from element matrices only. If nodal forces or accelerations are used to apply the load, this information should also be extracted individually.
- It is necessary to assemble the global matrices from the element matrices.

Later on we have discovered another problem. When different coordinates systems have been used during modeling, the element matrices must be accordingly transformed during assembly. However, at that point, the FULL file had all necessary information, and hence we have switched to using the FULL file instead of the EMAT file. Since ANSYS 6.0, the FULL file maintains all the original matrices, the load vector, the Dirichlet, and equation constraints in the file. ANSYS 8.0 allows us to make the assembly only and write the FULL file without a real solution phase with WRFULL. Since then it is also possible to extract the system matrices with HBMAT in the Harwell-Boeing format. Since ANSYS 13, there ANSYS APDL Math that allows us to work with system matrices directly through APDL (ANSYS Parametric Design Language) commands. The latter theoretically means that one can theoretically implement model reduction directly in APDL.

There are some problems with FULL file as well. In EMAT file, one can always find directly three system element matrices (mass, damping, and stiffness). In FULL file, this happens only after modal analysis and only in the case of a structural model that makes impossible to use it for an arbitrary multiphysics application. For other analysis, FULL file contains a linear combination of system matrices, in other words, a ready-to-solve linear system for a particular analysis (static, transient, or harmonic response). After several trials, the next method has been chosen in MOR for ANSYS.

Let us consider a linear system of equations to solve in harmonic response analysis

$$(-\omega^2 M + i\omega E + K)x = f \tag{18.2}$$

The FULL file in this case contains a complex matrix that is a combination shown in brackets in Eq. (18.2). The damping matrix is its imaginary part and the mass and stiffness matrices form its real part. If one extracts matrices from two FULL files are evaluated for two different frequencies then with simple transformation one can restore all three system matrices. This procedure has nothing to do with physical properties of the original system, and two frequencies to extract matrices from the FULL file should be chosen merely to reduce rounding errors during transformations.

18.3.2
Solvers

Each step of an iterative Krylov subspace algorithm requires us to compute a matrix-vector product, for example, for the first-order system

$$A^{-1}Eh \tag{18.3}$$

where h is some vector. The system matrices are high dimensional and sparse and one cannot afford to compute A^{-1} explicitly. The only feasible solution is to solve a linear system of equations for each step, and this constitutes the main computational cost.

There are two types of linear solvers, direct and iterative. In the case of model reduction, the direct solvers have an advantage as follows. The calculations (Eq. (18.3)) should be repeated for many vectors and we need to solve a system of linear equations with the same matrix with many right hand sides. In this case, in the case of a direct solver, one first factorizes the matrix and then uses a factor in the fast back substitution step to solve systems with many right hand sides. The size of the factor and thus the factorization time can be significantly reduced by matrix reordering for which METIS [14] has been employed.

We have started by using the TAUCS solver [15]. Yet, it is working well for positive definite matrices only and then we have added the UMFPACK solver [16] for unsymmetric matrices. Finally, we have switched to MUMPS [17] that can work with symmetric positive definite, symmetric indefinite, and unsymmetric matrices. Now MUMPS is the default solver in MOR for ANSYS. More information on practical use of solvers can be found at *http://MatrixProgramming.com*.

18.4
Open Problems

MOR for ANSYS has been employed in diverse engineering applications and some of them are reviewed in this book. A full list of MOR for ANSYS publications could be found at *http://ModelReduction.com*. In this section, however, the focus is not on successful applications of MOR for ANSYS in practice (to this end please see [2, 5–7, 18–23] and recent publications on the web site) but rather on sharing open problems that comes from practical applications. This will help to develop model reduction further.

When software based on numerical methods is used by engineers, most often they should develop certain know-how and even intuition to achieve reliable results with the software. For example, in finite elements, this would be a quality of the mesh and settings to converge a nonlinear problem. Model reduction is not an exception, and the main question here is the order of the reduced model. In addition, it may be necessary to choose several expansion points, and in this case, one should choose how many expansion points are needed and at what values of the Laplace variable and decide how many moments should be matched for each expansion point (Chapter 3).

From experience, It can be said that this does not cause big problems in real-life applications of MOR for ANSYS. It basically means that at the beginning an engineer should make a couple of tests where the optimal dimension and if necessary expansion points are chosen by error and- trial. After that such settings are usually valid for that particular class of problems. However, more research in this direction in order to incorporate more intelligence in model reduction software would be welcome. Promising in this direction are low-rank grammian methods (SVD-Krylov in [4]), as they theoretically could use global error estimates. There are many new papers from mathematicians along this way but what is missing is experience on how it is working for real-life applications. On the other hand, better theoretical understanding of error indicator experimentally found in [7] could also help even at least partly to solve such a problem.

Most applications in finite elements are related to structural mechanics. In this case, as it was mentioned earlier, there are model reduction methods that are already implemented in commercial software: mode superposition, Guyan reduction, CMS. In this respect, it is pretty instructive to compare two books: mathematical model reduction presented in [4] and engineering model reduction in [24]. Mathematicians make it the general case, and they prove theorems that allow us to understand what one can expect from model reduction algorithms. Engineers present their methods rather as a cookbook, in their view it is just evident that the lowest modes contain all necessary dynamic information, hence no need to discuss it from mathematical viewpoint.

This does not mean, however, that mathematical model reduction is automatically better in practical applications. Our experience shows that Krylov-based model reduction is somewhat faster and more accurate. Let us consider a hard disk drive actuator/suspension system from [25] (Figure 18.5a). In Figure 18.5b, there is a relative error shown for model reduction with the Arnoldi algorithm and mode superposition. One sees that with the Arnoldi algorithm, one reaches almost numerical accuracy and mode superposition gives the accuracy in the range of 0.1%. Yet, engineering model reduction does not depend on the input matrix, and because the nodes can be preserved, one can easily couple reduced models between each other. Engineers have also enormous know-how on how effectively employ their method in practical applications. Such experience is missing in the case of Krylov model reduction for structural mechanics. As for the accuracy, from the engineering viewpoint, Figure 18.5 does not show advantages of Krylov-based model reduction at all, as 1% accuracy is already good enough for engineering applications.

At present, these two different communities described in [4] and [24] remain separate. With an exception of [26], there are almost no papers where there is a comparison of mathematical and engineering model reduction for structural mechanics. This is a pity, as this could help to find better methods suited specifically for this area. It would be very interesting to understand why engineering model reduction is working in case of structural mechanics, as one can use this methods for such an application only. This means that structural model have some specific mathematical properties that distinguish them from other finite element models. It would be good to write these properties explicitly, as this could help us to develop a specific model reduction method for structural mechanics.

Finally, let us consider coupled problems with an example, an underwater electroacoustic (tonpilz) transducer [27] shown in Figure 18.6. There are three domains in the model, structural, piezoelectric, and acoustic coupled with each other, and the system matrices shown in Eq. (18.4) have a pretty interesting

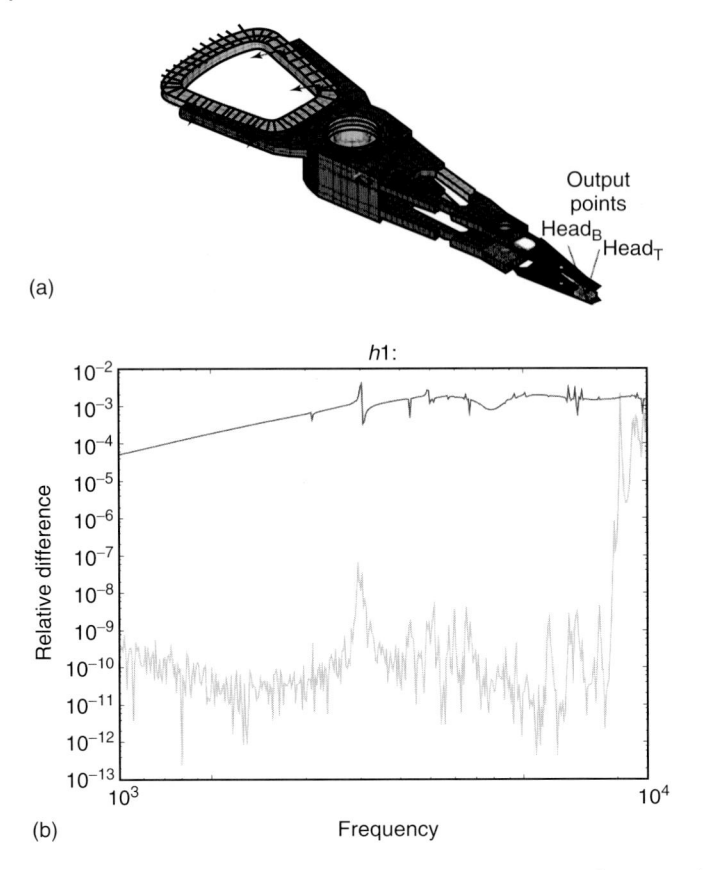

Figure 18.5 (a) Hard disk drive actuator/suspension system from [25]. (b) Relative error between the original and reduced system (dark grey line – mode superposition, light grey line – the Arnoldi algorithm).

structure.

$$\begin{bmatrix} M_s & 0 & 0 \\ 0 & 0 & 0 \\ M_{sa} & 0 & M_a \end{bmatrix} \begin{Bmatrix} \ddot{u} \\ \ddot{V} \\ \ddot{p} \end{Bmatrix} + \begin{bmatrix} E_s & 0 & 0 \\ 0 & 0 & 0 \\ 0 & 0 & E_a \end{bmatrix} \begin{Bmatrix} \dot{u} \\ \dot{V} \\ \dot{p} \end{Bmatrix} \\ + \begin{bmatrix} K_s & K_{se} & K_{sa} \\ K_{se} & K_e & 0 \\ 0 & 0 & K_a \end{bmatrix} \begin{Bmatrix} u \\ V \\ p \end{Bmatrix} = \begin{Bmatrix} f_s \\ f_e \\ f_a \end{Bmatrix} \quad (18.4)$$

Let us start first with a piezoelectric problem and ignore acoustics (remove matrices related to p in Eq. (18.4)). While mode superposition does not work, MOR for ANSYS has been used successfully by different engineering groups to reduce a finite element model of a piezoelectric device to use at system level simulation [18–20]. A coupled piezoelectric problem has interesting mathematical properties, as the stiffness matrix is symmetric indefinite and the mass matrix is singular.

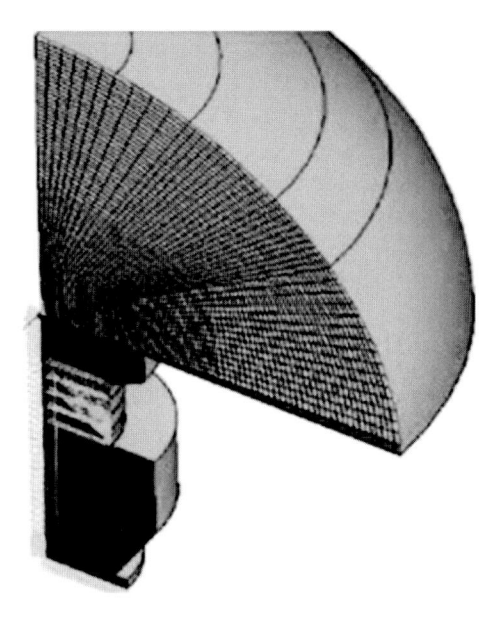

Figure 18.6 Model of a tonpilz transducer [27].

This, for example, requires special processing to convert a reduced model to the state-space form as the inverse of the M matrix does not exist.

Some fully coupled structural acoustic models (please ignore voltage-related members in Eq. (18.4)) have been also successfully reduced by MOR for ANSYS [21, 22]. However, here we had also a case shown in Figure 18.7 where the application of model reduction was not that successful. In this case of a coupled structural acoustic model of a loudspeaker [23], we have observed very slow convergence of the Arnoldi algorithm in the case of one expansion point. Until the order of the reduced model of 200, the generated reduced model did not match the response of the original model at all. Only after that the relative error went down but then we needed to generate 1000 Arnoldi vectors to obtain reasonable accuracy.

In this case, the structural acoustic coupling leads to unsymmetric system matrices and not proportional damping that increases computational requirements and at the same time, presumably adds unpleasant numerical properties of the dynamical system. It would be good to understand better from a mathematical viewpoint the difference of the loudspeaker model from structural acoustic in [21, 22].

Another problem with Eq. (18.4) is that owing to its structure, the Arnoldi algorithm does not preserve stability in the reduced model automatically, and it is necessary to develop engineering level tricks not only to choose the dimension of the reduced model but also to obtain it stable. In general, the Arnoldi and second-order Arnoldi algorithms do not take into account the structure of Eq. (18.4). This means that reduced matrices are completely dense and structural properties of Eq. (18.4) get lost. It could be an interesting research to check if structure-preserving model reduction will help to solve the problems that we encountered so far.

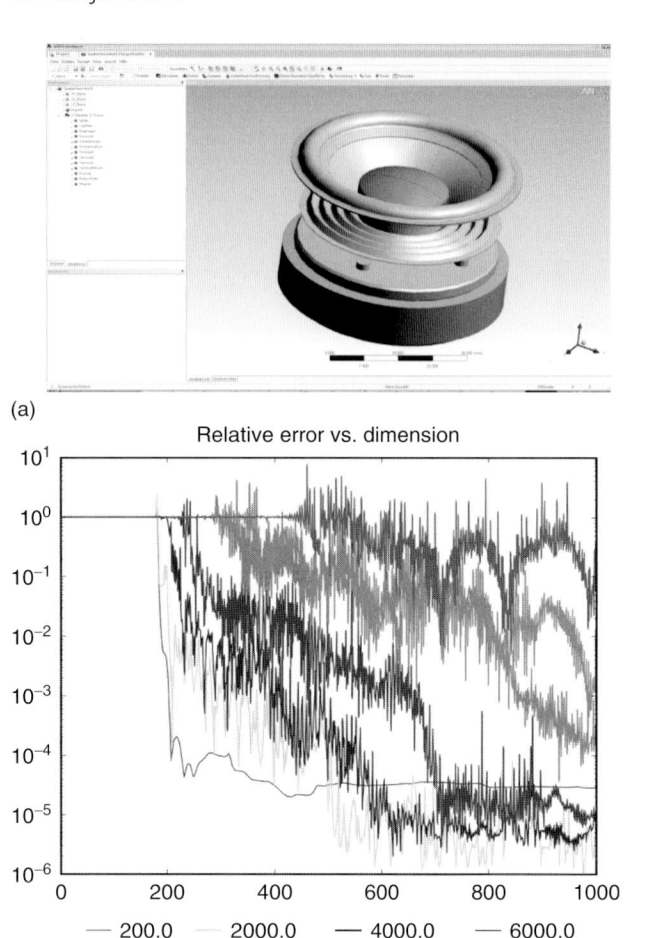

Figure 18.7 (a) Loudspeaker. (b) Relative error in the reduced model as a function of the model dimension for different frequencies [23].

18.5
Conclusion

The chapter demonstrates how the development of software can happen in the case when there is a wish to bring a new idea in practice. The most important is no doubt to find that particular theoretical idea that could improve the life of an engineer. A must here would be the understanding of the latest developments in mathematics and at the same time what is necessary in practice. At this level, rapid prototyping in Mathematica, Matlab, or similar environment is the best solution, as this allows us quickly to try different approaches and compare them with each other. Yet, when the way to proceed is clearer then it makes sense to go outside of the rapid prototyping environment and implement the software in such a way that it can be distributed independently. At this level, it is important to use available numerical libraries to reduce the development time. In general, the key is to find a compromise between research, programming, and numerics because the time, resources, and money are usually lacking. A good philosophy would be "worse is better" [28] that roughly speaking advices us not to try to implement everything in the best way. No doubt, one has to employ it wisely. Finally, a bit more informal view on this problem could be found in [29].

References

1. Rudnyi, E.B. and Korvink, J.G. (2006) *Lecture Notes in Computer Science*, **3732**, 349–356.
2. Rudnyi, E.B., Bechtold, T., Korvink, J.G., and Rossi, C. (2002) Solid Propellant Microthruster: Theory of Operation and Modelling Strategy, Nanotech 2002 – At the Edge of Revolution, AIAA Paper 2002-5755, September 9–12, Houston, TX.
3. Rudnyi, E.B. and Korvink, J.G. (2002) *Sensors Update*, **11**, 3–33.
4. Antoulas, A.C. (2005) *Approximation of Large-Scale Dynamical Systems*, Society for Industrial and Applied Mathematics, ISBN: 0898715296.
5. Bechtold, T., Rudnyi, E.B., and Korvink, J.G. (2006) *Fast Simulation of Electro-Thermal MEMS: Efficient Dynamic Compact Models*, Microtechnology and MEMS, Springer, ISBN: 3540346120.
6. Han, J.S., Rudnyi, E.B., and Korvink, J.G. (2005) *Journal of Micromechanics and Microengineering*, **15** (4), 822–832.
7. Bechtold, T., Rudnyi, E.B., and Korvink, J.G. (2005) *Journal of Micromechanics and Microengineering*, **15** (3), 430–440.
8. Bai, Z. and Su, Y. (2005) *SIAM Journal on Scientific Computing*, **26** (5), 1692–1709.
9. Eid, R., Salimbahrami, B., Lohmann, B., Rudnyi, E.B., and Korvink, J.G. (2007) *Sensors and Materials*, **19** (3), 149–164.
10. Benner, P., Feng, L., and Rudnyi, E.B. (2008) Using the superposition property for model reduction of linear systems with a large number of inputs. MTNS2008, Proceedings of the 18th International Symposium on Mathematical Theory of Networks and Systems (MTNS2008), Virginia Tech, Blacksburg, VA, July 28–August 1, 2008, p. 12.
11. Feng, L.H., Rudnyi, E.B., and Korvink, J.G. (2005) *IEEE Transactions on Computer-Aided Design of Integrated Circuits and Systems*, **24** (12), 1838–1847.
12. Rudnyi, E.B., Moosmann, C., Greiner, A., Bechtold, T., and Korvink, J.G. (2006) *5th MATHMOD, Proceedings*, Vol. 1: Abstract Volume, p. 147, Vol. 2: Full Papers CD, p. 8, February 8–10, Vienna University of Technology, Vienna. ISBN: 3-901608-30-3.
13. ANSYS Inc. (2010) Guide to Interfacing with ANSYS in Programmer's Manual.
14. Karypis, G. and Kumar, V. (1999) *SIAM Journal on Scientific Computing*, **20** (1), 359–392.
15. Rotkin, V. and Toledo, S. (2004) *ACM Transactions on Mathematical Software*, **30**, 19–46.
16. Davis, T.A. (2004) *ACM Transactions on Mathematical Software*, **30** (2), 196–199.
17. Amestoy, P.R., Duff, I.S., and L'Excellent, J.-Y. (2000) *Computer Methods in Applied Mechanics and Engineering*, **184**, 501–520.
18. Han, J.S. (2008) Krylov subspace-based model order reduction for piezoelectric structures. 2008 KSME CAE and Applied Mechanics Division's Spring Conference, KSME 08CA007, pp. 13–14.
19. Han, S.-O., Wolf, K., Hanselka, H., and Bein, T. (2009) Design and analysis of an adaptive vibration isolation system considering large scale parameter variations. SPIE Conference on Active and Passive Smart Structures and Integrated Systems, Proceedings of SPIE Vol. 7288, p. 728829.

20. Kurch, M., Klein, C., and Mayer, D. (2009) A framework for numerical modeling and simulation of shunt damping technology. The Sixteenth International Congress on Sound and Vibration, Krakow, July 5-9, p. 8.
21. Lippold, F. and Hübner, B. (2009) Application of MOR for ANSYS to hydro turbine runner dynamics. ANSYS Conference & 27. CADFEM Users Meeting, Congress Center Leipzig, November, 18–20.
22. Puri, R.S., Morrey, D., Bell, A.J., Durodola, J.F., Rudnyi, E.B., and Korvink, J.G. (2009) *Applied Mathematical Modelling*, **33** (11), 4097–4119.
23. Rudnyi, E.B., Moosrainer, M., and Landes, H. (2009) Efficient simulation of acoustic fluid-structure interaction models by means of model reduction. ICTCA 2009, 9th International Conference on Theoretical and Computational Acoustics, Dresden, September 7–11, 2009.
24. Qu, Z.-Q. (2004) *Model Order Reduction Techniques: with Applications in Finite Element Analysis*, Springer, ISBN: 1852338075.
25. Hatch, M.R. (2002) Vibration Simulation Using MATLAB and ANSY.
26. Koutsovasilis, P. (2009) Model order reduction in structural mechanics – coupling the rigid and elastic multi body dynamics. Dissertation. Technische Universität Dresden.
27. Clayton, L. (2010) *ANSYS Advantage*, **IV** (1), 17–19. http://www.ansys.com/About+ANSYS/ANSYS+Advantage+Magazine (accessed 2012).
28. Gabriel, R.P. (1991) Lisp: Good News, Bad News, How to Win Big, AI EXPERT.
29. Rudnyi, E.B. (2009) Engineering Computing: Mixing Knowledge Transfer, Programming, and Numerics, Case Study: Model Reduction, http://evgenii.rudnyi.ru/doc/misc/EngineeringComputing.html (accessed 2012).

19
SUGAR: A SPICE for MEMS
Jason V. Clark

19.1
Introduction

SUGAR is a modeling, design, and simulation tool that is one of the several efforts that pioneered the use of compact electromechanical models to simulate microelectromechanical system (MEMS) [1–5]. Leveraging off of some of the successful attributes of SPICE [6–8], SUGAR extends the utility of netlists and modified nodal analysis to accommodate electromechanical models, extends graphics to show flexures deforming in 3D alongside electrical circuits, and extends versatility by using the common MATLAB environment to invite user modifications and to allow unfettered access to the multitude of MATLAB functions and toolboxes.

Recent extensions to SUGAR include SugarCube [9], PSugar [10], iSugar [11], SugarX [12, 13], and SugarAid [14]. SugarCube adds a novice friendly interface to SUGAR that enables nonexperts to parametrically explore and layout the design/performance space of ready-made MEMS, which were previously created in SUGAR by experts. PSugar extends the modeling capabilities of SUGAR to include complex engineered systems with algebraic constraints. iSugar integrates SUGAR with SPICE for its extensive analog compact circuit models, with COMSOL for its finite element modeling, and with SIMULINK for its system-level modeling capabilities. SugarX bridges the gap between experiment and simulation by extracting geometric and material properties from true devices and importing the parameters into a corresponding SUGAR model. And SugarAid extends PSugar into the area of computer aided learning for students of science, engineering, technology, and mathematics. Owing to space constraints, we limit our discussions to SUGAR, SugarCube, PSugar, and iSugar.

19.2
SUGAR

In SUGAR, parameterized compact models are used to design and simulate MEMS. New models may be added through SUGAR's model function m-files. Material and environmental parameters are prescribed within a process m-file,

System-level Modeling of MEMS, First Edition. Edited by T. Bechtold, G. Schrag, and L. Feng.
© 2013 Wiley-VCH Verlag GmbH & Co. KGaA. Published 2013 by Wiley-VCH Verlag GmbH & Co. KGaA.

and geometry and connectivity are prescribed within a netlist file. The process file includes quantities such as Young's modulus, Poisson's ratio, thermal expansion coefficient, residual stress and strain gradient, temperature, and viscosity. The types of solvers include static, steady-state, modal, and transient analyses. SUGAR command lines are entered in the MATLAB workspace, and the graphical results are displayed in MATLAB figure windows. For example, the following MATLAB commands load a SUGAR netlist, perform static DC analysis, and display the deflected structure in 3D:

```
net = cho_load('comb_drive.net');   %Load netlist
q = cho_dc(net);                    %Solution vector
cho_display(net, q);                %Display deflection
```

where cho stands for the *carbon, hydrogen,* and *oxygen* elements in sugar.

SUGAR's netlist accommodates subnets, loops, and simple arithmetic operations. Coordinates of elements are not required, since positions are relative. Both mechanical and electrical elements are configured by branching off one node to another. The netlist syntax order follows *model, nodes,* and *parameters*. For instance,

```
uses process_file_polymumps.m
anchor     p1 [node2] [length = *, width = *, thickness = *]
beam3d     p1 [node2 node3] [length = *, width = *, thickness = *, ...]
combdrive  p1 [node3] [num_fingers = *, gap = *, finger_width = *, ...]
```

For a tangible example, we demonstrate how an advanced micromirror can be modeled and simulated in SUGAR. This example is chosen because the device is very difficult to simulate using finite element method (FEM) on a typical consumer PC. The true micromirror and its SUGAR emulation are shown in Figure 19.1. The micromirror consists of a circular recessed mirror plate, a thousand comb fingers, cosine-shaped flexures, and perforated flexures. The comb drive array converts an electric potential into a mechanical force that pulls the pair of tethers. The moment arm converts this translational force into a moment that rotates the circular mirror [5] (Figure 19.2).

Unlike FEM tools where components of a structure are created at the time of use, tools such as SUGAR [1–5] rely on the reuse of parameterized compact models. If the compact models do not exist then they can be created by the use of subnets, matrix condensation [5], or other reduced order modeling methods. Therefore, when creating netlist components, it is advantageous to make them suitable for general use by other designers. Table 19.1 depicts a practical choice of parameters for the necessary components for modeling the micromirror.

19.2.1
Equation of Motion

In SUGAR, the equation of motion has the form

$$M\ddot{q} + D\dot{q} + Kq = \sum F_{ext} \qquad (19.1)$$

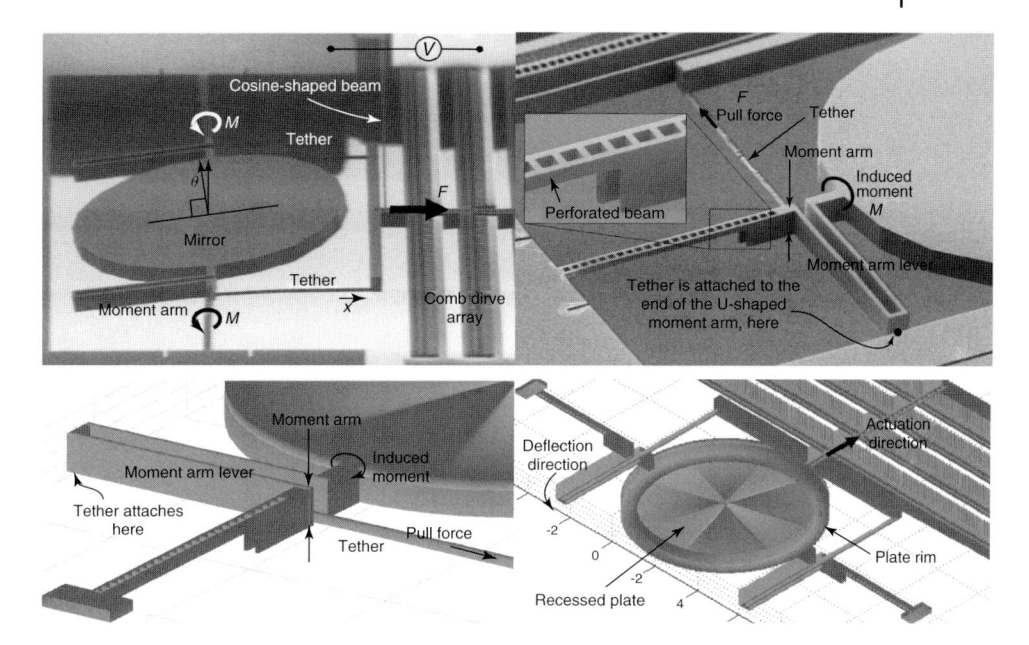

Figure 19.1 Scanning electron microscope image and SUGAR display of a micromirror. (*Source*: M. Last and V. Melanovic, personal communication.) [5].

where $\dot{q}$ and q are the flow and displacement state vectors and M, D, and K are the mass, damping, and stiffness matrices of the electromechanical elements, respectively [5]. A wide variety of phenomena may be modeled by including forces to the right-hand side of Eq. (19.1). An electrostatic force F_{Elec} applied on the tips of the comb drive fingers is most common. A few examples of other forces are as follows. Nonlinear deflections of structural components may be included by adding

$$F_{\text{stiffness}} = K_1 q + K_2 q^3 \tag{19.2}$$

where K_1 and K_2 are piecewise continuous matrix functions of displacement and q^3 is a vector of cubed displacements. Thermal expansion in the device may be included by adding

$$F_{\text{thermal}} = AE\alpha \left(T - T_0\right) \tag{19.3}$$

where T is the average temperature of a Joule-heated beam due to electric current, T_0 is the temperature of the ambient, A is the cross-sectional area, E is the Young modulus, and α is the coefficient of thermal expansion. Planar stress and strain gradients may be included by adding

$$F_{\text{stress}} = A\sigma_{\text{residual}} \tag{19.4}$$

where σ_{residual} is a tensile (positive) or compressive (negative) residual stress of the material; and

$$F_{\text{strain}} = EI_y \Gamma \tag{19.5}$$

/ # 19 SUGAR: A SPICE for MEMS

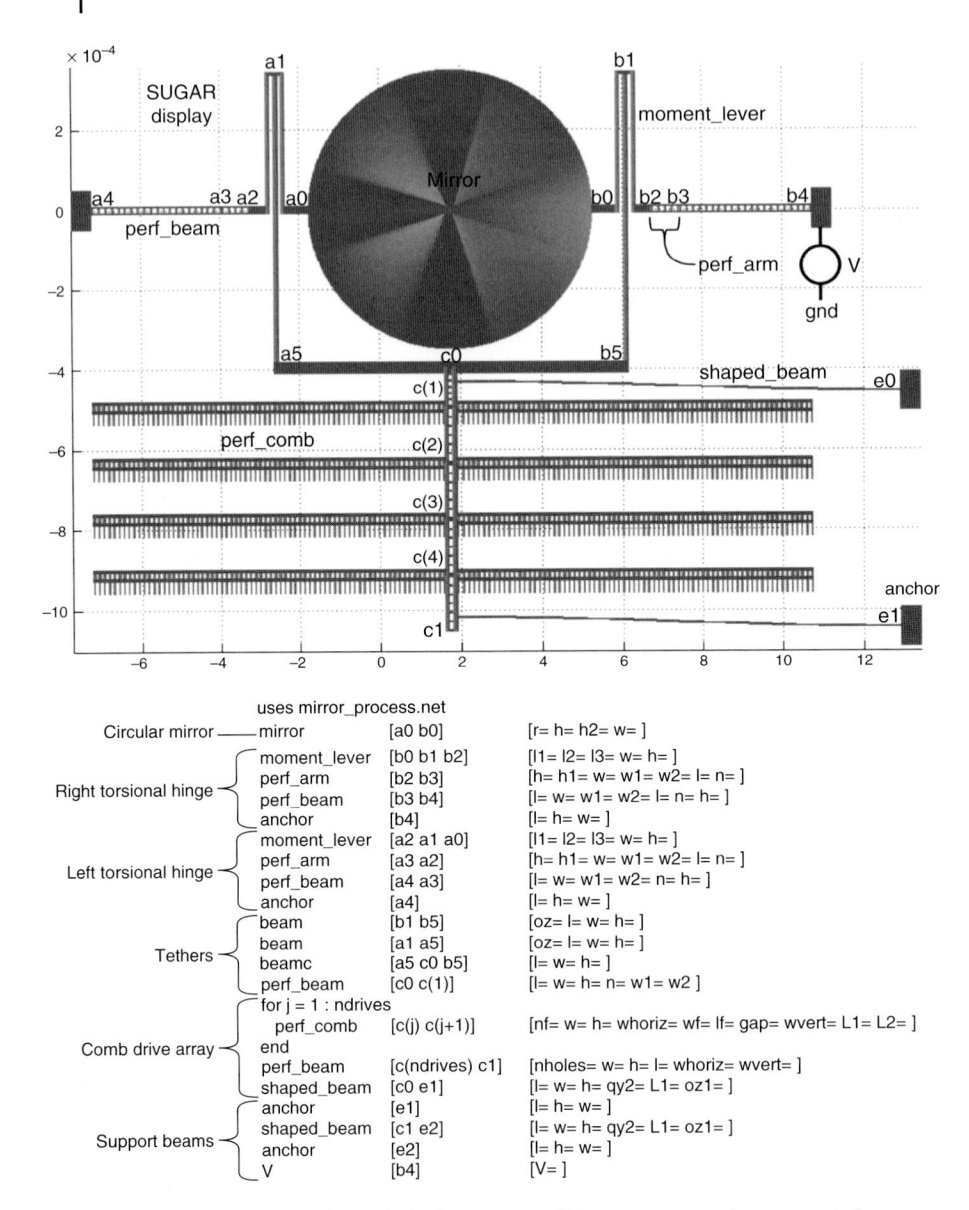

Figure 19.2 SUGAR netlist and display. For simplicity, parameter values are not shown.

```
                              uses mirror_process.net
        Circular mirror ——— mirror           [a0 b0]        [r= h= h2= w= ]
                         ┌ moment_lever   [b0 b1 b2]     [l1= l2= l3= w= h= ]
                         │ perf_arm        [b2 b3]        [h= h1= w= w1= w2= l= n= ]
    Right torsional hinge│ perf_beam       [b3 b4]        [l= w= w1= w2= l= n= h= ]
                         └ anchor          [b4]           [l= h= w= ]
                         ┌ moment_lever   [a2 a1 a0]     [l1= l2= l3= w= h= ]
                         │ perf_arm        [a3 a2]        [h= h1= w= w1= w2= l= n= ]
    Left torsional hinge │ perf_beam       [a4 a3]        [l= w= w1= w2= n= h= ]
                         └ anchor          [a4]           [l= h= w= ]
                         ┌ beam            [b1 b5]        [oz= l= w= h= ]
                         │ beam            [a1 a5]        [oz= l= w= h= ]
              Tethers    │ beamc           [a5 c0 b5]     [l= w= h= ]
                         └ perf_beam       [c0 c(1)]      [l= w= h= n= w1= w2 ]
                           for j = 1 : ndrives
                              perf_comb    [c(j) c(j+1)]  [nf= w= h= whoriz= wf= lf= gap= wvert= L1= L2= ]
         Comb drive array  end
                           perf_beam       [c(ndrives) c1] [nholes= w= h= l= whoriz= wvert= ]
                         ┌ shaped_beam    [c0 e1]         [l= w= h= qy2= L1= oz1= ]
                         │ anchor          [e1]           [l= h= w= ]
         Support beams   │ shaped_beam    [c1 e2]         [l= w= h= qy2= L1= oz1= ]
                         │ anchor          [e2]           [l= h= w= ]
                         └ V               [b4]           [V= ]
```

where Γ is the concave up (positive) or concave down (negative) strain gradient of the material, I_y is the second moment of area, and F_{strain} is the applied moment vector. As a final example, noninertial effects due to the micromirror operating in an accelerated reference frame may be included by adding

$$F_{\text{noninertial}} = M\ddot{R} - M\omega \times (\omega \times r) - 2M\omega \times \dot{r} - M\dot{\omega} \times r \qquad (19.6)$$

Table 19.1 Compact modeling building blocks for designing the micromirror (Figures 19.1 and 19.2).

Building block/component	Model name [node list]	Selected parameters
1	mirror (circular plate with rim) [a b]	r (radius) w (rim width) h (plate thickness) $h2$ (rim thickness)
2	moment_lever (moment arm lever) [a c b]	$l1$ (arm length) $l2$ (arm length) w (width) h (thickness)
3	perf_beam (perforated beam) [a b]	l (length) w (width) h (thickness) n (number of perforations) $w1$ (rail width) $w2$ (rung width)
4	perf_arm [a b]	l (length) w (width) $h, h2$ (thickness) n (number of perforations) w (main width) $w1$ (mirror width)
5	beam [a b] and beam c (with center node) [a c b]	l (length) w (width) h (thickness)
6	V (voltage source) [a b]	V (voltage) l, w (cosmetic length, width)
7	perf_comb (perforated comb drive) [a b]	n (number of fingers) l (finger length) w (finger width) h (thickness) g (gap) w_p (perforation width)
8	shaped_beam [a b]	l (length) w (width) h (thickness) $r_{x_1}, r_{y_1}, r_{z_1}$ (node-1 rotation) x_2, y_2, z_2 (node-2 translation) $r_{x_2}, r_{y_2}, r_{z_2}$ (node-2 rotation)

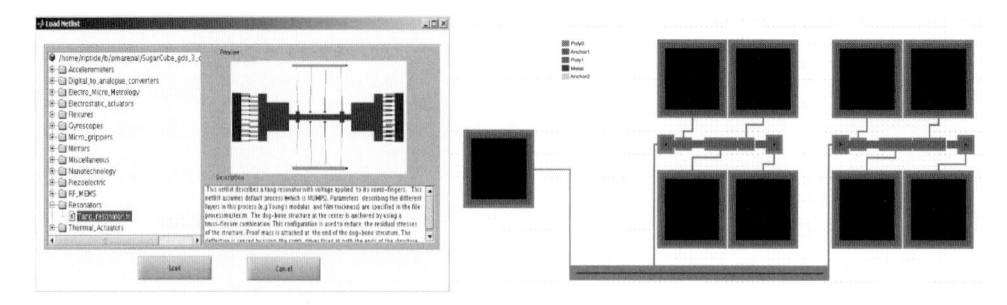

Figure 19.3 SugarCube library window and GDSII layout.

where the terms on the right-hand side are the translational force, centrifugal force, Coriolis force, and transverse force, respectively. The vector R is the position of the substrate, ω is the angular frequency vector of the substrate at R, and r is the position vector of all inertial nodes.

19.3
SUGAR-Based Applications

SugarCube provides a novice-friendly graphical user interface (GUI) to SUGAR, where the user is able to load ready-made parameterized microsystems that were previously programmed using SUGAR. The user is able to modify the key parameters with sliders that are bounded by practical limits. Single-button operations for static, modal, and transient analyses are available, where the deflected device is shown along with a 2D curve or 3D manifold for parameterized sweeps. In addition, the same configurations that are parametrically swept may be exported in GDSII layout format for subsequent fabrication. With a single button click, the layout feature in SugarCube automatically generates an array of parameterized devices with automatic etch holes, tracing lines, multilayer bonding pads, and so on. For example, a novice user is able to get online and load one of the many MEMS from a library, quickly explore its design space, and output its layout in GDSII format for fabrication (Figure 19.3). Such tasks can be done in a matter of minutes.

19.3.1
Library

SugarCube has a library of ready-made MEMS. Each MEMS file in the library is a parameterizable SUGAR netlist. In the library window, the MEMS are searchable by hierarchical directories such as accelerometers, gyroscopes, microgrippers, RF-MEMS, and thermal actuators. By highlighting a file in the library, an image of the device is immediately displayed along with a description of what it is, what it has been used for, what types of analyses can be done in SugarCube, and often a reference for more information about the device. New devices created by SUGAR experts may be imported into the SugarCube.

19.3.2
Design/Simulation

On selecting MEMS to explore from the library, the user may parametrically explore design and simulation (Figure 19.4). The parameters that appear are prescribed within the netlist, and therefore may be different from file to file. Each parameter value may be adjusted with a slider. Initial value bounds are given, which are intended for practical or common limits. However, the user may override any limits by typing numerical values in the fields. The lower right part of the window is for simulation. The user is able to choose which node and degree of freedom to inspect. The available solvers are static, sinusoidal, steady-state, and transient analyses.

19.3.3
Layout Generation

In SugarCube, the same parameters that are swept for parametric simulation create the same design configurations for layout; that is, every data point in SugarCube's 3D manifold corresponds to a device that can be laid out, whereby a similar manifold from experimental data can be subsequently created. To create layout arrays, a single button push is all that is required. To help with the tedious tasks that are often associated with MEMS, layout has been automated within SugarCube; for example, automatic common ground tracing, etch hole generation, and automated anchor and bond connects.

19.3.4
Common Ground Tracers

SugarCube is able to automatically configure common voltage tracers from each device in an array to a shared bond pad, which is useful for achieving a common ground between a multitude of devices. Such tracers are useful for reducing chip real estate by reducing the need for a large ground pad for each device. Such tracers are also useful for decreasing the time to probe device arrays, by requiring the repositioning of one probe instead of two. And common ground tracers are useful for reducing the possibility of undesirable voltage loops, which is beneficial for side by side comparisons of actuators. We show an example of automatically generated tracers attached to a common ground of the device in Figure 19.5. The common ground bond pad is configured to the left of each row. Figure 19.5 also demonstrates SugarCube's ability to conserve chip real estate. That is, the rows and columns are not equally spaced. Equally spaced arrays are also possible in SugarCube. The GDSII file from SugarCube has been imported into a free CleWin layout viewer [15] in Figure 19.5.

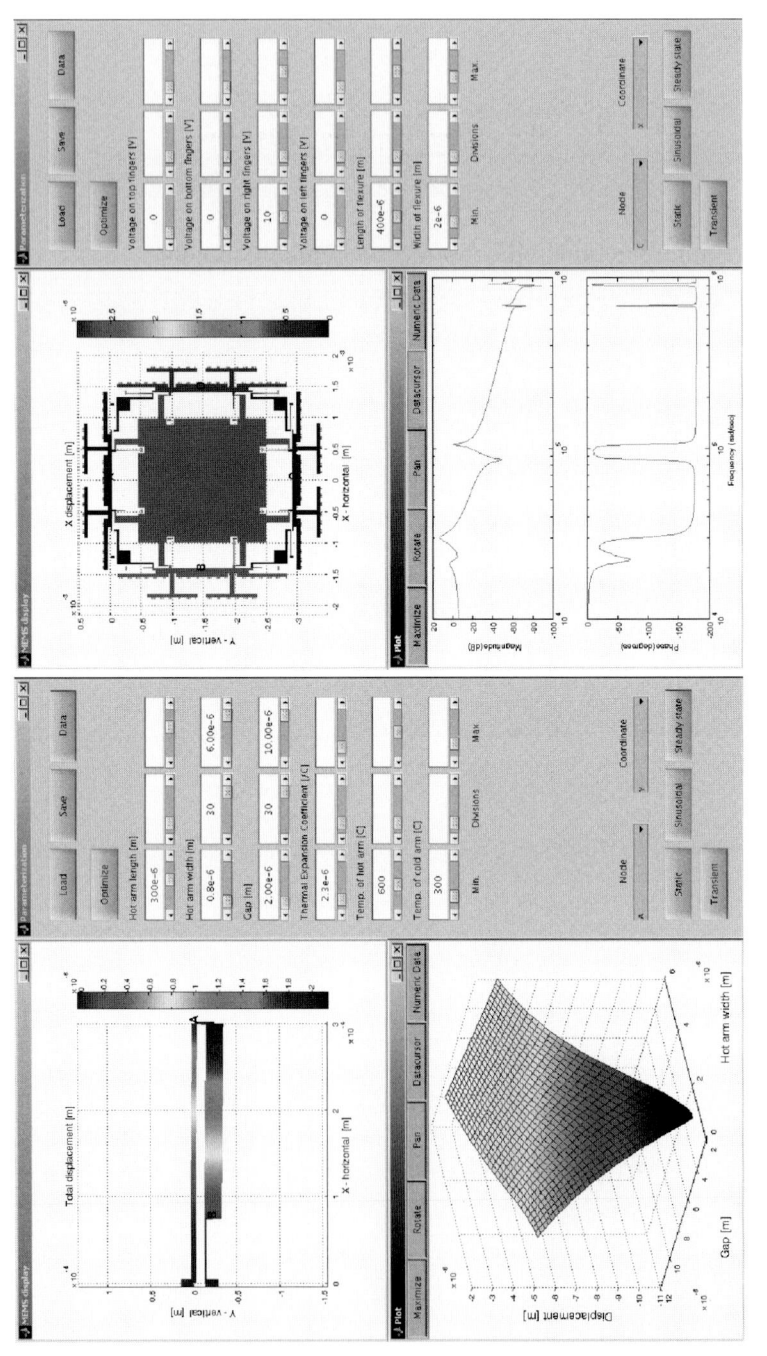

Figure 19.4 Static analysis of a thermal actuator and a frequency response of a gyroscope.

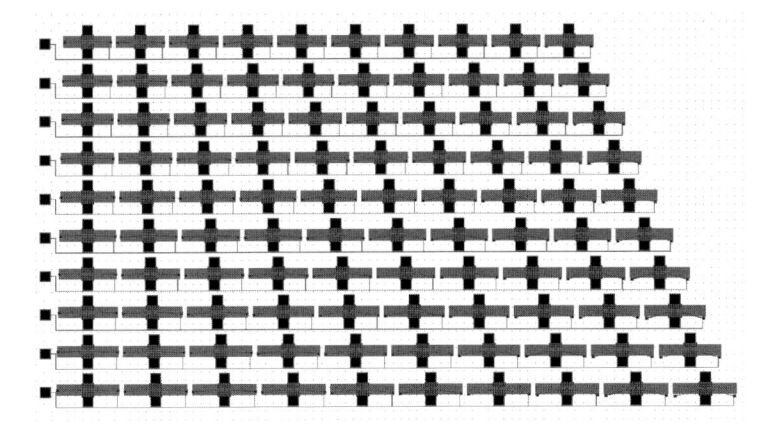

Figure 19.5 Parameterized GDSII layout array result from a single button click.

19.3.5
Etch Holes

SugarCube automatically generates etch holes in a given layout. These etch holes might be required for proper release of the oxide layer underneath the device layer. Generally, drawing etch holes in a layout is a time-consuming process. Etch hole errors could lead to a device that is not released. SugarCube's automatic etch hole generator can address these problems. To achieve this, the user just needs to specify those structures that need to be released. SugarCube identifies these elements and decides if the etch holes are required or not. Particularly, in case of SOIMUMPS [16], if the dimensions of a structure to be released is less than a specific tolerance, no etch holes are required. They will be released when exposed to an etchant. This type of information is specified in SugarCube's process file.

19.3.6
Multilayer Pads

SugarCube automatically generates multilayer bonding pads (or anchors) and common-ground pads that are often used in multilayer fabrication processes such as PolyMUMPs [16]. Such multilayered bonding pads are able to connect to any other structural layer in the process. They are topped by a metal layer for wire bonding or probing. Design rule layer size specifications for bonding pads are specified in the SUGAR process file.

19.3.7
Parameterized Arrays

Parameterized layout arrays of MEMS are easily created in SugarCube. Designers often layout an array of devices with slightly varying dimensions. This is often done to explore the dependence of particular design parameters on performance or to determine the limits of linearity, limits of fabrication, and so on. With conventional CAD, changing the dimensions of complicated device geometries can be difficult. It often requires the designer to recreate large portions of the design configuration, which can be tedious and error prone. If a large, varying array of devices must be configured, several hours to days may be spent on designing, debugging, and redesigning before the array is ready for layout submission. Using SugarCube, the user can reduce this time to seconds or minutes. In addition, SugarCube is able to simulate the entire array and plot the performance manifold, or optimize the design to achieve a particular performance metric such as resonance frequency.

19.3.8
Optimization

We implement a pattern search algorithm in SugarCube to solve an optimization problem. This function is provided by default in MATLAB's optimization toolbox. The objective function with constraints and bounds is fed into the patternsearch function and searches for the optimum geometry iteratively. The starting point for the search is the default parameters available in SugarCube. This process is repeated until the objective function attains the minimum possible value or any of the other stopping criteria are satisfied (Figure 19.6) [9, 12].

19.3.9
NEMS

Compact models of carbon nanotube (CNT) models, including zigzag and armchair chiralities, have also been implemented. We leverage the work of Li and Chou [17], who demonstrated that a network CNT model based on a hexagonal lattice structure comprising C-C bonds can be modeled using a structural flexure for each C-C bond. However, the model in [17] for a large number of C-C bonds can be computationally expensive. Using reduced order modeling, we have created two-node linear CNT compact model with 12 degrees of freedom. The degrees of freedom are independent of the number of C-C bonds. We have also created an efficient display routine for our reduced-order model that generates a flat 2D image and maps the image onto a deformed 3D cylindrical shape of required radius and length.

Such compact CNT models are advantageous for designing nanomechanical property testers, such as [18, 19]. This device is a microscale stress–strain tester for CNTs and nanowires. It applies a load on a specimen to measure its axial modulus. The displacement is applied by thermal actuators, a load is applied by microflexures, and displacement may be sensed or additional force may be applied with its interdigitated electrodes. The chosen stiffness of the device depends on the properties of the nanoscale specimen to be investigated. Another example is of the multiwalled CNTs for nanomotors, as in [20]. The nanomotor consists of an outer CNT sleeve surrounding a central CNT support. Figure 19.7 shows a parameterized CNT, a nanomechanical tester, and a nanomotor. At present, only small deflections and rotations are possible with our compact two-node CNT model.

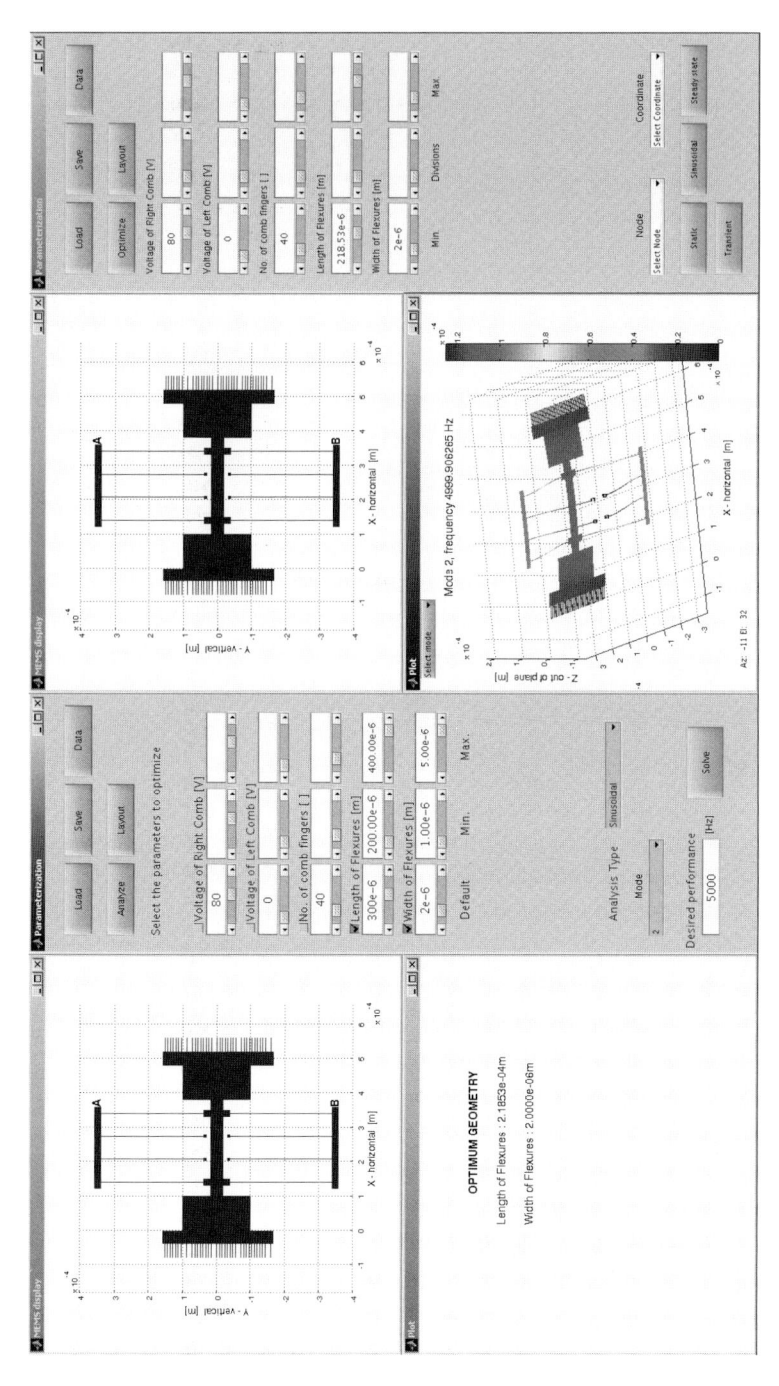

Figure 19.6 Optimization of the geometry for a particular resonant frequency.

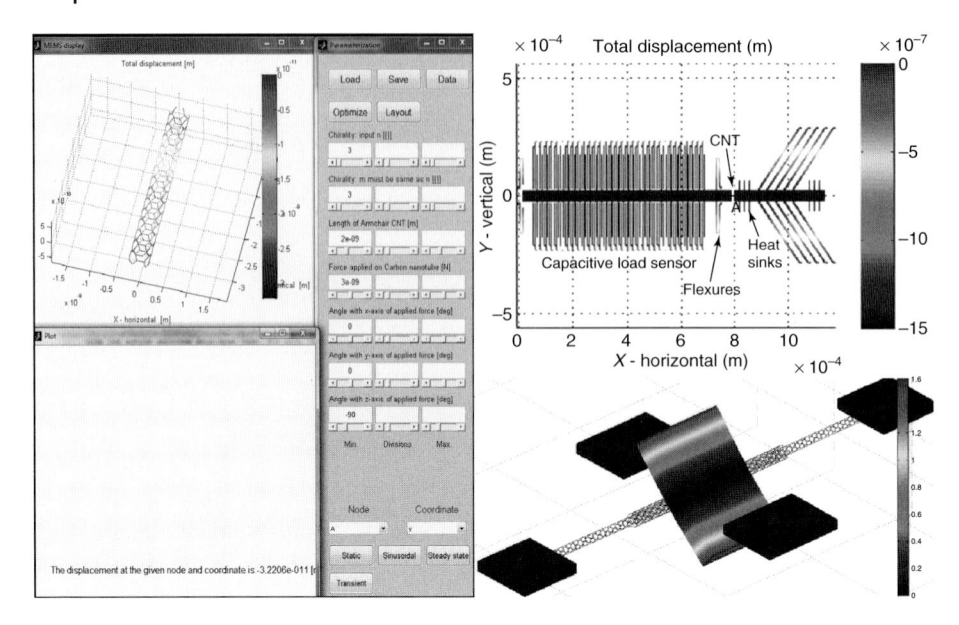

Figure 19.7 Parameterization of NEMS: CNT chirality, nanomaterial testing device, and nanomotor.

19.3.10
PSugar

PSugar is being developed to offer advances in designing and modeling complex microsystems with inequality constraints as follows. For design, we are developing a novel GUI that allows users to quickly configure complex systems in 3D using a computer mouse or pen at a faster rate than might be drawn with pencil and paper. We couple the GUI to a new and powerful netlist language for design flexibility. For modeling, we apply recent advances in analytical system dynamics and differential algebraic equations (DAEs) into a framework that facilitates the systematic modeling of multidisciplinary systems that may comprise static or dynamic constraints. For instance, with PSugar we appear to be the first CAD for MEMS tools that is able to efficiently dynamically simulate the most complex microsystem fabricated by Sandia National Laboratory (SNL). Such complex MEMS comprise gears, hinges, sliders, and so on, along with electronic components, comb drives, and electromechanical flexures.

19.3.11
GUI Configuration

In general, most GUIs are complicated, requiring a large number of button clicks which amounts to an undesirable amount of time to configure draw planes or objects in three-dimensional space. Such cumbersome tasks often break the natural flow of ideas during the design phase. To address this problem, we are exploring a methodology that facilitates the configuration of draw planes in device elements in three dimensions with minimal button clicks. Draw planes may be uniquely defined in 3D with a minimum of three coordinate points. These three points are defined by using the mouse or pen to click on three objects, such as an axis plane or a device component that has already been positioned on the screen. Since the position of a computer mouse or pen cursor is defined by two-dimensional coordinates, we project the cursor onto objects in three-dimensional space. For example, Figure 19.8a shows the positioning of the first of three coordinates used to define a draw plane. This first button click projects the cursor onto the *xz*-plane. Figure 19.8b shows the positioning of the second coordinate of the draw plane, and Figure 19.8c shows the positioning of the final coordinate that uniquely defines the draw plane. The three button sequence shown in Figure 19.8a–c took a fraction of a second. Figure 19.8d shows the positioning of an element on the draw plane. This two-node resistor element required two button clicks. Any additional resistors that may be configured end-to-end stemming from this initial element will only require one additional button click each. Draw planes may be repositioned by dragging the node of a plane along its interface with another plane, such as the *xz*-plane. The nodes of elements may be similarly repositioned. And to assist with free-hand positioning, snap-to-grid and snap-to-node features are implemented. The significant benefit of our methodology is that it allows systems to be configured faster than can be drawn on paper.

Although there are tools that are able to translate netlists into graphical images, there is a lack of tools that are able to go the other direction, that is, translate graphical images into netlists. In PSugar, the Graphical Window (GW) and Netlist Window (NW) are coupled as follows. Any element that is configured in the GW immediately appears in the NW as an additional line of text, and any element that is typed into the NW immediately appears in the GW. The benefit of this feature allows the best of both methods to be used when appropriate. For instance, the GW may be best for quickly configuring a filter device; however, creating a layout comprised of a parameterized array of devices that vary by one property along the *x*-direction and vary by a different property along the *y*-direction is best done using a nested for-loop within a netlist.

Elements are chosen for the GW by using the Element Menu Window (EMW). An element may be elementary, such as a molecule, resistor, operational amplifier, or flexure; or an element may be an assemblage of several elementary components, such as a CNT, a band-pass filter, or a micro gyroscope. The listing of selected elements in a particular EMW may be defined by a user. The EMW, the GW, and the NW are identified in Figure 19.8.

19.3.12
DAEs

ODE solvers are typically not equipped to solve a system of DAEs where the system mass matrix M may be singular or zero [21]. There are many complex microsystems that are amenable to DAEs, such as those comprising elements

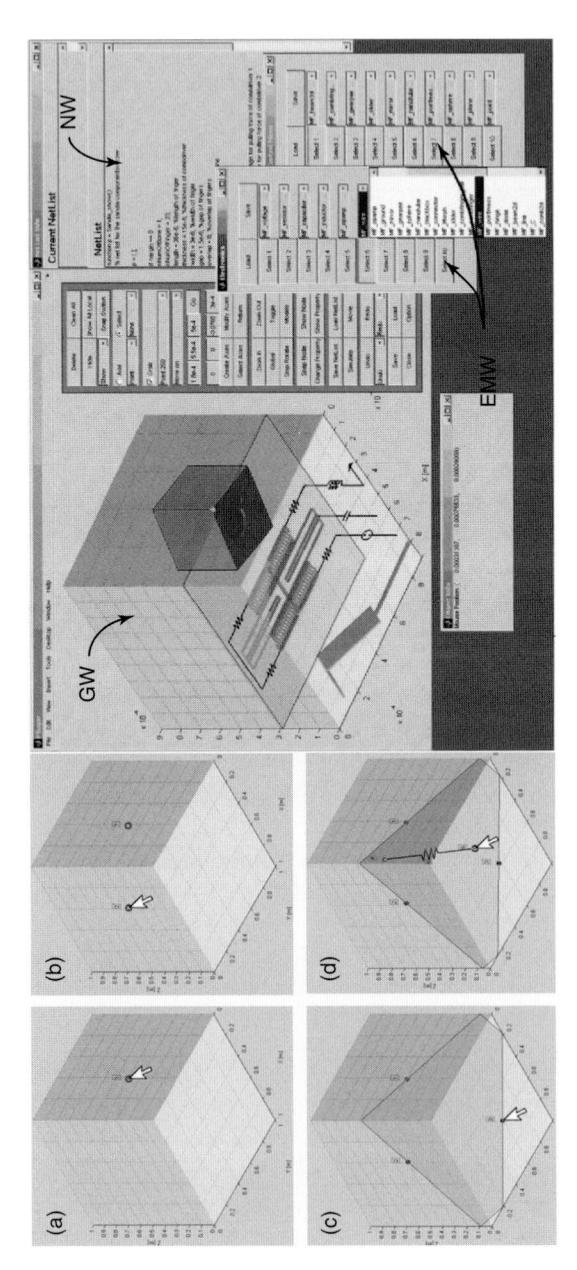

Figure 19.8 GUI design. (a–c) Define draw plane, three clicks. (d) Configure element, two clicks.

without inertia; elements with displacement, flow, dynamic variable, or effort constraints; or elements with inequality constraints. DAEs that represent a system may be expressed in several forms and may yield different differential indices. The DAE form that we use is derived from the first law of thermodynamics, which yields a differential index of at most three. Rigorous details for the derivation of this particular DAE form can be found in [22]. The DAE has the form

$$F = \begin{pmatrix} \dot{q} - f \\ M\dot{f} + \Phi_q^T \kappa + \Psi_q^T \mu - \Upsilon \\ \Phi \\ \Psi \\ \Gamma \\ \dot{s} - \Lambda \end{pmatrix} = 0 \qquad (19.7)$$

where κ and μ are the unknown Lagrange multipliers; $\Phi(q, t) = 0$ and $\Psi(f, q, t) = 0$ are the displacement and flow constraints; $\Gamma\left(e^y, s, \dot{q}, q, t\right) = 0$ is an algebraic vector of implicit effort constraints; s is the dynamic variable to account for phenomena where it is necessary to account for such as the derivative of flow $\dot{s} = df/dt$ or the integral of displacement $s = \int q \mathrm{d}t$; $M(f, q, t) = \nabla_f^2 T^*$ is the mass; and $\Upsilon = Q - \left(\nabla_f T^*\right)_q f - \left(\nabla_f T^*\right)_t + \nabla_q T^* - \nabla_q V - \nabla_f D$.

Systematic modeling in PSugar goes as follows. Each element has a representative parameterized model function containing its energy functions, constraints, and efforts. For instance, an electromechanical model function j returns the symbolic scalars $V_j = \frac{1}{2} q_j^T K_j q_j$ for its potential energy, $T_j = \frac{1}{2} f_j^T M_j f_j$ for its kinetic energy, and $D_j = \frac{1}{2} f_j^T R_j f_j$ for its power dissipation, where K, M, and R are multidomain matrices. The assembler collects energy functions from the N components, $V = \sum_{j=1}^N V_j$, $T = \sum_{j=1}^N T_j$, and $D = \sum_{j=1}^N D_j$; the externally applied effort vector Q; and the algebraic displacement and flow constraint vectors Φ and Ψ. The assembler then substitutes these functions into Eq. (19.7) for symbolic differentiation. For computational efficiency, the resulting equation of motion is automatically converted from symbolic form to a text m-file form. It is this m-file that the solver iterates. Hence, in PSugar, the modeler's effort is significantly reduced to simply providing energy functions, constraints, and efforts. So the traditional practice of rigorously manipulating a model into a particular form (as done with original SUGAR and SPICE) is not required with PSugar.

19.3.13
Simulation

In PSugar, the DAE in Eq. (19.7) is converted to this nonlinear algebraic counterpart, and the residues of the algebraic constraints are used for accuracy control. For brevity, here we describe a low-order approximation of Eq. (19.7) by replacing its differentials with linear finite differences, such as $\dot{q}_{n+1} \approx (q_{n+1} - q_n)/h_{n+1}$. In doing so, Eq. (19.7) has the form

$$\begin{pmatrix} M_{n+1} \dfrac{f_{n+1}-f_n}{h_{n+1}} + \Phi_q^T \Big|_{n+1} \kappa_{n+1} + \Psi_q^T \Big|_{n+1} \mu_{n+1} - \Upsilon_{n+1} \\ \Phi_{n+1} \\ \Psi_{n+1} \\ \Gamma_{n+1} \\ \dfrac{s_{n+1}-s_n}{h_{n+1}} - \Lambda_{n+1} \end{pmatrix} = 0 \quad (19.8)$$

where h_{n+1} is the $(n+1)^{th}$ time step and the jth column of the Jacobian of Eq. (19.8) is approximated as $[F(\hat{y}, t_n) - F(\check{y}, t_n)]/2\varepsilon$, where $\hat{y}_j = \left[f_j, q_j, e_j^\lambda, s_j, \mu_j, \kappa_j \right]^T + \varepsilon$, $\check{y}_j = \left[f_j, q_j, e_j^\lambda, s_j, \mu_j, \kappa_j \right]^T - \varepsilon$, $\varepsilon = 10^{-d}$ is a perturbation parameter, and d is equal to about half the number of significant digits [21].

There are several public domain solvers available for DAEs [21]. The choice of solver usually depends on the differential index of the DAE; that is, the minimum number of times that some or all the equations would need to be differentiated in time to determine its underlying ODE. However, using an ODE solver to solve the resulting underlying ODE is not preferred, because the solution trajectory often drifts from the solution manifold that is defined by the explicit constraints in the original DAE. Methods such as BDF (backward differentiation formula) and IRK (implicit Runge–Kutta) improve Euler's method by using higher-order approximations for $\dot{q}_{n+1}$, $\dot{f}_{n+1}$, and $\dot{s}_{n+1}$, and using variable step sizes. Since MATLAB's ode15s and ode23t solvers are valid only for index-1 DAE systems, we do not use them to solve Eq. (19.7), because its index can be as high as three. At present, we use a public domain high-index DAE solver.

In Figure 19.9, we show a complex engineered MEMS by SNL [23] and its PSugar counterpart. Both the real and simulated images show gears, hinges, slider, rack-and-pinion, comb drives, and flexures. For clarity, the PSugar components are labeled. We also show a corresponding simulation of gear rotation versus time. A pair of voltage ramps applied across the two sets of orthogonal comb drives rotates the smallest gear a quarter turn. The ramps end at 0.01 s, and the system settles back to its unactuated state. The family of curves shows parameterized responses of gear rotation as a function of time for a varying set of folded flexure widths, where shifts in frequency are clearly noticeable. SNL data was not available.

19.4
Integration of SUGAR + COMSOL + SPICE + SIMULINK

Our goal with iSugar is to develop an integrated systems design framework that integrates compact models, finite element models, and system-level analyses in a flexible MATLAB environment. For compact modeling, we use SUGAR for its ease of device configuration, parameterization, and layout capabilities; for FEM,

19.4 Integration of SUGAR + COMSOL + SPICE + SIMULINK

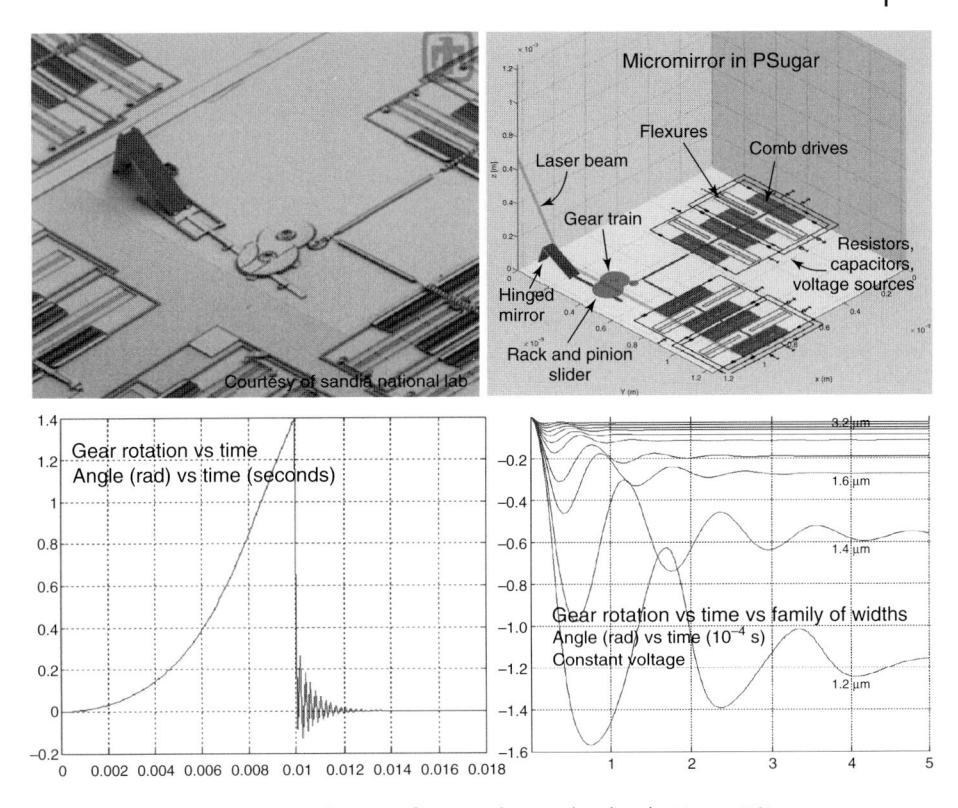

Figure 19.9 Complex engineered MEMS from Sandia simulated with PSugar [10].

we use COMSOL for its transparent interface and MATLAB integration; and for system analysis, we use SIMULINK for its simple graphical building-block style of modeling and MATLAB integration. We also accommodate SPICE circuit analysis syntax within iSugar, where SPICE syntax is recognized within a SUGAR netlist.

Modeling some systems may require the use of different numerical methods so that computational efficiency is optimized without sacrificing model accuracy. Although a few commercial tools have the ability to integrate with MATLAB and be controlled by SIMULINK, iSugar has the ability to control all aspects of the integration from within itself. That is SIMULINK, SPICE, and COMSOL are fully accessed and controlled from within iSugar, without the user having to directly start-up the other tools. Such control within iSugar facilitates a more holistic approach to design and analysis with greater utility. Moreover, it is not necessary for the user to know how to use the other tools that iSugar is integrated with to take advantage of their modeling capabilities. Although iSugar is readily available and open source, the tools that we have integrated with it (i.e., MATLAB, SIMULINK, and COMSOL) are available commercially. We show iSugar's framework and its integration with SIMULINK in Figure 19.10.

SIMULINK is a system-level simulation tool that is based in MATLAB. It uses graphical building blocks to configure systems. SIMULINK has a large library

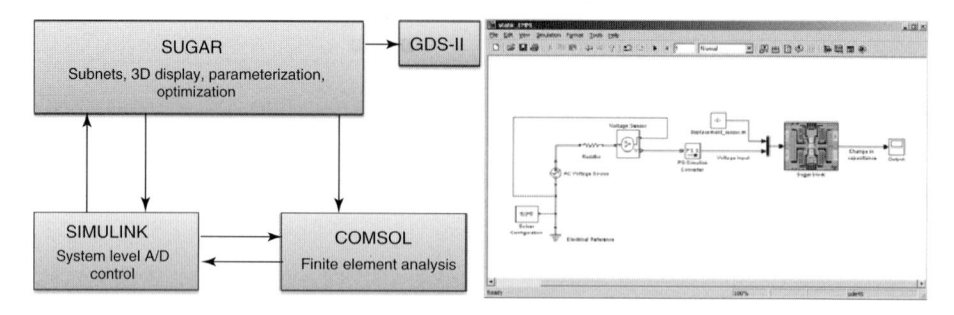

Figure 19.10 iSugar framework and an example of SUGAR integration with SIMULINK.

of building blocks that span a wide variety of modules including control theory, digital signal processing, COMSOL, and SUGAR. For instance, SIMULINK can be used to impart feedback and control signals, or environmental disturbances such as noninertial forces, temperature fluctuations, or noise. Similar to COMSOL, SIMULINK operations can also be carried out the MATLAB workspace, which we exploit with iSugar. The seamless integration of iSugar with SIMULINK allows for parametric optimization of the MEMS component as its performance is explored in a more complete system.

19.4.1
Integration

One goal for MEMS designers is to predict the performance of their devices under realistic interactive conditions. Modeling such systems more completely than convention includes interface electronics, packaging, temperature variations, external vibrations, electromagnetic radiation, and noninertial forces. A system-level simulation tool can be used to efficiently control such disturbances, since such sources do not require a detailed modeling as the MEMS structure. In iSugar, we integrate SUGAR with SIMULINK by implementing a SIMULINK SUGAR-block. These blocks can be used to perform different SUGAR operations such as simulating static, modal, and transient performance of MEMS and displaying the MEMS in their deflected states.

In use, the user is able to interconnect one or more SUGAR blocks of MEMS, one or more COMSOL blocks, and a host of other SIMULINK blocks to emulate a more complete system. In Figure 19.10, we show an example of a system-level configuration in SIMULINK that connects control circuitry to a MEMS SUGAR block. The output of the SUGAR block is defined by the user. For instance, the output might be the mechanical deflection of node, resonance amplitude, capacitance of a comb drive.

COMSOL is a CAE tool that is based on FEM. It has a wide range of capabilities to model and simulate multiple energy domains, which is especially important in fields such as MEMS. The accuracy of complicated models computed by COMSOL is usually better than those computed by SUGAR. A useful feature in COMSOL that

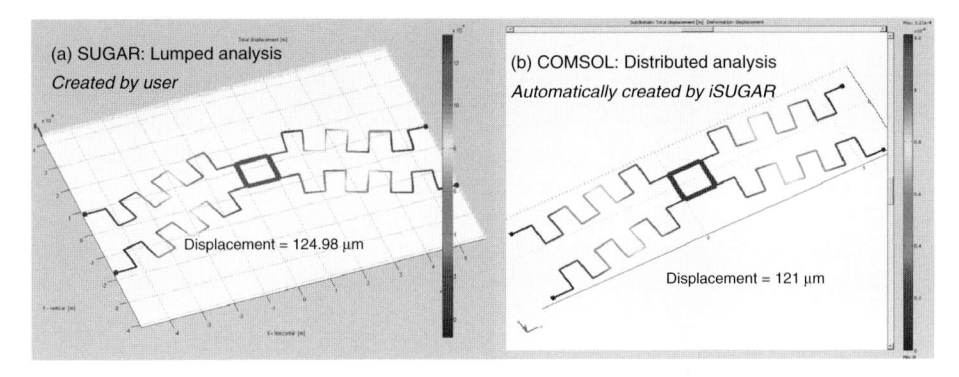

Figure 19.11 Automated verification of system of compact models using FEM.

we exploit is COMSOL Script, which is based in MATLAB. That is, every operation in COMSOL can be performed from MATLAB's workspace. This allows users to effectively control all COMSOL capabilities from within iSugar. This integration also allows parameterized designs that are difficult to configure within COMSOL to be easily configured in iSugar and then imported into COMSOL for detailed analysis. Subdomain and boundary conditions are also imported.

19.4.2
Verification

When developing compact models, it is important to verify them against another accepted form of modeling, such as analytical or FEM. Although compact modeling is much more computationally efficient than FEM, this is usually done at the cost of refined information. For instance, FEM often provides temperature, charge, and stress distributions on structures; yet, compact analysis is often limited to the effective equivalent information lumped at the nodes. Moreover, compact models are often created by reducing various types of physics involved in the problem to the bare minimum. So determining the accuracy and limits of compact models is often necessary; and even more so, determining the accuracy and limits of a system of compact models due to possible proximity effects is often necessary.

Such verification can be easily done using iSugar. Importing of geometry, material properties, and boundary conditions from SUGAR to COMSOL is automated. This design is automatically meshed and simulated. We show an example of iSugar automatic's verification in Figure 19.11. After a serpentine flexure is created in SUGAR (with nine lines of netlist text), it can be exported to COMSOL and simulated by a single MATLAB command.

19.5
Conclusion

In this chapter, we have discussed a few application areas of SUGAR and its extensions. SUGAR is a netlist-based simulation tool that uses compact models for designing and simulating MEMS. SUGAR has been used to efficiently, fairly, and accurately model advanced MEMS that are difficult to parametrically design or simulate using traditional FEM tools. SugarCube is a novice-friendly interface to SUGAR for easily exploring the design space of ready-made MEMS. SugarCube has been used to generate large arrays and wafer-level layout much faster than conventional tools, and SugarCube has been used to introduce students to MEMS. PSugar extends SUGAR's capabilities to include compact models with algebraic constraints. PSugar has been used to create the first dynamic simulation of Sandia's complex MEMS, and iSugar integrates the compact MEMS models of SUGAR, the compact analog circuit models of SPICE, and the finite element analysis methods of COMSOL with the system-level capabilities of SIMULINK.

References

1. Clark, J.V., Zhou, N., and Pister, K.S.J. (1998) MEMS simulation using SUGAR v0.5. Proceedings Transducer's Solid-State Sensor and Actuator Workshop, Hilton Head Island SC, June 8–11, 1998, pp. 191–196.
2. Zhou, N., Clark, J.V., and Pister, K.S.J. (1998) Nodal analysis for MEMS design using SUGAR v0.5. Technical Proceedings of the Fourth International Conference on Modeling and Simulation of Microsystems, Santa Clara CA, April 6–8, 1998, pp. 308–313.
3. Fedder, G.K. and Jing, Q. (1999) *IEEE Transaction on Circuits and Systems II, Analog and Digital Signal Processing*, **46** (10), 1309–1315.
4. Lorenz, G. and Neul, R. (1998) Network-type modeling of micromachined sensor systems. Proceedings of International Conference on MSM, Santa Clara CA, April 1998, pp. 233–238.
5. Clark, J.V. and Pister, K.S.J. (2007) *Journal of Microelectro-mechanical Systems*, **16** (6), 1524–1536.
6. Nagel, L.W. and Pederson, D.O. (1973) SPICE (Simulation Program with Integrated Circuit Emphasis), Memorandum No. ERL-M382. University of California, Berkeley, April 1973.
7. Nagel, L.W. (1975) SPICE2: A Computer Program to Simulate Semiconductor Circuits, Memorandum No. ERL-M520. University of California, Berkeley, May 1975.
8. Quarles, T.L. (1989) Analysis of Performance and Convergence Issues for Circuit Simulation, Memorandum No. UCB/ERL M89/42. University of California, Berkeley, April 1989.
9. Marepalli, P. and Clark, J.V. *Journal of Microelectromechanical Systems*, 408–410.
10. Zeng, Y. and Clark, J.V. (2010) Complex engineered MEMS simulation using PSugar v0.5. 18th Biennial IEEE UGIM (University Government Industry Micro/Nano) Symposium, June 28–July 1, 2010.
11. Marepalli, P. and Clark, J.V. (2011) Integration of Sugar, Comsol, Spice, and Simulink. Nanotech2011, International Nanotechnology Conference and Exhibition, Boston MA, June 13–16, 2011.
12. Marepalli, P. (2012) Advances in CAD for MEMS. MS thesis. Purdue University.
13. Marepalli, P., Magana, A., Taleyarkhan, M.R., Sambamurthy, N., and Clark, J.V. (2011). *Journal of Online Engineering Education*, **2** (1), 1–9.
14. Marepalli, P., Li, F., and Clark, J.V. (2012) SugarX: real-time online experimental control of MEMS. Nanotech 2011, International Nanotechnology Conference and Exhibition, Boston MA, June 13–16, 2011.
15. CleWin, WieWeb Software. (2012) Achterhoekse, Molenweg 76, 7556 GN Hengelo, The Netherlands,

http://www.wieweb.com/nojava/layoutframe.html.

16. Allen, C., Greg, H., DeMaul, M., Steve, W., and Busbee, H. *PolyMUMPS Design Handbook, a MUMPS Process* (2005), and *SOIMUMPs Design Handbook, a MUMPS Process* (2009). MEMSCAP Inc.

17. Li, C. and Chou, T.-W. (2003) *Applied Physics Letters*, **84** (1), 121–123.

18. Espinosa, H.D., Yong, Z., and Moldovan, N. (2007) *Journal of Microelectromechanical Systems*, **16** (5), 1219–1231.

19. Bansal, R. and Clark, J.V. (2011) *Sensors and Transducers Journal*, **13** (Special Issue), 408–410.

20. Fennimore, A.M., Yuzvinsky, T.D., Han, W.-Q., Fuhrer, M.S., Cumings, J., and Zettl, A. (2003) *Nature*, **424**. 408–410.

21. Brenan, K.E., Cambell, S.L., and Petold, L.R. (2012) *Numerical Solution of Initial-Value Problems in Differential-Algebraic Equations*, SIAM.

22. Layton, R.A. (2012) Analytical system dynamics. PhD thesis. University of Washington, Seattle.

23. Sandia National Laboratories (2012) http://mems.sandia.gov.

20
Model Order Reduction Implementations in Commercial MEMS Design Environment
Sandeep Akkaraju

20.1
Introduction

As the NEMS/microelectromechanical system (MEMS) industry matures, the design challenge continues to move from the microstructure design to the microsystem design. With the maturity of the process technology and the increase in computing power, designers are now looking to optimize MEMS from a system standpoint. Traditional N/MEMS CAD tools provide functionality to design at a microstructure level.

At present, MEMS modeling and simulation is performed at various levels of granularity by MEMS engineers working on different aspects of the manufacture. *Ab initio* models are based on atomistic, quantum mechanical or molecular dynamics. Such models are typically used in process modeling to predict materials behavior (such as physical properties or etch behavior). Component-level models can include lumped models and finite element representations of a component such as a plate or a comb drive. Device models represent the working of the microstructure or nanostructure under investigation. Algorithmic models are used to capture the behavior of a certain logic or control element within a system. Finally, system level models are used to model the entire microsystem.

20.1.1
Ab initio (First Principles) Simulations

First, principle simulations are typically based on atomistic or quantum mechanical molecular dynamics principles. However, this is a growing area of research, only a few tools have made it into the everyday repertoire of the MEMS design engineer. One such tool is IntelliSense's IntelliEtch that uses atomistic principles to simulate the etching of silicon (Figure 20.1).

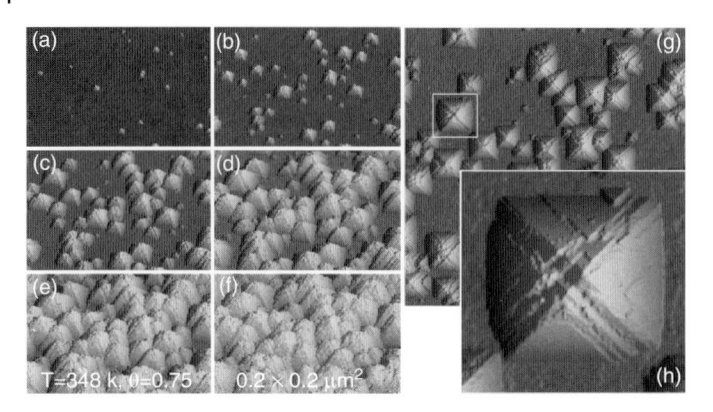

Figure 20.1 (a–h) Atomistic calculations are used to predict hillock formation and surface morphology during wet etching of silicon. *Ab initio* techniques allow the user to capture effects of micromasking, which can lead to hillock formation, preventing smooth etches.

20.1.2
Technology CAD (TCAD)

At this level, the microstructure is simulated at the process level. Simulators such as AnisE™ and RECIPE™ from IntelliSense and SUPREM™-based simulators from vendors such as CrossLight and Synopsys simulate the actual process flow based on process settings and physical simulation of the process, such as diffusion, growth, or etching. TCAD-based models are typically set up and run by process engineers.

These simulations are useful in understanding the effect of the process on the final physical geometry of the device. Since they are based on the actual physical models, they are often very time consuming. For instance, IntelliSense's RIE/ICP simulation tool RECIPE is based on the actual simulation of the plasma etching process and polymer deposition process. These tools are used to determine the influence of the process and mask set on the final geometry of the device (Figure 20.2).

20.1.3
Schematic or Component-Based Design (Top-Down Design)

One of the primary advantages of a hierarchical approach is that the design entry is done in terms of fundamental building blocks or components. This allows the user to enter a parameterized model of the device in terms of both layout and manufacturing data. Since the data entry is done in terms of parameterized abstract models, users can analyze the devices at different granularities. The element model can be represented in terms of lumped models, distributed models, or Rayleigh-Ritz-based finite element method (FEM) or boundary element method (BEM) models. The user can then easily perform an accuracy time trade-off.

Figure 20.2 Technology CAD (TCAD) tools are used to accurately predict the physical etching and processing of MEMS devices. A complex process of formation of microneedles being simulated in software.

One of the disadvantages of schematic design is that the user is limited to using components in the design library. Arbitrary geometries and new physical or material models are difficult to incorporate into the design. Since most schematic models are, to a degree, based on lumped models, they cannot accurately capture nonidealities. For instance, accurate capture of electrostatics, fluidics, or contact and postcontact physics in arbitrary geometries requires full 3D modeling. Similarly, packaging-level effects such as influence of viscoelastic overmolds, effects of die bumping, and substrate attach are difficult to capture in lumped element or compact models (Figure 20.3).

20.1.4
Layout-Based Design (Bottom-Up Design)

Three-dimensional design, typically the entry mode for the mechanical designer, is still the most popular methodology for MEMS design. Originally pioneered by IntelliSense in the early 1990s, this still remains the most popular methodology for MEMS design at present.

Layout-based design combines the 2D mask layout with the process flow to create 3D solid models of the MEMS device. These solid models are discretized and analyzed using 3D FEM/BEM methods. Layout-based design has been tremendously successful in microstructure design because it combines design intent with manufacturing.

Schematic synthesis tools, such as those incorporated in SYNPLE can convert component-based schematics into ready-to-use mask layouts, or hexahedral meshes that can be further used in 3D analysis of microsystems. In addition, bottom-up design can fully capture the complex multiphysics inherent in MEMS devices (Figure 20.4).

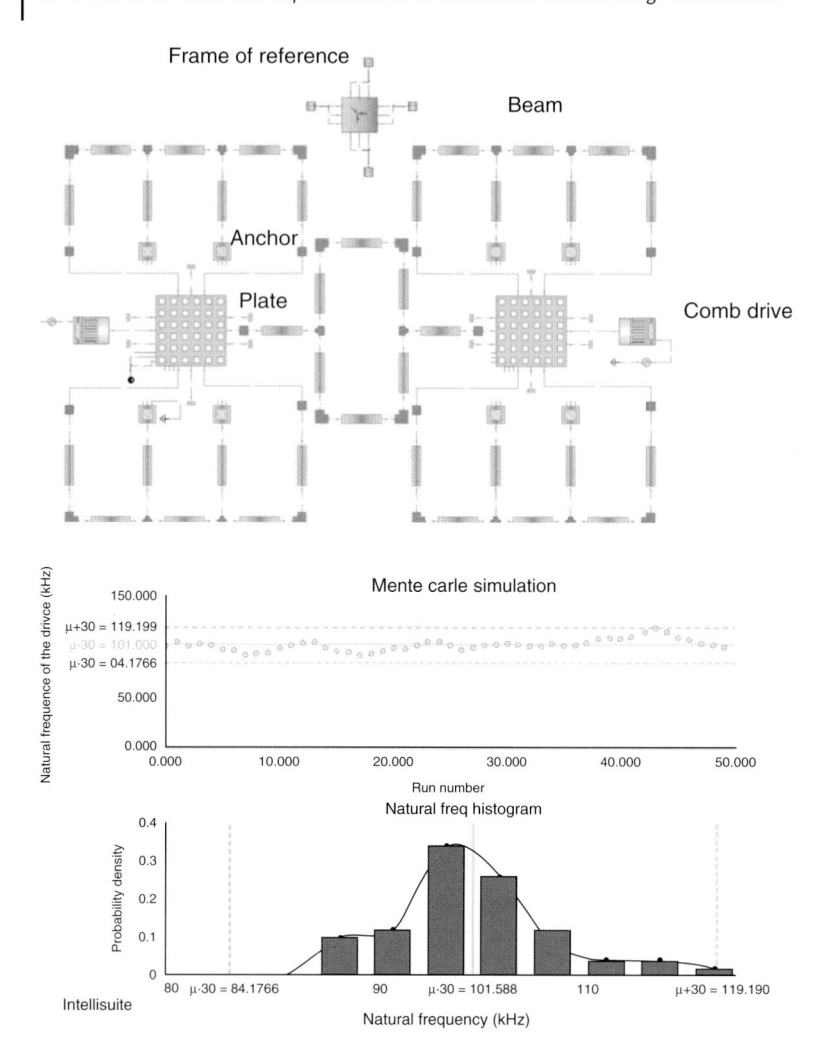

Figure 20.3 Schematic of a band-pass filter in SYNPLE. SYNPLE allows the user to quickly setup a parametric model of an MEMS device. This figure shows the results of a Monte Carlo simulation plotting the anticipated variation of natural frequency. Use of compact models allows users to develop robust designs that are inherently manufacturable.

20.1.5
System Model Extraction (SME)

The layout-based discretized 3D models need to be converted into system models. Typical MEMS 3D models can contain between 100 000 and 1 000 000 DOFs and system simulators are not designed to handle such complex problems.

Many designers use lumped model approximations to represent the microstructure in system simulations. While these are sufficient to proof-of-concept analysis, they simplify real-world effects such as etching effects, stresses in beams and

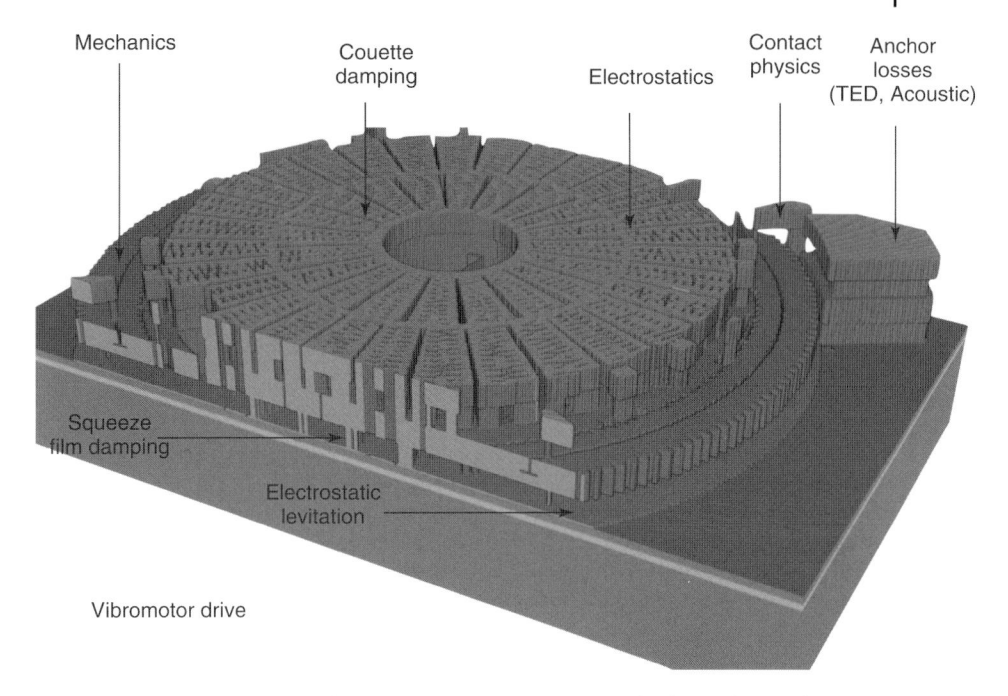

Figure 20.4 MEMS design is inherently multidomain in nature. The figure depicts the different kinds of models that need to be incorporated in the modeling of a vibro-drive. FEM/BEM tools are typically used to capture the multiphysical effects in MEMS devices.

suspensions, and levitation effects due to charge reflectance. For instance, uneven sidewall angle due to etching can lead to large quadrature errors in inertial devices and levitation effects in comb drives can lead to lowered sensitivity of devices.

A new class of numerical algorithms based on modal superposition and Krylov/Arnoldi subspace reduction techniques have been developed in the recent years to convert FEM models into arbitrary degree of freedom (NDOF) models. These algorithms are used to capture the total energy and energy dissipation in the system. On the basis of this, the FEM/BEM models can be reduced to efficient compact system models that can be incorporated into system simulators.

The advantage of using ROMs is that they accurately capture the device behavior across multiple energy domains (mechanical, thermal, electrostatics, fluidics, damping, etc.). The *ROMs* can be typically exported into a variety of hardware description languages (HDLs) for use with electrical or system simulators (Figure 20.5).

20.1.6
Verification

Verification in MEMS is quite different from that in the integrated circuit (IC) world. The IC world typically uses layout versus schematic (LVS) and design rule

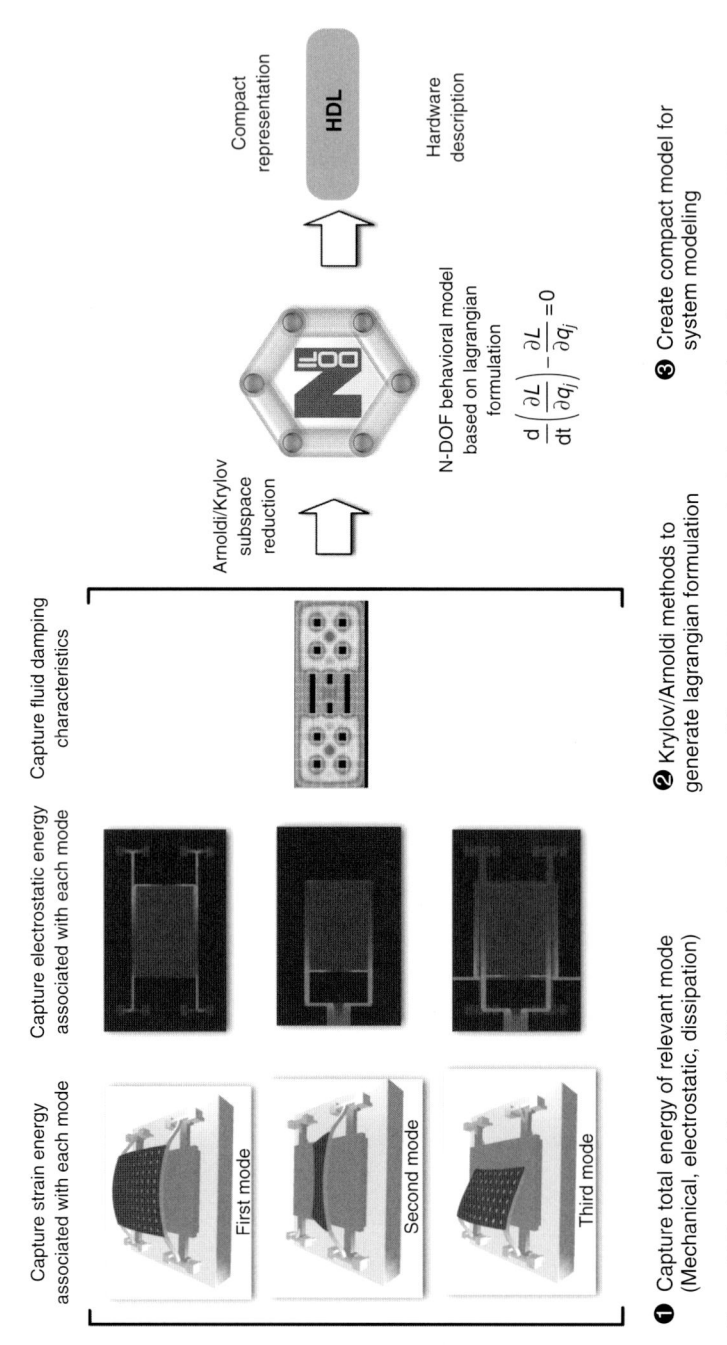

Figure 20.5 Methodology for efficient extraction of compact models. IntelliSuite creates the lookup-table-based Lagrangian ROMs by capturing the energy in each of the physical domains (i.e., mechanical, electrostatic, fluidic, etc.).

check (DRC) techniques. While DRCs can be used in the MEMS world, support for curves, beziers, and all-angle geometries are needed. LVS provides little benefit to the MEMS designer because of the inherent 3D nature of the design.

Schematic versus 3D comparisons (**SV3D**) are needed to make sure that the schematic capture has been accurately translated into a 3D design. In MEMS design, verification is the process of comparing the results from the schematic model (top-down approach) with the results of the 3D-based approach (bottom-up approach). This typically involves benchmarking the schematic, 3D finite element, and ROM results.

20.1.7
System Simulation

Accurate system simulation can be performed using ROMs described in the previous section. However, ROMs are not easily parameterized. While the process of creating a large number of ROMs can be automated, it is time consuming. Another alternative is to use lumped parameter models. However, considerable effort is required to derive realistic lumped models that take into account process-related effects, such as uneven etching and sidewall profiles, and complex physics such as electrostatic charge reflectance and levitation.

This poses a conundrum to the designer between choosing ROMs, which can capture complex physics and process nonidealities but cannot be easily parametrized, or lumped models, which can be parametrized but require significant effort to include nonidealities.

20.2
IntelliSense's Design Methodology

IntelliSense's software architecture is based on a unique combination of the best of bottom-up process-driven design and top-down synthesis. Top-down methodology allows the user to quickly explore a wide range of design options, while bottom-up design provides the accuracy to produce first-time-right silicon (Figure 20.6).

20.2.1
From the Top Down

State-of-the-art schematic capture and simulation tools allow the user to take a hierarchical approach to the design space. SYNPLE provides a large multidomain library of electrical, mechanical, thermal, digital and controls, and MEMS libraries. These elements may be combined in an effortless drag-and-drop manner and then wired to create schematics of multiscale multidomain systems. As a result, the design analyst can quickly survey a large design space before initiating a detailed analysis and verification process.

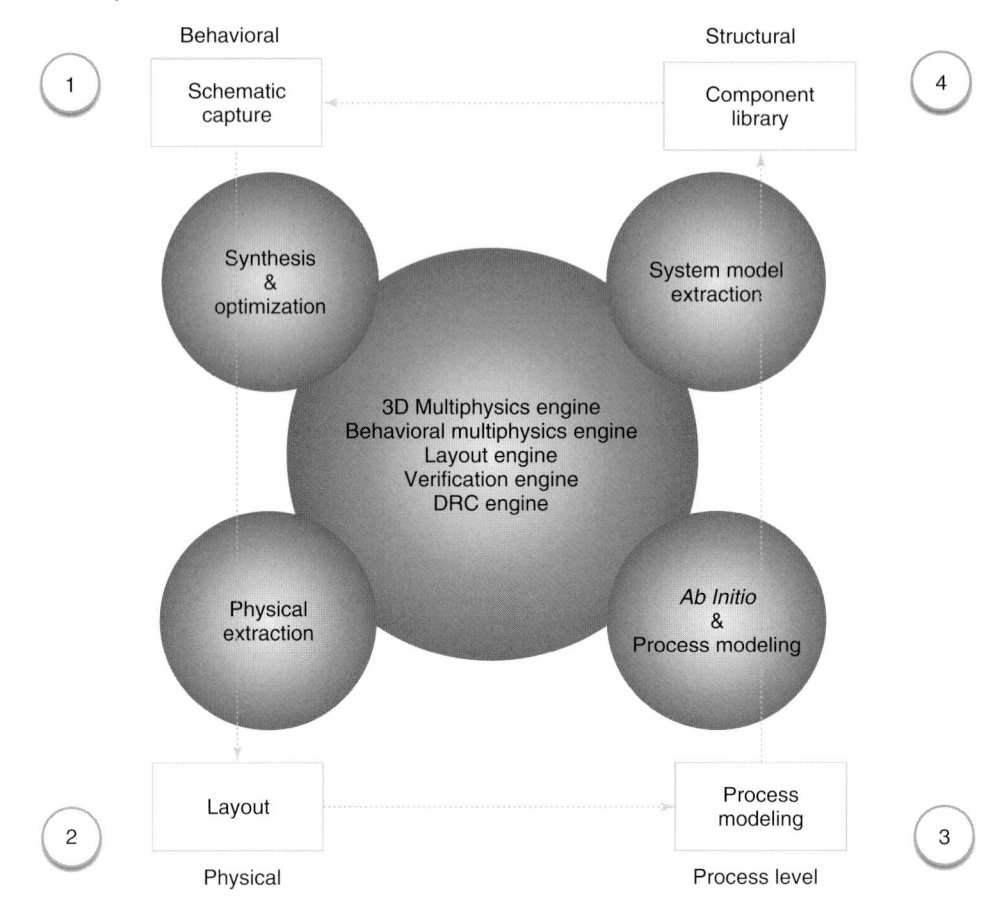

Figure 20.6 IntelliSuite design flow. Core computational engines and databases combined with synthesis, optimization, process modeling, physical, and system model extraction provide a friction-free and efficient workflow.

The top-down approach allows the user to combine readily available component blocks into a *netlist*. Owing to the simplified nature of the component models, users can perform device-level optimization using design of experiments (DoE), robust design, or other techniques. Users can start with component-based schematic capture and use optimization techniques to explore a vast design space. Built-in place and route algorithms can then be used to convert the schematic into a mask layout or an optimally meshed model ready for full 3D analysis (Figure 20.7).

20.2.2
One Step at a Time

IntelliSuite's bottom-up architecture is based on process elements – familiar process steps, such as photolithography, thin film deposition, and selective etching

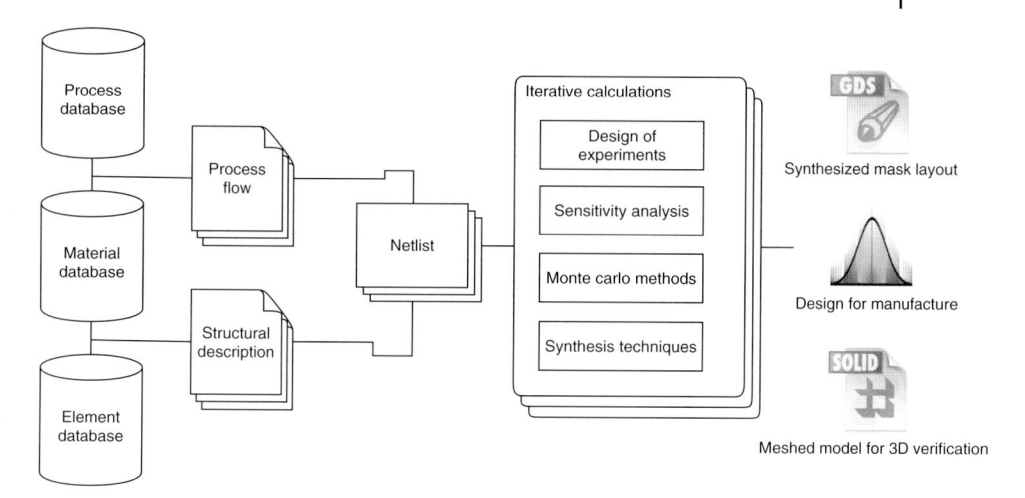

Figure 20.7 Top-down modeling is fast but of lower fidelity. It can be used for rapid exploration of design space and optimizing the design for performance and manufacturability.

form the basis of understanding the final device geometries. The process steps, combined with the mask geometries, can be used to build the final virtual device (power users can also import 3D geometries from popular CAD programs). In addition, the analysis modules (fully integrated thermoelectromechanical (TEM) analysis, high-frequency electromagnetic analysis, and microfluidics analysis) can be used to analyze the performance of MEMS models.

IntelliSuite features a comprehensive material and process database, allowing the analyst to understand materials properties such as conductivity, film stresses, and mechanical strength as a function of processing parameters. Subsequently, this enables the analyst to produce more realistic models (Figure 20.8).

20.2.3
Closing the Loop

IntelliSuite offers the analyst a number of tools to close the loop between top-down and bottom-up modeling. Synthesis and placement tools such as *MEMS-Synth* and *Hexpresso* can automatically transform the schematic into a ready-to-use layout or a meshed structure for FEM/BEM analysis. In addition, graphics tools allow the analyst to visualize the results of schematic level analysis in 3D, the natural context for MEMS design.

Similarly, system model extraction (SME) tools based on energy storage and dissipation in multiple physical domains can accurately capture the dynamics of the MEMS device. The reduced order models from SME can capture all of the device and packaging effects. The derived ROMs can be directly used in schematic-level cosimulation with the electronics or alternately exported into popular HDLs for use in simulators such as PSPICE, HSPICE, Cadence Virtuoso, Mathworks Simulink, and MentorGraphics SystemVision.

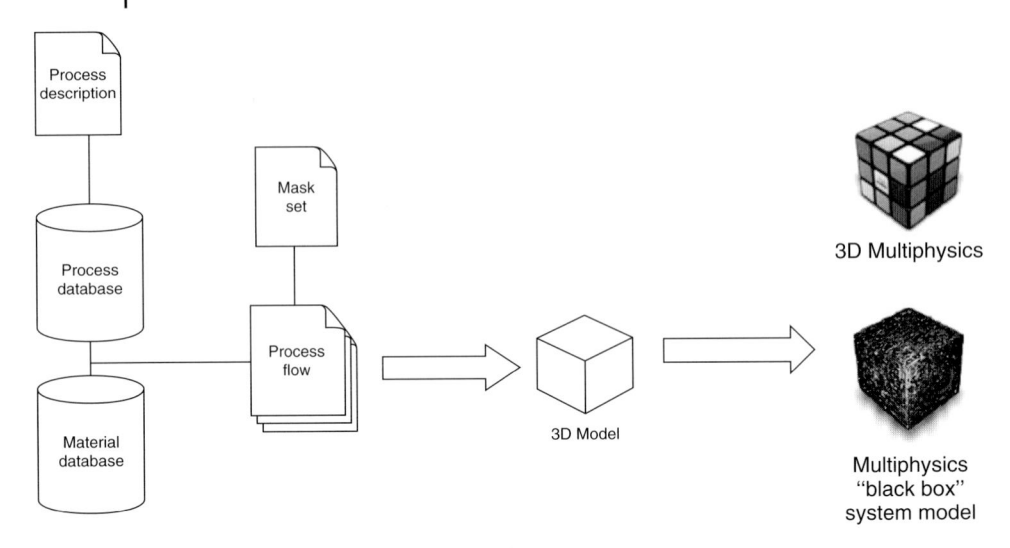

Figure 20.8 Bottom-up modeling is accurate but slower. Use bottom-up modeling for accurately capturing the device behavior and encapsulating it into a black-box system model.

By presenting a uniform framework for simultaneous top-down and bottom-up methodologies and toolsets to easily switch between methodologies, IntelliSuite allows information capture from the entire design team.

20.3
Implementation of System Model Extraction in IntelliSuite

20.3.1
High-LEVEL Overview

Lagrangian mechanics provides a structured approach that accounts for the total energy of all physical domains within an MEMS and automatically derives the equations of motion with full coupling effects (the reader is directed to Chapter 12 for a through background on the Lagrangian computation). The general procedure for calculating the Lagrangian function is as follows:

1) Choose the generalized coordinates of the system q_j. In the case of a discretized (meshed) MEMS device, the eigenshapes (Φ_j), eigenfrequency (ω_j), and generalized mass (m_j) of the device are easy to compute. The mode shapes are chosen as the basis functions or the generalized coordinates of the system.
2) The analyst is given the ability to intelligently identify the relevant/dominant modes of the system. This is accomplished by computing a "modal contribution factor" by performing a standard electromechanical relaxation analysis and solving for the initial deformed shape (derived from the residual stress without external loads) and the final deformed state (with mechanical loads and applied

voltages). A QR decomposition algorithm is then used to determine the contribution of each mode to the final deformed state of the structure. In most cases, 95% of the energy is typically contained within two to three dominant modes of the system.

3) Compute the kinetic energy (*T*) and potential energy (*V*) of each of the energy fields within the system. The Lagrange function is defined as: $L = T - V$. In the case of electrostatically actuated MEMS, the electrical potential energy is readily derived from the capacitance between various entities in the system. The electrostatic potential energy is computed from the system capacitance matrix (C_{ks}) derived using a boundary element formulation as previously reported by IntelliSense [1]. The mutual capacitance energy as a function of modal amplitudes is calculated and stored in a lookup table (LUT). In addition, the damping coefficient (ξ_j) of each mode is also calculated using the technique described in the next section.

4) Determine the generalized forces in the system, based on the displacements in the system arising from the nonconservative work (∂W) performed in the system. In the case of an MEMS or a purely mechanical device, this can be derived from the principal of virtual work. The strain energy function (∂W_{st}) of the system is determined for a user-defined maximum displacement of the device. A number of points are generated along the path of the maximum displacement, and the strain energy function is evaluated for each of the modes for each of these points. The strain energy function for each mode at every generated point of displacement is stored in a second LUT, known as the *strain energy versus modal amplitude LUT*.

5) The equations of motion can be derived as

$$m_j \ddot{q}_j + 2\xi_j \omega_j \dot{q}_j + \frac{\partial W_{st}}{\partial q_j} = \frac{1}{2} \sum_r \frac{\partial C_{ks}}{\partial q_j} \times (V_k - V_s)^2 + \sum_{i=1}^{n} \Phi_j^i \times F_i \quad (20.1)$$

Once these analyses are complete, the information is stored in a final LUT that can be used in a system level simulator. Extrapolation and interpolation within the LUT can be used to determine the energy associated with a particular excitation of the device. Within the system level simulator, different loading conditions can be applied to the system model to analyze how it will react, allowing optimization of the device and its control structure.

The algorithm for extracting the basis functions and the Lagrangian described earlier has been implemented in IntelliSuite's TEM analysis module. The equations of motion based on the Lagrangian are automatically translated into a SPICE/HDL model and are implemented in IntelliSense's SYNPLE software.

20.3.2
Capturing Residual Stress and Film Damping Effects

In general, the deformed state and dynamics of a mechanical system can be accurately described as a linear combination of modal shape functions. An equation describing this transformation is given below.

$$\Phi_{ext}(x, y, z, t) = \Phi_{initial}(x, y, z) + \sum q_i(t) \times \Phi_i(x, y, z) \quad (20.2)$$

where Φ_{ext} represents the deformed state of the structure, $\Phi_{initial}$ the initial equilibrium state (derived from the residual stress without external loads), $\Phi_i(x, y, z)$ the displacement vector of the *i*-th mode, and q_i the coefficient of the *i*-th mode, which is referred to as the *scaling factor* for the *i*-th mode.

Film damping arising from the viscous effects of the surrounding medium play an important role in the dynamics of MEMS. The reader is directed to Chapter 2.3 for a detailed discussion on film damping. Film damping force can be modeled as an implicit function of the motion.

$$F_{FD} = F(\Phi_{ext}) = F\left(\Phi_{initial}, q_1, q_2, \ldots, \Phi_1, \Phi_2, \ldots, \frac{\partial q_1}{\partial t}, \frac{\partial q_2}{\partial t}, \ldots\right) \quad (20.3)$$

where F_{FD} is the film damping force.

If the modal cross talk is negligible, the total force can be simplified to the sum of separate modal forces by ignoring the modal cross talk.

$$F_{FD} = \sum F_i\left(q_i, \Phi_i, \frac{\partial q_i}{\partial t}, \Phi_{initial}\right) = \sum G_i\left(q_i, \frac{\partial q_i}{\partial t}, \Phi_{initial}\right)\Phi_i \quad (20.4)$$

The damping force consists of two terms: one is the viscous effect in fluid flow, referred to as the *viscous force*, and the other reflects the compressibility of the flow medium, referred to as the *spring force*. It is assumed that the viscous force is proportional to the modal velocity, and the spring force is proportional to the modal displacement.

$$\langle F_i, \Phi_i \rangle = C_i \frac{\partial q_i}{\partial t} + K_i q_i \quad (20.5)$$

where $<F_i, \Phi_i>$ is the dot product of vectors F_i and Φ_i and C_i and K_i are the damping coefficient and stiffness coefficient of the *i*-th mode, respectively.

Typically, C_i and K_i are both nonlinear and frequency dependent.

$$C_i = C_i(q_i, \omega), K_i = K_i(q_i, \omega) \quad (20.6)$$

For small motions, a linearized model can be used, C_i and K_i take on values near the initial state $\Phi_{initial}$, and they are frequency dependent only.

$$C_i = C_i(q_i = 0, \omega) = C_i(\omega), K_i = K_i(q_i = 0, \omega) = K_i(\omega) \quad (20.7)$$

As can be seen, the damping and stiffness coefficients are still frequency dependent, so they cannot be directly used in macromodel simulation for transient analysis. Methods can be used to transform frequency-dependent parameters to frequency-independent ones. For the transformation method used in IntelliSuite, the reader is referred to the work by J Mehner, *et al.* in Chapter 12 and [2, 4].

20.3.3
Implementation

The modal superposition method is an efficient way for computation because only one equation is needed for one mode to describe a coupled system entirely. An overview of the modal superposition method implementation in fluid damping macromodel extraction is described below. Many aspects of the method are performed automatically and simultaneously without requiring user interaction.

1) **Create a FEM/BEM model.** IntelliSuite enables the creation of hybrid FEM/BEM models to capture the mechanical/fluidic and electrostatic domains, respectively.
2) **Simulate modal contributions.** Perform a frequency analysis to find the natural frequencies and modal shapes of the system. Then perform a nonlinear static FEM simulation or a full nonlinear coupled static FEM/BEM relaxation simulation. Solve for the initial deformed shape caused by residual stresses without external loads, and the final deformed shape, with mechanical loads and applied voltages. Project the simulated shape deformation on the space spanned by the calculated modal shapes by performing the operation of dot product in Eq. (20.5). Solve for the coefficients that determine the contribution of each mode to the shape deformation. Select which modes are significant in terms of the magnitude to the shape deformation and should be included in the calculation of the energy domains of the system.
3) **Calculate strain energy.** The strain energy function of the system is determined for a user-defined maximum displacement of the device. A number of points are generated along the path of the maximum displacement and the strain energy function is evaluated for each of the modes for each of these points. The location of each node of the system is determined from the expression derived in step 2, which describes the deformed state of the system as a linear combination of modal shapes. A single mechanical domain FEM analysis is performed at each generated point, and the strain energy is then calculated at each generated point. Then the strain energy for each relevant mode is stored in an LUT.
4) **Calculate capacitance (electrostatic energy).** Similar to step 3, evaluate the capacitance matrix of the system for each generated point. The electrostatic energy can then be calculated. A single electrostatic domain BEM analysis is performed at each generated point, and the mutual capacitance is then calculated at each point. Then the mutual capacitance for each relevant mode is stored in an LUT.
5) **Calculate film damping.** Apply sinusoidal displacement loads at the selected shape modes and perform a frequency analysis. Compute the pressure distribution and project it onto the space spanned by the calculated modal shapes. Then compute the frequency-dependent damping and stiffness coefficients. Fit the coefficients into a curve and transform them to frequency-independent coefficients.

With calculation results obtained in steps 3–5, the user can write the Lagrangian equations for the system. The equations of motion based on the Lagrangian are automatically translated into a SPICE/VHDL/Verilog-A/Simulink MEX model and are implemented in IntelliSense's SYNPLE software. IntelliSuite currently supports seamless integration with toolsets from Cadence (Spectre, PSPICE), Mentor Graphics (ADMS, System Vision), Synopsys (HSPICE, VSS), Mathworks (Simulink), and Tanner.

20.4
Benchmarks

20.4.1
Accelerometer

Figure 20.9a shows a finite element 3D model (IntelliSuite) of an silicon on insulator (SOI)-based capacitive accelerometer intended for microgram sensing of seismic applications. It is used to measure the acceleration, and the acceleration response is measured in terms of change in capacitance between the electrodes. The two rectangular dark gray structures in the middle are the electrodes. The capacitance change between the inertial mass and the electrodes is recorded to sense the acceleration. The macromodel was extracted using IntelliSuite's SME module and the system model is simulated in SYNPLE.

Figure 20.9b shows the dynamic response of the accelerometer to a 1 ms, 1g acceleration pulse. The difference between the FEM and the Lagrangian SME model is less than 2%. The FEM simulation took approximately 20 min to complete as opposed to 10 s for the SME model (Figure 20.10).

20.4.2
Inertial Gyroscope

The start-up response of the gyro is shown in the Figure 20.11a,b below. The x-displacement (drive motion) and the y-displacement (sense motion due to Coriolis force) are plotted as a function of time. The response of the gyro to a sinusoidal rotational input is shown in the Figure 20.3c.

The FEM/BEM multiphysics model took nearly 24 h to compute the first five cycles of the gyro response. Computing the first 50 cycles of the start-up response would take nearly 10 days. The Lagrangian model took about 4 h to extract. Once extracted, the Lagrangian results matched the coupled FEM/BEM to within 1% of displacement and took approximately 30 s to compute on a modern desktop PC. The response to the sinusoidal input took less than 1 min to compute.

One of the major advantages of this approach is the seamless and efficient integration of the device model with circuit simulation. The analyst is able to cosimulate the MEMS device and the associated application-specific integrated circuit (ASIC).

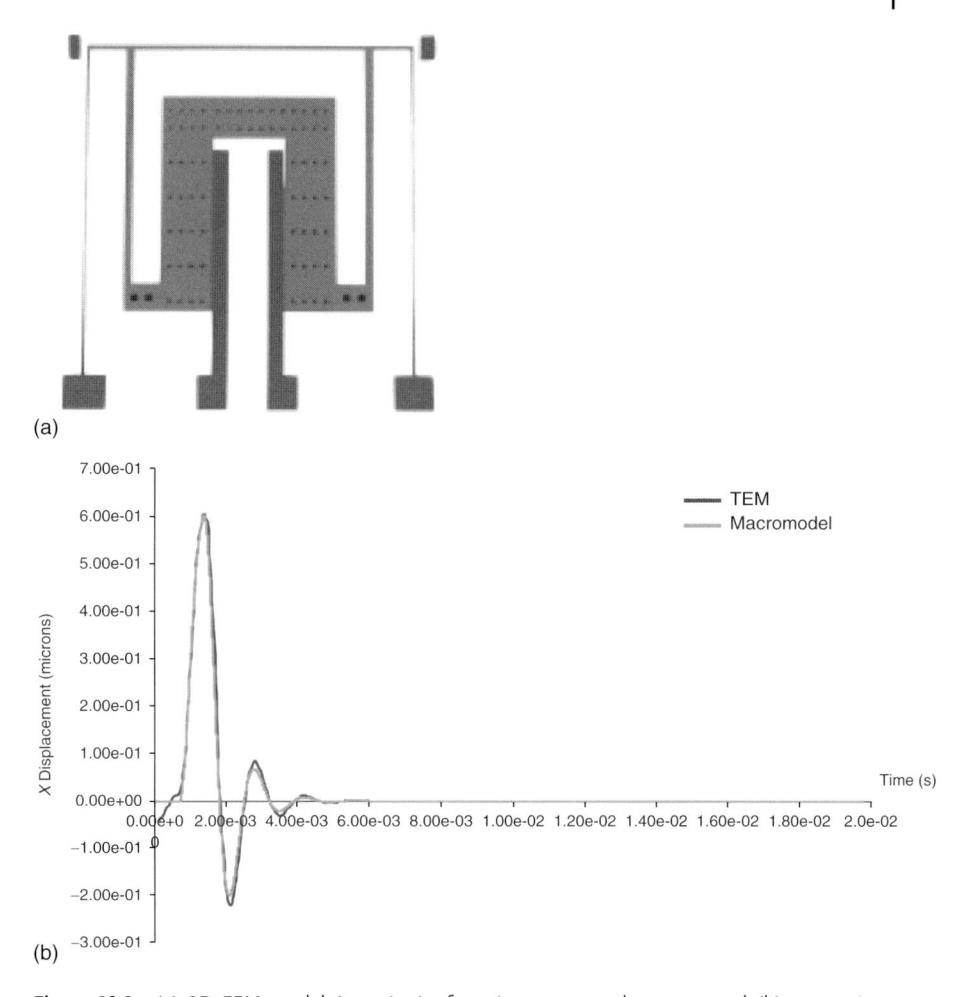

Figure 20.9 (a) 3D FEM model (top view) of a microgram accelerometer and (b) comparison of the dynamic response of the accelerometer, to a 1 ms 1g loading, between FEM and SME calculations.

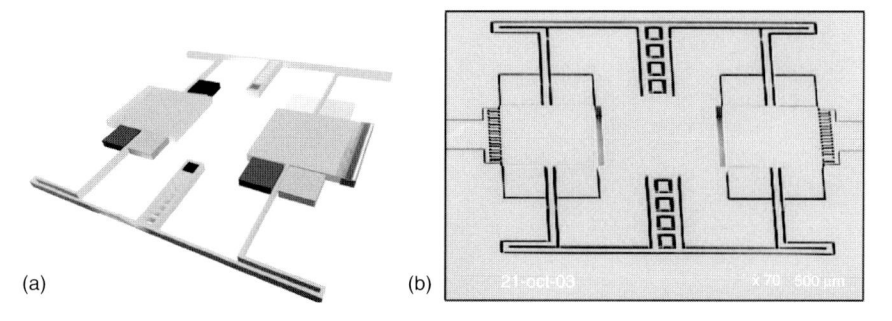

Figure 20.10 (a,b) Inertial grade gyro FEM model and fabricated device. (Source: Picture courtesy Zaman *et al*. Georgia Tech University.)

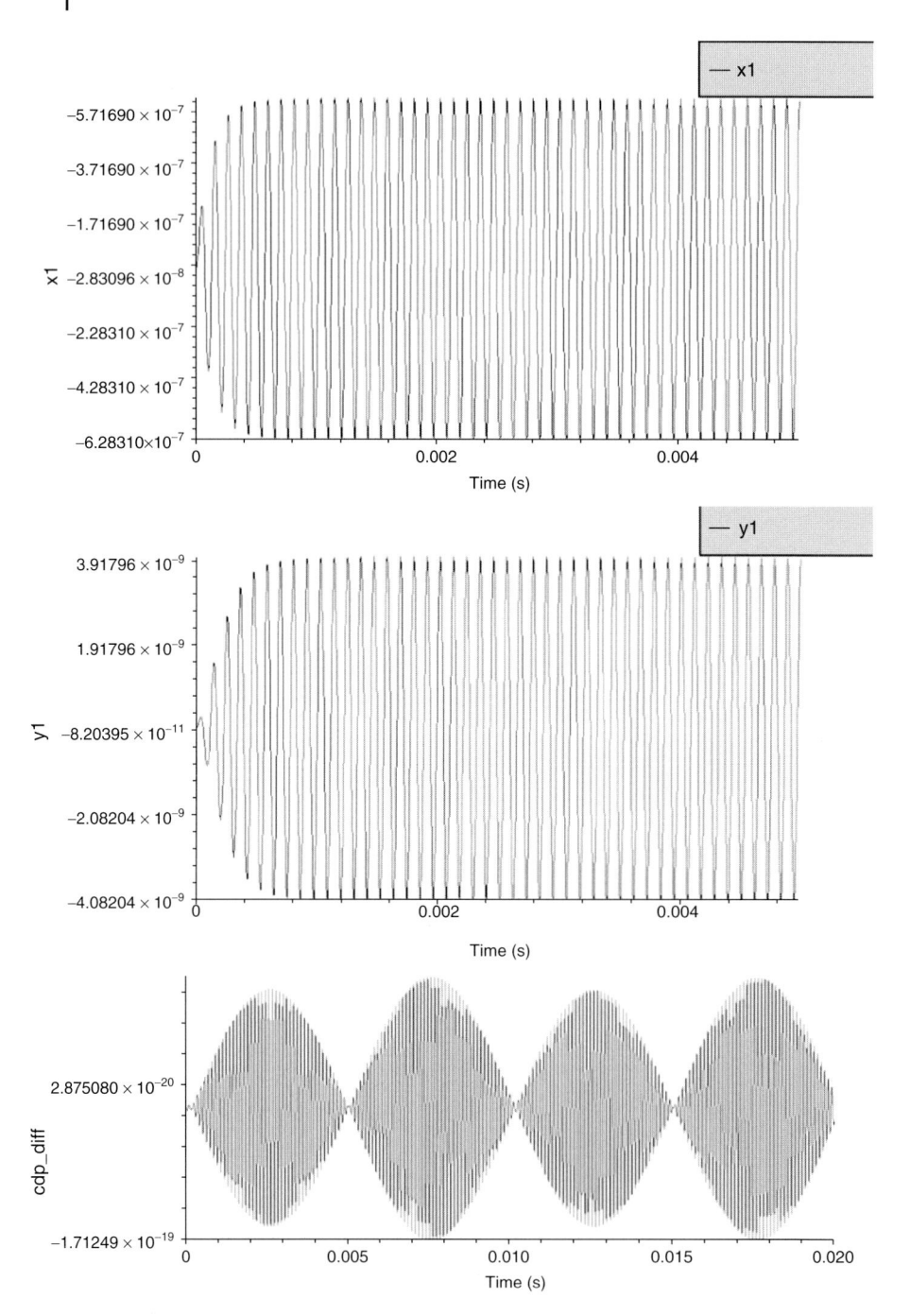

Figure 20.11 (a,b) Start-up response of the gyro and (c) gyro response to a sinusoidal rotational input.

20.4.3
Fluid Damping

Fluid damping macromodel extraction was verified for a variety of cases. A simple benchmark case is shown below whereby the damping between a moving plate and a fixed plate is extracted and compared with closed form solutions in Eq. (20.8). The reader is referred to [3] for a detailed derivation of the following equations.

$$C(\Omega) = \frac{64\sigma p_0 A}{\pi^6 d \Omega} \sum_{m=odd} \sum_{m=odd} \frac{m^2 + n^2 c^2}{(mn)^2 \left[\left(m^2 + n^2 c^2\right)^2 + \frac{\sigma^2}{\pi^4}\right]}$$

$$K(\Omega) = \frac{64\sigma^2 p_0 A}{\pi^8 d} \sum_{m=odd} \sum_{m=odd} \frac{1}{(mn)^2 \left[\left(m^2 + n^2 c^2\right)^2 + \frac{\sigma^2}{\pi^4}\right]}$$

(20.8)

where $C(\Omega)$ is the frequency-dependent damping coefficient; $K(\Omega)$, the squeeze stiffness coefficient; p_0, the ambient pressure; d, the thickness of the film; Ω, the response frequency; A, the area of the plate; $c = L/w$, the length-to-width ratio of the plate; and σ, the squeeze number of the system. L and w are the length and width of the plate, respectively, and η is the dynamic viscosity. The viscous and spring forces normalized by the transverse displacement of the rectangular plate are: $F_{vis} = C\Omega$ and $F_{spr} = K$.

Figure 20.12 shows the comparison of the viscous and spring force extracted by IntelliSuite macromodeling technique compared with the closed form solution.

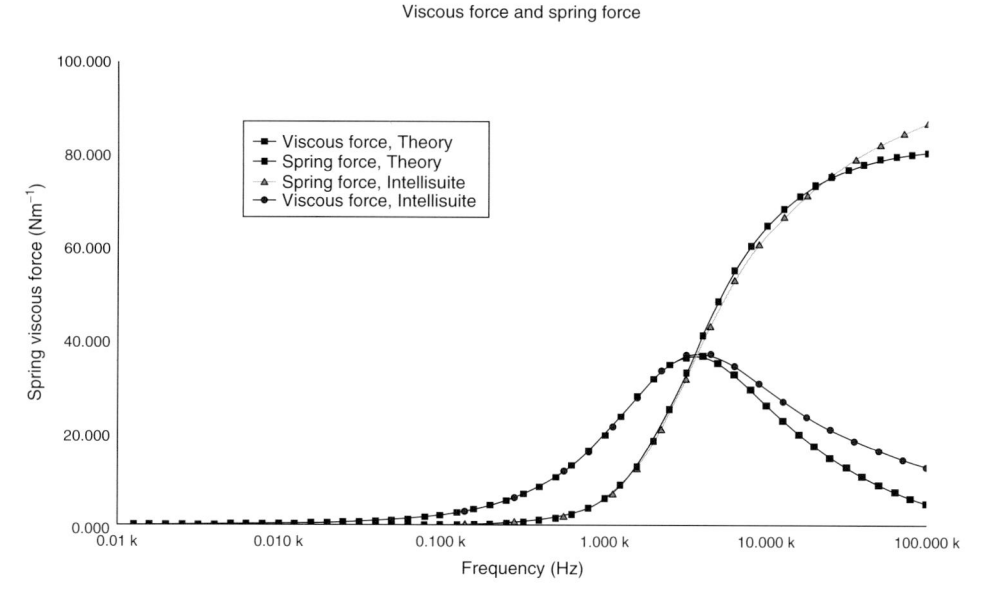

Figure 20.12 Comparison of viscous and spring force extracted by IntelliSuite and compared to theory. The following parameters were used in the model: $L = 2$ mm, $w = 1$ mm, $d = 20$ μm, $\eta = 2$ μPa·s, and $p_0 = 1000$ Pa.

20.4.4
Coupled Package-Device Modeling

The ambient environment, for example, temperature, affects the performance of packaged MEMS devices. So it is critical to understand the packaging effects to design an MEMS. Macromodel extraction of packaged devices is typically carried out in the following four steps.

1) Perform a static analysis to obtain the thermal stress and the deformation of the complete packaged device.
2) Extract the core MEMS device and refine its grid.
3) Load the stress and the deformation information from step 1 to the core device.
4) Perform the analysis of the core structure under the initial conditions as applied in step 2.

As a benchmark model, the thermomechanical performance of the inertial gyroscope described earlier is considered. The package considered in this case is a wafer-level chip-scale MEMS package (WLCSP) bumped onto an ASIC. Figure 20.13 shows the model of the packaged gyroscope.

The results of the package modeling are shown in Figures 20.14 and 20.15. Figure 20.15a shows the capacitance of the sense electrodes as a function of the package temperature while Figure 20.15b shows the spring softening of the drive mode as a function of temperature and voltage.

SME allows for an easy method to fully capture thermal, electrostatic, and mechanical effects. Figure 20.15 shows the transient start-up response of the packaged device is calculated in a matter of seconds using the macromodel solution. Macromodel response matched the finite-element-based response to

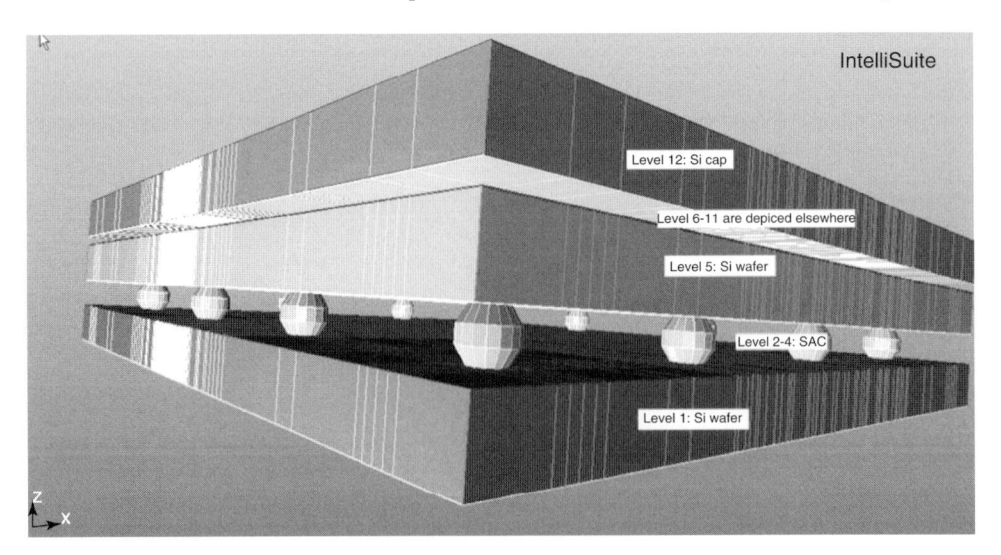

Figure 20.13 An MEMS gyro die wafer bumped onto an ASIC.

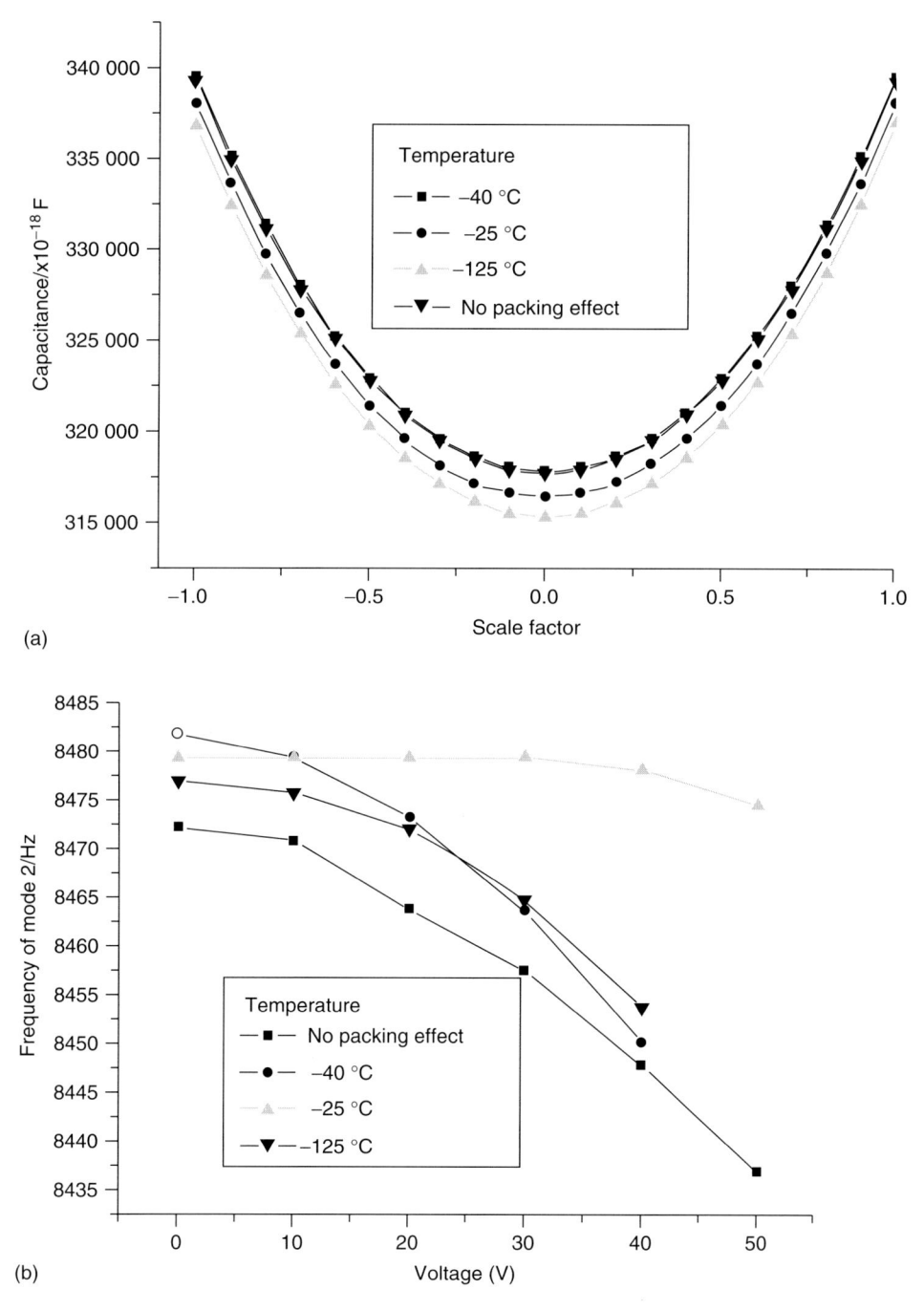

Figure 20.14 (a) Capacitance and (b) electrostatic spring softening (frequency) as a function of applied voltage and temperature.

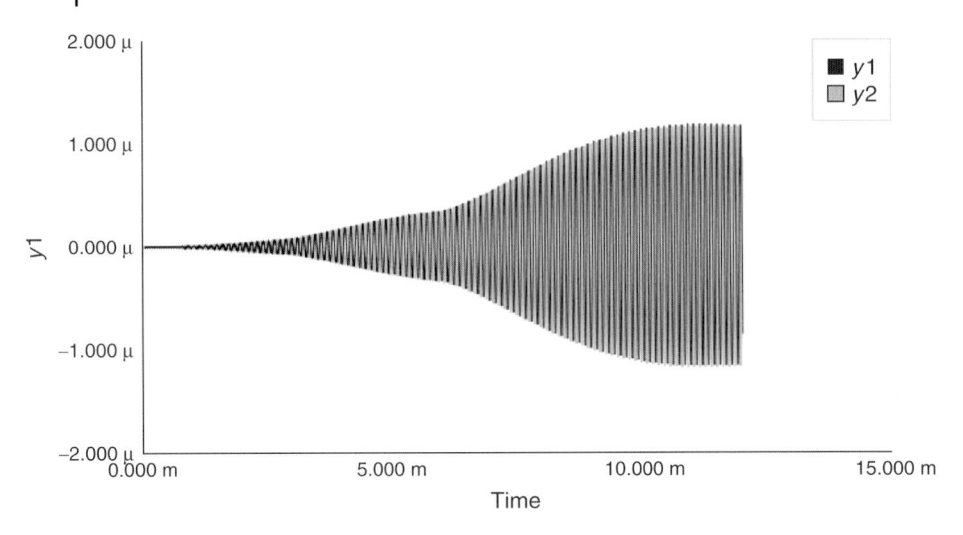

Figure 20.15 Schematic-level modeling of the packaged device Coriolis response.

within 3% with a 1000 × improvement computational performance for transient simulations.

20.5 Summary

IntelliSense provides tools that allow the user to simulate their devices at any stage of the design cycle.

SYNPLE is designed as a tool for initial design exploration. Parametric analysis and optimization features allow a designer to explore a large design space very quickly. Once a schematic has been constructed, it is easy to automatically extract a mask layout or meshed model to use for further analysis.

The TEM analysis module uses Exposed Face Meshing methods to decouple the mechanical and electrostatic meshes of devices. This allows our users to perform mechanical mesh refinement in areas with high stress gradients and electrostatic mesh refinement in areas of high charge density. This greatly simplifies the mesh needed for the device without sacrificing any accuracy. TEM is optimized for mixed domain analysis, and models can be exported to system level solvers for quick transient analysis as well as CMOS integration.

When one compare the time required to run a full dynamic finite element analysis to the time required to run an SME and dynamic analyses in SYNPLE, SME with SYNPLE is orders of magnitude faster. Running optimization analyses with SME and SYNPLE can take a few hours, while running multiple optimization analyses with a FE solver would take days, weeks, or possibly months depending on the extent of the optimization.

Overall, IntelliSuite has robust automated system model extraction routines that allow the user to create coupled macromodels that capture multiphysics including, electrostatics, mechanics, thermal behavior, and fluid damping. Future extensions to the tool will allow users to capture thermoelastic damping and structural-acoustic losses as well.

References

1. Mehner, J., Gabbay, L., and Senturia, S.D. (2000) *Journal of Microelectromechanical Systems*, **9**, 262–278.
2. He, Y., Marchetti, J., and Maseeh, F. (1997) An improved meshing technique and its application in the analysis of large and complex MEMS systems. Symposium on Micromachining and Microfabrication, Micromachined Devices and Components, 1997, Dallas, TX.
3. Blech, J.J. (1983) *Journal of Lubrication Technology*, **105**, 615–620.
4. Mehner, J.E., Doetzel, W., Schauwecker, B., and Ostergaard, D. (2003) Reduced order modeling of fluid structural interactions in MEMS based on modal projection techniques. Transducers'03, vol. 2, pp. 1840–1843.

21
Reduced Order Modeling of MEMS and IC Systems – A Practical Approach

Sebastien Cases and Mary-Ann Maher

21.1
Introduction

Recent advances have proven that microelectromechanical system (MEMS) developments are no longer an exotic application field of the semiconductor industry with the real success of devices such as accelerometers and gyroscopes in the gaming industry or pressure sensors in tire pressure monitoring systems (TPMS) in the automotive industry, to name a few.

Systems incorporating MEMS devices are growing with respect to their level of integration and complexity, often including multiple MEMS sensors/actuators, analog and digital circuitry, microcontrollers, and custom packaging. Many of the delays in deploying MEMS-based systems to market stem from errors made in integrating the MEMS with the rest of the system, causing costly redesigns. Codesign of the product enables designers to catch composition errors early and also enables designers to optimize the entire system, trading off requirements between the MEMS, electronics, and packaging. The results are higher product performance, lower manufacturing costs, and faster time to market.

SoftMEMS has created behavioral modeling software called the *Compact Model Builder* that enables codesign of MEMS-based systems by allowing users to create models of their sensors or actuators for use in simulation with electronics.

This chapter presents the Compact Model Builder as a solution to create an efficient link between the multiple energy domain 3D design and analysis environments based on finite element models (FEMs) or boundary element models (BEMs) and the electrical one based on circuit and systems simulators. We describe these environments and present how the Compact Model Builder can be used to create the models that unite these environments.

The Compact Model Builder uses the reduced order modeling techniques of dynamical systems (Chapter 3) to create a behavioral model, which can be *de facto* used by integrated circuit (IC) industry simulators. The goal of model order reduction is to find a low-dimensional but accurate approximation of the large-scale dynamic system. This way, one can drastically reduce time required for transient

System-level Modeling of MEMS, First Edition. Edited by T. Bechtold, G. Schrag, and L. Feng.
© 2013 Wiley-VCH Verlag GmbH & Co. KGaA. Published 2013 by Wiley-VCH Verlag GmbH & Co. KGaA.

and harmonic simulation and find a compact representation suitable for system level simulation [1].

Reduced order modeling is a general concept, not limited to MEMS design, used when the amount of data becomes so unwieldy that simulations require days to complete. The IC industry is also taking advantage of reduced order modeling algorithms. As netlists are becoming longer after parasitics back-annotation, for instance, reduction allows the extraction of a smaller netlist that require shorter simulation times.

By focusing on a limited set of load cases and degrees of freedom, and by applying methods such as substructuring using Guyan condensation and mode superposition (nonlinear dynamic problems) (Chapter 3), the initial problem can be reduced to smaller sized matrices (called *reduced matrices*), which can easily be reformulated by the Compact Model Builder into formats suitable for electronic simulations.

Another important use of the builder is to encapsulate intellectual property (IP), so that it may be used in system design by third parties. In this case, the designers do not wish to reveal the detailed physical model of their sensor or actuator, and the model enables their customers to evaluate whether the IP will suit their needs.

The Compact Model Builder is sold as part of SoftMEMS' popular MEMS design software MEMS Pro and MEMS Xplorer. Although the primary application field of the builder was the development of MEMS devices, it is also used to analyze and model multiple energy domain problems in the semiconductor, biotechnology, and solar industries.

Even though the tool is very flexible and allows users to target several physical domains with different levels of coupling between them (from basic electrostatic structure to more complex thermopiezoelectric structure coupling including fluid–structure interaction), the accuracy of the generated models relies on the knowledge of the user and especially on the validity of the original finite element model and its boundary conditions.

In this chapter, we describe the reduced order modeling techniques used by the Compact Model Builder. We begin by defining the modeling problem faced by the developers of commercial products based on multienergy domain systems. We continue by explaining the requirements of the modeling system, followed by a description of the modeling system itself. We end by describing future work in the area and open problems from an industrial perspective.

21.2
The MEMS Development Environment

To bring a sensor-based product to market on time and on budget involves a diverse group of people with various types of expertise. The team that designs these systems often includes materials scientists, process engineers, modeling experts, MEMS experts, circuit designers, and packaging experts. Large vertically integrated companies may have all the needed expertise in house. In small start-ups, there may be expertise lacking in certain areas. These start-ups must partner or use consultants to gain the needed expertise.

Modeling typically plays an important role in the design process. Phenomena occurring on multiple scales must be modeled from the nanoscale engineering of surfaces in a radio frequency (RF) switch, modeling of stiction, or bubble formation, to the packaging of multiple chips. Designers must also model phenomena on different timescales, as the electronics tend to operate on the nanosecond scale and the MEMS tend to operate at the microsecond to millisecond scale. Models must be composable and implement more accurate equations such as the nonlinearity induced by coupled field as the length scale is reduced.

At present, there is no one design flow for products incorporating MEMS, as there is in the semiconductor world where a top-down design predominates. Because of their various expertise and the jobs that they must do, engineers use a variety of CAD software. For example, the mechanical engineers typically use a finite or boundary element solver to analyze the MEMS or sensor in 3D. Electrical engineers may use a circuit simulation program such as SPICE [2]. Previously, designers would not share models. Each designer would use their own models for the part of the design they were working on. The mechanical designer would create a finite element model, and the electronics designer would redo the model from their perspective. In many companies, the communication between modeling teams shows a limited efficiency.

Creating a reduced order model is thus a part of a modeling hierarchy in between those performing detailed analysis to the circuit-level models used by electrical engineers and can ensure that there is consistency between the models. However, because the reduced order model is a bridge between the 3D and quasi 1D world, knowledge of the modeling techniques from both worlds is important as well as some knowledge of the reduction process itself. This knowledge is highly specialized. Mechanical designers have deep knowledge of thermomechanical considerations and how to run finite element programs. Circuit designers typically have knowledge of electrical considerations and circuit-level modeling but lack mechanical intuition. If a company has a modeling department, specialized engineers may be hired to create models but normally the designer must do their own analysis. An ideal flow may be shown (Figure 21.1). Hopefully, through the use of reduced order modeling, a consistent modeling hierarchy can be created so that high-level behavioral models become more refined and accurate, as they are simulated in FEM/BEM environments.

21.3
Modeling Requirements and Implementation within SoftMEMS Simulation Environment

Considering the constraints described earlier, and in order to be useful, the reduced order model must satisfy a number of requirements. The tool should support design environments used by the circuit designers and the sensor/MEMS designers. These

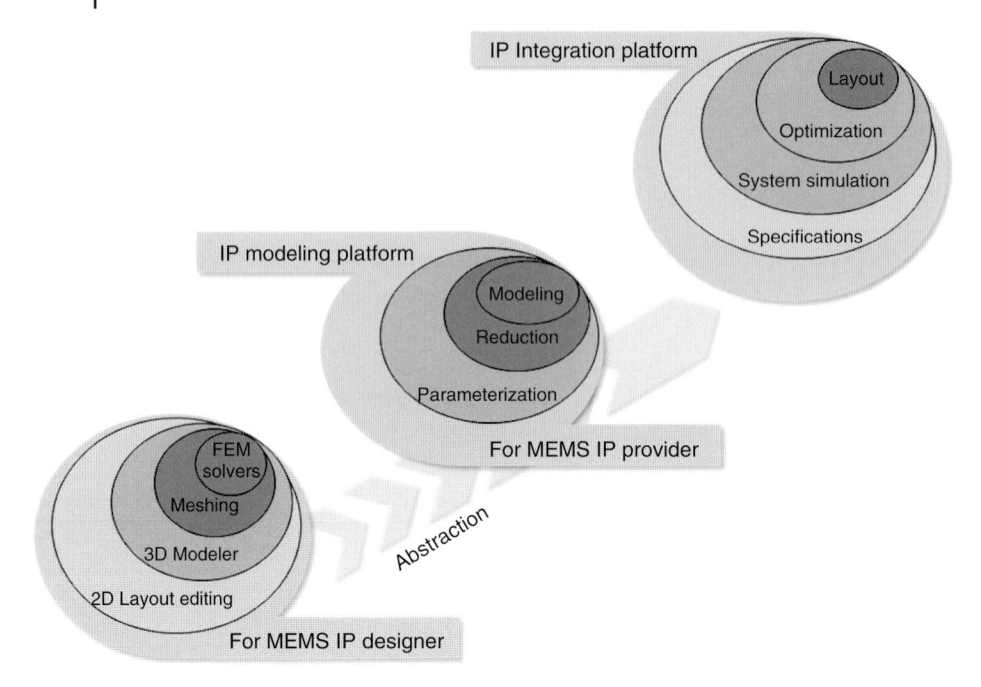

Figure 21.1 Modeling abstraction levels.

environments may include a wide set of IC-based layout, schematic, and simulations tools such as those from Cadence Design Systems (San Jose, CA), Mentor Graphics (Wilsonville, OR), Tanner EDA (Monrovia, CA), Agilent Technologies (Santa Clara, CA), Dolphin Integration (Meylan, France), and different finite element/boundary element solvers such as Ansys Multiphysics or HFSS from Ansys (Canonsburg, PA), Comsol Multiphysics from Comsol (Stockholm, Sweden), or Oofelie from Open Engineering (Angleur, Belgium).

Since these designers may be in different locations with different computer systems, support of multiple platforms is also important. Another constraint is that the tool should be easy to use. As stated earlier, the designers running the reduced order modeling programs are typically not modeling experts, so that it is important to add as much expert knowledge into the tool as possible.

The reduced order model approximates the finite element model. A designer needs to know when the model is accurate and when it is not. A very important issue that we see in industry is that the reduced order model is extracted by one team and passed on to another team without explaining or understanding the conditions under which the model will give accurate results. This is an extremely serious issue as the model user may drive the model with inputs outside its range of extraction or into a nonlinear range where it is not accurate. There are several important issues with approximation.

1) The spatial resolution, typically the signals at the point of master degrees of freedom chosen for the reduction fit well with the reduced order model. Making

sure that the users choose the appropriate number of degrees of freedom and their location on the model is important, an automatic choice can be given, but this option must be used with care. Missing phenomena that occur between the selected locations can be an issue such as bending and buckling.

2) *The temporal resolution*: can the model be used for transient analysis? Are the timescales that the model will react to appropriate for simulation with circuits?

3) *The range of validity of the model*: the models are typically valid for a range of input stimuli and certain stimuli conditions. Users may try to use the model outside its range of operational accuracy. Nonlinearity is also a concern.

4) The physical effects that the model can capture are also important. Some sources of error stem from the effects modeled by the finite element program in the first place and some sources of error occur from the model reduction process. A typical source of error that can be traced to the finite element model is the improper modeling of energy dissipation mechanisms such as anchors or damping. Incomplete geometric modeling of the finite element model can also lead to incorrect reduced order models. It is important that the physical effects included are made clear in the model documentation.

One typical example that shows the requirements is the transient simulation of a micromirror in an optical switch. Designers would like to be able to simulate the switching of the mirror in order to evaluate control circuitry. This simulation requires a time domain simulation of the mirror and its actuator including damping and thermal effects. Depending on the actuator type, this might be a coupled electrostatic mechanical model or some other coupled field reduction.

21.3.1
Models and Inputs

Reduced order modeling is a mechanism for creating behavioral models from finite element programs (Figure 21.2). The reduced order model is typically placed into a schematic or netlist as a part of a larger simulation.

This requires that the reduced order model has connection points (Figure 21.3) that represent the through and across variables to be used in the simulation. Depending on the formulation, position or velocity may be used to represent the mechanical across variable, while the applied force represents the through variable (Chapter 2).

Other conservative or dissipative energy domains [3] can be described using the across and through variables such as the electrostatic domain (with respectively, electrical potential and current) or magnetostatic domain (with magnetomotive force, or magnetic potential and magnetic flux).

However, depending on the model to be created, reduced order modeling may not be the right methodology, and other techniques may be used. An overview on alternative approaches is given in Chapters 2 and part II. The SoftMEMS Compact Model Builder allows users to create models from a variety of inputs.

➤ From a mechanical
finite element model

➤ To a circuit-level
behavioral model

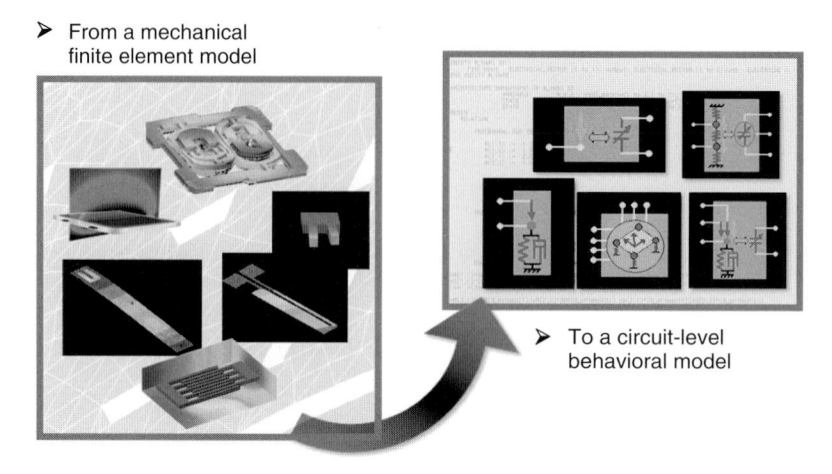

Figure 21.2 From FEM to circuit-level behavioral model.

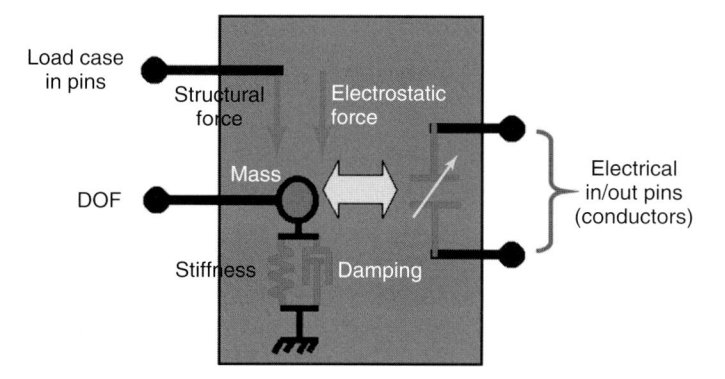

Figure 21.3 Input and output pins on a reduced order model.

1) FEM/BEM data – model order reduction
2) User-defined analytical equation
3) Experimental data
4) Combination of above models.

Knowing when to use each type of modeling process is important. If the geometry is very close to that used to create primitives in the library, a reduced order model may not be necessary. For example, users may select models from the SoftMEMS library, which are parameterized by process and geometry (Figure 21.4). These analytical models for beams, gaps, and comb drives may give excellent results for structures created using deep reactive-ion etching (DRIE is a highly anisotropic etch process used to create deep, steep-sided holes, and trenches in wafers). Reduced order models often are best for geometrically complex models where the analytical approximations based on simple geometry fail. However, often the

21.3 Modeling Requirements and Implementation within SoftMEMS Simulation Environment | 489

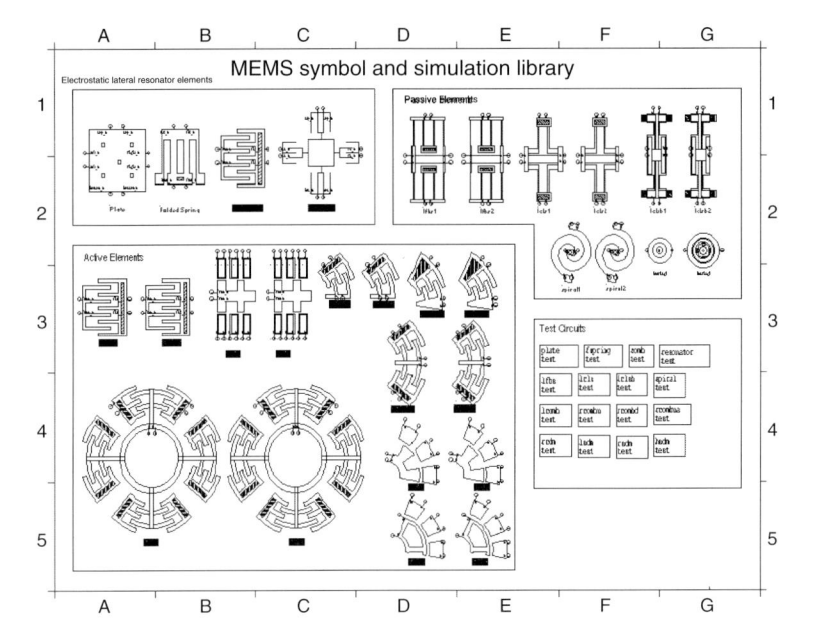

Figure 21.4 Libraries of models.

analytical models that can be calibrated to experimental data can be more accurate than reduced order models. This is due to the fact that the FEM model may not capture all the needed physical effects. This result is similar to the IC industry, where FEM/BEM simulations are done on new models, but SPICE models are used to characterize transistor models based on measured data.

A combination of techniques may also be required. For example, a coupled reduced order mechanical –electrostatic model may be supplemented by a proprietary damping equation to complete the model. SoftMEMS allows the flexibility to combine modeling techniques. Ideally, the reduced order model would create a reduced order model coupling a number of physical fields. However, this is often impractical. Supplementing the reduced order modeling technique often creates the best model.

For this purpose, the Compact Model Builder also supports the composition of models into larger models (Figure 21.5). Typically, a MEMS device may be built out of subparts each of which has a reduced order model. The builder allows users to correctly construct larger models.

The SoftMEMS Compact Model Builder allows the creation of system simulator ready models. The program formats the models, so that they are ready to be used directly in the simulator. For example, for Tanner's C-code modeling, the code is generated ready to link to the simulator. For Cadence and Agilent, Verilog-A [4] models are generated, and for Mentor and Dolphin, VHDL-AMS [5] models are created. It is necessary to understand the various implementations of the so-called hardware description languages (HDLs) (Chapter 4) as various companies create

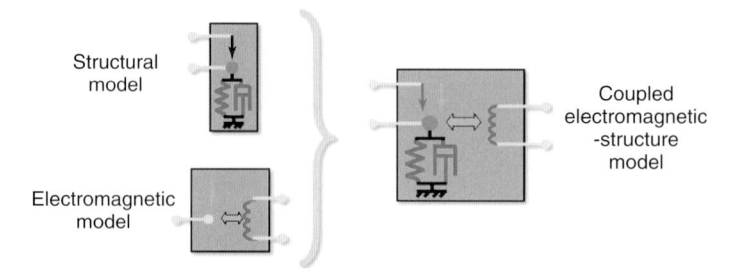

Figure 21.5 Combination of models.

their own – the various flavors of SPICE are well known. SoftMEMS' Compact Model Builder supports, for example, several flavors of VHDL-AMS, one for Mentor Graphics' ADVance-MS, and a different version for Dolphin's SMASH, for example.

21.3.2
Modeling Process

This section describes SoftMEMS' reduced order modeling techniques and software. As described earlier, users may not be modeling experts. A wizard is used to setup the needed inputs for the tools and the user is guided through the various steps.

1) In the first phase, the user will describe the type of model to be created; initially, the FE/BE model may include coupling between different types of solvers; so for a given set of coupled fields, the user can select which types of effects need to be captured in the reduced order model.
2) Specifying inputs and outputs for the model (Figure 21.6) is realized by choosing a set of degrees of freedom defined in the finite element model. This is an important step as each selected degree of freedom will be represented as one input/output pin in the behavioral model, and mis-selecting a degree of freedom can lead to missing data in the model, which can be a problem in case of strong coupling between different physics fields
3) Using load cases as the acceleration or the ambient temperature can also be of interest when the user wants to apply environmental factors to their simulations. In such cases, the load case will be defined as an input pin only.
4) Applying the reduction algorithm.
5) Once the algorithm completes, the tool compares the results of the finite element model with the newly generated reduced order model and reports on any accuracy issues that were found in generating the model. After the structural reduction of the FE model, the accuracy of the reduced mass and the resonant frequency can be extracted to measure the validity of the model with respect to the FEM results (Figure 21.7). The same kind of verification procedure can be performed after an electrostatic reduction, where the algorithm compares the values of the reduced capacitance against the ones extracted from the FEM/BEM solver (Figure 21.8).

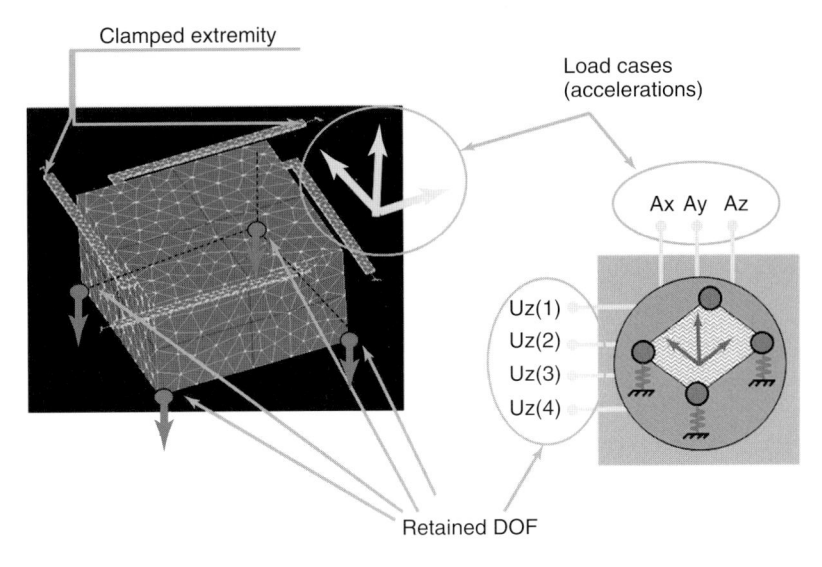

21.3 Modeling Requirements and Implementation within SoftMEMS Simulation Environment

Figure 21.6 Specifying inputs and outputs of a model.

Expected eigen frequency = 860.5332519165376
Approximated eigen frequency = 860.5359254900691 (3.106879978861649E-4% shift)
Reduced mass = 1.151938403721327E-6
Corrected mass = 1.151945561601173E-6 (6.213769610471114E-4% shift)

Figure 21.7 Mass and eigenfrequency accuracy report.

Capa [1,1] accuracy estimation:

Point	Reference	Approximation	Difference
1	1.9490860e-15	1.9490860e-15	0.0000000e+00
2	1.7708000e-15	1.7708000e-15	7.8886091e-31
3	1.6326938e-15	1.6326938e-15	0.0000000e+00

Mean absolute value = 1.7841933e-15
Maximum absolute difference = 7.8886091e-31 (0.0000%)

Figure 21.8 Capacitances accuracy report.

21.3.3
Model Output

As introduced at the beginning of Section 21.3, the design environment (simulator used, operating system, etc.) may differ from one group to another. In this context, the reduced order model extracted previously must be saved under an independent representation. The SoftMEMS Compact Model Builder saves its data in a format called *BML* (see Figure 21.9). Storing models in this format allows the user to recall the model from a different operating system without data loss and easily reformat it in a different behavioral language.

The ability of the tool to generate a model that has the same behavior on different simulation tools is also important. Toward this goal, SoftMEMS supports various EDA simulators and modeling languages such as Verilog-A or VHDL-AMS where the translation is quite obvious. But some users prefer that the model be interpreted as a SPICE equivalent circuit where a mapping of the data onto SPICE elements is necessary (Figure 21.10). Some others use simulators that have their own Application Programming Interface (APIs) and the model can be written as C-code.

For simulators such as Mathwork's Simulink, output may be in the form of a block diagram using simulink blocks (Figure 21.11). The terms of the reduced order model equation are translated into Simulink blocks.

21.3.4
Parameterization

The Compact Model Builder can handle several geometries and support different coupling fields. Generating static models can be useful to simulate the global behavior of the device, but having access to parameterized models can open new

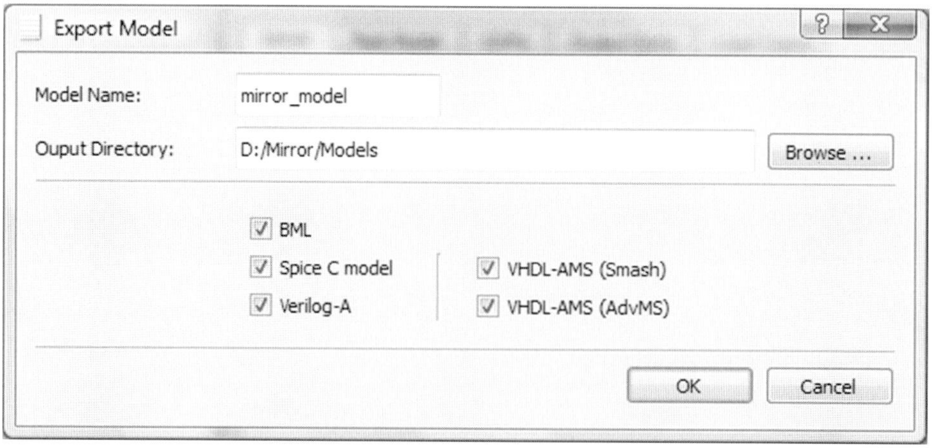

Figure 21.9 Selection of output language.

21.3 Modeling Requirements and Implementation within SoftMEMS Simulation Environment | 493

Figure 21.10 Model using SPICE elements.

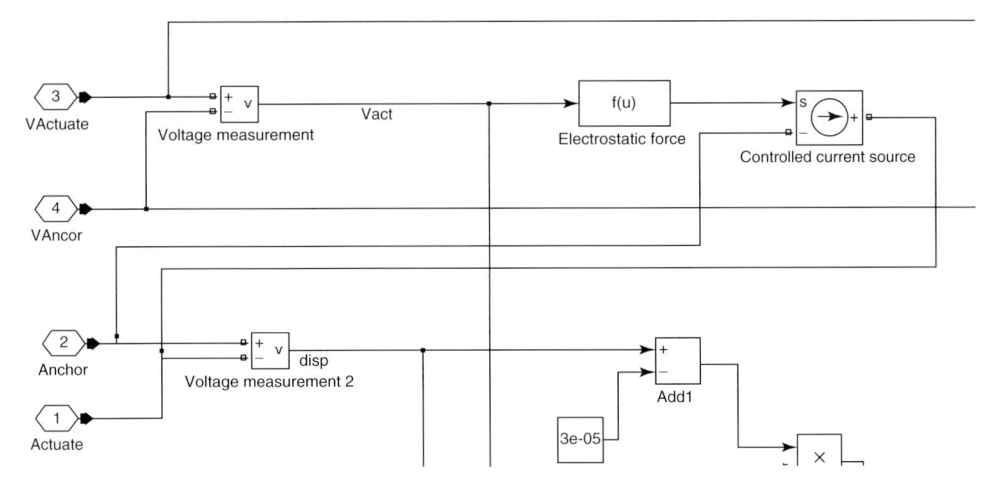

Figure 21.11 Simulink block diagram representation.

fields of investigation that could be difficult to implement in a finite element solver because of limitations on the elements used to describe the model (elements valid for small displacements) or simply because of the amount of time required to simulate the different problems.

In this context, an outer loop may be put around the model generation process to produce statistical and/or parameterized models. We have also had some success to use symbolic analysis in conjunction with reduced order modeling techniques to create model parameterizations.

The simplest example to illustrate this feature is the comb drive. It can be interesting to extract a parameterized reduced order model where the number of fingers of the comb can be a parameter for the simulation.

Two FEMs are created (Figure 21.12), one with an odd number of fingers and the second one with an even number of fingers.

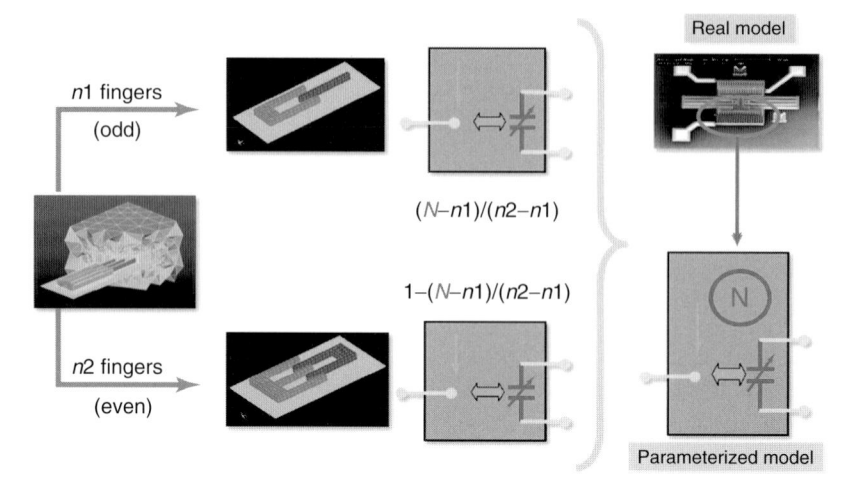

Figure 21.12 Parameterized reduced order model: models decomposition.

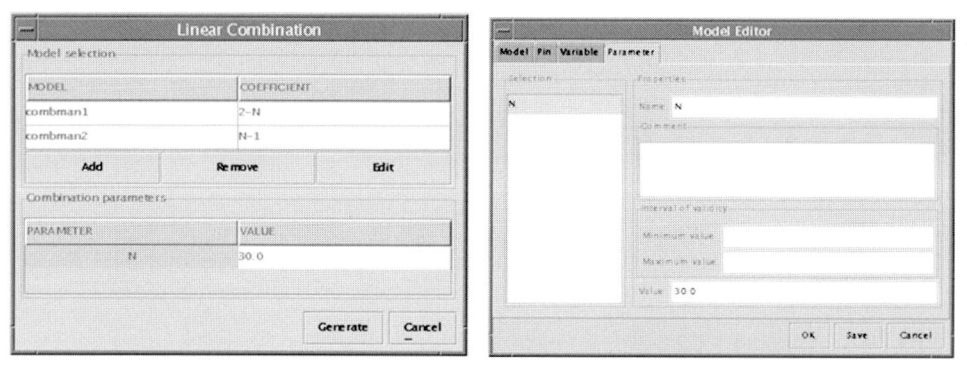

Figure 21.13 Linear combination and parameterized reduced order model.

The two FEMs represent small problems that can be easily reduced with the Compact Model Builder. By using the combination of models, the two reduced order models will be combined to generate the parameterized model (Figure 21.13).

21.4
Applications

Reduced order modeling techniques are often used to model sensors/MEMS with electronics. The underlying idea is the cosimulation of an accurate representation of the device to refine the electrical design that will be used to translate a physical

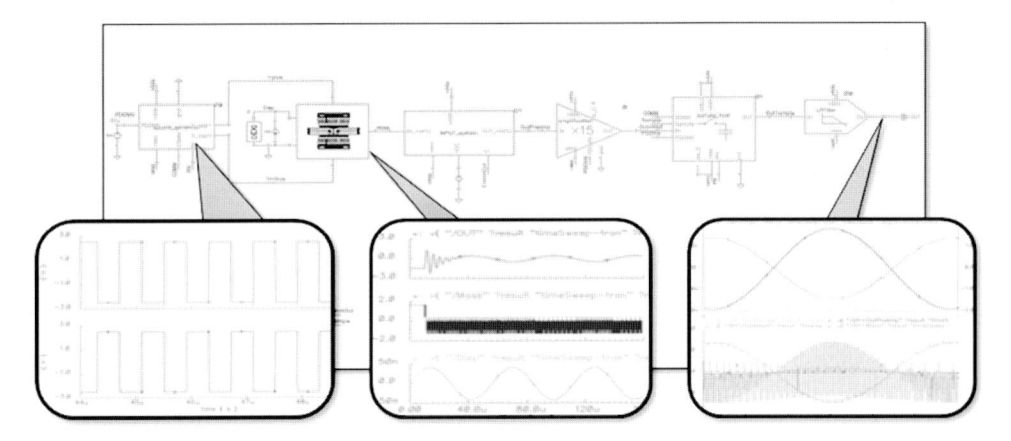

Figure 21.14 Reduced order model used in an electronic schematic.

quantity (acceleration, pressure, magnetic fields, etc.) into an analog or numerical signal or, for actuation, which signal must be applied to operate the device (switches, inkjet heads, etc.) in the given specification range.

Initially users wanted to extract models of their inertial sensors (Figure 21.14), where the predominant effect was the electrostatic–structural coupling.

The complexity of the device makes it difficult to write a behavioral model that captures all the effects that can be extracted from the FEM simulations. The structural reduced order modeling algorithm of the Compact Model Builder allows capturing a model taking into account the different displacements of the structure as a function of external loads. In addition, the electrostatic-structure algorithm is used to extract the capacitance behavior of the comb drives. Finally, the combination algorithm is used to generate the final electrostatic-structure model of the device.

Then, fluid–structure interaction algorithms [6] have been implemented to capture the harmonic response of the device and provide more accurate damping models, which were requested to design the active control of the device (Figure 21.15).

Reduced order modeling is also useful to describe semiconductor devices where modeling of thermal effects is important. Thus, users have also made use of the Compact Model Builder for creating thermomechanical models of packages [7] to be simulated with the electronics and their sensor models (Figure 21.16).

For example, in a pressure sensor package, the ambient temperature change can be transmitted through the package and the materials used for chip attachment expand at different rates because of their different coefficients of thermal expansion. This expansion can cause stress that is transmitted to the sensor dies. Heat from any ICs packaged with the sensor dies may also cause stress changes.

In our simulations, several models have been created to capture the different physical effects using different approaches. Thermal FEM analysis has been used to study the temperature repartition across the different materials of the package and to extract a lumped RC model of the temperature distribution up to the sensor die.

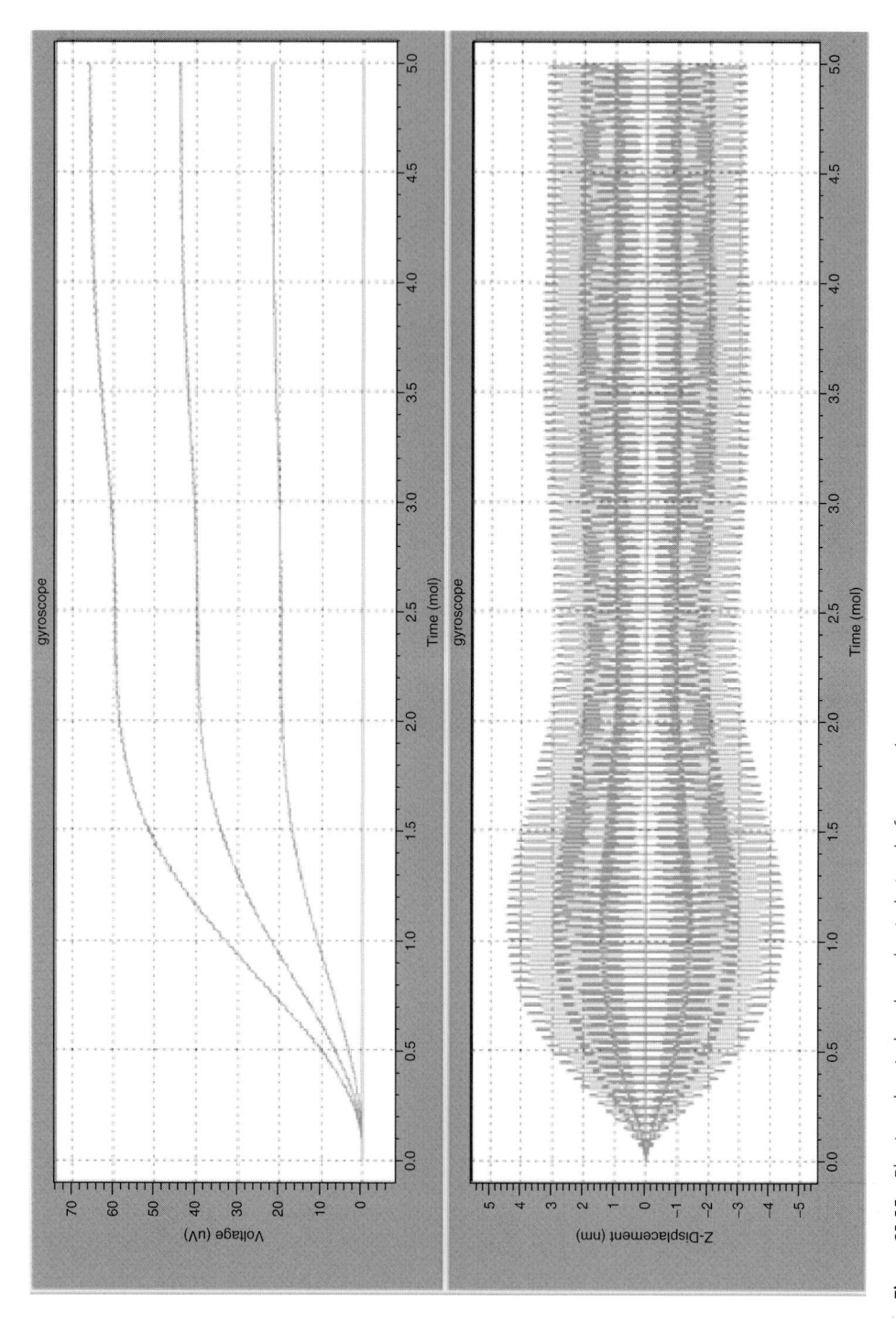

Figure 21.15 Plotting electrical and mechanical signals from a schematic.

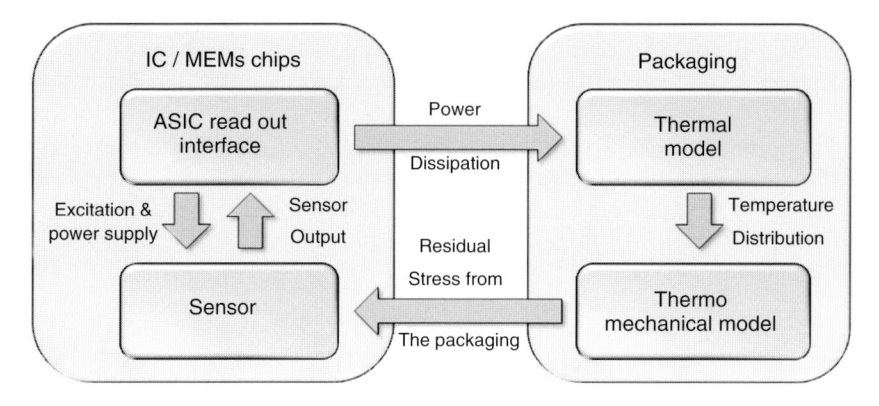

Figure 21.16 Synopsis of the thermomechanical problem.

Then, analytical models of the stress repartition close to the piezoresistive sensor elements as a function of the temperature variations have been implemented. This intermediate model allows us to use the Compact Model Builder to create reduced order models of the piezoresistive sensor elements and then apply the generated thermal stress of the pressure sensor's membrane directly to these reduced order models.

Simulation of the packaging, electronics, and sensors die (Figure 21.17) allows the designers to understand the interactions of components and determine errors. In order to cosimulate the system, models of sensors, package, and electronics must be created and simulated together.

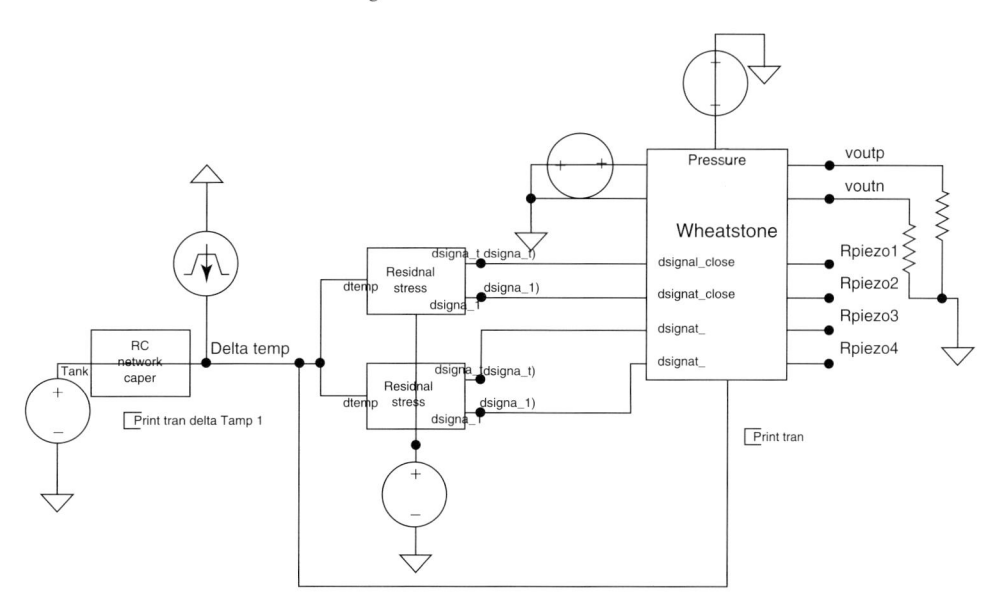

Figure 21.17 Schematic of the package and sensor including thermal effects.

Thus, engineers can place temperature signals on the package; simulate the package, sensor, and electronics; and view the output of the electronics while examining the behavior of the sensor device.

Besides the system level cosimulation aspects, reduced order models can find other application fields, such as testing of MEMS devices. Thus, the model can be used in defining test routines to be performed on the real device. Parameters and conditions to be tested can be explored during manufacturing steps. During the tests, unusual results can be compared to simulation to try to understand what is a test artifact and what is a real problem.

The goal is not yet to replace costly and difficult testing environment but to optimize the use of these equipments by identifying problems in design earlier and then to reduce the testing burden.

21.5
Conclusions and Outlook

In this chapter, we have introduced some of the problems faced by the MEMS design community to codevelop mechanical devices and electronic circuitry. We have presented some of the functionalities of our Compact Model Builder to illustrate how it can be used to solve some of these problems by bridging the gap between the mechanical (and 3D multiple energy domains) and the electrical worlds.

Industrial users are now being driven by time to market needs to quickly create system level models for use by electrical engineers to codesign electronics and sensors. Reduced order modeling is one of the important tools to enable the creation of these models. This is why the Compact Model Builder has followed the requirement trend starting from simple mechanical models to support complex devices and include various field couplings such as magnetostatic–structure interactions. But, simply adding new features will not make a tool better than another one. A number of factors, such as the ergonomics, the accessibility of key data, in addition to the basic mathematical algorithms, are involved in making a usable tool for the end user.

References

1. Rudnyi, E.B. and Korvink, J.G. (2006) *Lecture Notes in Computer Science*, **3732**, 349–356.
2. Nagel, L.W. and Pederson, D.O. (1973) SPICE (Simulation Program with Integrated Circuit Emphasis), Memorandum No. ERL-M382, University of California, Berkeley, April 1973.
3. Senturia, S.D. CAD for microelectromechanical systems. International Conference on Solid-State Sensors and Actuators (Transducers'95), Stockholm, June 26–29.
4. Open Verilog International (1996) Verilog-A Language Reference Manual, Version 1.0, August 1, 1996.
5. IEEE Standard 1076-1993.
6. Mehner, J.E., Dötzel, W., Schauwecker, B., and Ostergaard, D. Reduced order modeling of fluid structural interactions

in MEMS based on modal projection techniques. International Conference on Solid-State Sensors and Actuators, (Transducers'03), Boston, MA, June 8–12.
7. Rencz, M., Székely, V., Kohári, Zs., and Courtois, B. A method for thermal model generation of MEMS packages. International Conference on Modeling and Simulation of Microsystems (MSM'00), San Diego, March 27–29.

22
A Web-Based Community for Modeling and Design of MEMS

Peter J. Gilgunn, Jason V. Clark, Narayan Aluru, Tamal Mukherjee, and Gary K. Fedder

22.1
Introduction

The range of topics in this book and the international character of the contributors testify to the diversity of the community modeling and designing microelectromechanical system (MEMS) devices and microsystems today. The activities of the community rest on mathematical foundations that yield tools for the hierarchical simulation of lumped elements arranged in generalized Kirchoffian networks, system-level MEMS modeling through model order reduction (MOR), behavioral modeling, and detailed algorithmic and mixed-level approaches. The particular requirements of modeling linear and nonlinear microsystems ranging from inertial to electrothermal and from radio-frequency (RF) to microfluidic are described in this book through relevant system-level examples, and the software used to implement and run the model simulations are also presented. In this chapter, we describe a web-based community for modeling and design of MEMS called *Serendi-CDI* [1] that serves as a nexus for global access to these tools and the documentation and content needed to employ them effectively. The goal of the Serendi-CDI web-based design community is to foster the widespread use of structured design methodologies and tools to reduce the cost of entry into the field and unleash the creative talents of a larger pool of designers. The approach is illustrated using primitive behavioral models, or simply primitive models, and hierarchical system simulation.

22.2
The MEMS Modeling and Design Landscape

Microsystem modeling and design is a challenging endeavor. It spans many physical domains, and the physical behavior of systems at the microscale is far removed from everyday life. Deep mathematical and physical intuition must be developed in order to obtain the expertise to efficiently and economically realize working microsystems. Knowledge of the various software tools used in the design

System-level Modeling of MEMS, First Edition. Edited by T. Bechtold, G. Schrag, and L. Feng.
© 2013 Wiley-VCH Verlag GmbH & Co. KGaA. Published 2013 by Wiley-VCH Verlag GmbH & Co. KGaA.

and simulation process and the skill to recognize and overcome obstacles to achieving simulation convergence must be learned first hand. This expertise is difficult to come by and takes years to develop through common routes such as academic and industrial research. Mentors for aspiring microsystem designers are concentrated in a small number of academic institutions and high-technology companies clustered in amenable locations spread around the world.

The topography of this landscape has not been conducive to delivering human intellectual design and modeling resources to the burgeoning microsystems industry. As traditional semiconductor manufacturers recognize the functionality enhancements possible through the integration of microsystems with their circuit intellectual property (IP) [2–4], the need for capable microsystems designers and modelers is increasing, and a transformative change in the way they are developed is required.

The internet provides a means for bringing about transformative change in the topography of the microsystem design field by compressing the time and distance that separates experts from engaged and enthusiastic newcomers whose technical strength lies outside MEMS.

22.3
Leveraging Web-Based Communities

In 2012, a circuit designer who wants to integrate a gyro in his system can discuss the scale of Coriolis force with an inertial MEMS specialist on the other side of the world who can direct him to a device characterization expert in a third country, who can connect him with a fabrication expert in a fourth country, and who knows the best person in a fifth country to run the coupled simulation. This kind of web-based community networking has wrought amazing changes in human society and promises even more transformative changes in how technology develops.

The internet increases accessibility to data and enhances a person's ability to find like-minded individuals with whom to share their ideas and knowledge. Specialized online communities cover a vast number of fields and enable entry into previously marginal disciplines. These communities offer information to their members, a chance for interaction with other members to discuss and resolve issues of interest to the community, and the opportunity to contribute to the community. For the interested novice, a web-based portal can be a less intimidating passage to the workings of hard engineering fields in which the knowledge gap between expert and novice is very large.

Several established web-based communities exist with a focus on technological implementations. For example, (i) The Designer's Guide Community [5] serves as a nexus for the simulation, modeling, and design of analog, mixed-signal, and RF circuits and (ii) Nanohub [6] is a web-based resource for research, education, and collaboration in nanotechnology. Serendi-CDI is an addition to technological online communities and has a particular focus on systematic hierarchical design of microsystems and nanosystems. These communities provide source content to aid their members in understanding how the models and tools work so that the individual can develop independence and contribute improved content back to the community. This is contrasted with the way in which companies offering design and simulation materials retain source content and deter robust community engagement and collaboration.

While online communities can be leveraged to benefit the progress of a field, it must be recognized that it is only a facet to building and sustaining a successful technological community. Limitations and disadvantages to an open source, online-only collaborative model must be considered and managed by the community and its administrators. Opportunities for face-to-face off-the-record interactions between community members at events such as technical conferences are still required. Social networking and online community research shows mixed online–offline interaction improves the sociability of an online community and enhances knowledge sharing among its members [7, 8]. It is too early too provide data on the progress of the Serendi-CDI community, but the growth of the community and its level of activity will provide useful data for determining the correct interaction mix for its members.

22.3.1
Concepts of Web-Based Design

The Serendi-CDI community concept is illustrated in Figure 22.1. The tools and knowledge of the community is represented by the space in the center of the image and comprises

1) collaboratively generated wiki-style content,
2) simulation tools such as SUGARCube [6],
3) downloadable source code of models, schematics, and simulation templates, and
4) a forum for members to raise questions and discuss the community resources.

The MEMS modeling and design community member space is defined to encompass all engaged individuals from novices in MEMS design and simulation to MEMS experts that have built up a deep reserve of knowledge and experience. The developmental goals of the community are represented by the curved arrows in the user space that show novices developing into experts by absorption of material from

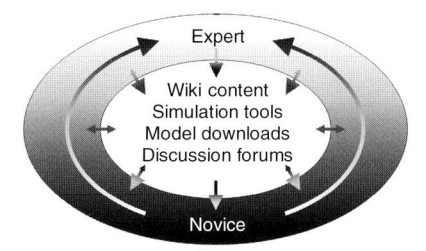

Figure 22.1 MEMS modeling and design web-based community concept.

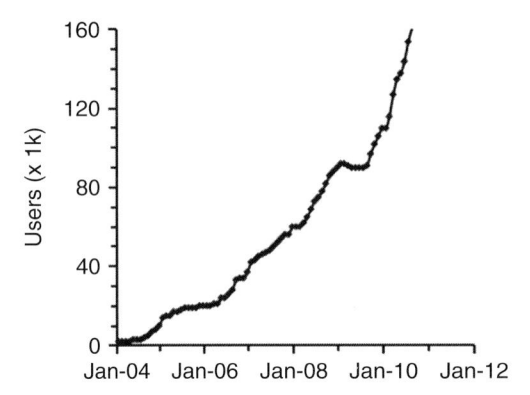

Figure 22.2 Web-based technologically focused community growth. (*Source*: Data extracted from Nanohub.)

the community and a process of gradually increasing reciprocity. Initial growth in community member numbers is fueled by word of mouth transmission. Community development and growth is later powered by the interaction of its members and their identification of areas of interest that generate activity among its members. In the ideal case, the community reaches a critical mass of engaged members at which point it becomes self-sustaining and self-directed. Figure 22.2 shows an example of web-based community growth taken from Nanohub. A very slow initial period is followed by highly nonlinear growth in member numbers, representing the viral nature of community development in the age of internet-based communications and networking.

22.3.2
Design Community Constitution

The web-based community portal is designed to facilitate member access to the content and features of interest to them and to be extensible to include new tools and features as the community grows. The key features and characteristics of the community are

1) **Open source philosophy**: Contents, such as articles, models, and code, are made available to community members under the terms of the permissive 3-clause Berkeley Software Distribution license (BSD-new or BSD-3) [9]. The goal of this licensing approach is to enable the members to make free use of the content on the site to further their design and modeling research and development while ensuring visibility of the source of the content in the web-based community is maintained. Community members are encouraged to upload their modified content through the portal to share their improvements and enhancements with the rest of the community. Community member uploads are licensed to the community under the terms of the BSD-3 license, which ensures recognition for their contributions.

2) **Wiki-style collaborative authoring**: The documentation available through the portal is authored in a collaborative way by the members of the community. Model user guides and tutorials explaining the physics of the models and the modeling process and detailed modeling information are provided in a wiki article format. Any member can create and edit an article and the edit history is maintained for review and rollback, if necessary. Article-specific discussion is provided so that the editors in remote locations can validate content changes and provide feedback to other editors. The crowd-sourcing approach to content generation and maintenance assures that topics of interest to the community remain fresh and relevant. Detailed access statistics provide a high-level snapshot of content viewing to identify hot topics in the community.
3) **Online simulation tools**: The portal provides access to tools such as SUGARCube, which allows users to select from a suite of MEMS structures and simulate them over a parameter space range. Ideally, the tools are standalone with backend processing functionality provided through the web server. This opens the power of MEMS simulation to members in situations without access to the computing and software infrastructure needed to do detailed simulations.
4) **Discussion forums**: A discussion forum is provided for the community members to bring their questions and suggestions to the community for review and development. The discussion forum provides a dynamic environment for raising and resolving issues in the community. Topics that produce widespread engagement are shunted to the wiki for further development as a collaboratively authored article.

22.4 MEMS Modeling and Design Online

Design is the process through which the materials, geometry, orientation, and interconnection of a set of components are identified to create a system that meets a set of specifications defined by a customer (Figure 22.3) [8, 10]. In the early 1980s, the digital integrated circuit (IC) industry demonstrated how the design process could be executed in a structured manner to achieve functional devices with a minimal number of fabricated prototypes [11], thus reducing time-to-market and nonrecurring engineering costs and accelerating yield ramps to volume production. The analog IC industry adopted this approach with success, and since 1995, MEMS designers have tried to follow suit [12]. However, the wide range of disciplines and the distributed and nonlinear nature of many MEMS components has made this implementation difficult. The web-based design community addresses the analysis aspect of the design flow through circuit-based design methodologies and schematic system representations as delineated by the dashed box in Figure 22.3.

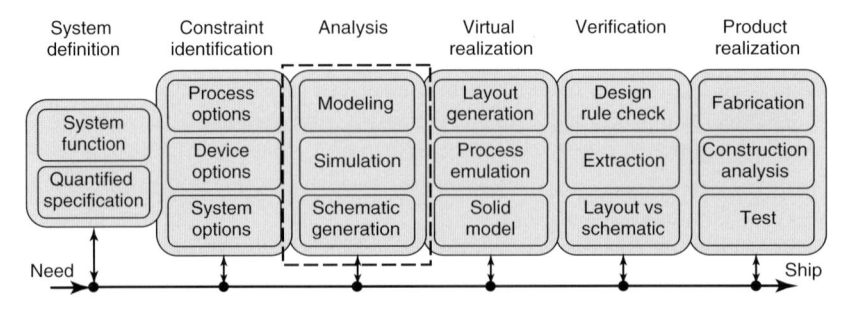

Figure 22.3 Iterative MEMS design flow with dense feedback, from customer requirements to finished product. The dotted line delineates the scope of the web-based community for MEMS design.

22.4.1
System-Level Modeling of MEMS

Circuit-based simulation of MEMS devices in a hierarchical system-level model requires the behavior of the MEMS devices to be codified in a way that is readable and executable by a simulator and interoperable with electrical components for cosimulation. The preferred means of doing this is through an analog behavioral model written in an analog hardware description languages (AHDLs) such as Verilog-AMS [13] or SystemC [14]. The MEMS model can be built up of composable primitive models of lumped elements through macromodels derived from finite element analysis (FEA) or by semianalytical reduced order models (ROMs) [15]. Owing to the specificity of FEA-based macromodels and ROMs, there is less room for model reuse and more issues with interoperability, as described in Chapter 1. FEA-based models and ROMs also defy hierarchical modeling, a central tenet of electronic circuit design, and the parameterization necessary for design space exploration; so these approaches have been eschewed in favor of parameterized primitive models, for the purposes of the web-based design community.

Hierarchical design allows a system to be broken down into smaller and smaller components until a level is reached, the primitive-level (the lowest level in the hierarchy), at which only a handful of multiply instantiated components are found. Primitive models can still be very complex but are more tractable and can be more physically complete than behavioral models of higher level components which, by the nature of their derivation, are coarse approximations that neglect more physical effects and characteristics.

The primitive models referred to in this chapter are from NODAS (NOdal Design of Actuators and Sensors), a library of parameterized models developed at Carnegie Mellon University [15] and currently encoded in Verilog-AMS. NODAS models span multiple physical domains and have been formulated as interoperable, composable and conservative systems to allow accurate system simulation given arbitrary interconnection in Kirchoffian networks such as those described in Chapter 2. NODAS comprises behavioral models for anchor, beam, electrostatic gap, and plate and sources for each nature. Linear, nonlinear, 2D, and 3D versions of elements exist and phenomena such as damping [16], contact physics [17], and the multimorph effect [18] have been modeled. The models have been validated against FEA and experimental data in a number of systems [16, 17]. Composable models are also available using SUGAR [19] and MEMS+ [20], and MEMS Pro/MEMS Xplorer [21].

22.4.2
Web-Based Community Conventions

Models uploaded to the web-based design community must follow nomenclature and sign conventions to ensure interoperability and ease of use and reuse. These are briefly explained. In Verilog-AMS, signal types, called "natures," that are related through the type of physics used to define them, are grouped together in a "discipline." For each discipline, there is a flow nature, representing the through variable (current in the electrical discipline), and a potential nature, representing the across variable (voltage in the electrical discipline). The value of a nature at a port (or terminal) of a module (i.e., an instance of a behavioral model) is read out by the modeler using an access function. Standard Verilog-AMS conservative disciplines, natures, and access functions (DNA) are used [22], but a MEMS-specific extension set (Table 22.1) is used in addition. In 2010, a survey of Verilog-AMS behavioral modelers in different institutions was made to define a base for the MEMS-specific DNA of the web-based community. To convert models from one nomenclature convention to another, only textual replacements are required. A detailed explanation for the choice of the potential nature for kinematic disciplines is given in Chapter 1.

In the behavioral models, displacement and angular displacement are used as the potential natures at module ports as explained by Fedder and Jing [15]. Velocity and acceleration natures are used internally in the models. The use of a velocity nature allows acceleration to be computed as the derivative of velocity that improves scaling and reduces the condition number of the system matrix and the number of equations thereby speeding up convergence and increasing its probability as explained by Iyer et al. [23]. Kinematic translational ports with acceleration natures and kinematic rotational ports with angular velocity natures are needed for inertial system simulations. The definition of conservative disciplines for these natures, rather than signal flow, follows Iannacci [24]. As the web-based design community

Table 22.1 Verilog-AMS behavioral model DNA of the web-based design community.

Discipline	Potential nature (access)	Flow nature (access)
mems_kinematic_translational	MEMS_Displacement (d)	MEMS_Force (f)
mems_velocity	MEMS_Velocity (vel)	MEMS_Force (f)
mems_acceleration	MEMS_Acceleration (acc)	MEMS_Force (f)
mems_kinematic_rotational	MEMS_Angular_Displacement (phi)	MEMS_Moment (m)
mems_angular_velocity	MEMS_Angular_Velocity (omega)	MEMS_Moment (m)
mems_angular_acceleration	MEMS_Angular_Acceleration (aacc)	MEMS_Moment (m)

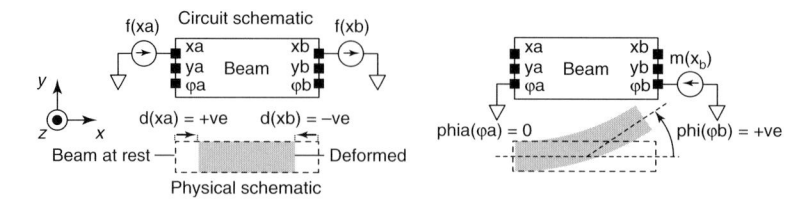

Figure 22.4 Brief examples of sign conventions for flows and potentials at module ports. Positive flows into a port act to produce positive potentials with respect to the local reference.

grows, this will expand through consensus. Other possible disciplines are fluidic [25], chemical [26], and optical [27], to name a few.

The definition of reference directions for MEMS modules is critical to interoperability but not straightforward. Unlike electrical discipline modules that have associated reference directions [28], MEMS modules have multiple ports that span disciplines and energy that can move internally between disciplines (for example, a force at one end of a beam can lead to a moment at the other end). It is not possible to associate a positive flow with an input port defined as positive, and a negative flow with an output port defined as negative. For behavioral models of the web-based community, the convention is used that positive flows into a port act to produce a positive potential with respect to the local reference (rotations follow right hand rule). More details on this convention is given in Chapter 1, but some examples of forces and moments acting on beams are provided in Figure 22.4 to support the remaining content of this chapter.

Definitions of structure orientation in space and reference frame must also be consistent among behavioral models to ease user deployment. The chip on which the devices are laid out is used as the reference frame for the behavioral models described here. The structures are oriented so the longitudinal axis of a beam is aligned along the x-axis with the "a" port at the left edge of the beam. Electrostatic gaps are oriented so that the gap is traversed in the positive y-direction as one moves from the lower plate to the upper plate, the "a" ports are at the left edge of the gap. Designers should check model documentation for structure orientation before using a model. Each 2D model has one orientation angle and each 3D model has three angles (the Euler angles) that describe structure orientation.

22.5
Encoding MEMS Behavioral Models

The goal of this section is to illustrate the Serendi-CDI makes primitive models available to its members and to show how the primitive models are composed into a complex structure with emergent behavior that becomes apparent through simulation. The open source philosophy of Serendi-CDI allows the members to see inside the model black box to the physics encoded there and modify or add to it, if necessary. The primitive model example of a nonlinear elastic beam is presented

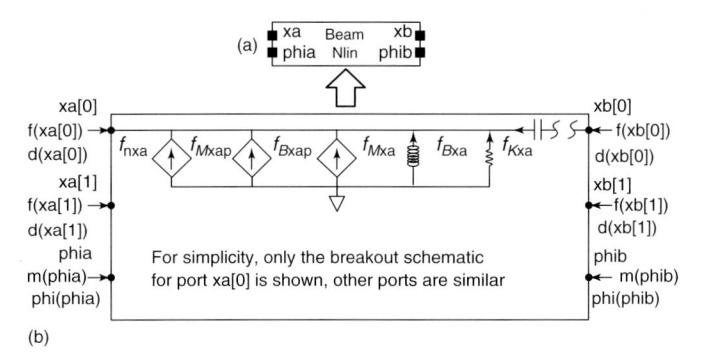

Figure 22.5 (a) NODAS circuit symbol and (b) breakout schematic showing the physics elements encoded in a 2D nonlinear beam behavioral model for the x-directed port on the a side of the beam. The "$f \rightarrow v$ convention" is used so familiar electrical symbols represent mass, damping, and spring terms. Note, there are analogous terms to f_{nxa}, f_{Mxap}, and so on, connected to the x-directed port on the b side of the beam, but they are not shown here because of space considerations.

in some detail and used along with other primitive models (plate, comb drive, anchor) to build a coupled microresonator. The potential to include other physical behaviors through model extensions is discussed briefly.

Behavioral models map the stimulus at the port of a module to the response at that port. Either port nature can be the stimulus and either can be the response, it is up to the modeler to determine which is which based on the characteristics of the system. The constitutive relations are formulated and translated into contribution statements

$$\text{response} < + g(\text{stimulus}) \tag{22.1}$$

that encode the mapping. Detailed information about the Verilog-AMS modeling syntax can be found in Kundert and Zinke [22] and Fitzpatrick and Miller [29]. Details of the structured modeling process for MEMS can be found in Jing [16] and Wong [18]. The 2D nonlinear beam is used to illustrate the modeling process.

22.5.1
Inside the Black Box – Nonlinear Beam

From the designer's perspective, a module is a black box, and all that is required is knowledge of how to connect the ports and the significance of the parameters at their disposal. On the other hand, the modeler must have intimate knowledge of the internal structure of the model, so we begin with a breakout schematic that makes explicit all the physics to be encoded. More physics typically leads to more accurate simulation results, but may lead to convergence and simulation time issues, so the modeler must consciously balance these tradeoffs.

Figure 22.5 shows a breakout schematic for the *x*-directed port on the "a" side of a 2D, homogeneous and nonlinear mechanical beam with rectangular cross-section.

The stimuli in the model derivation are the translational and rotational displacements of the ports (d and phi, respectively) which lead to the forces and moments (f and m, respectively) flowing into the ports as the responses. The model incorporates responses due to inertia, f_M and m_M, Couette damping due to relative motion in fluid, f_B and m_B, elasticity, f_K and m_K, and axial stress due to beam bending f_n, which leads to variable bending stiffness that drives the nonlinearity of the beam. The "p" subscript denotes responses coupled from the displacements at other ports. The "$f \rightarrow v$ convention" [30] is used so that masses are represented as inductors and springs as capacitors. The kinematic translation ports are encoded as buses so that they are represented by x[0] in the x-direction and x[1] in the y-direction. As the model is 2D, only rotation about the z-axis is included and only one pair of kinematic rotational ports phia and phib are needed. All external and internal forces and moments are modeled as flowing into a port.

At the xa[0] port (i.e., x-directed kinematic translation port on the "a" side of the beam)

$$\sum f = f(xa[0]) + f_{nxa} + f_{Kxa} + f_{Bxa} + f_{Mxa} + f_{Bxap} + f_{Mxap} = 0 \qquad (22.2)$$

which gives the behavioral response to encode in Verilog-AMS as

$$f(xa[0]) <= -f_{nxa} - f_{Kxa} - f_{Bxa} - f_{Mxa} - f_{Bxap} - f_{Mxap}. \qquad (22.3)$$

Expressions for the behavioral response are derived similarly for the other ports. The form of the expressions for the various internal forces and moments are determined by considering the response of the beam to displacements at its ends using matrix structural analysis applied to the Euler–Bernoulli beam equation with no distributed load

$$EI_z \frac{\partial^4 u_y}{\partial x^4} - \frac{f_{axial}}{wt} \frac{\partial^2 u_y}{\partial x^2} = 0 \qquad (22.4)$$

where,

$$u_y(x) = s_1(x)\, d(xa[1]) + s_2(x)\, d(xb[1]) + s_3(x)\, \text{phi}(\text{phia}) + s_4(x)\, \text{phi}(\text{phib}). \qquad (22.5)$$

is the transverse displacement of the beam neutral axis from its rest position at a distance x from the beam reference edge for a given set of shape functions s_i, w is the beam width, t is the thickness, E is the Young's modulus, $I_z = w^3 t/12$ is the bending moment of inertia about the z-axis, and $f_{axial} = f_{Kxa} + f_{nxa}$ is the axial stress. Derivation of the expressions using a set of shape functions s_i is beyond the scope of this work; however, details can be found in Przemieniecki [31], Senturia [30], and Jing [16]. A subset of the nonlinear-beam stimulus-response mappings are stated here for the purpose of illustration.

$$f_{Kxa} <= \frac{Ewt}{l}\left(d(xb[0]) - d(xa[0])\right), \qquad (22.6)$$

$$f_{Bxa} <= -\frac{\eta l(w+b)}{3 g_z}\text{vel}(xa[0]), \qquad (22.7a)$$

$$f_{Bxap} <= -\frac{\eta l(w+b)}{6 g_z}\text{vel}(xb[0]), \qquad (22.7b)$$

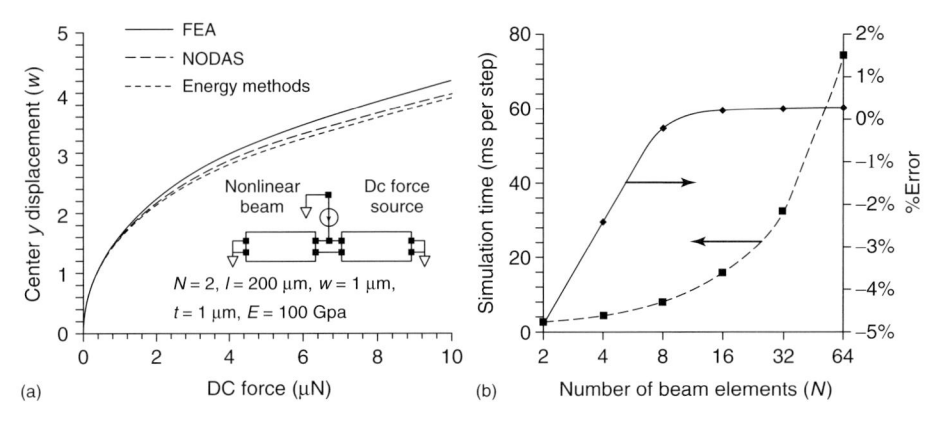

Figure 22.6 (a) Two-dimensional, homogeneous, nonlinear and fixed-fixed beam simulations and theoretical prediction comparison. (b) Error trend as a function of the number of beam elements N.

$$f_{Mxa} <= -\frac{\rho wtl}{3} \text{acc}(xa[0]), \tag{22.8a}$$

$$f_{Mxap} <= -\frac{\rho wtl}{6} \text{acc}(xb[0]), \tag{22.8b}$$

$$f_{nxa} <= \frac{Ewt}{2l} \int_{d(xa[0])}^{1+d(xb[0])} \left(\frac{\partial u_y}{\partial x}\right)^2 \partial x = \frac{Ewt}{2l}\Delta, \tag{22.9}$$

where l is the length, b is the bloat or effective width increase due to damping edge effects, ρ is the density of the beam and η is the viscosity of air.

The nonlinear response of the beam arises from the dependence of the bending stiffness on axial stress. Responses are computed in the local beam frame and transformed to the chip frame using a Euclidean rotation transformation. For example, the local frame force responses at the y-directed port on the a side of the beam is

$$f_{Kya} <= -\left(\frac{12EI_z}{l^3} + \frac{f_{axial}}{5l}\right)(d(xb[1]) - d(xa[1])) + \left(\frac{6EI_z}{l^2} + \frac{f_{axial}}{10}\right)(\text{phi}(\text{phib}) - \text{phi}(\text{phia})). \tag{22.10}$$

22.5.2
Behavioral Model Performance – Nonlinear Beam

NODAS DC simulation results for a fixed–fixed beam modeled using two of the nonlinear beam elements described earlier, and a DC force source are shown in Figure 22.6a. The trend in error percentage, (NODAS – FEA)/FEA, and simulation time for various numbers of elements is shown in Figure 22.6b. Consistent with trends in the number of mesh elements for FEA accuracy, the NODAS error

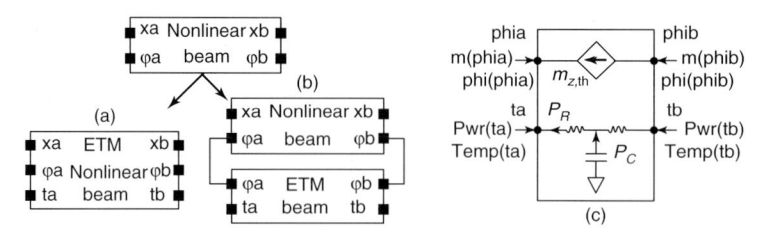

Figure 22.7 Electrothermal multimorph model extension options. (a) New combined module and (b) new ETM module connected in parallel to nonlinear beam. (c) ETM breakout model.

decreases as the number of elements N increases, but the simulation time per step increases. This tradeoff must be respected by the designer who has to have an understanding of the degree of simulation accuracy needed to determine how well design specifications are met.

22.5.3
Behavioral Model Extensions

The member community is encouraged to modify or add extensions to the models available through the web portal to foster community growth and to maintain its relevance. Extensions are needed for new physics and phenomena (e.g., Casimir, electrothermal, or chemisorption effects) or to accommodate new processes or material systems (e.g., inhomogeneous multimorph beams or piezoelectric materials). Extensions to the model physics can be done in two ways: (a) new ports can be added to the module and new branches can be added internally to the breakout model, as shown in Figure 22.7a or (b) new models can be created and connected in parallel to the main model through the appropriate ports, as shown in Figure 22.7b. As an example, an electrothermomechanical (ETM) multimorph extension to the 2D nonlinear beam model is described.

Heat power P flowing through a beam with inhomogeneous transverse cross-section due to temperature gradients Temp(tb) − Temp(ta), typically generated by Joule heating in other elements of the device, give rise to bending moments $m_{z,th}$ in the beam about the z-axis. This situation is described by the breakout model in Figure 22.7c. The bending moment was first described for bimorphs by Timoshenko [32], expanded to MEMS multimorphs by Iyer [33], and encoded in a behavioral model by Wong [18]. The thermally generated bending moment is

$$m_{z,th} = \beta \left(\frac{\text{Temp (ta)} + \text{Temp (tb)}}{2} - T_{\text{gnd}} \right), \tag{22.11}$$

where T_{gnd} is the ground temperature and β is a gain parameter that depends on the material properties, the geometric parameters, and the relative positions of the materials in the beam [34].

The decision whether to follow path (a) from Figure 22.7 or path (b) depends on the general relevance of the extension. In a schematic such as that representing

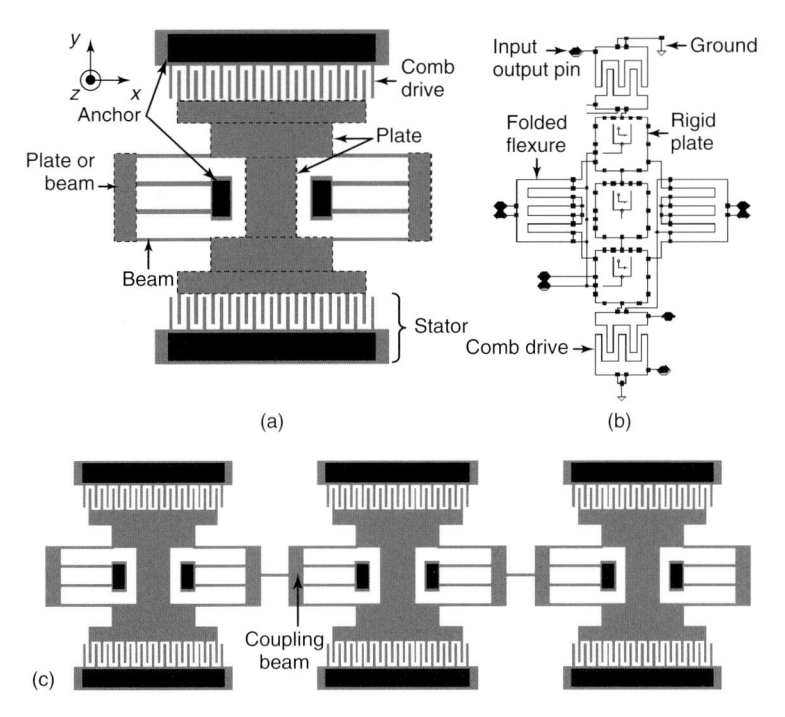

Figure 22.8 (a) Plan view schematic of a microresonator partitioned into four unique primitive models whose mechanical behavior is encoded for simulation using Verilog-AMS. (b) NODAS schematic of a microresonator comprised of composed comb drive and folded flexure models and rigid plate primitive models. (c) Plan view schematic of a micromechanical filter made from three microresonators coupled by an elastic beam.

the two-dimensional resonant mass structure shown in Figure 22.8a, the beams have a constant cross-section for which $\beta = 0$ and therefore $m_{z,th} = 0$. If all beam elements had thermal ports, the complexity of interconnection would be increased for no added benefit to simulation accuracy, so a global change to the nonlinear beam model would be undesirable and the creation of an ETM module would be more sensible.

22.5.4
Behavioral Model Composability

The power of composability of behavioral models is presented in this section using the example of an in-plane folded flexure microresonator in a band-pass filter. A single microresonator is shown in plan view schematic in Figure 22.8a with individual plate elements delineated by dashed lines. The microresonator is actuated capacitively by a pair of comb drives that each consist of two sets of laterally offset parallel beams, called *fingers* that are separated in an orthogonal direction. Typically, one set of fingers in a comb drive is attached to a mechanical anchor to form the stator, while the other set is part of a rotor or moving element.

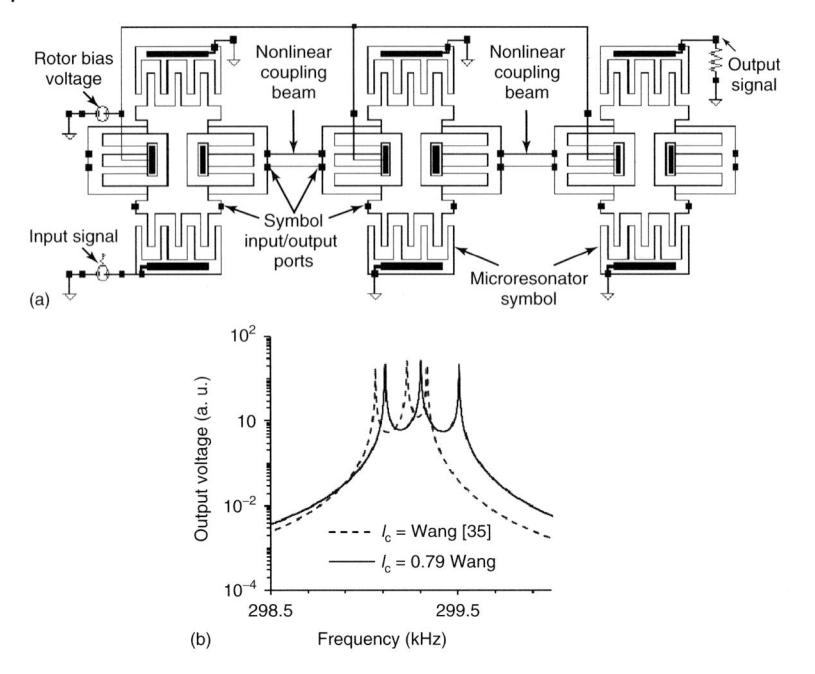

Figure 22.9 (a) Circuit schematic of a band-pass filter formed from composed resonator models coupled by 2D nonlinear beam primitive models. (b) NODAS simulated band-pass frequency response for coupling beams with lengths l_C as given by Wang and Nguyen [35] and 0.79 times the length given by Wang.

The rotor consists of a set of plates and beams and is suspended above a substrate by some form of flexure connected to an anchor. Voltages are applied to the microresonator through the anchor connections. Two matched folded flexures are used in this microresonator and are designed to give the greatest in-plane motion in the y-direction. This complex structure can be viewed as being made of beams, plates, comb drives, and anchors. The comb drives can be considered as primitive models, but could also be decomposed further to beams and electrostatic gaps.

There are many ways to compose a circuit schematic of the microresonator. One approach is to compose the comb drive and folded flexure of primitive models and generate symbol views of these composed models. The comb drive model and the folded flexure models are then instantiated with rigid plate primitive models that form the shuttle proof mass as shown in Figure 22.8b. It is also possible to make a composed model of the shuttle from rigid plate primitive models. The choice is in the hands of the designer, but should depend on the reusability of the composed model. The designer can generate a parameterized symbol view of the microresonator and quickly couple together three microresonators with elastic beams to simulate the band-pass filter structure shown in Figure 22.8c.

The circuit schematic of the band-pass filter is shown in Figure 22.9a. The symbolic view of the composed microresonator model is chosen so the circuit schematic has a form reminiscent of the actual structure, which serves a mnemonic function. The filter and its behavior were reported by Wang and Nguyen [35] and simulated using behavioral modeling by Fedder and Jing [15]. The microresonators are modeled using 2D nonlinear beam elements that form the flexures, the comb fingers, and the coupling beams. No constraints are placed on the motion or behavior of the primitive models. The composed models are easily parameterized to enable a sweep of the design space or to estimate the effects of process variation on device performance, for example. The simulated passband behavior of the filter and the resonant peaks of each microresonator are shown in Figure 22.9b. The dimensions of the microresonator are taken from Wang and Nguyen [35] but the output is for indication only as it is simplified in comparison by taking the voltage across a 1 MΩ resistor. Responses are shown for two cases: (i) coupling beam length $l_c = 75.2$ µm as given by Wang and (ii) $l_c = 59.7$ µm, which represents a coupling spring with twice the spring constant as that used by Wang. The second case shows a more symmetric passband. Composability allows simple elements to be arbitrarily interconnected to form myriad complex devices whose emergent behaviors, like the form of the separation of the resonances in the coupled microresonator example, may be difficult to predict before simulation.

22.6
Conclusions and Outlook

Structured hierarchical design of MEMS using composable, interoperable and primitive behavioral models of beams, electrostatic gaps, plates, and anchors can be delivered effectively to a global community of designers with varying levels of expertise through a web-based design community. An open-source collaborative environment can transform the field by lowering the barrier of entry to the field, providing the resources for novices to develop into expert designers. Such a web-based community is a valuable facet of the community-building process that also includes personal contact between members in the real world. This chapter has detailed the structure of the Serendi-CDI web-based community and the design and modeling resources available to members of the community. The scope of disciplines applied in MEMS design and the range of physics utilized for describing MEMS have by necessity constrained this chapter to provide only an overview of the process of modeling and designing them. However, the shape of the process has been outlined and the key considerations in building behavioral models for MEMS design have been presented in a way to provide confidence to the reader to join this exciting field.

References

1. Serendi-CDI *www.serendi-cdi.org* (accessed 25 May 2011).
2. Johnson, R.C. (2010) IBM, WiSpry teamed on tunable RF MEMS, *EE Times*, June 28 ed.
3. Ramanathan, R.N. and Willoner, R. (2006) *Silicon Innovation: Leaping from 90 nm to 65 nm*, White Paper, Intel Corporation, 7 pp.
4. International Technology Roadmap for Semiconductors (2009) Annual Report International Technology Roadmap for Semiconductors, Executive Summary.

5. The Designer's Guide, www.designers-guide.org/index.html (accessed 31 January 2011).
6. Nanohub www.nanohub.org (accessed 31 January 2011).
7. Matzat, U. (2010) *American Behavioral Scientist*, **53** (8), 1170–1193.
8. Haythornthwaite, C. and Kendall, L. (2010) *American Behavioral Scientist*, **53** (8), 1083–1094.
9. Open Source Initiative – the BSD License www.opensource.org/licenses/bsd-license.php (accessed 31 January 2011).
10. Mukherjee, T., Fedder, G.K., Ramaswamy, D., and White, J. (2000) *IEEE Transactions on Computer-Aided Design of Integrated Circuits and Systems*, **19** (12), 1572–1589.
11. Mead, C. and Conway, L. (1980) *Introduction to VLSI Systems*, Addison-Wesley Publishing Company, Reading.
12. Fedder, G.K. (2006) in *System-Level Simulation of Microsystems, in MEMS: A Practical Guide to Design, Analysis and Applications* (eds J. Korvink and O. Paul), William Andrew, Inc., Norwich, pp. 187–228.
13. Accellera Verilog Analog Mixed-Signal Group www.vhdl.org/verilog-ams (accessed 31 January 2011).
14. Open SystemC Initiative www.systemc.org (accessed 31 January 2011).
15. Fedder, G.K. and Jing, Q. (1999) *IEEE Transactions on Circuits and Systems II: Analog and Digital Signal Processing*, **46** (10), 1309–1315.
16. Jing, Q. (2003) Modeling and simulation for design of suspended MEMS, Carnegie Mellon University Dissertation.
17. Wong, G.C., Tse, G.K., Jing, Q., Mukherjee, T., and Fedder, G.K. (2003) Accuracy and composability in NODAS. Proceedings of the 2003 International Workshop on Behavioral Modeling and Simulation San Jose, CA, pp. 82–87.
18. Wong, G. (2004) Behavioral modeling and simulation of MEMS electrostatic and thermomechanical effects, Carnegie Mellon University Masters Report.
19. Clark, J.V. and Pister, K.S.J. (2007) *Journal of Microelectromechanical Systems*, **16** (6), 1524–1536.
20. Coventor www.coventor.com/mems-ic/mems-product-design-platform.html (accessed 31 January 2011).
21. softMEMS www.softmems.com/products.html (accessed 23 January 2012).
22. Kundert, K. and Zinke, O. (2004) *The Designer's Guide to Verilog-AMS*, Kluwer Academic Publishers, Boston, MA.
23. Iyer, S., Jing, Q., Fedder, G.K., and Mukherjee T. (2001) Convergence and speed issues in analog HDL model formulation for MEMS Technical Proceedings of the 2001 International Conference on Modeling and Simulation of Microsystems, Hilton Head, SC, pp. 590–593.
24. Iannacci, J. (2007) Mixed-domain simulation and hybrid wafer-level packaging of RF-MEMS devices for wireless applications. Università degli Studi di Bologna Dissertation.
25. Wang, Y., Lin, Q., and Mukherjee, T. (2006) *IEEE Transactions on Computer-Aided Design of Integrated Circuits and Systems*, **25** (2), 258–273.
26. Cenni, F., Mir, S., and Rufer, L. (2009) Behavioral modeling and simulation of a chemical sensor with its microelectronics front-end interface. IWASI 2009. 3rd International Workshop on Advances in Sensors and Interfaces, Bari, Italy, pp. 92–97.
27. Briere, M., Carrel, L., Michalke, T., Mieyeville, F., O'Connor, I., and Gaffiot, F. (2004) Design and behavioral modeling tools for optical network-on-chip. Design, Automation and Test in Europe Conference and Exhibition, Paris, France, pp. 738–739.
28. Desoer, C.A. (1969) *Basic Circuit Theory*, McGraw-Hill, New York.
29. Fitzpatrick, D. and Miller, I. (1997) *Analog Behavioral Modeling with the Verilog-A Language*, Kluwer, Boston, MA.
30. Senturia, S. (2000) *Microsystem Design*, Kluwer, Boston, MA.
31. Przemieniecki, J.S. (1985) *Theory of Matrix Structural Analysis*, Dover Publications, Mineola.
32. Timoshenko, S.P. (1925) *Journal of the Optical Society of America*, **11** (3), 233–255.

33. Iyer, S. (2003) Modeling and simulation of non-idealities in a z-axis CMOS-MEMS gyroscope, Carnegie Mellon University Dissertation.
34. Gilgunn, P.J. (2010) SOI-CMOS-MEMS electrothermal micromirror arrays, Carnegie Mellon University Dissertation.
35. Wang, K. and Nguyen, C.T.-C. (1997) High-order micromechanical electronic filters. Proceedings of the IEEE MEMS Workshop, Nagoya, Japan, pp. 25–30.

Index

a

Ab initio (first principles) simulations 461–462
Acceleration methods of computing Gramians 82–83
Acceleration sensitivity 304
Accelerometer 148–149, 360, 474
– seismic perturbations detection of using 363–370
– – battery modeling 368–369
– – digital controller in SystemC 365–366
– – embedded software, cross-compiled GNU GCC application for MIPS 369
– timed dataflow model of 359–361
– – electrical linear network model of 361–363
Across quantities 26
Additive white Gaussian noise (AWGN) 366
Algebraic equations 95, 97–102
Analog and mixed signal (AMS) 357–374
Analog hardware description languages (AHDLs) 4, 11–12, 506
– simulation capabilities 13–14
ANSYS®, MOR for 343, 425–437
– Arnoldi algorithm model 427
– open problems 432–436
– – tonpilz transducer 435
– practice-oriented research 426–429
– programming issues 429–432
– – obtaining system matrices from ANSYS 430–431
– – partial solve (PSOLVE) 430
– – solvers 431–432
– Second Order ARnoldi (SOAR) 428
Antiresonance 327
Application Programming Interface (APIs) 492

Arnoldi algorithm 64, 216, 279, 427–435
Arrayed bent-beam thermal actuator 128
Asymptotic waveform evaluation (AWE) 60
Automated model order reduction methods 9

b

Balanced realization 70
Balanced system 80
Balanced truncation, motivation of 78
Balancing transformation 78–80
– computation 81–82
Band Arnoldi process 65
Basic electrothermal actuator 127
Basis 57–59
– definition 58
– equivalent definition of 59
BCA (Bit-Cycle Accurate) compliant 365
Beam
– coupled electrothermal model of 138
– electrothermal model of 131–134
– thermomechanical model of 134–136
Behavioral compact models 30
Behavioral model composability 513–515
Behavioral model performance 8
– encoding MEMS behavioral models 508–515
– – behavioral model composability 513–515
– – behavioral model extensions 512–513
– – nonlinear beam 511–512
Bent-beam actuators 128
Biot number 132
Block–Arnoldi algorithm 213, 216–220
Black Box, nonlinear beam 509–511
Block Krylov Subspace 59–60
Block moments 74
Body load contribution vector 297–298

System-level Modeling of MEMS, First Edition. Edited by T. Bechtold, G. Schrag, and L. Feng.
© 2013 Wiley-VCH Verlag GmbH & Co. KGaA. Published 2013 by Wiley-VCH Verlag GmbH & Co. KGaA.

Boolean equations 96, 102–104
Boundary model 173–174
Branch 12

c

Cadence simulation 191
Cadence Virtuoso's plotting capabilities 420
Cantilever model 277
Carbon nanotube (CNT) models 448–450
Chip/adhesive/package model 153–155
Circuit model or network model 8
Circuit-level design for RF MEMS devices 335–355
– extraction of reduced order model 340–345
– – extraction procedure 343–345
– – input function, handling nonlinearities in 341–343
– – second order ODE systems 341
– microswitch 351–354
– reduced order models application in 335–355, See also Vibrating devices
– – model equations for 337–340
– – Rayleigh mode preserving damping 338
– – stress stiffening 338
Circuit-level simulation 348
Closed loop force-feedback loop 378
Coefficient of thermal expansion (CTE) 130–131, 152
Combdrive 265
Commercial MEMS design environment 461–481
– accelerometer 474
– benchmarks 474–476
– coupled package-device modeling 478–480
– fluid damping 477
– inertial gyroscope 474–476
– IntelliSense's design methodology 467–470
– IntelliSuite 470–474, See also individual entry
– model order reduction implementations in 461–481
– – Ab initio (first principles) simulations 461–462
– – layout-based design (bottom-up design) 463–464
– – schematic or component-based design (top-down design) 462–463
– – system model extraction (SME) 464–465
– – system simulation 467
– – technology CAD (TCAD) 462
– – verification 465–467
Communicating synchronous finite state machines (CSFSMs) 366
Compact model builder 483
Compact model derivation 29–32
– behavioral compact models 30
– compact modeling of MEMS 30–31
– equivalent network models 30
– finite network (FN) models 29–30
– mathematical model order reduction (MOR) 32
– mixed-level modeling (MLM) 31
– Physics-based compact models 31
Compact modeling of MEMS 8, 30–31, 192–194, See also RF-MEMS devices, compact modeling of
– behavioral compact models 30
– equivalent network models 30
– Physics-based compact models 31
Complementary variables 117
Complex matching 328–330
– consequence of matching 331–332
– real matching 330–331
Complex matrix-valued function 62
Complex Navier-Stokes equations 170
Component mode synthesis (CMS) 427
Component, terminology 8
Composable model libraries 14–15
COMSOL 454–457
Constraint mechanical systems (CMS) 99
Controllability 68–69
– controllability (of a state) 68
– controllability Gramian 68
Coriolis forces 293, 299–300, 397
Correction factors 175
Cosimulation 110–112
– cosimulation interface 112
– coupling algorithms 111–112
Cosimulation aspects 357–374
Coupled electrothermal model, equivalent circuit of 137–138
– of a beam 138
Coupled multiphysics microsystems 4–6
– multiscale modeling and simulation hierarchy 6
– sampling of 5
Coupled package-device modeling 478–480
Coupling algorithms 111–112
CoventorWare™ 149–150
Current 101

d

3D parametric-library-based MEMS/IC design 407–423
– 3D parametric library for MEMS design–MEMS+® 409–415
– – 3D design entry 409–410
– – integration with EDA tools 414–415
– – integration with MATLAB and Simulink 413–414
– – integration with Simulink 413–414
– – integration with system simulators 412–413
– – MEMS model library 410–412
– micromirror array design example 419–422
– obstacles to widespread adoption 408
– toward manufacturable MEMS designs 415–419
– – free parameter in MEMS design 415
– – parameterization of process and material properties 415–418
– – process design kits 418–419
De facto standard 14
Deep reactive-ion etching (DRIE) 488
Deflation 64
Degrees of freedom (DOFs) 7, 164
Delays 358
Delta cycles 110
'Device level' continuous-field models 19
Differential algebraic equations (DAEs) 53, 96, 450–453
Differential equations of MEMS models 97–102
Digital controller in SystemC 365–366
Digital light processing (DLP) projection system 419
Digital micromirror device (DMD) 419
Dirichlet boundary condition 215
Discontinuous forcing functions 114–116
Discrete empirical interpolation method (DEIM) 237, 250–255
Display 360
Distributed effects in microsystems, modeling of 163–187, *See also* Squeeze film damping (SQFD) in MEMS modeling
– mixed-level approach for 163–187, *See also* Mixed-level modeling (MLM)
Distributed nodal method (DNM) 151
Dominant modes 302
Doubly fixed beams 148
Dynamic pull-in voltage 268

e

Electrical linear networks (ELNs) 357, 361–363
Electrical stiffness 342
Electrofluidic Kirchhoffian network 28
Electromechanical sigma–delta modulators for inertial MEMS sensors 377–401, *See also under* Inertial MEMS sensors
Electrostatic driving principles for MEMS 263–287
– dynamic pull-in voltage 268
– examples
– – electrostatic micropump diaphragm 281–285
– – IBM scanning-probe data storage device 275–281
– – plate model formulation 282–284
– linear MOR for 263–287
– model order reduction methods 269–274
– – for nonlinear systems 272–274
– – parametric systems 272
– – polynomial projection 273–274
– – representation of nonlinearities 270
– – second-order linear systems 270–272
– – systems with few nonlinearities 272
– – systems with many nonlinearities 273
– – systems with nonlinear inputs 272
– – training input 274
– – trajectory piecewise-linear (TPWL) method 274
– nonlinear MOR for 263–287
– phase space 268
– pull-in 266–268
– pull-in voltage 267
– regimes of the trajectory 268–269
Electrostatically actuated micropump
– physics-based electrofluidic compact model of 32–41
– – membrane drive 34–36
– – micropump 39–40
– – model calibration 40–41
– – parameter extraction 40–41
– – tubes 38–39
– – valves 36–38
Electrostatically actuated RF MEMS switch 41–48
Element integration 156–157
Element Menu Window (EMW) 451
Embedded Linux file (ELF) 366
Embedded Software, Cross-Compiled GNU GCC Application for MIPS 369
Energy harvesting modules, system-level simulation of 313–333
– experimental results 318

Energy harvesting modules, system-level simulation of (contd.)
- micropower module 315–316
- modeling and simulation 318–327
- – finite element method (FEM) model 320–321
- – harmonic MEMS circuit cosimulation 325–327
- – lumped element modeling 319–320
- – model order reduction 322–323
- – piezoelectric energy harvester 319
- – transient MEMS circuit cosimulation 323–325
- piezoelectric harvester, maximum power point for 327–332
- – complex matching 328–330
- – consequence of matching 331–332
- – real matching 330–331
- vibrational harvesters 316–317
- wireless autonomous sensor nodes 314–315

Energy storage system (ESS) 313–315
Equation of motion 440–444
Equivalent circuit method 136
Equivalent network models 30
Etch holes 447
Euler–Bernoulli beam theory 154
Events 113
Expansion points, different choices of 75–76

f

Feedback control system 378
Feedback linearization 395–396
Fill_table(...) function 369
Fingers 513
Finite element analysis/method (FEA/FEM) 126, 147, 320–321, 337
- FEM-based spatial discretization 215
Finite networks (FNs) 29–30, 163–164, 170–173
Finite state machines (FSMs) 96, 102–104
Fluid damping 477
Fluidic systems 99–100
Fraunhofer EMFT 32
Function block 389
Functional mock-up interface (FMI) 112

g

Galerkin method 73, 84, 134, 243
'Gappy-least-squares' approximation 250
Gas rarefaction effects 175–177
Gaussian pulse module 360–361
Gauss-Seidel iteration 111

Generalized Kirchhoff's current law (GKCL) 25–26
Generalized Kirchoffian network 43–44
Generic modeling approach for microdevices and systems 21–23
Gramian-based MOR methods 77–83, 239
- acceleration methods of computing the Gramians 82–83
- balanced system 80
- balanced truncation, motivation of 78
- balancing transformation 78–80
- – computation 81–82
- error bound of 84–85
- extension to more general systems 83
- passivity of 84–85
- square-root algorithm (SR-method) 82
- stability of 84–85
- truncation 80–81
Gram-Schmidt process 63–64
Graphical user interface (GUI) to SUGAR 444, 450–451
Graphical Window (GW) 451
Guyan reduction 427
Gyros 149
Gyroscopic matrix 300

h

Hall sensors 150
Hankel singular values (HSVs) 70
Hardware description language (HDL) 20, 213, 344, 489
Harel state charts 103
Harmonic distortion 258–259
Harmonic MEMS circuit cosimulation 325–327
Heterogeneous modeling with SystemC AMS 358–363, See also under SystemC AMS
Hexpresso 469
Hierarchically nested states 104
High-dimensional equation systems 108
Higher order electromechanical $\Sigma\Delta M$ 391–400
- band-pass EM$\Sigma\Delta M$ approach 399
- for MEMS accelerometer 391–397
- – advanced model 396–397
- – design methodology 391–392
- – example design 392–395
- – feedback linearization 395–396
- for MEMS gyroscopes 397–400
Hole model 174–175
Homotopy 108
Hybrid systems 113

i

Index reduction   109
Induced norm   62
Inertial gyroscope   474–476
Inertial MEMS sensors   377–401, *See also* Higher order electromechanical $\Sigma\Delta M$; Second order electromechanical $\Sigma\Delta M$
– electromechanical sigma–delta modulators for, system level modeling of   377–401
– – feedback control system   378
Initial state (z_0)   103
Initial state of the system   67
Initialize() function   361
In-plane actuators   127–129
Input alphabet (X)   103
Integrated circuit (IC) systems   483–498
IntelliSense's design methodology   467–470
– closing the loop   469–470
– one step at a time   468–469
– from the top down   467–468
IntelliSuite, system model extraction in   470–474
– high-LEVEL overview   470–471
– implementation   473–474
– lookup table (LUT)   471
– residual stress and film damping effects, capturing   471–472
Intellisuite™   150
Internal variables   106

k

Kirchhoff's current law (KCL)   25, 101, 171
Kirchhoff's flow law (KFL)   12
Kirchhoff's mesh rule   12
Kirchhoff's node rule   12
Kirchhoff's potential law (KPL)   12
Kirchhoff's voltage law (KVL)   101
Kirchhoffian networks, system-level modeling using   19–48
– basic principles   19–48
– compact model derivation   29–32
– from continuous-field level to compact models   23–29
– generalized Kirchhoff's current law (GKCL)   25
– generalized networks for tailored system-level modeling of microsystems   20–32, *See also individual entry*
Knudsen numbers   176
Kronecker product   240
Krylov-subspace methods   59, 215, 219, 239, 271, 314, 431

l

Lagrange equations   98
Lanczos algorithm   216
Laplace transform   60
Laplace variable   229
Laser Doppler vibrometer (LDV) system   159
Limiter block   389
Linear independence   57–60
Linear mechanical systems   295–297
Linear model order reduction, for MEMS electrostatic actuators   263–287, *See also* Electrostatic driving principles for MEMS
Linear system theory   66–71
– balanced realization   70
– controllability   68–69
– observability   68–69
– passivity of a system   70–71
– realization theory   69–70
– stability of a system   70–71
– transfer function   66–67
– two different LTI systems, measure of the difference between   67–68
Linear time-invariant (LTI) systems   56
Lithographie, Galvansformung, Abformung (LIGA) technology   125
Long-short beam actuator   128–129
Long-short-beam thermal microactuator   128
Lumped element models   8, 30–31, 101, 198–201, 319–320
Lyapunov equations   81

m

Macromodeling   8, 30, 104–105
Markov parameters   75
Mass flow rate   100
Master algorithms   112
Mathematical model order reduction (MOR)   32
Mathematical structure of MEMS models   96–104, *See also* Numerical methods for system-level simulation
– algebraic equations   97–102
– Boolean equations   102–104
– differential equations   97–102
– finite state machines   102–104
– fluidic systems   99–100
– hierarchically nested states   104
– Lagrange equations   98
– mass flow rate   100
– model behavior   97
– multibody system (MBS)   98–99
– networks   101–102
– solution of the DAE system   97

Mathematical structure of MEMS models (*contd.*)
– terminal behavior 97
– terminal variables 97
MATLAB 413–414
Matlab/Simulink 304–306
Matrix function norms 62–63
Matrix norms 61–62
Matrix 57
Maximum power point (MPP) principle 313
McMillan degree of the system 69
Mealy machines 103
Membrane drive 34–36
MEMS-Synth 469
Microbeam model 152–153
Microhotplates 214, 220–223
Micromirror array design 419–422
– Cadence Virtuoso's plotting capabilities 420
Micropower module 315–316
Micropump, physics-based system-level model of 39–40
Microswitch 351–354
Microsystems modeling, issues in 3–17
– analog hardware description languages 11–12
– automated model order reduction methods 9
– composable model libraries 14–15
– – *de facto* standard 14
– coupled multiphysics microsystems 4–6
– handling complexity, following the VLSI paradigm 10–11
– – view 10
– model validation 15–16
– model verification 15–16
– multiscale modeling and simulation 6–7
– parameter extraction 15–16
– system-level model
– – general attributes of 12–13
– – terminology 7–9
– system-level models for microsystems 3–4
Minimal realization of the system 69
Mixed-level modeling (MLM) 15, 31, 163–164
– alternative damping models, comparison with 185–186
– evaluation 179–180
– – numerical evaluation 179–180
– – experimental evaluation 180–185
– of squeeze film damping in MEMS 169–179
– – automated model generation 178–179
– – coupling with mechanical models 177–178
– – finite network-based evaluation of the Reynolds equation 170–173
– – gas rarefaction effects 175–177
– – motivation for using 168–169
– – physics-based lumped element models 173–175
– – total damping force calculation 177
Mixed-mode simulation 7
Mixed-signal systems 109–110, 113
Modal-superposition (MSUP)-based nonlinear MOR for MEMS gyroscopes 291–308, *See also* Reduced order model generation pass algorithms
– extraction of capacitances 301–302
– – analytical approach 301
– – for comb cell conductors and platelike capacitors 301–302
– – hybrid approach 301
– – numerical approach 301
– flow chart of 294–295
– – expansion pass 294
– – generation pass 294
– – master nodes 294
– – use pass 294
– multivariable capacitances, data sampling and function fit procedures for 302–304
– system simulations of MEMS based on modal superposition 304–307
– – based on Kirchhoffian networks 306
– – full order FEM models 306–307
– – Matlab/Simulink 304–306
– theoretical background of 295–299
– – body loads of capacitive sensors 297–298
– – linear mechanical systems 295–297
– – nonlinear electromechanical interactions 297–298
– – parametric reduced order models for packaging interactions 298–299
– vibratory gyroscope 293–294
Model behavior 97
Model description languages 105–107
Model order reduction (MOR) 4, 108, 238, 269–274, 322–323, *See also* ANSYS®, MOR for; Electrostatic driving principles for MEMS; Gramian-based model order reduction
– basic idea of 71–73
– expansion points, different choices of 75–76
– moment-matching 73–77
– – moments and moment vectors 73–74

– – projection matrices W and V, computation of 74–75
Model order reduction (MOR), system-level modeling of MEMS by means of 53–87
– applied to electrothermal simulation 54
– Gramian-based MOR methods 55
– linear system theory 66–71
– mathematical background 53–87
– mathematical preliminaries 56–66
– – basis 57–60
– – Block Krylov subspace 59–60
– – Krylov subspace 59
– – Laplace transform 60
– – linear independence 57–58
– – matrix 57
– – norms 61–63
– – orthogonality of the vectors 58
– – rational function 60–61
– – scalar 57
– – subspace 57–58
– – vector 57
– – vector space 57–58
– model order reduction, basic idea of 71–73
– moment-matching MOR 73–77
– nonlinear systems 86
– nonzero initial condition 85–86
– numerical algorithms 63–66
– parametric systems 86
– reduced model, stability, passivity, and error estimation of 84–85
– for second-order systems 86
Model validation in microsystems modeling 15–16
Model verification in microsystems modeling 15–16
Modeling formalisms (MFs) 357
Models of computation (MoCs) 357
Modified Gram-Schmidt process 63–64
Moment-matching-based linear MOR 213–234
Moment-matching model order reduction 73–77, See also Model order reduction (MOR)
– development of 76–77
– error bound of 84
– Padé approximation 76
– passivity of 84
– stability of 84
Moments of transfer function 73
Moore machines 103
Multibody system (MBS) 98–99
Multidimensional moment matching 229

Multifeedback and local resonators (MFLRs) architecture 398
Multilayer pads 447
Multiparameter momentmatching 229
Multiscale modeling 6–7, 108

n
Nanohub 502
Navier-Stokes equations 166–167, 170, 175
NEMS 448–450
Netlist Window (NW) 451
Nètwork method 98
Network on Chip (NoC) 366
Networks 101–102
– current 101
– Kirchhoff's current law (KCL) 101
– Kirchhoff's voltage law (KVL) 101
– lumped elements 101
– voltage 101
Newton iteration 111
Newton-Raphson iterations 108
NOdal Design of Actuators and Sensors (NODAS) 506
Nodal model of microbeam element 152
Node voltages 101
Nonlinear DAEs, solution of 107–109
Nonlinear model order reduction, for MEMS electrostatic actuators 263–287, See also Electrostatic driving principles for MEMS
Nonlinear Navier-Stokes equations 170
Nonlinear transmission-line model 246–247
Nonzero initial condition 85–86
Norms 61–63
– induced norm 62
– matrix norms 61–62
– operator norm 62
– subordinate matrix norm 62
– vector norms 61–62
Numerical algorithms 63–66
– Arnoldi algorithm 64–65
– Band Arnoldi process 65
– deflation 64
– Gram-Schmidt process 63–64
– modified Gram-Schmidt process 63–64
Numerical methods for system-level simulation 107–112
– mixed-signal simulation cycle 109–110
– nonlinear DAEs, solution of 107–109
– – choice of tolerances 108
– – high-dimensional equation systems 108
– – homotopy 108
– – index reduction 109
– – model order reduction 108
– – stiffness and multiscale systems 108

o

Observability 68–69
– observability Gramian 68
– observability (of a state) 68
One-end fixed beams 148
Open circuit condition 327
Open Core Protocol (OCP) 365
Operator norm 62
Optical-based microsystems 6
Order of the Krylov subspace 59
Order of the reduced model 250
Ordinary differential equations (ODEs) 53, 95, 131, 164, 215, 337
Orthogonality of the vectors 58
Out-of-plane actuators 129
Output alphabet (Y) 103
Output function (g) 103
Oversampling ADC 379

p

Packaging effects of MEMS devices, system-level modeling of 147–160
– element integration 156–157
– FEM and experimental validation 157–160
– impact on typical MEMS devices 148–150
– – accelerometers 148–149
– – doubly fixed beams 148
– – fine meshing 150
– – gyros 149
– – hall sensors 150
– – one-end fixed beams 148
– – pressure sensors 149
– – solid modeling 150
– – thermal actuators 149
– single substructures, behavioral modeling of 152–156
– support model 155–156
– – nodal model 156
– system partitioning 151–152
Packaging interactions, parametric reduced order models for 298–299
Padé approximation 76, 271
Padé via Lanczos (PVL) 77
Parameter extraction in microsystems modeling 15–16
Parametric electrothermal MEMS models 213–234
Parameterization 13, 492–494
Parameterized Arrays 447–448
Parametric model order reduction (pMOR) 213, 228–230, 295

– moment-matching-based linear MOR for 213–234
– – application of 223–227
– – application to extraction of thin-film thermal parameters 227–232
– – DOT® (Design Optimization Tools) 231
– – parameter extraction methodology 230–232
– for packaging interactions 298–299
Partial differential equation (PDE) 53, 95, 131, 164, 337
Partial solve (PSOLVE) 430
Passivity of a system 70–71
Periodic AC analysis 14
Periodic steady-state analysis 14
Periph_io_get() 369
Periph_io_set() 369
Petrov-Galerkin projection 72, 84
Phase space 268
Physical design kits 13
Physics-based compact models 8, 19, 31–41, *See also under* Electrostatically actuated micropump
Physics-based lumped element models 173–175
– boundary model 173–174
– hole model 174–175
Piezoelectric energy harvesters 316–317
Piezoelectric generator 317–318
– design of 317–318
– fabrication of 317–318
Piezoelectric harvester, maximum power point for 327–332
Poiseuille flow 172
Polynomial projection 273–274
Power spectral density (PSD) of output bitstream 385
Pressure sensors 149
Primitive behavioral model 8
Probe-based microsystems 6
Process design kits (PDKs) 418–419
Processing() function 358, 361, 365
Projection-based nonlinear model order reduction 237–260
– discrete empirical interpolation method (DEIM) 250–255
– – thermal analysis 251–253
– evaluation cost for 239–240
– harmonic distortion 258–259
– problem specification 238
– projection principle 239–240
– pull-in effect 256–258
– – generalizing from training inputs 258
– Taylor series expansions 240–245

– – microfluidic channel example   241–242
– – model reduction via quadratic Taylor expansion   242–244
– – stability issues   244
– trajectory piecewise-linear method   245–250
– – nonlinear transmission-line model   246–247
Proper orthogonal decomposition (POD) methods   55, 239
Pseudo bimorph electrothermal actuator   127
Pseudo-linear relation   22
PSugar   450
Pull-in effect   256–258

q

Quadratic Taylor expansion, model reduction via   242–244
Quadrature phase shift keying (QPSK)-modulated   364–367
Quantization noise loop transfer function (QNTF)   385–386
Quantizer block output   381

r

Radio-frequency (RF) switches   3–5
Rates   358
Rational function   60–61
Rational interpolation   75
Rayleigh mode preserving damping   338
Real matching   330–331
Realization theory   69–70
Reduced order modeling (ROM) of MEMS   8, 19, 483–498, See also SoftMEMS simulation environment
– applications   494–498
– error estimation of   84–85
– generation pass algorithms   299–304
– – extraction of body load contribution vectors for   299–301
– MEMS development environment   484–485
– passivity   84–85
– stability   84–85
Reference node   101
Relevant modes   302
Resistance Inductance Capacitance (RLC)   361
Resistivity   130
Reynolds equation-based modeling strategies   166–167
– based on decomposition into cells   167
– based on FEM   167
– based on GKN theory   167
– based on modified Reynolds equation   167
– finite network-based evaluation of   170–173
RF-MEMS devices, compact modeling of   191–208
– RF-MEMS multistate attenuator parallel section   194–205
– – lumped element network   198
– whole RF-MEMS multistate attenuator network   205–207

s

Scalar variable   57
Scaling factor   472
Second Order ARnoldi (SOAR)   428
Second order electromechanical $\Sigma\Delta M$
– for a MEMS accelerometer   380–391
– – advanced model   386–391
– – basic model   380–386
– – function block   389
– – limiter block   389
– – power spectral density (PSD) of output bitstream   385
– – switch block   382
Second-order linear systems, MOR methods for   270–272
Second order ODE systems   341
Seismic perturbations detection using accelerometer   363–370
Serendi-CDI   501–504
Sigma–delta modulators ($\Sigma\Delta M$)   378–379, See also Second order electromechanical $\Sigma\Delta M$
Signal-flow model (or block-diagram model)   8
Silicon-based microhotplate   220–223
SIMPLORER®   223–225
SIMULINK   413–414, 454–457
Single substructures, behavioral modeling of   152–156
– chip/adhesive/package model   153–155
– microbeam model   152–153
Single-input multiple-output (SIMO) system   56, 75
Single-input single-output (SISO) system   56, 217
Singular value decomposition (SVD)   239
SoftMEMS simulation environment   485–494
– implementation within   485–494
– issues with   486–487
– – range of validity of the model   487

SoftMEMS simulation environment (contd.)
– – spatial resolution   486
– – temporal resolution   487
– model output   492
– modeling abstraction levels   486
– modeling process   490–491
– modeling requirements within   485–494
– models and inputs   487–490
– parameterization   492–494
Solution of the DAE system   97
Solvers   431–432
Source stepping   116
Sparse Lyapunov equations   83
Specific heat   130
Spectre-Verilog A model   279
SPICE (Simulation Program with Integrated Circuit Emphasis)   104–105, 137, 454–457
– essential features of   105
– subcircuits in   105
Spring force   472
Square-root algorithm (SR-method)   82
Squeeze film damping (SQFD) in MEMS modeling   165–169
– mixed-level modeling of   169–179, See also individual entry
– Reynolds equation-based modeling strategies   166–167
– – based on decomposition into cells   167
– – based on FEM   167
– – based on GKN theory   167
– – based on modified Reynolds equation   167
Stability of a system   70–71
State charts   96
State diagrams   102
State of the system   67
State set (Z)   102
State-space representation   56, 215
State-space transformation   69
Static random access memory (SRAM)   419
Stiffness   108
Stress stiffening   338
Structured model   8
Subcircuits in SPICE   105
Subordinate matrix norm   62
Subspace   57–58
SUGAR, for MEMS   439–458
– compact modeling building blocks for designing   443
– differential algebraic equations (DAEs)   450–453
– equation of motion   440–444
– model   440
– nodes   440
– parameters   440
– simulation   453–454
– SUGAR, COMSOL, SPICE, SIMULINK, integration of   454–457
– SUGAR-based applications   444–454
– – accelerometers   444
– – common ground tracers   445–447
– – design/simulation   445
– – etch holes   447
– – GUI configuration   450–451
– – gyroscopes   444
– – layout generation   445
– – library   444
– – microgrippers   444
– – multilayer pads   447
– – NEMS   448–450
– – optimization   448
– – parameterized arrays   447–448
– – PSugar   450
– – RF-MEMS   444
– – thermal actuators   444
– verification   457
Superposition   427
Support model effect on MEMS devices   155–156
Surface micromachined beamlike electrothermal microactuators   125–143
– classification and problem description   127–131
– – arrayed bent-beam thermal actuator   128
– – bent-beam actuators   128
– – in-plane actuators   127–129
– – long-short beam actuator   128–129
– – out-of-plane actuators   129
– coupled electrothermal model, equivalent circuit of   137–138
– material properties   129–131
– – coefficient of thermal expansion   130–131
– – resistivity   130
– – specific heat   130
– – thermal conductivity   129–130
– modeling   131–136
– – electrothermal model of a beam   131–134
– – thermomechanical model of beam   134–136
– solving   136–139
– system-level modeling of   125–143
Switch block   382
System in package (SiP)   147

System model extraction (SME) 464–465
System partitioning 151–152
SystemC AMS 357–374
– heterogeneous modeling with 358–363
– – 2.4 GHz RF transceiver 366–368
– – SystemC AMS timed dataflow (TDF) 358–359
– – timed dataflow model of accelerometer 359–361
System-level modeling of MEMS 53–87, 506–507, See also under Model order reduction (MOR); Packaging effects of MEMS devices; Surface micromachined beamlike electrothermal microactuators
– algorithmic approaches for 95–118, See also Mathematical structure of MEMS models
– – advanced simulation techniques 113–118
– – complementary variables 117
– – discontinuous forcing functions 114–116
– – emerging problems 113–118
– – hybrid or mixed-signal systems 113
– – model equations, structural changes in 116–118
– description 104–107
– – general approaches for 104–107
– – macromodeling 104–105
– – model description languages 105–107
– – SPICE (Simulation Program with Integrated Circuit Emphasis) 104–105
– general attributes of 12–13
– for microsystems 3–4
– terminology 7–9
– – behavioral model 8
– – circuit model or network model 8
– – compact model 8
– – component 8
– – lumped-element model 8
– – macromodel 8
– – physics-based model 8
– – primitive behavioral model 8
– – reduced-order model 8
– – signal-flow model (or block-diagram model) 8
– – structured model 8
– using generalized Kirchhoffian networks 19–48, See also Kirchhoffian networks

t

Tailored system-level modeling, generalized Kirchhoffian networks for 20–32

– generic modeling approach for microdevices and systems 21–23
– pseudo-linear relation 22
Tanner's C-code modeling 489
Taylor series expansions 240–245
– quadratic Taylor expansion, model reduction via 242–244
TDF cluster 358
Technology CAD (TCAD) 462
Terminal behavior 97
Terminal variables 97
Thermal actuators 149
Thermal conductivity 129–130
Timed dataflow model of accelerometer 359–361
Timestep 358
Tire pressure monitoring systems (TPMS) 483
Total damping force calculation 177
Training input 274
Trajectory piecewise-linear (TPWL) 237, 245–250, 264, 274, 285
– nonlinear transmission-line model 246–247
– stability issues 247–249
Transfer function 66–67
Transient MEMS circuit cosimulation 323–325
Transition function (f) 103
Truncated-balanced realizations (TBRs) 244
Truncation 80–81
Tubes, system-level model of 38–39

u

UML behavioral state machines 103
Unsymmetric Lanczos process 74
U-shaped actuator 127
U-shaped principle 125

v

Valves, compact model of 36–38
Variable gap parallel plate capacitor 264–269
Vector function norms 62–63
– real valued vector function 62
Vector norms 61–62
Vector space 57–58
Vector 57
VerilogA compact models 196–201
VerilogA model 352
Very high speed integrated circuit hardware description language–analog and mixed-signal extensions (VHDL-AMS) 106
Very-large-scale integration (VLSI) 4, 10–11
– in handling complexity 10–11

Vibrating devices 345–351
Vibrational harvesters 316–317
Vibratory gyroscope 293–294
Viscous force 472
Voltage 101
Volterra method 240

w

Wafer level package (WLP) 147
Web-based community for MEMS modeling and design 501–515
– concepts of web-based design 503–504
– design community constitution 504–505
– – discussion forums 505
– – online simulation tools 505
– – open source philosophy 504
– – wiki-style collaborative authoring 505
– encoding MEMS behavioral models 508–515
– – nonlinear beam 509–511
– leveraging web-based communities 502–505
– – Designer's Guide Community 502
– – Nanohub 502
– – Serendi-CDI 502
– MEMS modeling and design online 505–508
– system-level modeling of MEMS 506–507
– web-based community conventions 507–508
White Noise block 381
Wireless autonomous sensor nodes 314–315
Wireless sensor network (WSN) 357